VOLUME ONE HUNDRED AND FOUR

ADVANCES IN VIRUS RESEARCH

Virus Entry

VOLUME ONE HUNDRED AND FOUR

Advances in VIRUS RESEARCH

Virus Entry

Edited by

MARGARET KIELIAN
Albert Einstein College of Medicine, Bronx, New York, United States

THOMAS C. METTENLEITER
Friedrich-Loeffler-Institut, Federal Research Institute for Animal Health, Greifswald – Insel Riems, Germany

MARILYN J. ROOSSINCK
Department of Plant Pathology and Environmental Microbiology, Center for Infectious Disease Dynamics, Penn State University, University Park, PA, United States

ELSEVIER

ACADEMIC PRESS
An imprint of Elsevier

Academic Press is an imprint of Elsevier
50 Hampshire Street, 5th Floor, Cambridge, MA 02139, United States
525 B Street, Suite 1650, San Diego, CA 92101, United States
The Boulevard, Langford Lane, Kidlington, Oxford OX5 1GB, United Kingdom
125 London Wall, London, EC2Y 5AS, United Kingdom

First edition 2019

ISBN: 978-0-12-818394-6
ISSN: 0065-3527

Publisher: Zoe Kruze
Acquisition Editor: Ashlie M. Jackman
Editorial Project Manager: Peter Llewellyn
Production Project Manager: James Selvam
Cover Designer: Greg Harris

Typeset by SPi Global, India

Contents

Contributors

Aurélie Albertini
Institute for Integrative Biology of the Cell (I2BC), CEA, CNRS, University Paris-Sud, Université Paris-Saclay, Gif-sur-Yvette, France

Marija Backovic
Institut Pasteur, Unité de Virologie Structurale, UMR3569 (CNRS), Paris, France

Laura Belot
Institute for Integrative Biology of the Cell (I2BC), CEA, CNRS, University Paris-Sud, Université Paris-Saclay, Gif-sur-Yvette, France

Kartik Chandran
Department of Microbiology and Immunology, Albert Einstein College of Medicine, Bronx, NY, United States

Jia Chen
Department of Microbiology and Immunology, Feinberg School of Medicine, Northwestern University, Chicago, IL, United States

Yuting Chen
Department of Entomology and Plant Pathology, North Carolina State University, Raleigh, NC, United States

Moshe Dessau
Azrieli Faculty of Medicine, Bar Ilan University, Safed, Israel

Maria Eugenia Dieterle
Department of Microbiology and Immunology, Albert Einstein College of Medicine, Bronx, NY, United States

Yves Gaudin
Institute for Integrative Biology of the Cell (I2BC), CEA, CNRS, University Paris-Sud, Université Paris-Saclay, Gif-sur-Yvette, France

Mara C. Harwood
Department of Cell and Developmental Biology; Cellular and Molecular Biology Program, University of Michigan Medical School, Ann Arbor, MI, United States

Angela R. Howard
Department of Molecular Biology and Microbiology, Tufts University School of Medicine, Boston, MA, United States

Rohit K. Jangra
Department of Microbiology and Immunology, Albert Einstein College of Medicine, Bronx, NY, United States

Lara M. Kleinfelter
Department of Microbiology and Immunology, Albert Einstein College of Medicine, Bronx, NY, United States

Barbara G. Klupp
Institute of Molecular Virology and Cell Biology, Friedrich-Loeffler-Institut, Greifswald – Insel Riems, Germany

Richard Longnecker
Department of Microbiology and Immunology, Feinberg School of Medicine, Northwestern University, Chicago, IL, United States

Thomas C. Mettenleiter
Institute of Molecular Virology and Cell Biology, Friedrich-Loeffler-Institut, Greifswald – Insel Riems, Germany

Eva Mittler
Department of Microbiology and Immunology, Albert Einstein College of Medicine, Bronx, NY, United States

Britta S. Möhl
Institute of Virology, School of Medicine, Technical University of Munich/Helmholtz Zentrum München, Munich, Germany

Yasuko Mori
Division of Clinical Virology, Center for Infectious Diseases, Kobe University Graduate School of Medicine, Kobe, Japan

James B. Munro
Department of Molecular Biology and Microbiology, Tufts University School of Medicine, Boston, MA, United States

Jose A. Navarro
Instituto de Biología Molecular y Celular de Plantas (IBMCP), Universidad Politécnica de Valencia-Consejo Superior de Investigaciones Científicas, Valencia, Spain

Mitsuhiro Nishimura
Division of Clinical Virology, Center for Infectious Diseases, Kobe University Graduate School of Medicine, Kobe, Japan

Vicente Pallas
Instituto de Biología Molecular y Celular de Plantas (IBMCP), Universidad Politécnica de Valencia-Consejo Superior de Investigaciones Científicas, Valencia, Spain

David A. Rasmussen
Department of Entomology and Plant Pathology, North Carolina State University, Raleigh, NC, United States

Felix A. Rey
Institut Pasteur, Unité de Virologie Structurale, UMR3569 (CNRS), Paris, France

Dorith Rotenberg
Department of Entomology and Plant Pathology, North Carolina State University, Raleigh, NC, United States

Jesus A. Sanchez-Navarro
Instituto de Biología Molecular y Celular de Plantas (IBMCP), Universidad Politécnica de Valencia-Consejo Superior de Investigaciones Científicas, Valencia, Spain

Megan M. Slough
Department of Microbiology and Immunology, Albert Einstein College of Medicine, Bronx, NY, United States

Chelsey C. Spriggs
Department of Cell and Developmental Biology, University of Michigan Medical School, Ann Arbor, MI, United States

Billy Tsai
Department of Cell and Developmental Biology; Cellular and Molecular Biology Program, University of Michigan Medical School, Ann Arbor, MI, United States

Melina Vallbracht
Institute of Molecular Virology and Cell Biology, Friedrich-Loeffler-Institut, Greifswald – Insel Riems, Germany

Anna E. Whitfield
Department of Entomology and Plant Pathology, North Carolina State University, Raleigh, NC, United States

CHAPTER ONE

Key checkpoints in the movement of plant viruses through the host

Jose A. Navarro, Jesus A. Sanchez-Navarro, Vicente Pallas*
Instituto de Biología Molecular y Celular de Plantas (IBMCP), Universidad Politécnica de Valencia-Consejo Superior de Investigaciones Científicas, Valencia, Spain
*Corresponding author: e-mail address: vpallas@ibmcp.upv.es

Contents

Abstract

Plant viruses cannot exploit any of the membrane fusion-based routes of entry described for animal viruses. In addition, one of the distinctive structures of plant cells, the cell wall, acts as the first barrier against the invasion of pathogens. To overcome the rigidity of the cell wall, plant viruses normally take advantage of the way of life of

Advances in Virus Research, Volume 104
ISSN 0065-3527
https://doi.org/10.1016/bs.aivir.2019.05.001

different biological vectors. Alternatively, the physical damage caused by environmental stresses can facilitate virus entry. Once inside the cell and taking advantage of the characteristic symplastic continuity of plant cells, viruses need to remodel and/or modify the restricted pore size of the plasmodesmata (channels that connect plant cells). In a successful interaction for the virus, it can reach the vascular tissue to systematically invade the plant. The connections between the different cell types in this path are not designed to allow the passage of molecules with the complexity of viruses. During this process, viruses face different cell barriers that must be overcome to reach the distal parts of the plant. In this review, we highlight the current knowledge about how plant RNA viruses enter plant cells, move between them to reach vascular cells and overcome the different physical and cellular barriers that the phloem imposes. Finally, we update the current research on cellular organelles as key regulator checkpoints in the long-distance movement of plant viruses.

1. Introduction

Plants can be considered, in a strict sense, not as multicellular but as supracellular organisms, as their cells are not physically separated from each other but connected via plasmodesmata (PD) channels to form a highly dynamic symplastic entity (Baluska, 2009; Lucas et al., 1993). While in animal cells, extracellular matrix components offer a platform for virus recognition and entry, the rigidity of plant cell walls constitutes a physical barrier that makes virus entry difficult (Stavolone and Lionetti, 2017; Underwood, 2012). To overcome this first restraining barrier, plant viruses take advantage of the food needs of certain insect vectors, fungi or nematodes. Alternatively, abiotic factors such as wind, rain, grafting or even grazing animals, can cause mechanical damage of sufficient magnitude to facilitate virus entry without having to break or remodel the cell wall. Once inside the cell and unlike animal viruses, cell-to-cell cytosolic connectivity is exploited by plant viruses for the transport of genomes and the need to release particles into the extracellular space is bypassed. This intercellular transport can be achieved by modifying the size exclusion limit of the tightly controlled PD, or alternatively, by remodeling its structure and breaking the ER continuity between cells (Benitez-Alfonso et al., 2010; Peña et al., 2012).

Once a given virus has moved to adjacent cells through the plasmodesmata, it must move through the vasculature to reach the distal parts of the plant. Despite lacking an equivalent to the animal nervous system, vascular plants have a dedicated integrative system for the accurate connection of distal organs. Vascular tissues perform essential roles in integrating the physiological (transport of water and nutrients), developmental (transport of signaling molecules) and structural (physical support) processes of higher plants. Thus, plant viruses usurp these endogenous routes to systemically invade a plant. From a practical viewpoint, this is the most frequent scenario in which plant disease is revealed.

Seminal discoveries revealed that plant viruses follow the pathway of photoassimilate translocation to invade the distal parts of the plant (see Nelson and van Bel, 1998: Pallas et al., 2011; Vuorinen et al., 2011 for review). First, photoassimilate flow requires transport from photosynthetic cells, located in source tissue, to the conducting sieve tubes in a process known as phloem loading (Zhang and Turgeon, 2018). After being transported through phloem, nutrients and signaling molecules must be unloaded into the sink tissues/organs to alleviate their great photoassimilate demand (Milne et al., 2018). Thus, viral long-distance movement involves the passage of viruses through various cellular barriers including the bundle sheath, vascular parenchyma, and companion cells for virus loading into sieve elements (Fig. 1). Viruses are then transported through sieve elements to distant tissues. The invasion of distant cells requires virus unloading from sieve elements into companion cells, followed by cell-to-cell movement into the bundle sheath and mesophyll cells.

In the first part of this review, we briefly summarize the different ways viruses overcome the first defense line that plants have against pathogen invasion. Once inside the plant cell, viruses, as biotrophic pathogens, must usurp either cell routes or host factors to accomplish their infection cycle and succeed in their perpetuation. The increasing number of these host factors and their mode of action in the different steps of the viral cycle has been recently reviewed (Folimonova and Tilsner, 2018; Harries and Ding, 2011; Hipper et al., 2013; Nagy, 2016; Wang, 2015) and will only be treated here in regards to their demonstrated roles in overcoming cellular physical barriers. In the second and third parts of this review, emphasis is placed on how viruses overcome the different cellular barriers, the CW at the intercellular transport or specific cell boundaries at the vascular tissue, to systemically invade the plant. In the last part of the review, we discuss the emerging view ascribed to essential organelles, chloroplasts and nuclei, as regulator points of long-distance plant RNA virus movement.

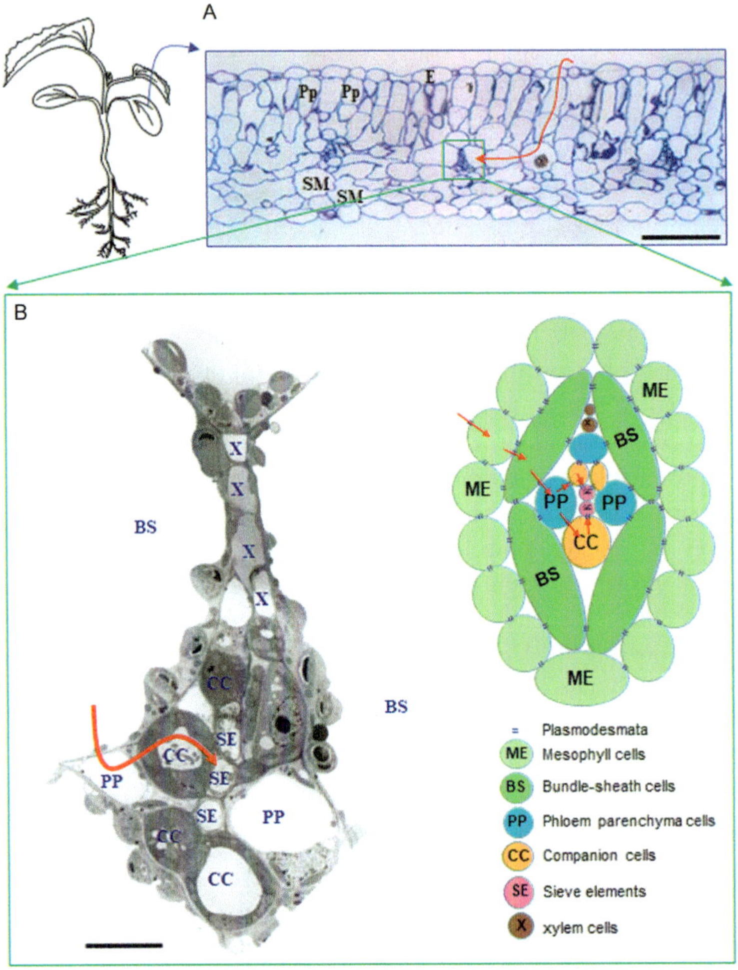

Fig. 1 Schematic representation of the pathway followed by plant viruses to reach the vascular tissue. (A) Transversal section of a *Chenopodium quinoa* leaf in which red arrow represents a putative itinerary of a plant virus to reach vascular tissue from the initially infected cell. After the initiation of the infection in the epidermis cells (E), viruses move to the palisade parenchyma (Pp) and spongy mesophyll (SM) cells to reach the vascular tissue. (B), Transmission electron microscopy of the major vein observed in the square of (A) showing the different cell types present at the vascular tissue. Viruses infecting the mesophyll cells reach the conductive sieve elements (SE) trough bundle sheath (BS) cells, the phloem parenchyma (PP) and companion (CC) cells. Bars correspond to 50 μm (A) or 5 μm (B). On the right, a schematic representation of the different cell types.

2. The different ways viruses enter a plant cell

Unlike animal viruses, those infecting plants are unable to autonomously enter their host cells. The presence of a rigid cell wall that impedes the diffusion of macromolecules larger than 60 kDa, blocks the free access of viral particles from the outside to the plasma membrane. Plant viruses have developed no mechanism to degrade the cell wall and, consequently, they cannot exploit any of the membrane fusion-based routes described for animal virus entry (Helenius, 2018). For this reason the presence of a lipid envelope having tethering factors, an essential feature for endocytosis subversion in animal cells, is lacking in the vast majority of plant viruses. Only a few plant-infecting species belonging to the families *Rhabdoviridae*, *Peribunyavirae* and *Fimoviridae* have a surrounding lipid membrane, and in these cases the envelopes appear to be an adaptation to their insect vectors rather than to their host plants (Buchmann and Holmes, 2015). Another major difference between plant and animal virus entry relies on the symplastic movement of plant viruses through the plasmodesmal continuum that potentially exists over the whole plant (Roberts and Oparka, 2003). The entry process for a plant virus is completed after the first cell is successfully colonized, since plant cuticle and cell wall breaching is no longer required. Thereafter, plant virus movement proteins (MPs) take the leading role in managing intracellular and symplastic intercellular trafficking by dilating the PD channels (see Section 3).

How plant viruses begin the colonization of an uninfected host is largely dependent on their lifestyles and transmission modes (Roossinck, 2010). Vertical transmission does not rely on vectors and occur directly from parents to their offspring. In a systemically infected plant viruses take advantage of symplastic movement to reach gametes or the embryo, thus the plant progeny initiates its life already infected. This is the only way that some cryptic asymptomatic viruses, causing infections that last for longer periods and do not move between cells, use for spreading (Sastry, 2013). In contrast, virus entry to another plant of the same generation, or horizontal transmission, is most frequently assisted by vectors that somehow breach the cell wall. To a lesser extent, very stable viruses can directly get in and out of plants through wounds and scratches caused either by contact among neighboring plants, which rub their leaves together, or through abiotic factors such as wind, rain, grafting, pruning shears and handling tools, or by grazing animals. In addition, cross-pollination can provide a pathway for a direct form of horizontal transmission between and within generations. More often than not, acute plant viruses are transmitted horizontally and vertically (Jones, 2018). Given the complexity and the extremely wide scope of this issue, we restrict our focus in the section below to review the virus entry mechanisms and pathways associated with vertical and assisted-horizontal transmission.

2.1 Virus entry into the embryo in vertical transmission

Some plant viruses have developed the ability to infect the embryo via two distinct but sometimes co-existing routes. On the one hand, the embryo can start its development infected if at least one of the two gametes that take part in the fertilization process has previously been infected. On the other hand, some plant viruses are capable of infecting the embryo after fertilization.

2.1.1 Virus entry into the zygote during fertilization

Virus transmission by pollen has been reported in more than 45 virus species, the majority belong to the genera *Alphacryptovirus*, *Potyvirus*, *Ilarvirus*, *Nepovirus* and *Tobravirus* but also to single species from up to 12 further genera including *Alfamovirus*, *Hordeivirus*, *Cucumovirus* and others (Card et al., 2007; Hull, 2014). However, only a few studies published to date have shown how these viruses infect male and female gametes at the microscopic level. For example, an examination of barley flower tissues by electron microscopy revealed that the seed- and pollen-transmitted isolate MI-1 of the *Barley stripe mosaic virus* (BSMV) invaded the floral primary meristem early and, subsequently, pollen mother cells and sperm and megaspore mother cell including the egg (Carroll and Mayhew, 1976a, b). Immunohistochemistry studies in pea infected with *Pea early browning virus* (PEBV) showed that PEBV was present in the synergids, the egg cells and the polar nuclei from unfertilized ovules and in mature pollen from unopened flower anthers (Wang and Maule, 1997). Ultrastructural images of developing anthers from young lettuce flowers infected with *Lettuce mosaic virus* (LMV) showed that detection of LMV in pollen mother cells and the surrounding tapetum was consistently correlated. Using in situ RT-PCR, *Prune dwarf virus* also was detected in the generative and vegetative cells of the pollen grains of almond (Silva et al., 2003). More recently, using in situ hybridization and transmission electron microscopy, Amari et al. (2007) showed that *Prunus necrotic ringspot virus* (PNRSV) was detected early in apricot pollen mother cells but later, it was distributed uniformly within the cytoplasm of uni- and bicellular pollen grains including generative cells.

To infect male gametes, plant viruses must enter pollen mother cells from the tapetum cells before the commencement of early meiotic prophase I. Thereafter, a fast deposition of callose between the plasma membrane and cell wall of microsporocytes destroys the cytoplasmic connections with the tapetum and seals off the pollen mother cells. Although the callose wall is dissolved after meiosis by the callase activity of the tapetum, the free microspores are released into the pollen sac excluding the possibility of further infection. Virus infection can be eluded during megasporogenesis, since embryo sac and megaspore tetrad are also callose-isolated from maternal tissues. However, callose isolation in some plants lasts a relatively short time or is incomplete. In any case, the functional megaspore remains in contact with the nucellus, when the callose disappears after the completion of meiosis, being exposed to infection again. Accordingly, PNRSV RNA was detected in young ovule tissues but not in megaspores. In a more advanced developmental stage, PNRSV RNAs were occasionally detected inside the mature embryo sac, specifically in the fused polar cells. To explain this, the authors suggested that this later infection came from the highly infected surrounding nucellus (Amari et al., 2009) (see Fig. 2A and B for a detailed illustration).

The reason some viruses and not others can infect gametes appears to be directly associated with their unusual capacity to invade primary meristems. The majority of plant viruses are excluded from meristem cells but the specific mechanism behind this finding is far from being understood (Gosalvez-Bernal et al., 2006; Sudarshana et al., 1998). It is assumed that the absence of vascular bundles coupled with the high cell multiplication rate, which requires an intense metabolic activity with high cytokinin and auxin levels (Azizi et al., 2015), restricts virus access to the meristem. More recently, it has been shown that an active RNA silencing mechanism cooperates in controlling the selective unloading of RNAs into the meristem (Foster et al., 2002). Some viruses produce a systemic silencing signal that spreads through the plant either with the virus or ahead of the infection front. Therefore, blocking the propagation of this systemic silencing signal, or relaying host mechanisms would increase not only viral replication but also the likelihood of meristem infection (Schwach et al., 2005).

The above hypothesis is certainly consistent with recent experimental evidence. Although the functional characterization of PEBV 12K protein has not been reported to date, the deletion of its corresponding gene completely abolished male and female gamete infection and seed transmission of PEBV (Wang et al., 1997). Nevertheless, 16K of the *Tobacco rattle virus* (TRV), an ortholog of PEBV 12K, is a suppressor of gene silencing that

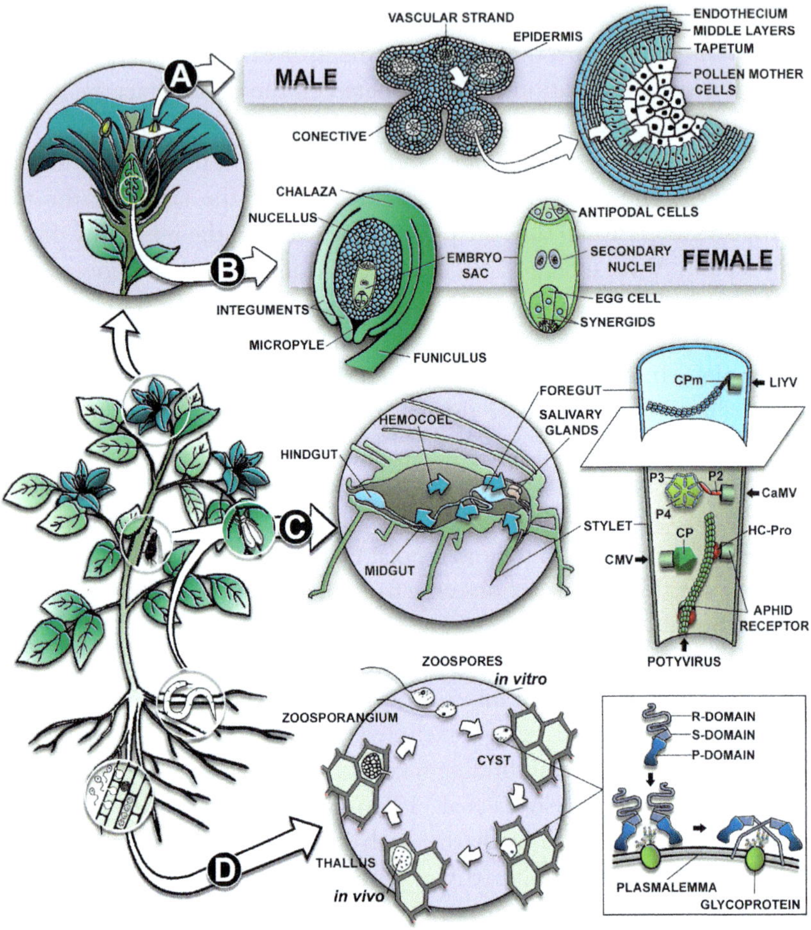

Fig. 2 Schematic overview of the different ways that plant viruses use to entry into a plant cell. (A) Virus entry into male gametes during vertical transmission. A transverse section of a young anther (left) and an enlarged view of a microsporangium before the microsporogenesis (right) is shown. Plant viruses may reach male gametes by symplastic movement (white arrows) from vascular strands to pollen mother cells before callose deposition. (B) Virus entry into female gametes during vertical transmission. A longitudinal section of the megasporangium or ovule (left) and an enlarged view of the mature embryo sac (right) showing the two synergids together with the female gamete (egg) is shown. Egg or zygote infection may occur by symplastic movement from the funiculus vascular strands through chalaza, integuments and nucellus but also during fertilization with infected pollen. (C) Virus entry assisted by piercing and sucking non-vertebrate vectors (aphids, whitefly, nematodes among others) during horizontal transmission. A schematic representation of an aphid and the route used by propagative and circulative persistent viruses is shown. The ingested virions must pass through the gut and the hemocoel to reach the salivary glands (blue arrows). The inset at the

(Continued)

allows TRV to enter the shoot apical meristems and invade the reproductive organs in *Nicotiana benthamiana* (Martín-Hernández and Baulcombe, 2008). In addition, the seed transmission determinant of BSMV, named γb, was characterized as a post-transcriptional gene silencing suppressor (Edwards, 1995; Yelina et al., 2002). This type of RNA-based surveillance may apply to some but not all virus species, e.g. *Tobacco mosaic virus* (TMV), suggesting the existence of other modes of meristem exclusion (Kim et al., 2005b; Schwach et al., 2005).

Plant viruses can also reach the megasporocyte from the cytoplasm of the pollen vegetative cell via the pollen tube. In apricot triporate pollen grains, PNRSV was located mainly at the apertures from which the pollen tube emerges. After pollen tube elongation, virus RNA was particularly localized at its end, in the growth zone. Interestingly, this viral distribution pattern is similar to that of hydroxyproline-rich glycoproteins, required to provide nutritional and structural support for pollen tube growth (Amari et al., 2007). From this observation, it could be hypothetically possible that PNRSV distributions resulted from viral hijacking of the intense traffic of exocytic vesicles that release new wall materials at the growing end. It is critical to note that during elongation of the pollen tube, transverse plugs of callose and cellulose are regularly deposited keeping the two sperm cells enclosed at the growing tip (Newbigin et al., 2009). Thus, early positioning of PNRSV at tube emerging apertures and, later, at growing tips could increase the transmission opportunities of this virus during fertilization.

2.1.2 Virus entry directly into the embryo

Direct colonization of the embryo has been described as an alternative entry mode for BSMV, but it represents the only entry for *Pea seed borne mosaic virus* (PSbMV) transmission in pea (Roberts et al., 2003; Wang and Maule, 1994). Direct infection of the embryo occurs by virus entry into seed, from the

Fig. 2—cont'd right represents the foregut epithelium (top) or the food/salivary canal located at the tip of an aphid stylet (bottom). Non-circulative semi- and non-persistent viruses bind at specific vector receptors located at the tip of the stylet and foregut epithelium using the capsid (LYIV and CMV) or the helper (CaMV and potyviruses) strategy. (D) Virus entry assisted by zoosporic fungi during horizontal transmission. A schematic representation of the asexual cycle of *Olpidium* sp. is shown. The virus in vitro and in vivo acquisition points are indicated. The inset at the right represents the proposed conformational change that viral CP undergoes when it binds to the zoospore plasmalemma. R domain could mediate the passage of viral RNA from the outside to zoospore inside and from thallus inside to the plant cell cytoplasm.

surrounding maternal testa, to the endosperm, suspensor and finally the embryo. This movement must occur symplastically during early embryogenesis while the suspensor is still functional, i.e., prior to its programmed cell death (apoptosis). For some time now, the main constraints against this pathway have been the unclear presence of symplastic connections crossing the maternal-filial endosperm boundaries (i.e., testa/endosperm and endosperm/suspensor or embryo). Roberts et al. (1998) showed that PSbMV cell-to-cell movement could be mediated by the cylindrical inclusion protein, a viral factor that modifies plasmodesmal apertures by inducing the formation of pinwheel inclusion bodies. A later work by the same authors revealed the presence of these inclusion bodies perpendicularly arranged to the testa cell walls at the endosperm junction in young developing pea seeds. These findings suggested that PDs were certainly present at this interface, although further investigation is necessary to evaluate if they existed constitutively or were created de novo during the course of the infection (Roberts et al., 2003).

Since viral replication is not supported in the endosperm, the timing of PSbMV arrival to the base of the suspensor exclusively depends on the amount of viruses reaching the maternal testa. In this scenario, the helper component-proteinase (HC-Pro) plays a relevant role in enhancing virus replication and long-distance movement. Actually, HC-Pro has been identified as one of the factors determining the difference between the seed transmitted and non-transmitted isolates of PSbMV (Johansen et al., 1996). How PSbMV transverses the boundary between the endosperm and suspensor cells represents the last controversial question in virus entry into the embryo. This is because an extracellular embryonic sheath composed of cell wall material, actively secreted by the endosperm, surrounds the embryo (Moussu et al., 2017). Nevertheless, a close examination of this sheath, especially at the base of the suspensor, revealed the existence of pore-like structures that provide symplastic continuity between the endosperm and suspensor cells and consequently, a way for virus entry into the embryo (Roberts et al., 2003).

2.2 Assisted virus entry in horizontal transmission

The entry of a virus into a plant from another plant of the same generation (horizontal transmission) is most frequently assisted by non-vertebrate biological vectors, which breach the cell wall when feeding or collecting nectar, but also by fungal cell wall degrading enzymes.

2.2.1 Virus entry assisted by piercing and sucking non-vertebrate vectors

Many plant viruses are transmitted by biological vectors that feed on the plant (Fig. 2C). However, the mechanism controlling their release into another plant is yet to be understood (Dietzgen et al., 2016). In the most simplistic mode, viruses with stable particles and high titers, such as tobamoviruses and potexviruses, would associate in a non-specific manner with the mouthparts of chewing insects or with the distal tips of sucking insect stylets during feeding. In this scenario, viruses are simply inoculated into the next plant as the vector starts feeding. Unfortunately, no experimental evidence of this mechanism has been provided (Jones, 2018).

A most plausible ingestion-egestion or ingestion-salivation mechanism has been proposed in which viral particle adhesion would take place along the anterior part of the insect's alimentary canal, i.e., the proximal tip of the maxillary stylet and foregut. The virion would later be released into a new plant during regurgitation and/or salivation. This mechanism is potentially used by non-persistent and semipersistent viruses that are commonly acquired by piercing and sucking non-vertebrate vectors. The needle-like mouthparts of these vectors can mechanically penetrate the cell wall and plasma membrane, assisted or not by salivary and gut enzymes. During feeding, vector secreted proteins can elicit a plant wound response, preventing severe damage of the cell or phloem clogging (Zhang et al., 2017). This is essential for successful virus infection, since viruses need to replicate and produce MPs in the first inoculated cell before they can move to neighboring cells.

The molecular interactions that occur between viruses and vectors implicate only the viral coat protein (CP) or additional non-structural virus-encoded proteins named helper components. The first situation is found in the aphid-transmitted cucumovirus, alfamovirus and carlavirus, nematode-transmitted nepovirus and whitefly-transmitted crinivirus. For example, *Cucumber mosaic virus* (CMV) CP contains a conserved βH-βI loop that is involved in electrostatic interactions with unknown aphid stylet components. Crinivirus particles are composed of four or five virus-encoded proteins but only the minor CP (CPm) is involved in the transmission of *Lettuce infectious yellows virus* by the whitefly *Bemisia tabaci* (Ng and Falk, 2006a). In the second option, the best characterized examples are definitely those of aphid-transmitted potyvirus and caulimovirus. The association of potyviruses with their aphid vectors is assisted by the HC-Pro bridges between the CP and the aphid stylet. The particular amino acids involved in the interaction between HC-Pro and CP have been mapped in the N-terminus of the CP (DTVDAGK motif) and in the C-terminus of the HC-Pro (PKT region). Another HC-Pro relevant domain near the N-terminus, the KITC region, is involved in stylet retention. In contrast, *Cauliflower mosaic virus* (CaMV) requires two helper proteins (p2 and p3) and the major CP (p4); in this case p3 acts as a bridge between p2, which binds the aphid stylet, and p4 (Ng and Falk, 2006b).

Finally, there is a wide variety of pathosystems where viruses can circulate, replicating (propagative) or not (circulative) inside their vectors, hence the term persistent. They include viruses transmitted by aphids (luteovirus, umbravirus, enamovirus and nanovirus), leafhoppers or planthoppers (geminivirus, rhabdovirus, phytoreovirus, marafivirus, tenuivirus), whiteflies (begomovirus), thrips (tospovirus), nematodes (nepovirus and tobravirus) and some others. In many of these systems, the ingested virions must pass through the gut and the hemocoel to reach the salivary glands. Finally, during salivation, the viruses are inoculated through the salivary ducts. In order to gain access to this way, the viral nucleic acids must be packaged since unencapsidated RNAs are not acquired. Therefore, a specific interaction between the CP and some vector membrane components is absolutely required. Since the molecular and cellular processes taking place within these vectors and the viral localization during circulation have been extensively reviewed (Blanc et al., 2014; Bragard et al., 2013; Dietzgen et al., 2016; Hull, 2014), this subject will not be further addressed in this review.

2.2.2 Virus entry assisted by fungi

Virus entry into plant roots can be assisted by two zoosporic fungi in the class *Chytridiomycetes* (*Olpidium bornovanus* and *O. brassicae/virulentus*). Virus acquisition varies depending on whether viruses have icosahedral (termed in vitro) or rod-shaped/filamentous (termed in vivo) virions (Fig. 2D). In vitro acquisition has been reported in 10 species belonging to 5 genera in the family *Tombusviridae*. These viruses must be acquired by zoospores in the soil/water medium surrounding the plant roots, since they can infect neither the resting spores nor the sporangia. A highly specific interaction between the viral CP and a zoospore glycoprotein receptor allows the adsorption of virus to the zoospore membrane (plasmalemma) (Ohki et al., 2010). However, the mechanism by which these viruses cross the plasmalemma to gain access to the zoospore cytoplasm is unknown.

It was proposed that virus internalization could take place during the withdrawal of the flagellum at the initial stage of zoospore encystment. Besides, it was observed that plasmalemma association induces a conformational change in the CNV CP that superficially exposes its R domain. Since this CP domain is involved in RNA genome interaction, conformational changes could mediate the passage of viral RNA through the plasmalemma. In contrast, the *Mirafiori lettuce big vein virus* and several varicosa viruses can be directly acquired in vivo by resting spores, sporangia and emerging zoospores as fungus grows in the root (Rochon et al., 2004).

Regardless of the way the viruses are acquired, the mechanism of release must be similar. Zoospore penetration in the root cell involves attachment to the root surface, encystment, enzymatic degradation of the cell wall, and the mechanical introduction of the cyst protoplasm. Once there, the carried viruses could reach the root cell cytoplasm before the appearance of the new plasmalemma or, alternatively, they could cross it by means of the above-mentioned CP conformational-dependent mechanism. At that moment, the viral infection of plant tissues can be initiated but the means by which in vivo acquisition occurs is still unknown (Rochon et al., 2004).

3. The plasmodesmata as a regulatory checkpoint in cell-to-cell movement

3.1 Plasmodesmata and movement proteins: Condemned to perpetual understanding

Plasmodesmata are intercellular pores connecting adjacent plant cells allowing membrane and cytoplasmic continuity and are essential routes for intercellular trafficking, communication and signaling in plant development and defense (Ganusova and Burch-Smith, 2019). PD establishes a direct connection of the plasma membrane, endoplasmic reticulum and cytoplasm between neighbor cells, forming a symplastic network that allows for the direct molecular exchange between cells. The inner part of PD consists of the desmotubule or a strand of endoplasmic reticulum (ER) that is continuous between neighbor cells. The presence of the desmotubule generates a space between the plasma membrane and the ER filled with the cytoplasm cells, called the cytoplasmic sleeve, in which there are some proteins or spokes located that divide this space into smaller domains (Overall and Blackman, 1996). These proteins are located on the surface of both membranes to regulate molecule movement (Cilia and Jackson, 2004). Different types of membrane proteins have been characterized including PD-located protein, PDLP1 (Thomas et al., 2008), PD callose-binding proteins localized to the neck region of PD (Simpson et al., 2009), remorins (nanodomain-organized proteins located in the plasma membrane and involved in cellular responses in plants) (Raffaele et al., 2009) and also actin and myosin (White and Barton, 2011) (Fig. 3A). PD can be classified according to their origin into primary PD formed during cell division when the ER strands are trapped in the forming cell plate, and secondary PD that are generated de novo after cell division (Burch-Smith et al., 2011). Both primary and secondary PD are structurally diverse, having a simple, twinned or branched type. Communication through PD involves both selective and nonselective movement of various molecules that function in diverse biological processes. PD are dynamic structures and the deposition of the β-1,3-glucan callose in the apoplast around PD modulates the aperture of the pore, affecting the size exclusion limit (SEL) and the size of the molecules that can freely pass through the PD. Callose accumulation at PD is regulated by the antagonistic enzymes responsible for the synthesis and degradation of callose, having activity that has been associated with a range of both biotic and abiotic stress responses (Brunkard and Zambryski, 2017; Tilsner et al., 2016). Molecules move from cell-to-cell through the cytoplasmic sleeve, along the membranes of the ER (Guenoune-Gelbart et al., 2008) or between the ER lumen via the desmotubule lumen (Barton et al., 2011). Closed PD inhibit all traffic, whereas open PD enable the movement of solutes of less than 1 kDa, and dilated PD are permeable to larger molecules, such as proteins or mRNAs. In growing tissues (e.g. embryo), the SELs are generally between 30 and 50 kDa (Kim et al., 2005a), although the PD permeability does not always correlate with the protein molecular weight. Large molecules (protein, mRNA, etc.) considered too large to pass the PD passively, require an active transport mechanism, using specific proteins that increase the SEL (e.g. MP of plant viruses) or/and unfold the transported molecule for translocation via PD.

Apart from the critical PD function in trafficking endogenous molecules and signals, PD are also critical for the establishment of pathogen infections and, in consequence, plants have evolved to limit pathogen transport via PD. Thus, upon infection the PD SEL is reduced by the accumulation of callose, a process mediated by defense hormones (Wang et al., 2013) or reactive oxygen species (Cui and Lee, 2016), with essential roles in the systemic distribution of signaling molecules that induce defense or systemic acquired resistance (SAR) in distal tissues to initiate defense responses

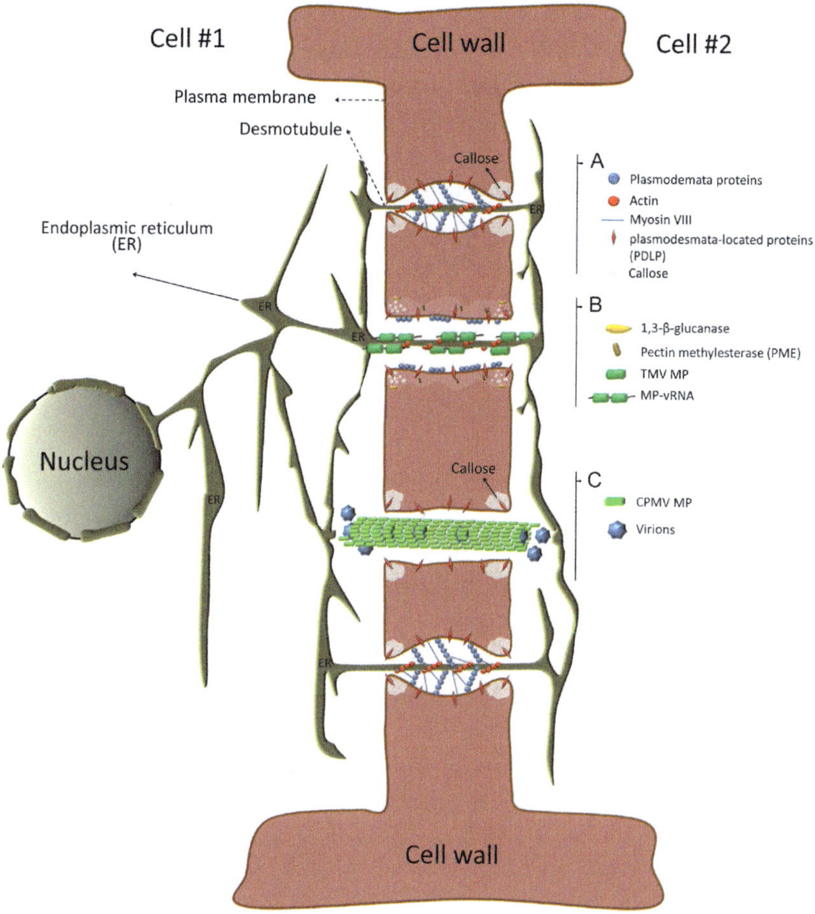

Fig. 3 Schematic representation of the main models of viral intercellular transport based on the movement proteins (MP) of Tobacco mosaic virus (TMV) or Grapevien fanleaf virus (GFLV). (A) Structure of plasmodesmata (PD) in a relaxed state, composed of the desmotubule or prolongation of the endoplasmic reticulum (ER), specific PD proteins, components of the cytoskeleton and callose deposits. (B) Modifications suffered by the PD structure during transport mediated by the MP of TMV. During the TMV infection, the 1,3-β-glucanase enzyme is released in the cell wall degrading the callose depositions at the PD, allowing its dilation. In addition, the transport of ions in the plasma membrane activates other cell wall modifying enzymes that reduce its rigidity. Finally, the structural disorganization of the PD protein components permits the transport of the ribonucleoprotein by diffusion through the membrane of the desmotubule. (C) Modifications generated by the MP of tubule-forming viruses, such as CPMV. In this model, the MP is assembled into tubular structures through its interaction with proteins located in the PD, PDLP. The tubules replace the desmotubule within the PD and allow the transport of virions between the cells, which are transported following the direction of assembly and disassembly of the tubule.

(Ganusova and Burch-Smith, 2019). Accordantly, PD represents a bottleneck for any of the following: plant development, defense and pathogen infection. Conversely, plant pathogens have evolved to use PD structures to facilitate their infections. Plant viruses encode MPs that are essential for either local or systemic transport. Regarding the peculiar PD structure, some common properties have been described for most viral MPs including: (i) the capacity to localize and to move through PD, (ii) the capability to increase the SEL necessary for the translocation of viral complexes, (iii) RNA-binding properties and (iv) association with the host cytoskeleton and the ER network to move viral complexes along the desmotubule. Viral MPs can be classified into two types: MPs that increase the PD SEL without affecting its structure and MPs that self-interact to form tubular structures that modify the PD pore by replacing the PD desmotubule with tightly packed MP molecules (Fig. 3). The first type mediates virus spread from cell-to-cell as ribonucleoprotein complexes (vRNPs), which is exemplified by the TMV (Liu and Nelson, 2013; Luo et al., 2018; Peña and Heinlein, 2012) (Fig. 3B). The second type allows the transport of encapsidated virus particles from the site of infection to neighbor cells, which is exemplified by the MP of *Cowpea mosaic virus* (CPMV) (Fig. 3C) (Ritzenthaler and Hofmann, 2007). In the second mechanism, both CP and MP are required for virus transport. Other viruses require the CP for virus transport but as part of the vRNP, where the formation of virus particles is not a prerequisite for virus translocation (e.g. Kaplan et al., 1998; Sanchez-Navarro and Bol, 2001).

For viruses transported as virions, it remains unclear if they could transport other complexes different from the encapsidated forms. The results obtained with the *Alfalfa mosaic virus* (AMV) model system, that permits the functional exchange of MPs assigned to the 30K family (Melcher, 2000; see next section), showed that the MPs of RNA viruses transported as virus particles (e.g. CPMV or *Grapevine fanleaf virus*, GFLV) are competent to transport vRNP in the AMV system. These results indicated that probably both vRNP and virions could be transported with the same MP (Fajardo et al., 2013; Peiro et al., 2014a, b; Sanchez-Navarro et al., 2006; Sánchez-Navarro et al., 2010). However, the MPs of DNA viruses (e.g. CaMV) that transport virus particles through tubular structures were not competent to transport vRNV in the AMV system (Sánchez-Navarro et al., 2010), suggesting that the transport of vRNP is specific for RNA viruses, independently of whether they form tubular structures or are transported as virus particles.

3.2 Intracellular transport of viruses encoding a single MP: The 30K superfamily concept

Over the last 2 decades, a very large number of MPs have been frequently grouped under the name of the 30K superfamily, as a result of their biochemical and structural similarities with the 30 kDa-MP of the TMV (Mushegian and Koonin, 1993). The "30K" superfamily includes an unexpected diversity of viruses ranging from those with positive- (*Bromoviridae, Comovirinae, Betaflexiviridae, Regressovirinae, Virgaviridae, Tombusviridae, Betaflexiviridae*) and negative-sense (*Rhabdoviridae, Tospoviridae, Phenuiviridae*) ssRNA genome to those with ssDNA (*Geminiviridae*) and dsDNA genome (*Caulimoviridae*). Despite this wide variety of viruses, all MPs belonging to the 30K family share a computer-predicted core domain that is composed of a series of β-elements framed by one α-helix at each side. In addition, they harbor an RNA-binding domain (Melcher, 2000).

With the exception of the bipartite begomovirus, all the above-mentioned viruses encode a single MP. In some of them, such as AMV, CMV and most likely PNRSV, the CP but not the virus particle is also required for virus movement (Aparicio et al., 2010; Kaplan et al., 1998; Sanchez-Navarro et al., 1997; van der Vossen et al., 1994). To enable virus transport, it is known that the C-terminal region of AMV and CMV MP needs to be compatible with the cognate CP (Nagano et al., 1999; Salanki et al., 2004; Sanchez-Navarro and Bol, 2001). Interestingly, deletion of the 29 C-terminal amino acids of the CMV MP increases its RNA-binding affinity (Andreev et al., 2004; Kim et al., 2004) and eliminates the CP requirement (Nagano et al., 2001). Similar CP-independent movement was observed when the 44 and the 38 C-terminal amino acids of the AMV and PNRSV MP, respectively, were removed (Aparicio et al., 2010; Sanchez-Navarro and Bol, 2001). Considering these findings, a model was proposed wherein the CP induces a conformational change in the MP which results in increased protein affinity for the viral RNA. At that point, the ribonucleoprotein complex (RNP) formed acquires a movement-competent state that allows viral transport (Andreev et al., 2004).

In support of this model, the interaction between either the AMV or PNRSV MP and their cognate CP has been shown to be mediated by the 44 and the 38 C-terminal amino acids of the corresponding MP (Aparicio et al., 2010; Sanchez-Navarro et al., 2006). Additionally, experimental evidence has illustrated the strong coordination between the MP RNA-binding properties and viral movement. For example, in the case of PNRSV, small variations of the RNA-binding affinity of the PNRSV MP make this protein unable to allow virus transport (Herranz et al., 2005).

3.2.1 Intracellular transport of plant viruses within the 30K superfamily is not associated with tubule formation: The TMV prototype

TMV has been considered as the model for non-tubule-mediated transport of viral ribonucleoprotein complexes (vRNPs) (Liu and Nelson, 2013; Waigmann et al., 2007). It is well established that TMV MP binds ssRNA (Citovsky et al., 1990), localizes to the PD (Tomenius et al., 1987), increases the PD SEL (Wolf et al., 1989), interacts with microtubules (Heinlein et al., 1995) and with different host cell factors (Niehl and Heinlein, 2011) and can be phosphorylated (Lee and Lucas, 2001). The manner in which the TMV MP interacts with ER membranes has become a new subject of discussion. The first topological model proposed that the TMV MP is an integral membrane protein with two α-helical transmembrane segments. The N- and C-termini are exposed to the cytoplasm and the opposite short loop to the ER lumen (Brill et al., 2000; Fujiki et al., 2006). However, Peiro et al. (2014a, b) showed that the two hydrophobic regions of the TMV MP do not span the biological membranes but are intimately associated with the cellular membranes, allowing maximum exposure of the MP hydrophilic domains to the cytoplasmic cellular components.

Despite all these functional properties, the TMV MP is indispensable but insufficient for cell-to-cell movement. The 126 kDa TMV replicase also is involved in the process (Guenoune-Gelbart et al., 2008) suggesting that viral replication and movement can be tightly linked in TMV (Heinlein, 2015) (Fig. 4A). Very early after TMV capsid disassembly, the vRNA is attached through its 5′ methylguanosine cap to specific endoplasmic reticulum (ER) sites, developing granular structures (Christensen et al., 2009). It also was suggested that the TMV MP associates with ER membranes and binds newly replicated vRNA forming a membrane-associated complex that most likely represents the TMV viral replication complex (VRC) (Asurmendi et al., 2004; Mas and Beachy, 1999; Reichel and Beachy, 1998). Immunolabeling fluorescent and electron microscopy revealed that these VRCs are localized at the opening of PDs and contained the proteins involved in TMV genome replication (126 and 183 kDa) and large amounts of MP (Saito et al., 1987; Szécsi et al., 1999). It is important to indicate that using in situ hybridization techniques (Mas and Beachy, 1999, 2000) and an original live-cell imaging method that combined human RNA-binding protein Pumilio1 and

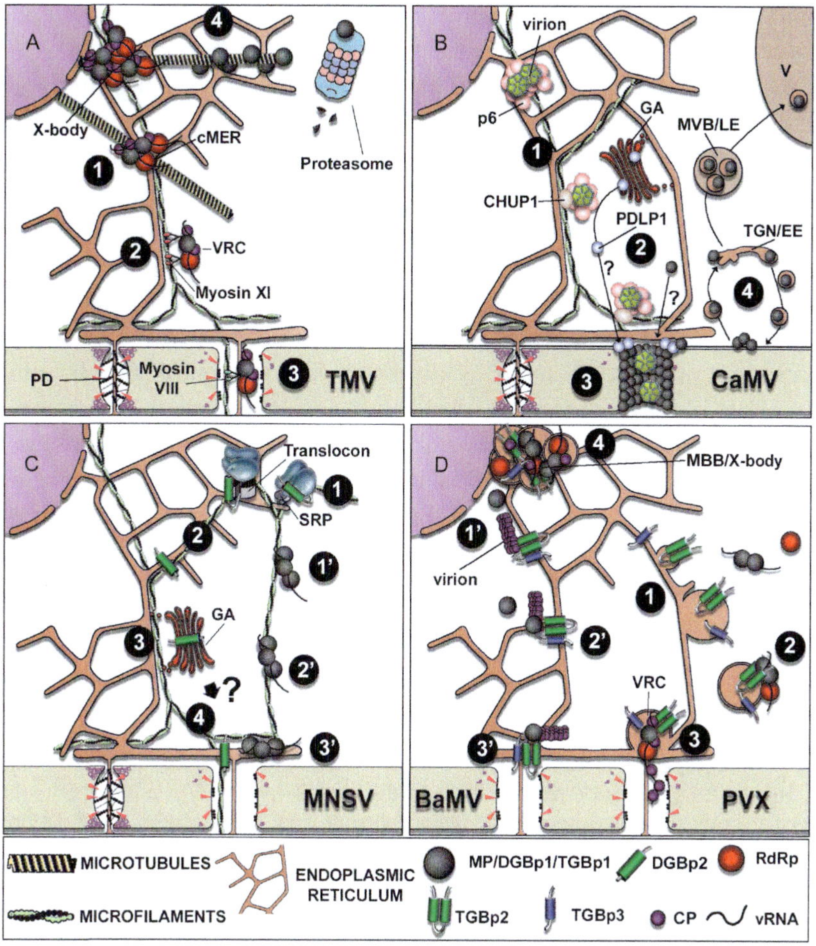

Fig. 4 Four models for cell-to-cell movement of plant viruses. (A) Tobacco mosaic virus, TMV. Early in the infection TMV RNA and MP associate with the endoplasmic reticulum (ER) at C-MERs. After recruitment of essential components for replication, viral replication complexes (VRCs) are thereby established for initial replication (1). Some VRCs may be detached from C-MERs and transported through the actin-ER network (2). Once in the plasmodesmata (PD), the aperture is enlarged by the MP and VRCs move to adjacent cells (3). At middle and late stages of the infection, MPs accumulate at VRCs and microtubules before degradation. Besides, VRCs becomes large viral factories named X-bodies (4). (B) Cauliflower mosaic virus, CaMV. After aphid mediated entry of CaMV into the plant cell, translation of viral proteins occurs at p6 inclusion bodies (IB) generating virion factories. At some stage of the infection, p6 IBs move to PD through actin microfilaments by recruiting CHUP1 from chloroplasts (1). CaMV MP also moves to the PD, independently of p6 IB, by using an unknown pathway and there, MP and p6 interact with PDLP1, a host protein that is targeted to the PD via the Golgi apparatus (GA) (2). CaMV MP modifies the PD by replacing the desmotubule by a tubule of tightly packed

(Continued)

bimolecular fluorescence complementation (Tilsner et al., 2009), the vRNA also localized at the VRCs confirming their role as replication sites. Moreover, ectopically expressed TMV MP localized to PD together with its cognate mRNA transcript (Sambade et al., 2008).

Kawakami et al. (2004) measured the rate at which TMV VRCs, labeled with a MP-green fluorescence protein fusion, move from cell-to-cell. Interestingly, cell-to-cell movement from secondary to tertiary infected cell was 5–10 times faster than required to reach the PD in the primarily infected cell. To explain that difference, it was proposed that the viral movement-competent complex (VMC) traversing the plasmodesmatal orifice should contain not only vRNA and MPs but also the factors required for viral replication. It is still unknown whether VRCs and VMCs are the same complexes, but in any case, VRC assembly in newly infected cells would not start from the early beginning of the replication process, thereby enhancing infection and leading to a faster movement. Taken together, these findings indicate that TMV replication can be spatially linked to movement, raising questions for discussion such as: how do VRCs move within cells to reach the PD? or what role does the MP play in this process?

Fig. 4—cont'd MP molecules that allows the passage of CaMV virions (3). CaMV MP is recycled through TGN/EE (trans-Golgi network/early endosome), plasma membrane (MP) and PD probably to regulate the correct uptake of the protein to PD for tubule formation (4). The excess MP may be recycled back to the central vacuole (V) via MVB/LE (multivesicular bodies/late endosome). (C) Melon necrotic spot virus, MNSV. MNSV DGBp1 and DGBp2 may use two independent pathways to reach the cellular periphery and PDs. Self-interacting DGBp1s bind vRNA to form RNPs (1′) that associate with microfilaments (2′) and move to cellular periphery in an actin-dependent manner (3′). Besides, after cotranslational/signal recognition particle (SRP)-dependent and translocon-assisted insertion into ER membranes (1), DGBp2 move along the ER in an actin-dependent manner (2) and is exported to the Golgi apparatus (GA) (2). Finally, DGBp2 is targeted to PD via an unidentified post-Golgi route (4). (D) Potato virus X, PVX. PVX TGBp2 induces the budding of ER vesicles containing TGBp3 and other replication components such as the viral replicase (RdRp)/vRNA complex (1). This putative VRCs reach the PD traveling along the ER-actin network driven by the PD targeting properties of TGBp2 and TGBp3 (2). Once there, the vRNA is introduced into the PD orifice while replication is still going on (3). In late stages of the infection, a perinuclear and intricate viral factory or X-body is generated (4). Alternatively, at the viral factories of Bamboo mosaic virus, BaMV, (VRC or ER-derived membrane-bound body, MBB), newly translate TGBp2 and TGBp3 form a membrane-associated complex that recruit viral particles through TGBp2-CP interaction (1′). Subsequently, TGBp2/3-virion complex recruits TGBp1 from VRCs (1′) or from the cytosolic pool (2′). The ternary complex TGBp1/2/3-virion, driven by TGBp3, then moves alongside the ER network toward the tightly curved desmotubules (3′).

Using either low or high concentrations of brefeldin A, an inhibitor of the secretory pathway, it was demonstrated that the intracellular trafficking of TMV MP to the PD is not dependent on a functional secretion pathway but requires intact ER (Tagami and Watanabe, 2007; Wright et al., 2007). The existence of direct ER membrane continuities between different cells through the desmotubule suggests that TMV employs a mechanism that involves the lateral diffusion of VRCs along ER membranes (Pitzalis and Heinlein, 2017).

A much-discussed subject has commonly turned around the driving force involved in VRC transport along the ER membrane and the relative contributions of actin microfilaments and microtubules in MP targeting to PDs. Some authors have considered that VRCs are transported directly by an actomyosin-dependent mechanism (Kawakami et al., 2004; Liu and Nelson, 2013; Wright et al., 2007). Although MP does not align with microfilaments (Hofmann et al., 2009), the link between the cytoskeleton and the VRCs could be provided by the 126kDa replicase, which co-aligns with and traffics along microfilaments in a myosin XI-2-dependent manner (Harries et al., 2009; Liu and Nelson, 2013). The association of the 126kDa replicase with the cytoskeleton is probably facilitated by interaction with the host elongation factor 1A (EF-1A), that forms complexes with tubulin and actin and, in turn, interacts with the vRNA 3′ UTR (Yamaji et al., 2006; Zeenko et al., 2002). The active role of specific class VIII and XI myosin family members in TMV movement has also been described (Amari et al., 2014). Class XI myosins could support the ER-associated movement of the VRCs but class VIII, which links desmotubule to the plasma membrane by forming radial spoke-like extensions, could assist intercellular transport through PD. Besides, TMV MP interacts with SYT1, a contractile calcium-binding synaptotagmin located in the contact sites between the ER and the plasma membrane at PD. SYT1 can change the distance between tethered membranes and forms complexes with the membrane tethering factor VAP27, the actin-binding protein NETC3, and the ER-associated actin and microtubule networks. It has been hypothesized that SYT1 could be involved in PD SEL modification assisted by myosin VIII. Therefore, TMV MP binding to SYT1 could interfere with its contracting ability, leading to a constitutive increase of the PD SEL (Pitzalis and Heinlein, 2017). Recent work has shown the presence of plasmodesmal localization signal (PLS) in the MP of TMV (Yuan et al., 2016) or Sugarcane mosaic virus (SCMV) (Xu et al., 2017), indicating the presence of putative cellular proteins that mediated the

MP targeting or association with PD. In this sense, SYT1, has been shown to regulate the capacity of diverse viral MPs to alter PD to promote virus transport (Levy et al., 2015; Lewis and Lazarowitz, 2010) and to interact with the TMV MP through the PLS (Yuan et al., 2018). A similar observation showed that the localization of potyvirus MP (P3N-PIPO) to the PD was affected by a direct interaction with a membrane raft-docking protein (AtPCaP1) (Vijayapalani et al., 2012), a developmentally regulated plasma membrane protein of *N. benthamiana* (NbDREPP) (Geng et al., 2015) or with a plasma membrane-associated Ca^{2+}-binding protein 1 (ScPCaP1) (Xu et al., 2017). It is most likely that such interactions represent one step in the PD targeting pathway of MP. The PLS has been associated with the transport of viral MPs to the PD, but fluorescent proteins carrying only the PLS portion remain associated with the plasma membrane rather than being incorporated into the PD (Yuan et al., 2016). This indicates that other proteins are required for the incorporation of MPs in the PD structure.

Finally, the role of microtubules in TMV movement is supported by evidence from a wide variety of studies (Ashby et al., 2006; Boyko et al., 2007; Ferralli et al., 2006; Kragler et al., 2003; Ouko et al., 2010; Seemanpillai et al., 2006). Similar to the stop-and-go movement of the Golgi stacks, the early TMV VRCs halt their movement at sites where the cortical microtubules intersect with the ER-actin network (Peña and Heinlein, 2012). These specific sites, which have been denominated as cMERs (cortical MT-associated ER sites), may function as centers for macromolecular complex assembly and subsequent delivery to specific organelles (Peña and Heinlein, 2013). Therefore, microtubules are involved in anchoring VRCs to cMERs, allowing the recruitment of host membranes and essential components for replication and movement (Niehl et al., 2012, 2013). In the early stages of the infection, microtubule polymerization may release the VRCs for acto-myosin mediated movement. However, in a controversial hypothesis, microtubule polymerization may also impulse the VRCs along the ER to another cMER. It was proposed that the interaction of TMV MP with the microtubule end-binding protein 1 (EB1) may control the dynamics of the process (Brandner et al., 2008). There could be an overlap between cMERs and ER-plasma membrane contact sites, connecting intracellular and intercellular movement at the edge of the infection (Pitzalis and Heinlein, 2017). In contrast, as the infection progresses, the VRCs coming from different cMERs could fuse to finally become mature viral factories known as X-bodies.

3.2.2 *Intracellular transport of plant viruses within the 30K superfamily associated with tubule formation*

The MPs of some members of the 30K superfamily (Melcher, 2000), predominantly of the families *Comoviridae*, *Bunyaviridae*, *Caulimoviridae* and *Bromoviridae*, work by producing tubules that protrude through the PD. Most of these viruses move from cell-to-cell as intact virus particles that are incorporated into the tubules during the assembly of these structures (Ritzenthaler and Hofmann, 2007). In order to make this happen, MPs and viral particles have to converge at the entry of the PD, independently of whether they are using the same or different intracellular pathways.

Among tubule-forming viruses, studies on intracellular transport essentially have focused on CaMV (Fig. 4B), GFLV and CPMV. The use of cytoskeletal inhibitors like latrunculin B, that prevents de novo actin polymerization and disrupts microfilament organization, and oryzalin, that inhibits microtubules assembly, has little, if any, impact on the targeting and tubule formation of CaMV and CPMV MPs that are ectopically expressed in protoplasts (Huang et al., 2000; Pouwels et al., 2002) and GFLV MP in Bright Yellow 2 (BY-2) cells (Laporte et al., 2003). However, cytoskeleton-specific drugs treatment induces an unusual localization of the GFLV MP tubules within BY-2 cells. When oryzalin was used, numerous small GFLV MP tubules were formed next to, but never going through the cell wall. Upon the disruption of microtubules and microfilaments, GFLV MP tubules were observed within the cytosol close to the nucleus (Laporte et al., 2003). These observations suggest that proper targeting of the GFLV MP to cross walls is strictly dependent on microtubules but, in its absence, microfilaments may target MP to the cell surface. Further investigation is required to determine if this situation is a general rule because in the case of CPMV and CaMV, it was not observed as a result of the use of cell wall-less protoplasts.

The removal of the desmotubule during the intercellular transport of these viruses does not necessarily imply that the endomembrane system is not required for their intracellular transport. In this sense, although the MPs of CaMV and CPMV accumulated in punctate structures in the plasma membrane after protoplast treatment with brefeldin A, tubule formation was drastically impaired (Huang et al., 2000; Pouwels et al., 2002). Once more, GFLV MP behaves differently since brefeldin A interfered with tubule formation and also with MP targeting (Laporte et al., 2003). Carluccio et al. (2014) demonstrated that CaMV MP harbors three redundant endocytosis motifs (Tyr signals) that mediate the recruitment of plasma membrane-derived vesicles or early endosomes probably via interaction with plasma membrane-located adaptins. CaMV MP recycling through early endosomes probably regulates the correct uptake of the protein to PD for tubule formation (Fig. 4B). Therefore, the brefeldin A effect on tubule formation by CaMV, CPMV and GFLV MPs could be related to the endocytic pathway inhibition. In an alternative but not mutually exclusive explanation, treatment with brefeldin A would result in retention in the ER of host proteins that also are required for tubule formation.

Some interesting questions arise from these findings: does a host receptor exist that is responsible for the precise targeting and tethering of tubule-forming MPs to the plasma membrane? Is that host factor also involved in the nucleation of the MP tubule? Since tubules are also formed in non-host plant protoplasts (Wellink et al., 1993) and, unexpectedly, in insect cells (Kasteel et al., 1996; Liu et al., 2011; Storms et al., 1995), the host factor acting as receptor might be conserved among plant and animal species. There is currently insufficient evidence to accept or reject this proposal but some results are pointing in the first direction. Laporte et al. (2003) demonstrated that, after plasmolysis and even in the presence of oryzalin, GFLV MP tubules were not randomly distributed but they projected from specific calreticulin-labeled foci related to membranous structures, intimately linked to the PD that anchor the plasma membrane to the cell wall (Hechtian strands). Fluorescence recovery after photobleaching experiments showed that an indirect interaction via a host protein probably occurs between the CPMV MP tubule and the surrounding plasma membrane (Pouwels et al., 2004). Finally, Amari et al. (2010) found that CaMV and GFLV MPs interact at the base of tubules with the PD-located protein 1 (PDLP1), a member of a redundant family of proteins that controls trafficking between cells (Thomas et al., 2008). The *Arabidopsis thaliana* triple mutant pdlp1/2/3 causes a decrease in the number of cells showing tubules, delayed infection and reduced symptoms. Interestingly, PDLP1 was targeted to the PD using the secretory pathway in a COPII-dependent way (Thomas et al., 2008) and a specialized post-Golgi route driven primarily by class XI myosins (Amari et al., 2011). PDLP1 also interacts with the MP of CPMV in plants but tubule assembly in protoplasts does not require an interaction with PDLP. This suggests that physical interaction between MPs and PDLPs is not required for assembly of the movement tubule, and that the beneficial role of PDLPs in virus movement is confined to the structural context of the PD (den Hollander et al., 2016). Recently the MP of the *Citrus psorosis virus*, the type member of the *Aspiviridae* family, genus *Ophiovirus*, has been shown to interact with PDLP1, and assemble tubule-like structures at PD by a mechanism dependent on the secretory pathway (Robles Luna et al., 2018).

Regarding the intracellular transport of viral particles, it is not clear exactly how it occurs since there is little or no information available. In the case of CaMV, the interaction between CaMV MP and virions may be mediated by the CaMV P3 protein but it has been suggested that the P3/virus complex travels to the PD independently from the MP (Stavolone et al., 2005). Instead, CaMV virus particles are located in cytoplasmic inclusion bodies composed of the CaMV P6 that move within cells along the microfilaments (Harries et al., 2009) and interact with Chloroplast Unusual Positioning 1 protein a plant protein which moves chloroplasts on actin microfilaments (Angel et al., 2013). Interestingly, CaMV P6 is also capable of interacting with the cognate CP and MP (Hapiak et al., 2008). In a possible scenario, CaMV P6 is not only capable of delivering virions to the plasma membrane in a MP-independent manner but it could also favor their transfer to the MP present in the tubule assembling sites. Furthermore, P6 associates with *Arabidopsis thaliana* Soybean Response to Cold 2.2 protein and PDLP1 at the plasmodesmata supporting a model in which P6 functions to transfer CaMV virus particles directly to MP at the PD.

The high concentration of viral MPs at the PD permits their oligomerization (Leastro et al., 2015; Leshchiner et al., 2008; Robles Luna et al., 2018), in which the capacity to form or not to form tubular structures should be an inherent property of the specific MP. Thus, a common sequence feature of the tubule-forming subgroup in the 30K family was the extra length of the N-terminal domain (Melcher, 2000) that was considered to be responsible for polymerization into tubules. The association of MP-derived tubules with PD correlated with structural alterations and cell wall modification of PD (Xie et al., 2016). However, the observation that MP-derived tubules can extend from the PD into the cytoplasm is consistent with the observations that MP-derived tubules can form on the surface of infected protoplasts, where the CW and PD do not exist (Xie et al., 2016). In this sense, the observation of tubular structures on the surface of insect cells in which only the MP is expressed (Kasteel et al., 1996; Liu et al., 2011; Storms et al., 1995; Wellink et al., 1993), implies that MP is the only viral component of the tubule, and that the PD is not required for MP tubule assembly.

Another aspect associated with the MP-derived tubular structures is the capacity to translocate the virus particles through the tubule. In this sense, deletion mutant analysis of the MPs revealed a large C-terminal region required for the assemblage of the structure (Melcher, 2000) that has been postulated to be oriented toward the tubular lumen, interacting with virus particles during its translocation through the tubule (Thomas and Maule, 1999). Thus, direct or indirect interaction between the C-terminal region of the MP and the cognate CP has been reported for different viruses, independently of the MP capacity to generate tubular structures (Aparicio et al., 2010; Carvalho et al., 2003; Sanchez-Navarro et al., 2006; Stavolone et al., 2005).

Other proteins associated with the viral intercellular movement are related to chaperons/co-chaperon complexes that may help in the delivery of viral movement protein to PD or help to unfold and refold viral movement and replication complexes to facilitate virus trafficking between plant cells through PD. Thus, some host factors required for virus infection (e.g. 41 kDa heat shock protein, DnaJ) (Du et al., 2013) are known to interact with Hsp70 or with Hsp90 and RAR1, a signaling component of resistance gene-mediated plant innate immune responses, to form co-chaperon complexes (Liu et al., 2004). The molecular co-chaperone SGT1 is critical for cell-to-cell movement and systemic infection of Tomato spotted wilt virus (Qian et al., 2018) that interacts with Hsp70 (Spiechowicz et al., 2007) and Hsp90 (Liu et al., 2004). Molecular chaperon proteins (HSP60, ATPase-associated with diverse cellular activities -AAA-proteins-) also have been characterized as part of the plasma membrane-tubule complexes isolated from the CPMV-infected protoplast (den Hollander et al., 2016). The critical role of chaperon proteins in viral transport is also reinforced by the observation of some viral genera (*Closterovirus*) encoding an Hsp70 that is critical for cell-to-cell movement (Peremyslov et al., 1999).

3.3 Viruses encoding the double gene block (DGB) movement proteins

Cell-to-cell movement of *Turnip crinkle virus* (TCV), *Melon necrotic spot virus* (MNSV), and *Pelargonium flower break virus* (PFBV) requires the coordinated action of two small proteins encoded in the central region of their genomes (Genoves et al., 2006; Li et al., 1998; Martinez-Turino and Hernandez, 2011). This couple of proteins, initially referred to as the double gene block movement proteins (DGBp1 and DGBp2) (Hull, 2014), is commonly found among members of the genera *Alphacarmovirus*, *Betacarmovirus*, *Gammacarmovirus*, *Alphanecrovirus*, *Betanecrovirus*, *Gallantivirus* and *Panicovirus*, all of which belong to the family *Tombusviridae*. The main components of the movement-competent complex during intracellular transport of these viruses and their mechanism of transportation through the PD is not known. However, CP deletion mutants of TCV and MNSV move from cell-to-cell in *A. thaliana* and melon (*Cucumis melo*) plants, respectively, indicating that the intra- and intercellular transport of these viruses could take place as ribonucleoprotein complexes (Genoves et al., 2006; Hacker et al., 1992).

Comparison of all DGBp sequences available in the database to date revealed that the highest degree of similarity was restricted to secondary structure elements, implying a functional role (Navarro et al., 2006). Structural and molecular studies with *Carnation mottle virus* (CarMV) and MNSV DGBp1s revealed the presence of a basic central region folded in an α-helix that was involved in RNA binding (Marcos et al., 1999; Navarro et al., 2006; Vilar et al., 2001, 2005). In addition, a putative β-sheet conformation at the C-terminus could be involved in host protein interaction and self-interaction (Genoves et al., 2009). MNSV and TCV DGBp1 RNA-binding activity was also shown to be directly associated with cell-to-cell movement. Subcellular fractionation experiments showed that, later in infection, CarMV DGBp1 was mainly associated with the cell wall fraction (Garcia-Castillo et al., 2003). Ectopically expressed MNSV DGBp1 generated cytoplasmic bodies that moved along microfilaments and concentrated at the cellular periphery close to the PD (Genoves et al., 2009) (Fig. 4C). All these findings suggest that DGBp1 could assist the intracellular transport of movement-competent complexes to PD by genome interaction.

The second MP, DGBp2, has been classified as CarMV-like or MNSV-like DGBp2 depending on whether it has one or two potential transmembrane domains (TMD), respectively (Serra-Soriano et al., 2014). CarMV-like DGBp2s, such as those of CarMV and TCV, and MNSV DGBp2 are targeted and co-translationally inserted in vitro into ER-derived microsomes following a signal recognition particle (SRP)-dependent and translocon-assisted process (Martinez-Gil et al., 2007, 2010; Navarro et al., 2006; Sauri et al., 2005). The membrane topology of these proteins was initially determined in vitro. The N- and C-terminal regions of the CarMV DGBp2 are on the opposite sides of the cytoplasm whereas TCV DGBp2 showed an orientation with both terminals exposed to the ER lumen (Martinez-Gil et al., 2007; Vilar et al., 2002). The authors did not rule out the possibility that the protein had other orientations in vivo. Genoves et al. (2011) demonstrated that MNSV DGBp2 has a type II topology in plants as it was previously described using a prokaryotic system. However, topological rules, such as the positive inside rule, were different for each situation (Genoves et al., 2011; Martinez-Gil et al., 2007).

Contemporary work has confirmed that MNSV and CarMV-like PFBV DGBp2s associate with plant ER membranes (Genoves et al., 2010; Martinez-Turino and Hernandez, 2011), although only MNSV DGBp2 has been shown to be targeted to PD via the Golgi apparatus (GA) in a COPII-dependent pathway (Genoves et al., 2010) (Fig. 4C). A mutational analysis showed that ER-to-GA transport of MNSV DGBp2 was essential but insufficient to promote MNSV cell-to-cell movement that required the final localization of the MP in the PD (Genoves et al., 2011). Although single-pass membrane proteins containing TMDs of about 19–20 amino acids in length are commonly targeted to the GA (Brandizzi et al., 2002), the TMD of MNSV DGBp2 by itself was unable to promote the export of the green fluorescence protein out of the ER to the GA, resulting in ER retention. Instead, ER export of MNSV DGBp2 relied on specific sorting motifs located in the cytoplasmic domain (DSSP), as expected for a COPII-dependent cargo, but also in the luminal domain (K49). An essential condition for ER export was the K49-dependent lateral diffusion of MNSV DGBp2 along the ER membranes, probably driven by the ER-associated actin system. Therefore, the authors suggested that connection with the actin-ER network must involve interaction with an actin-associated transmembrane adaptor (Serra-Soriano et al., 2014).

An early model was proposed in which the DGBp2 cytosolic regions would interact with DGBp1/vRNA. This ternary complex would be transported from the mitochondrial associated-replication sites to PD through the endomembrane system in a DGBp2-driven process (Navarro et al., 2006; Vilar et al., 2002). Although this was a striking and simple model, it presents certain inconsistencies. As mentioned above, the DGBp1 C-terminus could be involved in host protein interaction and self-interaction, and PFBV DGBp2 has a leucine zipper-like motif, which is possibly involved in protein-protein interactions. However, to date there is no reported interaction between DGBp1 and DGBp2. Moreover, MNSV DGBp1 and DGBp2 can autonomously reach the cellular periphery and the PD, respectively, following different and independent routes. Compounding the issue is the fact that neither of the TCV DGBps appears to be associated with cell periphery or with the cytoskeleton. TCV DGBp1 has a nuclear localization mediated by two nuclear localization signals (NLSs) that are necessary for cell-to-cell movement. TCV DGBp2 is found in both the nucleus and cytoplasm (Cohen et al., 2000). Thus, the mechanism governing carmovirus movement could be more complex than originally thought and, possibly, different models should be considered depending on specific viruses.

3.4 Viruses encoding the triple gene block (TGB) movement proteins

A number of positive ssRNA viruses share a conserved genetic module of three overlapping genes formerly known as the "triple gene block" (TGB). Sequence comparison revealed that the TGB module is present in a wide range of genera including *Potexvirus*, *Carlavirus*, *Allexivirus*, *Foveavirus*, *Hordeivirus*, *Benyvirus*, *Pomovirus* and *Pecluvirus*. TGB-encoded proteins (TGBp1, TGBp2 and TGBp3) are required for virus movement as they act in concert to deliver viral genomes to and through the PD into adjacent cells. The TGB modules were initially classified into two different groups depending on the structural properties of the proteins: potex-like and hordei-like. However, two novel classes corresponding to the TGB proteins of Beet necrotic yellow vein virus (BNYVV) and some related benyvirus, and the TGB proteins of Hibiscus green spot virus (HGSV) have also been described (Morozov and Solovyev, 2003; Solovyev et al., 2012; Verchot-Lubicz et al., 2010).

Most of the studies indicate that TGBp1 self-interacts, binds to RNA, ATP and Mg^{2+} and possesses helicase and NTPase activities (Verchot-Lubicz et al., 2010). In addition, potexvirus TGBp1 increases the PD SEL (Howard et al., 2004), promotes translation of virus-derived RNA (Atabekov et al., 2000; Rodionova et al., 2003), suppresses the silencing spread through the plant by promoting Argonaute 1 (AGO1) degradation (Chiu et al., 2010; Voinnet et al., 2000) and, in some cases, inhibits intracellular silencing (Senshu et al., 2011). TGBp1s are commonly found in the cytoplasm and nucleus as inclusion bodies (Chang et al., 1997; Samuels et al., 2007). Hence, it is thought that TGBp1 is responsible for remodeling or formation of movement-competent RNPs.

In contrast, TGBp2 and TGBp3 are small integral membrane proteins localized in the ER and/or ER-derived granular vesicles (see Park et al., 2014; Solovyev et al., 2012 for review). TGBp2 exhibits the highest degree of sequence conservation among the TGB proteins. It contains two TMDs with both its N- and C-terminal ends facing the cytosol. In some particular cases, it has been reported that TGBp2 increases PD SEL (Haupt et al., 2005; Tamai and Meshi, 2001) and binds vRNA in a non-specific manner in vitro (Hsu et al., 2009). Unlike TGBp2, TGBp3s forms two main groups depending on the presence of a single (genera *Potexvirus*, *Carlavirus*, *Foveavirus* and *Allexivirus*) or two (genera *Hordeivirus*, *Pomovirus* and *Pecluvirus*) TMDs. It is generally accepted that both TGBp2 and TGBp3 form homologous and heterologous interactions that are required to deliver TGBp1 to the PD.

In particular, it has been postulated that TGBp3 is the driving force for the intracellular transport of the RNPs (Lim et al., 2008, 2009; Samuels et al., 2007; Schepetilnikov et al., 2008; Zamyatnin Jr. et al., 2004).

3.4.1 Potex-like movement

Two different models have been proposed to describe the transportation mode of potex-like viruses within cells. The first model, typified by *Potato virus* X (PVX), is based on the presence of TGBp2 vesicles and proposes that replication and movement are tightly connected at the PD (Tilsner et al., 2013) (Fig. 4D). Early in PVX infection, TGBp2 and TGBp3 are inserted into ER membranes but TGBp2 rapidly induces the budding of membranous vesicles containing TGBp3 from the ER (Ju et al., 2005, 2007). Ribosomes, virions, and viral replicases are also included in these granular vesicles that possibly represent early VRCs where replication and translation occur. The replicase/vRNA complex may be indirectly recruited to VRCs by the TGBp2 RNA-binding capability (Bamunusinghe et al., 2009; Tilsner et al., 2013). Due to the PD targeting properties of TGBp2 and TGBp3, VRCs may reach the cellular periphery traveling along the ER-actin network. Once there, some of the VRCs anchor to the PD entry where nascent 5′-terminal vRNA may be diverted by the helicase activity of TGBp1, from replication/translation toward encapsidation with locally synthesized CP (Karpova et al., 2006; Zayakina et al., 2008). The 5′ genome region, that it is being encapsidated, is introduced into the PD orifice while replication in the 3′ end of the viral genome is still progressing. Such a mechanism called "coreplicational insertion," involves heterologous interactions among TGBps, CP, replicase and vRNA and TGBp1/2 PD gating (Tilsner et al., 2013). In late stages of the infection, TGBp1 may reorganize actin and endomembranes (ER and Golgi), recruiting TGBp2/3 vesicles to finally generate a perinuclear intricate viral factory or X-body. A detailed analysis of the X-bodies has shown that it contains large amounts of vRNAs and TGBp2/3 vesicles inside and viral particles at the periphery (Linnik et al., 2013).

An alternative model has been proposed for the potexvirus *Bamboo mosaic virus* (BaMV) based on a TGBp3-driven process (Fig. 4D). BaMV TGBp3 interacts with TGBp2 in the ER to be further targeted to the periphery of the cell where they accumulate in specific structures. In contrast to what was observed in PVX, the cytoskeleton and Golgi-dependent secretory pathway are dispensable for TGBp2/3 transport; however, several conserved residues located at the C-terminal end are required (Harries et al., 2009). It was shown that peripheral structures are subdomains of the cortical ER, enriched in ER-shaping proteins known as reticulons (Lee et al., 2010; Wu et al., 2011). Since the desmotubule is a tightly curved strand of the ER, Tilsner et al. (2013) proposed that ER-shaping reticulons may be involved in plasmodesmata biogenesis. TGBp3, which exhibits a high affinity to desmotubules, may hijack the plasmodesmata biogenesis machinery or cause the reticulons to move within cell and through the PD. It was recently demonstrated that the TGBp2/3-based structure is capable of forming a stable minimal complex with the BaMV viral particles. Therefore, TGBp2/3 recruits virus particles from perinuclear X-bodies through CP-TGBp2 interaction. Subsequently, newly translated TGBp1 is transferred from X-bodies to TGBp2/3-virus complex via interactions with TGBp2 and/or viral CP. The complex formed is delivered to PD in a manner dependent on the TGBp3 targeting signal (Chou et al., 2013). During intercellular movement, CP phosphorylation by casein kinase 2α may assist VRC disassembly, allowing the release of the vRNA at the PD. In addition, two host membrane-associated proteins, a Rab GTPase activation protein 1 and a serine/threonine kinase-like protein of *N. benthamiana*, may be involved in BaMV cell-to-cell movement. The former is possibly related to the intracellular transport of BaMV VRCs associated with membranes, targeting cargo to the PD. The serine/threonine kinase-like protein enhances the movement of BaMV; however, the underlying mechanisms remain unclear (Liou et al., 2015).

3.4.2 Hordei-like movement

In contrast to potex-like viruses, hordei-like virus movement does not require the viral CP (Verchot-Lubicz et al., 2010). Instead, the vRNAs of these viruses are "encapsidated" by TGBp1 in replication sites (Lim et al., 2008). In *Barley stripe mosaic virus* (BSMV) and *Poasemi latent virus* (PSLV) (genus *Hordeivirus*), the TGBp3 interacts with TGBp2 and TGBp1/vRNA in the ER to form a virus movement complex that is directed to PD (Lim et al., 2008; Schepetilnikov et al., 2008; Zamyatnin Jr. et al., 2004). The C-terminal TMD of PSLV TGBp3 has been identified as the PD targeting element driving this transport (Shemyakina et al., 2011a). The specific role of TGBp2 is unknown but BSMV TGBp2 appears to increase the efficiency of the TGBp3-mediated transport of TGBp1 to the CW. The intracellular transport of PSLV TGBp3 requires neither cytoskeleton (microtubules and microfilaments) nor the secretory pathway, and most probably uses a passive lateral diffusion in ER membranes (Jackson et al., 2009; Schepetilnikov et al., 2008). In contrast, actin cytoskeleton interactions are relevant in BSMV movement since Latrunculin B treatments reduces the amount of TGBp3 appearing at the CW and delays the local spread of BSMV (Lim et al., 2013).

In Potato mop-top virus (PMTV) (genus *Pomovirus*), TGBp3 forms ER-derived motile granules that accumulate at the PD, using the ER-actin network (Haupt et al., 2005; Tilsner et al., 2010). TGBp2 are recruited to these motile granules by direct interaction with TGBp3 but TGBp1 is thought to be indirectly incorporated into the TGBp2/3 complex most likely through TGBp2-vRNA-TGBp1 association (Cowan et al., 2002). Additionally, it has been reported that ectopically expressed TGBp1 requires transient association with microtubules and MT-associated bodies to be targeted to PD. The interaction of TGBp1-microtubules may be important for the formation of transport-competent vRNP bodies as well as degradation of the excessive TGBp1 (Shemyakina et al., 2011b). When motile granules containing TGBp1/2/3/vRNA multimeric complex contacts the PD, TGBp2 and TGBp3 could collaborate in PD gating, thus promoting the transport of vRNA through the PD pore. At this point TGBp2/3, which does not move from cell-to-cell, may associate with endosomal vesicles probably to be reutilized in new rounds of transport (Haupt et al., 2005).

3.5 Viruses encoding multiple movement proteins

Virion assembly is a prerequisite for cell-to-cell movement of closteroviruses like *Beet yellows virus* (BYV). These viruses have filamentous virus particles composed of two distinct parts: a long body formed by the major capsid protein (CP) and a short tail formed by the minor capsid protein (CPm) (Dolja, 2003). BYV involves five proteins in its cell-to-cell movement (Alzhanova et al., 2000; Peremyslov et al., 1999), four of which are virion components: HSP70h, a homolog of the cellular heat shock protein HSP70; a 64kDa protein; CPm and CP. The only non-structural BYV MP is a 6-kDa hydrophobic protein (p6). CPm, p64, and Hsp70h act cooperatively in the assembly of viral tails (Alzhanova et al., 2001, 2007; Napuli et al., 2003).

BYV p6 is an integral type III protein that inserts into rough ER membranes. In infected plants, it forms dimers via a disulfide bridge near its N-terminus. Although p6 does not contribute to virion assembly and its specific role is unknown, it is essential for viral transport (Peremyslov et al., 2004; Zamyatnin Jr. et al., 2006). BYV HSP70h has an N-terminal ATPase domain and a C-terminal domain harboring a peptide-binding pocket. It can move autonomously along the microfilaments in a class VII myosin-dependent manner and be finally located in the PD (Avisar et al., 2008; Medina et al., 1999; Prokhnevsky et al., 2005). Inhibition of HSP70h translation (Peremyslov et al., 1999) and MF disassembly (Prokhnevsky et al., 2005) completely arrests translocation of BYV from cell-to-cell. Since BYV HSP70h binds tightly to the short tail of viral particles (Napuli et al., 2000), it has been proposed that short tails might represent a movement device empowered by the ATPase activity of Hsp70h (Dolja, 2003).

3.6 Is viral movement coupled to viral replication process?

It is clear that a tight connection between virus replication and movement represents an efficient way of enhancing virus movement, since it facilitates the initiation of replication in newly colonized cells. An increasing number of studies have shown this correlation between the presence of mobile membrane-associated inclusions containing VRCs and viral cell-to-cell movement. The above-described mechanisms, used by TMV or PVX for cell-to-cell movement at the edge of the infection are clear examples of this.

The association between replication and movement can be particularly relevant in multipartite viruses. This is because they must translocate two or more genome segments before their replication can be restarted in recipient cells. One such example is typified by the dianthovirus *Red clover necrotic mosaic virus* (RCNMV). The RCMNV genome is divided into two RNA molecules, RNA1 and RNA2. RNA1 encodes an auxiliary replicase, the RdRp and the CP, while RNA2 only encodes the MP. Unlike TMV, the ER membrane arrangements involved in RCMNV VRCs formation are triggered by two replicases. RCMNV VRCs also include some chaperones (HSP70 and HSP90) and small GTPases (Arf1 and Sar1) involved in vesicle trafficking (Hyodo et al., 2013; Mine et al., 2012). The auxiliary replicase recruits the RNA2 template for replication and other VRC components but also function as an ER membrane anchor (Kusumanegara et al., 2012). Similar to what was observed in TMV, RCMNV VRCs change their localization and size from early cortical punctate bodies to later bigger perinuclear aggregates.

RCNMV MP (35k) belongs to the 30k superfamily and thus it shares some characteristics with the TMV MP. It binds single-stranded nucleic acids and increases the PD aperture (Park et al., 2013). However, it is exclusively located at PD when it is expressed alone, but during infection it is also recruited to VCRs in the ER membranes. This binding is facilitated by the glyceraldehyde 3-phosphate dehydrogenase, which also interacts with the auxiliary replicase (Kaido et al., 2014). Interestingly, MP recruitment to VRC only involves the negative strand and/or replication of RNA1 but not RNA2. This seems to be a strategy of keeping the MP close to RNA1, thus ensuring the formation of transport complexes containing RNA1. Since RNA2 encodes the MP, it will certainly be close to the MP. Thus, MPs may simply attach RNA2 in their vicinity after or during translation (Kaido et al., 2009). It remains unsolved whether RNA1 and RNA2 form independent transport complexes or whether they travel in a single transport complex stabilized by the interaction among proteins or RNAs.

For potyviruses it seems that there is no other option other than engaging in simultaneous movement and replication. Members of the genus *Potyvirus*, the largest known plant virus genus, do not code specific MPs. Nevertheless, accumulated experimental evidences in recent years, mainly using *Turnip mosaic virus* (TuMV) and *Tobacco etch virus* (TEV), indicate that the intracellular movement of these viruses mainly rely on VRCs (Heinlein, 2015). When TuMV infects a cell, the viral protein 6K2 induces the formation of ER-derived motile vesicles at ER exit sites, in a COPII-COPI-dependent manner. Very recently, it was shown that 6K2 interacts with Root Hair Defective 3 protein (RHD3), an ER fusogen required for efficient ER fusion (Movahed et al., 2019). The emerging vesicles are indeed VRCs as they contain vRNAs, dsRNAs, replication-related viral proteins, and host factors (Cotton et al., 2009). These 6K2-induced vesicles move intracellularly along microfilaments and ER membranes but controversial data have been reported about the route that they later take for virus replication. The 6K2 vesicles can migrate from the ER to the GA and from there to chloroplasts, where they induce additional membrane rearrangements (Wei et al., 2010a). However, it has also been published that 6K2 vesicles can circumvent the GA by taking an unconventional route that involves pre-vacuolar compartment/multivesicular bodies (Cabanillas et al., 2018). In any case, the intercellular movement of TuMV requires at least two viral PD-located proteins, P3N-PIPO and CI. The PD targeting of P3N-PIPO requires a 19 amino acid region at the N-terminus of PIPO and a functional COPII-COPI-dependent system but not microfilaments (Vijayapalani et al., 2012). In addition, P3N-PIPO targeting to the PM or PD may be assisted by the Ca^{2+}-binding protein 1(ScPCaP1), but this does not determine its final location (Cheng et al., 2017). In contrast, the PD localization of CI is possibly dependent on the physical interaction with the PD. P3N-PIPO can modify the PD SEL (Wei et al., 2010b) while CI can function as a helicase.

The 6K2 interacts with CI but not P3N-PIPO; this interaction is essential for intercellular movement of 6K2 vesicles. Thus, CI could be acting as a docking point between P3N-PIPO and 6K2-coated VRCs in cell-to-cell movement (Movahed et al., 2017).

4. The bottleneck at the phloem cells to systemically invade the plant

4.1 The phloem as a robust but dynamic conductive tissue required for virus transport

The phloem consists of enucleate conducting sieve elements (SEs) in close association with the non-conducting companion cells (CCs), the phloem parenchyma (PP) and the bundle sheath (BS) cells (Lucas et al., 2013; Zhang and Turgeon, 2018) (Fig. 1). Phloem cells interface vary broadly in terms of PD frequency and morphology, and these specialized PD connections can regulate entry and exit to phloem, and condition the strategies utilized by each plant species for phloem-mediated long-distance transport (Lee and Frank, 2018). With the exception of viruses that use xylem vessels (e.g. Moreno et al., 2004; Opalka et al., 1998; Wan et al., 2015), most of the plant viruses usurp the phloem distribution routes to systemically invade the plant (Pallas et al., 2011; Vuorinen et al., 2011). Although a significant part of the translational machinery is present in the phloem of some angiosperms (Lin et al., 2009), there is no proven evidence yet for protein translation (Kragler, 2010). Thus, the inability to be expressed in this conductive tissue is a key bottleneck for the success of plant viruses in a systemic infection. This is particularly relevant to multipartite viruses that must coordinate phloem transport steps to maintain genome integrity and address at least one of each genome segment per cell to recapitulate the viral cycle (Gilmer et al., 2018). Recently, however, Sicard et al. (2019) have shown that distinct viral genome segments of a multipartite plant DNA virus do not necessarily co-localize in individual infected cells. These authors propose a viral multicellular way of life for these viruses where the infection operates at a level above the individual cell level (Sicard et al., 2019).

Although viruses need phloem to translocate long distances (Vuorinen et al., 2011), this conductive tissue does not remain invariable and unmodified under virus infection. Thus, compelling evidence has demonstrated that the proteome of phloem sap is dynamic and under developmental control. CMV induced the up-accumulation of a limited number of phloem proteins associated with stress responses (Malter and Wolf, 2011).

Serra-Soriano et al. (2015) identified 19 different proteins whose accumulation in the phloem sap was altered during a compatible plant virus infection including redox and hypersensitivity response-related proteins. The levels of many amino acids and organic acids as well as sugar acids were higher in the phloem sap of Citrus tristeza virus-infected plants while glucose and sugar alcohols were decreased as a result of CTV infection (Killiny et al., 2017). Viral infection can even cause a quantitative shift in the mode by which photoassimilates are loaded into the sieve tube (Gil et al., 2011). More recently, Collum and Culver (2017) used vascular-specific promoters and a translating ribosome affinity purification strategy to identify phloem-associated translatome responses to infection by TMV, demonstrating that the number of translated gene alterations that occurred in response to infection is at least fourfold higher in phloem-specific translatomes than in non-phloem translatomes. The same authors observed that TMV reprograms phloem transcription by interfering with AUX/IAA proteins, resulting in favorable conditions for viral phloem loading and systemic transport specifically in mature tissue (Collum et al., 2016). At the transcriptomic level, very few data exist on the influence of viral infection on the phloem mRNA population. Ruiz-Medrano et al. (2007) observed that the effect of CMV infection on the composition of the phloem sap transcriptome was minor in nature and that only 91 highly diverse mRNA species, including enzymes involved in general metabolism, transcription factors and signaling agents, were upregulated in infected-phloem tissues. In very few cases, it has been shown that phloem modification could even lead to macroscopic alterations. Thus, a phloem-limited fijivirus induces the formation of neoplastic phloem tissues that house virus multiplication in the host plant (Shen et al., 2016).

Pioneering studies on the long-distance movement of plant viruses led to consolidation of the paradigmatic view that phloem does not only transport sugars, hormones, amino acids and small signaling molecules but that proteins and different types of nucleic acids, including mRNAs, used the same systemic source–sink pathway (Kehr and Kragler, 2018; Lucas et al., 2001, for review). Indeed, recent data obtained by either studying the exchange of mRNAs between a parasitic plant and its hosts (Kim et al., 2014) or by using heterografts of closely related Arabidopsis ecotypes (Thieme et al., 2015) have revealed that mRNA transport can take place on a genomic scale. This unpredictable large number of potentially mobile RNAs would make it unlikely that viral genomes would require a specific structural mRNA motif in order to be transported through the phloem. In contrast to this scenario, Zhang et al. (2016) provided strong evidence that tRNA-related sequences trigger systemic mRNA transport in plants and Banerjee et al. (2009) demonstrated that a sequence motif of StBEL5 RNA, a member of the BEL1-like transcription factor family, located at either of its 3′ or 5′UTR can enable movement of StBEL14, a transcript without long-distance movement capabilities. The mere presence of RNA molecules within the phloem, however, does not guarantee that these RNAs are capable of moving through the phloem without the aid of host proteins and/or a particular structural motif. Phloem can be considered either as a communications integrator or as a checkpoint and surveillance post that limits the presence and movement of certain molecules based on their size and/or specific structural requirements. The manner of movement of plant viruses through the phloem is still a matter of controversy (Folimonova and Tilsner, 2018; Hipper et al., 2013). Well-documented examples exist that demonstrate that viral particles are the viral entities transported through vascular tissues (e.g. Simon-Buela and Garcia-Arenal, 1999; Vaewhongs and Lommel, 1995). However, forms other than particles are also feasible for phloem virus transport since CP can be dispensable or required in a non-particle form (see Gilmer et al., 2018; Hipper et al., 2013; Scholthof, 2005; Waigmann et al., 2004). In the last two cases, viral genomes could be transported with the help of phloem proteins with RNA-binding capabilities (Hipper et al., 2013; Pallas and Gomez, 2013; Spiegelman et al., 2013) as demonstrated for viroid RNA transport (Gómez and Pallás, 2004), for specific phloem mRNAs that are carried through the phloem translocation stream within a polypyrimidine tract binding protein (RBP50) ribonucleotide complex (Ham et al., 2009) or for the systemic delivery of siRNA in pumpkin by a plant PHLOEM SMALL RNA-BINDING PROTEIN 1–ribonucleoprotein complex (Ham et al., 2014). Alternatively, the ribonucleotide complex could be formed between the genomic RNA and proteins of viral origin as demonstrated for CMV that traffics into minor vein sieve elements as a ribonucleoprotein complex that contains the viral RNA, CP and 3a MP, with a subsequent viral assembly occurring in the parietal layer of the sieve element (Blackman et al., 1998). No matter the manner in which plant viruses move through the phloem, to be loaded or to be unloaded, it is obvious that they would interact with a wide variety of host factors that could either promote or restrict their long-distance movement. Since plant virus-host factors interactions have been reviewed during the last few years (Folimonova and Tilsner, 2018; Harries and Ding, 2011; Hipper et al., 2013; Pallas et al., 2011), they are not described here. Restriction of the systemic movement is one of the developed responses of plants to counteract virus infection. In most cases, this restriction is mediated by host factors related to resistance genes that preclude or modulate virus accession to the phloem. Environmental factors, such as temperature and/or light, can also restrict virus access to the phloem. The next section presents a review of the scenarios for which a specific phloem boundary has been determined.

4.2 The bundle sheath phloem interface

Minor veins are the preferred sites for photosynthate and virus loading, since this class of veins accounts for 95% of the total vein surface in the source leaves (reviewed in Nelson and van Bel, 1998). The minor veins of herbaceous dicot plants are surrounded by the BS cells which enclose the VP cells, CC, SE, and tracheary elements of xylem (X) (Fig. 1). Non-necrotic resistance in soybean line PI 346304 to Cowpea chlorotic mottle virus (CCMV) was related to the inability of the virus to gain access to the BS/phloem interface (Goodrick et al., 1991). Using mild and severe strains of TMV (Ding et al., 1995) or TMV mutants defective in the phloem-dependent movement that infected VP and CC less efficiently than the wild-type TMV, but infected the BS cells with similar efficiency (Ding et al., 1996), these authors also found that the BS-phloem interface was a barrier for virus movement. Using transgenic tobacco exhibiting replicase-mediated resistance, Wintermantel et al. (1997) demonstrated that CMV is restricted from entering minor veins in these resistant plants. The restriction in long-distance virus movement resulted from the inability of CMV to pass from bundle sheath cells into vascular parenchyma and companion cells, preventing access to the conducting elements of the phloem and thus restricting long-distance movement. Thompson and Garcia-Arenal (1998) used a symplastic phloem loading plant, cucumber, and reassortants of CMV and Tomato aspermy virus (TAV) to demonstrate that the BS-phloem interface is a boundary for the systemic movement of these viruses.

In the melon exotic accession PI 161375, the gene *cmv1*, confers recessive resistance to CMV strains of subgroup II (e.g. CMV-LS). Immunogold labeling at transmission electron microscopy level showed that CMV-LS remains restricted to the bundle sheath (BS) cells in the resistant line, and does not invade the vascular parenchyma or intermediary cells, whereas, in the susceptible line "Piel de Sapo" (PS), the virus invades all vein cell types (Guiu-Aragonés et al., 2015). *Cmv1* was identified as a Vacuolar Protein Sorting 41 gene (CmVPS41) conserved among plants, animals and yeast, and required for post-Golgi vesicle trafficking toward the vacuole (Giner et al., 2017). Since *cmv1* is capable of blocking CMV-LS in the BS cells and VPS41 is involved in intracellular membrane trafficking, this suggests that CmVPS41 is involved in CMV transport in the boundary BS cells/phloem, rather than in viral replication. Lartey et al. (1998) identified an *A. thaliana* mutation (*vsm1*) that restricts the systemic movement of tobamoviruses by blocking the step of viral entry into the host plant vasculature from the infected leaf tissue, although the exact cell boundary has not been determined.

Although temperature is a critical factor for phloem-dependent virus movement (e.g. Garcia-Castillo et al., 2001; Xiong et al., 1993), there is no available information about the precise restriction point. Through immunocytochemistry and in situ hybridization, Ding et al. (1999) determined that BMV infection in barley is restricted predominantly to cells within and associated with vasculature, predominantly BS cells, when plants are grown at 24/20 °C (day/night).

4.3 The companion cells-sieve elements complex

The CC-SE complex can be considered as the traffic control center of the phloem (Oparka and Turgeon, 1999). It must function in several roles, including phloem loading and unloading, long-distance transport, and as a quality control center for monitoring the nature of substances passing into and out of the phloem. The size and the number of plasmodesmata favor connections and interactions between both cell types. The plasmodesmal complex is characteristically branched in the CC wall, with a single pore opening into the SE lumen. This functionally important complex has been demonstrated to be a key regulatory checkpoint to restrict virus entry by avoiding systemic dissemination.

Ding et al. (1998) studied the accumulation of tobamoviruses and potyviruses in minor veins of inoculated leaves from representatives of the Solanaceae and Fabaceae plant families. The authors observed that vascular parenchyma cells were the predominant, and sometimes the only cells, to become visibly infected among the cells surrounding the SE suggesting that fully susceptible host infection of CCs is a limiting step in systemic infection. Thus, in apoplastic loading, some species of viruses can bypass the infection of the CC and enter the midvein phloem stream directly from the surrounding parenchyma tissue (Cheng et al., 2000).

Another example in which the complex CC-SE was recalcitrant to virus invasion was the infection of CPMV in cowpea (*Vigna unguiculata*) plants

(Silva et al., 2002). As stated above, this virus uses a tubule-guide mechanism to move from cell-to-cell. Tubular structures, indicative of virus particle movement, were never found in PD connecting phloem parenchyma cells and CC or CC and SE, but they were present in other phloem cells. The authors suggested that CPMV could employ a mechanism for phloem loading and unloading that is different from the typical tubule-guided cell-to-cell movement in other cell types.

Cucumber green mottle mosaic virus (CGMMV) was undetected by immunocytochemical procedures in the intermediary cell–sieve element complex in inoculated cotyledons, where photoassimilate loading takes place (Moreno et al., 2004); although in this case it was shown that the virus can alternatively move systemically via the xylem. The absence of viral infection in the CC-SE has previously been reported for other virus-host combinations such as Sunn-hemp mosaic virus (SHMV) in beans (*Phaseolus vulgaris*) (Ding et al., 1998) and MNSV in melon plants (Gosalvez-Bernal et al., 2008), with the VP–CC boundary acting as an important barrier for systemic movement (see Ueki and Citovsky, 2007; Waigmann et al., 2004). In these situations, it was suggested that viruses could exploit the plasmodesmata between the SE and PP cells to gain access to the phloem rather than entering the CCs directly.

The mechanism by which the long-distance movement of tobacco etch virus (TEV) is blocked in Arabidopsis ecotype Col-0 plants relies on the function of at least three genes: RTM1 (restricted TEV movement 1), RTM2, and RTM3. Analysis of the activity of RTM1 and RTM2 regulatory sequences and localization of RTM1 and RTM2 fusion proteins revealed that components of the RTM system function in the phloem, very probably within the sieve elements (SEs), to restrict TEV long-distance movement (Chisholm et al., 2001). In recent times, the RTM1 jacalin and the RTM2 small heat shock protein have been located at the sieve tubes of Arabidopsis, revealing that they can be used as landmarks for exploring the organization and dynamics of phloem cells in vivo (Cayla et al., 2015).

A small non-AUG-initiated ORF, P3a, that is required for long-distance movement, has recently been discovered in poleroviruses and luteoviruses (Smirnova et al., 2015). In this case, it seems that this small ORF acts by allowing the virus to overcome the CC-SE barrier. The authors have suggested that P3a functions only in the cell where it is expressed and assists the virus in its exit from the infected-phloem cells and loading into the phloem. This could be achieved by increasing the size exclusion limit of the specialized PD that connect the phloem companion cells and the sieve tube, the so-called pore plasmodesmal unit (PPU).

5. Cellular organelles as key checkpoints for viral long-distance movement

Although it is not intuitive to think about how intracellular organelles can modulate virus phloem transport, a large body of evidence has unquestionably demonstrated their critical participation as key checkpoints for viral long-distance movement. In this section, we review different virus-hosts systems in which these organelles, mainly the nucleus/nucleolus and chloroplasts, are essential for successful virus transport through the phloem.

5.1 Nucleus/Nucleolus

Although most (+)ssRNA viruses replicate in the cytoplasm, the nucleus and/or nucleolus have emerged as essential organelles in their infection cycle, and particularly in their long-distance movement (see Solovyev and Savenkov, 2014; Taliansky et al., 2010 for review). One of the best examples in which this requirement has been clearly demonstrated is the umbravirus, Groundnut rosette virus (GRV). Umbraviruses do not code for a CP and, consequently, do not form conventional virus particles in infected plants. They are, however, able to systemically infect their hosts by forming cytoplasmic ribonucleoprotein (RNP) particles between the viral genome and the product of ORF 3. These filamentous GRV RNP complexes also contain fibrillarin, one of the major proteins of the nucleolus whose main function is in rRNA processing and modification. This nuclear protein is recruited and relocalized to the cytoplasm by GRV ORF 3 (Taliansky et al., 2010). Remarkably, silencing of fibrillarin prevents long-distance movement without affecting viral replication or cell-to-cell movement. The RNP particles accumulate as inclusions at the cytoplasm and are thought to move through the phloem facilitating the movement of the viral genome to distal parts of the plant. The requirement of fibrillarin in the promotion of the viral systemic movement has also been demonstrated for rice stripe tenuivirus (RSV). The weak silencing suppressor p2 protein encoded by RNA2 moves to and through the nucleus and interacts with fibrillarin in this organelle (Zheng et al., 2015). Fibrillarin depletion affects the nucleolar localization and cytoplasmic distribution of p2 and abolishes the systemic movement of RSV in *N. benthamiana* plants (Zheng et al., 2015).

Another virus that can move systemically in plants in the absence of the CP is PMTV. Pomoviruses contain the conserved triple gene block of movement proteins. The N-terminus of TGBp1 appears to be important for long-distance movement of viral RNA by an unknown mechanism, and such movement is correlated with the association of TGB1 with the nucleolus (Torrance et al., 2011). It was demonstrated that the TGBp1 N-ter domain is required for importin α-mediated nucleolar localization of the virus being this step an essential checkpoint in establishing the efficient systemic infection of the entire plant (Lukhovitskaya et al., 2015). Recently, Li et al. (2018) showed that the TGBp1 of BSMV localizes to the nucleus and the nucleolus, and that TGBp1 possesses both a nucleolar localization signal (NoLS) and a nuclear localization signal (NLS). Neither the NoLS nor NLS mutant viruses could infect *N. benthamiana* systemically. BSMV TGBp1 interacts directly with the glycine–arginine-rich (GAR) domain of the nucleolar protein fibrillarin. These results support a model whereby fibrillarin is hijacked by TGBp1 to form a Fib–RNP movement complex that is required for BSMV cell-to-cell and long-distance movement (Li et al., 2018). Fibrillarin is also able to interact with the P20 protein of Bamboo mosaic potexvirus satRNA (satBaMV) which in turn interacts with the satellite RNA to form a ribonucleoprotein complex that is required for helper virus-independent long-distance trafficking of the satRNA (Chang et al., 2016).

In addition to the fibrirallin-interacting proteins, other evidence demonstrates that the involvement of the nucleus/nucleolus on the viral phloem movement relies on the presence and function of nuclear/nucleolar localization signals in viral proteins. The AMV CP is a multifunctional protein required for diverse processes during infection (Bol, 2005; Pallas et al., 2012). Herranz et al. (2012) identified an N-terminal lysine-rich NoLS in the AMV CP required both for entry and accumulation in the nucleolus of infected cells, and a C-terminal leucine-rich domain that might function as a nuclear export signal. Remarkably, virus particle formation and systemic movement were completely abolished in plants infected with a mutant AMV RNA 3 unable to target the nucleolus. Similarly, basic amino acid mutations in the nuclear localization signal of hibiscus chlorotic ringspot virus p23 prevented it from entering the nucleus, resulting in restriction of virus long-distance movement, which in turn abrogated symptom expression in the newly emerged leaves (Gao and Wong, 2013). Unexpectedly, the 30-K MP of a tobamovirus, the turnip vein clearing virus (TVCV), in addition to localizing to the ER membrane and PD, targeted the nucleus in a NLS-dependent manner, where it localized to novel F-actin-containing filaments that were associated with chromatin (Levy et al., 2013). The 30K NLS was necessary for efficient TVCV cell-to-cell movement and systemic infection in *N. benthamiana* and *A. thaliana* plants, suggesting that nuclear targeting of MP encoded by TVCV and potentially other subgroup 3 tobamoviruses, may play a role in modulating host defense or cellular physiology to favor virus movement and infection (Levy et al., 2013).

5.2 Chloroplasts

Chloroplasts carry out photosynthesis and, thus, are the main metabolic energy sources for the plant cell. In addition, since these organelles serve as sites for the synthesis of important phytohormones, they can mediate relevant plant defense responses against viruses and other phytopathogens. The interaction of viruses with the chloroplasts machinery has been known from the 70s, first by identifying the remarkable ultrastructural modifications caused by virus infection and later on by associating the replication of some important plant RNA viruses with the rearrangements observed in the chloroplast membranes (see Bhattacharyya and Chakraborty, 2018; Budziszewska and Obrepalska-Steplowska, 2018; Li et al., 2016; Zhao et al., 2016).

Although the critical involvement of the chloroplast in viral symptoms (Li et al., 2016), viral replication (Bhattacharyya and Chakraborty, 2018; Budziszewska and Obrepalska-Steplowska, 2018), and cell-to-cell movement (Bhat et al., 2013; Burch-Smith and Zambryski, 2012; Zhao et al., 2016) has been well documented, evidence for the potential role of these organelles in long-distance virus movement is limited to very few examples. *Alternanthera mosaic virus* (AltMV) TGBp3 accumulates preferentially in mesophyll cells. AltMV TGBp3 has an internal signal essential for chloroplast localization (Lim et al., 2010). Deletion of the TGBp3 start codon or alteration of the chloroplast localization signal limited cell-to-cell movement to the epidermis and the mesophyll cells and eliminated long-distance virus movement (Lim et al., 2010). Li et al. (2005) identified a tobacco protein designated IP-L that interacts with Tomato mosaic virus (ToMV) CP using a Gal4-based yeast two-hybrid system. Using virus-induced gene silencing (VIGS), it was shown that IP-L facilitated the long-distance movement of ToMV. Using immunoblotting, Zhang et al. (2008) showed that both the IP-L and a majority of the ToMV CP are co-localized in the chloroplast thylakoid membranes. This led the authors to suggest that this interaction could affect the function and stability of the chloroplast, and thus lead to chlorosis. In addition, the ToMV MP interacts with the Rubisco small subunit (RbCS) of *N. benthamiana* in vitro and in vivo (Zhao et al., 2013). ToMV and TMV were able to establish efficient local infection but their systemic movements were significantly reduced in NbRbCS-silenced plants indicating that this abundant and essential chloroplastic protein can act as a key regulatory checkpoint in the long-distance movement of tobamoviruses.

An Arabidopsis cDNA library generated from companion cells was screened with the CP of LMV and the RTM proteins earlier described in restricting the long-distance movement of different potyviruses, using the yeast two-hybrid system (Sofer et al., 2017). Subcellular studies in *N. benthamiana* plants revealed that most of the viral and cellular protein pairs co-localized at the periphery of chloroplasts, which suggests a putative role for plastids in this process. The authors suggested that one of the interacting proteins, BCAT4, which is a phloem protein, might act as a bridge between RTM2 and the RTM1/RTM3 complex in SE (Sofer et al., 2017).

A remarkable interplay between the PD and chloroplasts in mediating the outcome of virus infection recently has been described. An Arabidopsis mutant with abnormal PD function was identified as the chloroplast-resident DEAH-type RNA helicase named INCREASED SIZE EXCLUSION LIMIT2 (ISE2) for its loss-of-function phenotype of increased PD-mediated trafficking (Burch-Smith and Zambryski, 2010). The expression of ISE2 was induced by viral infection and, remarkably, increased or decreased ISE2 expression led to faster viral systemic spread of TMV and TuMV in *N. benthamiana* plants (Ganusova et al., 2017). These unexpected results revealed a complex relationship between PD and chloroplasts, especially with respect to virus susceptibility and the outcomes of infection. Another significant example in which chloroplast function can modulate long-distance virus movement is the infection of *potato leaf roll virus* (PLRV) in *N. benthamiana.* Disruption of chloroplast function through downregulation of phytoene desaturase promotes the systemic accumulation of the aphid-borne and phloem-limited PLRV (DeBalsio et al., 2018). In addition, the fecundity of the aphid vector was increased when fed on PDS-silenced plants relative to PLRV-infected plants. Again, these results revealed that normal chloroplast function can negatively regulate the systemic accumulation of plant viruses. Finally, it was shown that the Lolium latent virus (LoLV) major CP sequence contains a typical chloroplast transit peptide (cTP) domain that interacts with an ANK repeat (ANKr) protein (Vaira et al., 2018). The interaction of LoLV major CP with NbANKr redirects it to the chloroplasts. The silencing of NbANKr expression in *N. benthamiana* through VIGS experiments significantly compromises systemic infection by LoLV.

6. Conclusions and future trends

The study of plant virus transport has led to seminal discoveries in different areas of plant molecular biology. Since the observations that: (i) the TMV MP increased the size exclusion limit (SEL) of the PD allowing the ribonucleoprotein complex to overcome the rigid physical barrier of the CW (Wolf et al., 1989) and that (ii) a phloem protein parallels the function and properties of a viral movement protein (Xoconostle-Cazares et al., 1999), an increasing number of relevant findings have revealed that viral and cellular RNAs share common mechanisms of transport among cells and through the vascular system. Significant progress has been made on the mechanistic insights dealing with the way viral RNAs reach and cross the plasma membrane and vasculature, but how these processes occur in a specific manner is far from being understood. Although the mobility of RNA is a more widespread property than initially thought, we are still far from understanding the structural requirements that confer this property to viral RNAs. Some issues that still need to be understood in greater depth are: (i) Is there any common structural signature to the viral movement proteins that make them localize in PD? (ii) Are there sequence motifs in viral RNAs that give them a greater ability to move? (iii) What is the nature of the physical entity that moves through the phloem? (iv) What is the intrinsic mechanism by which some cellular organelles are essential for the systemic movement? A study of the translocation of viral RNAs both at the intercellular and vascular levels can continue to help elucidate the mechanisms underlying these processes. Plant virology will continue to be an essential discipline to understanding the development and stress processes of plants.

Acknowledgments

This work was funded by grant BIO2017-88321-R from the Spanish Ministerio de Ciencia, Innovacion y Universidades and the Prometeo Program GV2015/010 from the Generalitat Valenciana. We apologize to those whose work could not be cited due to space limitations.

References

Alzhanova, D.V., Hagiwara, Y., Peremyslov, V.V., Dolja, V.V., 2000. Genetic analysis of the cell-to-cell movement of beet yellows closterovirus. Virology 268, 192–200.

Alzhanova, D.V., Napuli, A.J., Creamer, R., Dolja, V.V., 2001. Cell-to-cell movement and assembly of a plant closterovirus: roles for the capsid proteins and Hsp70 homolog. EMBO J. 20, 6997–7007.

Alzhanova, D.V., Prokhnevsky, A.I., Peremyslov, V.V., Dolja, V.V., 2007. Virion tails of beet yellows virus: coordinated assembly by three structural proteins. Virology 359, 220–226.

Amari, K., Burgos, L., Pallas, V., Sanchez-Pina, M.A., 2007. Prunus necrotic ringspot virus early invasion and its effects on apricot pollen grain performance. Phytopathology 97, 892–899.

Amari, K., Burgos, L., Pallas, V., Sanchez-Pina, M.A., 2009. Vertical transmission of Prunus necrotic ringspot virus: hitch-hiking from gametes to seedling. J. Gen. Virol. 90, 1767–1774.

Amari, K., Boutant, E., Hofmann, C., Schmitt-Keichinger, C., Fernandez-Calvino, L., Didier, P., Lerich, A., Mutterer, J., Thomas, C.L., Heinlein, M., Mely, Y., Maule, A.J., Ritzenthaler, C., 2010. A family of plasmodesmal proteins with receptor-like properties for plant viral movement proteins. PLoS Pathog. 6, e1001119.

Amari, K., Lerich, A., Schmitt-Keichinger, C., Dolja, V.V., Ritzenthaler, C., 2011. Tubule-guided cell-to-cell movement of a plant virus requires class XI myosin motors. PLoS Pathog. 7, e1002327.

Amari, K., Di Donato, M., Dolja, V.V., Heinlein, M., 2014. Myosins VIII and XI play distinct roles in reproduction and transport of tobacco mosaic virus. PLoS Pathog. 10, e1004448.

Andreev, I.A., Hyon Kim, S., Kalinina, N.O., Rakitina, D.V., Fitzgerald, A.G., Palukaitis, P., Taliansky, M.E., 2004. Molecular interactions between a plant virus movement protein and RNA: force spectroscopy investigation. J. Mol. Biol. 339, 1041–1047.

Angel, C.A., Lutz, L., Yang, X., Rodriguez, A., Adair, A., Zhang, Y., Leisner, S.M., Nelson, R.S., Schoelz, J.E., 2013. The P6 protein of Cauliflower mosaic virus interacts with CHUP1, a plant protein which moves chloroplasts on actin microfilaments. Virology 443, 363–374.

Aparicio, F., Pallás, V., Sánchez-Navarro, J., 2010. Implication of the C terminus of the Prunus necrotic ringspot virus movement protein in cell-to-cell transport and in its interaction with the coat protein. J. Gen. Virol. 91, 1865–1870.

Ashby, J., Boutant, E., Seemanpillai, M., Groner, A., Sambade, A., Ritzenthaler, C., Heinlein, M., 2006. Tobacco mosaic virus movement protein functions as a structural microtubule-associated protein. J. Virol. 80, 8329–8344.

Asurmendi, S., Berg, R.H., Koo, J.C., Beachy, R.N., 2004. Coat protein regulates formation of replication complexes during tobacco mosaic virus infection. Proc. Natl. Acad. Sci. U. S. A. 101, 1415–1420.

Atabekov, J.G., Rodionova, N.P., Karpova, O.V., Kozlovsky, S.V., Poljakov, V.Y., 2000. The movement protein-triggered in situ conversion of potato virus X virion RNA from a nontranslatable into a translatable form. Virology 271, 259–263.

Avisar, D., Prokhnevsky, A.I., Dolja, V.V., 2008. Class VIII myosins are required for plasmodesmatal localization of a closterovirus Hsp70 homolog. J. Virol. 82, 2836–2843.

Azizi, P., Rafii, M.Y., Maziah, M., Abdullah, S.N., Hanafi, M.M., Latif, M.A., Rashid, A.A., Sahebi, M., 2015. Understanding the shoot apical meristem regulation: a study of the phytohormones, auxin and cytokinin, in rice. Mech. Dev. 135, 1–15.

Baluska, F., 2009. Cell-cell channels, viruses, and evolution. Ann. N. Y. Acad. Sci. 1178, 106–119.

Bamunusinghe, D., Hemenway, C.L., Nelson, R.S., Sanderfoot, A.A., Ye, C.M., Silva, M.A., Payton, M., Verchot-Lubicz, J., 2009. Analysis of potato virus X replicase and TGBp3 subcellular locations. Virology 393, 272–285.

Banerjee, A.K., Lin, T., Hannapel, D.J., 2009. Untranslated regions of a mobile transcript mediate RNA metabolism. Plant Physiol. 151, 1831–1843.

Barton, D.A., Cole, L., Collings, D.A., Liu, D.Y., Smith, P.M., Day, D.A., Overall, R.L., 2011. Cell-to-cell transport via the lumen of the endoplasmic reticulum. Plant J. 66, 806–817.
Benitez-Alfonso, Y., Faulkner, C., Ritzenthaler, C., Maule, A.J., 2010. Plasmodesmata, gateways to local and systemic virus infection. Mol. Plant Microbe Interact. 23 (11), 1403–1412.
Bhat, S., Folimonova, S.Y., Cole, A.B., Ballard, K.D., Lei, Z., Watson, B.S., Sumner, L.W., Nelson, R.S., 2013. Influence of host chloroplast proteins on tobacco mosaic virus accumulation and intercellular movement. Plant Physiol. 161, 134–147.
Bhattacharyya, D., Chakraborty, S., 2018. Chloroplast, the Trojan horse in plant–virus interaction. Mol. Plant Pathol. 192, 504–518.
Blackman, L.M., Boevink, P., Cruz, S.S., Palukaitis, P., Oparka, K.J., 1998. The movement protein of cucumber mosaic virus traffics into sieve elements in minor veins of Nicotiana clevelandii. Plant Cell 10, 525–538.
Blanc, S., Drucker, M., Uzest, M., 2014. Localizing viruses in their insect vectors. Annu. Rev. Phytopathol. 52, 403–425.
Bol, J.F., 2005. Replication of alfamo- and ilarviruses, role of the coat protein. Annu. Rev. Phytopathol. 43, 39–62.
Boyko, V., Hu, Q., Seemanpillai, M., Ashby, J., Heinlein, M., 2007. Validation of microtubule-associated tobacco mosaic virus RNA movement and involvement of microtubule-aligned particle trafficking. Plant J. 51, 589–603.
Bragard, C., Caciagli, P., Lemaire, O., Lopez-Moya, J.J., MacFarlane, S., Peters, D., Susi, P., Torrance, L., 2013. Status and prospects of plant virus control through interference with vector transmission. Annu. Rev. Phytopathol. 51, 177–201.
Brandizzi, F., Frangne, N., Marc-Martin, S., Hawes, C., Neuhaus, J.M., Paris, N., 2002. The destination for single-pass membrane proteins is influenced markedly by the length of the hydrophobic domain. Plant Cell 14, 1077–1092.
Brandner, K., Sambade, A., Boutant, E., Didier, P., Mely, Y., Ritzenthaler, C., Heinlein, M., 2008. Tobacco mosaic virus movement protein interacts with green fluorescent protein-tagged microtubule end-binding protein 1. Plant Physiol. 147, 611–623.
Brill, L.M., Nunn, R.S., Kahn, T.W., Yeager, M., Beachy, R.N., 2000. Recombinant tobacco mosaic virus movement protein is an RNA-binding, alpha-helical membrane protein. Proc. Natl. Acad. Sci. U. S. A. 97, 7112–7117.
Brunkard, J.O., Zambryski, P.C., 2017. Plasmodesmata enable multicellularity: new insights into their evolution, biogenesis, and functions in development and immunity. Curr. Opin. Plant Biol. 35, 76–83.
Buchmann, J.P., Holmes, E.C., 2015. Cell walls and the convergent evolution of the viral envelope. Microbiol. Mol. Biol. Rev. 79, 403–418.
Budziszewska, M., Obrepalska-Steplowska, A., 2018. The role of the chloroplast in the replication of positive-sense single-stranded plant RNA viruses. Front. Plant Sci. 9, 1776.
Burch-Smith, T.M., Zambryski, P.C., 2010. Loss of increased size exclusion limit ISE1 or ISE2 increases the formation of secondary plasmodesmata. Curr. Biol. 20, 989–993.
Burch-Smith, T.M., Zambryski, P.C., 2012. Plasmodesmata paradigm shift, regulation from without versus within. Annu. Rev. Plant Biol. 63, 239–260.
Burch-Smith, T.M., Stonebloom, S., Xu, M., Zambryski, P.C., 2011. Plasmodesmata during development: re-examination of the importance of primary, secondary, and branched plasmodesmata structure versus function. Protoplasma 248, 61–74.
Cabanillas, D.G., Jiang, J., Movahed, N., Germain, H., Yamaji, Y., Zheng, H., Laliberte, J.F., 2018. Turnip mosaic virus uses the SNARE protein VTI11 in an unconventional route for replication vesicle trafficking. Plant Cell 30, 2594–2615.
Card, S.D., Pearson, M.N., Clover, G.R.G., 2007. Plant pathogens transmitted by pollen. Australas. Plant Pathol. 36, 455–461.

Carluccio, A.V., Zicca, S., Stavolone, L., 2014. Hitching a ride on vesicles: cauliflower mosaic virus movement protein trafficking in the endomembrane system. Plant Physiol. 164, 1261–1270.
Carroll, T.W., Mayhew, D.E., 1976a. Anther and pollen infection in relation to the pollen and seed transmissibility of two strains of barley stripe mosaic virus in barley. Can. J. Bot. 54, 1604–1621.
Carroll, T.W., Mayhew, D.E., 1976b. Occurrence of virions in developing ovules and embryo sacs of barley in relation to the seed transmissibility of barley stripe mosaic virus. Can. J. Bot. 54, 2497–2512.
Carvalho, C.M., Wellink, J., Ribeiro, S.G., Goldbach, R.W., van Lent, J.W.M., 2003. The C-terminal region of the movement protein of Cowpea mosaic virus is involved in binding to the large but not to the small coat protein. J. Gen. Virol. 84, 2271–2277.
Cayla, T., Batailler, B., Le Hir, R., Revers, F., Anstead, J.A., Thompson, G.A., Grandjean, G., Dinant, S., 2015. Live imaging of companion cells and sieve elements in arabidopsis leaves. PLoS One 102, e0118122.
Chang, B.Y., Lin, N.S., Liou, D.Y., Chen, J.P., Liou, G.G., Hsu, Y.H., 1997. Subcellular localization of the 28 kDa protein of the triple-gene-block of bamboo mosaic potexvirus. J. Gen. Virol. 78, 1175–1179.
Chang, C.H., Hsu, F.C., Lee, S.C., Lo, Y.S., Wang, J.D., Shaw, J., Taliansky, M., Chang, B.Y., Hsu, Y.H., Lin, N.S., 2016. The nucleolar fibrillarin protein is required for helper virus-independent long-distance trafficking of a subviral satellite RNA in plants. Plant Cell 28, 2586–2602.
Cheng, N.H., Lin, C.L., Carter, S.A., Nelson, R.S., 2000. Vascular invasion routes and systemic accumulation patterns of tobacco mosaic virus in Nicotiana benthamiana. Plant J. 23, 349–362.
Cheng, G., Dong, M., Xu, Q., Peng, L., Yang, Z., Wei, T., Xu, J., 2017. Dissecting the molecular mechanism of the subcellular localization and cell-to-cell movement of the sugarcane mosaic virus P3N-PIPO. Sci. Rep. 7, 9868.
Chisholm, S.T., Parra, M.A., Anderberg, R.J., Carrington, J.C., 2001. Arabidopsis RTM1 and RTM2 genes function in phloem to restrict long-distance movement of tobacco etch virus. Plant Physiol. 127, 1667–1675.
Chiu, M.H., Chen, I.H., Baulcombe, D.C., Tsai, C.H., 2010. The silencing suppressor P25 of Potato virus X interacts with Argonaute1 and mediates its degradation through the proteasome pathway. Mol. Plant Pathol. 11, 641–649.
Chou, Y.L., Hung, Y.J., Tseng, Y.H., Hsu, H.T., Yang, J.Y., Wung, C.H., Lin, N.S., Meng, M., Hsu, Y.H., Chang, B.Y., 2013. The stable association of virion with the triple-gene-block protein 3-based complex of Bamboo mosaic virus. PLoS Pathog. 9, e1003405.
Christensen, N., Tilsner, J., Bell, K., Hammann, P., Parton, R., Lacomme, C., Oparka, K., 2009. The 5' cap of tobacco mosaic virus (TMV) is required for virion attachment to the actin/endoplasmic reticulum network during early infection. Traffic 10, 536–551.
Cilia, M.L., Jackson, D., 2004. Plasmodesmata form and function. Curr. Opin. Cell Biol. 16, 500–506.
Citovsky, V., Knorr, D., Schuster, G., Zambryski, P., 1990. The P30 movement protein of tobacco mosaic virus is a single-strand nucleic acid binding protein. Cell 60, 637–647.
Cohen, Y., Qu, F., Gisel, A., Morris, T.J., Zambryski, P.C., 2000. Nuclear localization of turnip crinkle virus movement protein p8. Virology 273, 276–285.
Collum, T.D., Culver, J.N., 2017. Tobacco mosaic virus infection disproportionately impacts phloem associated translatomes in Arabidopsis thaliana and Nicotiana benthamiana. Virology 510, 76–89.
Collum, T.D., Padmanabhan, M.S., Hsieh, Y.C., Culver, J.N., 2016. Tobacco mosaic virus-directed reprogramming of auxin/indole acetic acid protein transcriptional responses enhances virus phloem loading. Proc. Natl. Acad. Sci. U. S. A. 113, 2740–2749.

Cotton, S., Grangeon, R., Thivierge, K., Mathieu, I., Ide, C., Wei, T., Wang, A., Laliberte, J.F., 2009. Turnip mosaic virus RNA replication complex vesicles are mobile, align with microfilaments, and are each derived from a single viral genome. J. Virol. 83, 10460–10471.

Cowan, G.H., Lioliopoulou, F., Ziegler, A., Torrance, L., 2002. Subcellular localisation, protein interactions, and RNA binding of Potato mop-top virus triple gene block proteins. Virology 298, 106–115.

Cui, W., Lee, J.Y., 2016. Arabidopsis callose synthases CalS1/8 regulate plasmodesmal permeability during stress. Nat. Plants 2, 16034.

DeBalsio, S.L., Rebelo, A.R., Parks, K., Gray, S.M., Heck, M.C., 2018. Disruption of chloroplast function through downregulation of phytoene desaturase enhances the systemic accumulation of an aphid-borne, phloem-restricted virus. Mol. Plant Microbe Interact. 31, 1095–1110.

den Hollander, P.W., Kieper, S.N., Borst, J.W., van Lent, J.W.M., 2016. The role of plasmodesma-located proteins in tubule-guided virus transport is limited to the plasmodesmata. Arch. Virol. 161, 2431–2440.

Dietzgen, R.G., Mann, K.S., Johnson, K.N., 2016. Plant virus–insect vector interactions: current and potential future research directions. Viruses 8, 303.

Ding, X.S., Shintaku, M.H., Arnold, S.A., Nelson, R.S., 1995. Accumulation of mild and severe strains of tobacco mosaic virus in minor veins of tobacco. Mol. Plant Microbe Interact. 8, 32–40.

Ding, X.S., Shintaku, M.H., Carter, S.A., Nelson, R.S., 1996. Invasion of minor veins of tobacco leaves inoculated with tobacco mosaic virus mutants defective in phloem-dependent movement. Proc. Natl. Acad. Sci. U. S. A. 93, 11155–11160.

Ding, X.S., Carter, S.A., Deom, C.M., Nelson, R.S., 1998. Tobamovirus and potyvirus accumulation in minor veins of inoculated leaves from representatives of the Solanaceae and Fabaceae. Plant Physiol. 116, 125–136.

Ding, X.S., Flasinski, S., Nelson, R.S., 1999. Infection of the Brome mosaic virus is restricted predominantly to sells in and associated with veins through a temperature dependent mechanism. Mol. Plant Microbe Interact. 12, 615–623.

Dolja, V.V., 2003. Beet yellows virus: the importance of being different. Mol. Plant Pathol. 4, 91–98.

Du, Y., Zhao, J., Chen, T., Liu, Q., Zhang, H., Wang, Y., Hong, Y., Xiao, F., Zhang, L., Shen, Q., Liu, Y., 2013. Type I J-domain NbMIP1 proteins are required for both tobacco mosaic virus infection and plant innate immunity. PLoS Pathog. 9, e1003659.

Edwards, M.C., 1995. Mapping of the seed transmission determinants of barley stripe mosaic virus. Mol. Plant Microbe Interat. 8, 906–915.

Fajardo, T.V., Peiro, A., Pallas, V., Sanchez-Navarro, J.A., 2013. Systemic transport of Alfalfa mosaic virus can be mediated by the movement proteins of several viruses assigned to five genera of the 30K family. J. Gen. Virol. 94, 677–681.

Ferralli, J., Ashby, J., Fasler, M., Boyko, V., Heinlein, M., 2006. Disruption of microtubule organization and centrosome function by expression of tobacco mosaic virus movement protein. J. Virol. 80, 5807–5821.

Folimonova, S.Y., Tilsner, J., 2018. Hitchhikers, highway tolls and roadworks, the interactions of plant viruses with the phloem. Curr. Opin. Plant Biol. 43, 82–88.

Foster, T.M., Lough, T.J., Emerson, S.J., Lee, R.H., Bowman, J.L., Forster, R.L.S., Lucas, W.J., 2002. A surveillance system regulates selective entry of RNA into the shoot apex. Plant Cell 14, 1497–1508.

Fujiki, M., Kawakami, S., Kim, R.W., Beachy, R.N., 2006. Domains of tobacco mosaic virus movement protein essential for its membrane association. J. Gen. Virol. 87, 2699–2707.

Ganusova, E.E., Burch-Smith, T.M., 2019. Plant-pathogen interactions through the plasmodesma prism. Plant Sci. 279, 70–80.

Ganusova, E.E., Rice, J.H., Carlew, T.S., Patel, A., Perrodin-Njoku, E., Hewezi, T., Burch-Smith, T.M., 2017. Altered expression of a chloroplast protein affects the outcome of virus and nematode infection. Mol. Plant Microbe Interact. 30, 478–488.

Gao, R., Wong, S.M., 2013. Basic amino acid mutations in the nuclear localization signal of hibiscus chlorotic ringspot virus p23 inhibit virus long distance movement. PLoS ONE 89, e74000.

Garcia-Castillo, S., Marcos, J.F., Pallas, V., Sanchez-Pina, M.A., 2001. Influence of the plant growing conditions on the translocation routes and systemic infection of Carnation mottle virus in Chenopodium quinoa plants. Physiol. Mol. Plant Pathol. 58, 229–238.

Garcia-Castillo, S., Sanchez-Pina, M.A., Pallas, V., 2003. Spatio-temporal analysis of the RNAs, coat and movement (p7) proteins of Carnation mottle virus in Chenopodium quinoa plants. J. Gen. Virol. 84, 745–749.

Geng, C., Cong, Q.Q., Li, X.D., Mou, A.L., Gao, R., Liu, J.L., Tian, Y.P., 2015. Developmentally regulated plasma membrane protein of Nicotiana benthamiana contributes to potyvirus movement and transports to plasmodesmata via the early secretory pathway and the actomyosin system. Plant Physiol. 167, 394–410.

Genoves, A., Navarro, J.A., Pallas, V., 2006. Functional analysis of the five melon necrotic spot virus genome-encoded proteins. J. Gen. Virol. 87, 2371–2380.

Genoves, A., Navarro, J.A., Pallas, V., 2009. A self-interacting carmovirus movement protein plays a role in binding of viral RNA during the cell-to-cell movement and shows an actin cytoskeleton dependent location in cell periphery. Virology 395, 133–142.

Genoves, A., Navarro, J.A., Pallas, V., 2010. The intra- and intercellular movement of Melon necrotic spot virus (MNSV) depends on an active secretory pathway. Mol. Plant Microbe Interact. 23, 263–272.

Genoves, A., Pallas, V., Navarro, J.A., 2011. Contribution of topology determinants of a viral movement protein to its membrane association, intracellular traffic, and viral cell-to-cell movement. J. Virol. 85, 7797–7809.

Gil, L., Yaron, I., Shalitin, D., Sauer, N., Turgeon, R., Wolf, S., 2011. Sucrose transporter plays a role in phloem loading in CMV-infected melon plants that are defined as symplastic loaders. Plant J. 66, 366–374.

Gilmer, D., Ratti, C., Michel, F., 2018. Long-distance movement of helical multipartite phytoviruses, keep connected or die? Curr. Opin. Virol. 33, 120–128.

Giner, A., Pascual, L., Bourgeois, M., Gyetvai1, G., Rios, P., Picó, B., Troadec, C., Bendahmane, A., Garcia-Mas, J., Martín-Hernández, A.M., 2017. A mutation in the melon vacuolar protein sorting 41 prevents systemic infection of cucumber mosaic virus. Sci. Rep. 7, 10471.

Gómez, G., Pallás, V., 2004. A long distance translocatable phloem protein from cucumber forms a ribonucleoprotein complex in vivo with Hop stunt viroid RNA. J. Virol. 78, 10104–10110.

Goodrick, B.J., Kuhn, C.W., Hussey, R.S., 1991. Restricted systemic movement of Cowpea chlorotic mottle virus in soybean with nonnecrotic resistance. Phytopathology 81, 1426–1431.

Gosalvez-Bernal, B., Garcia-Castillo, S., Pallas, V., Sanchez-Pina, M.A., 2006. Distribution of carnation viruses in the shoot tip, exclusion from the shoot apical meristem. Physiol. Mol. Plant Pathol. 69, 43–51.

Gosalvez-Bernal, B., Genoves, A., Navarro, J.A., Pallas, V., Sanchez-Pina, M.A., 2008. Pathway for phloem-dependent movement of Melon necrotic spot virus in melon plants. Mol. Plant Pathol. 9, 447–461.

Guenoune-Gelbart, D., Elbaum, M., Sagi, G., Levy, A., Epel, B.L., 2008. Tobacco mosaic virus (TMV) replicase and movement protein function synergistically in facilitating TMV spread by lateral diffusion in the plasmodesmal desmotubule of Nicotiana benthamiana. Mol. Plant Microbe Interact. 21, 335–345.

Guiu-Aragonés, C., Díaz-Pendón, J.A., Martín-Hernández, A.M., 2015. Four sequence positions of the movement protein of cucumber mosaic virus determine the virulence against cmv1-mediated resistance in melon. Mol. Plant Pathol. 16, 675–684.

Hacker, D.L., Petty, I.T., Wei, N., Morris, T.J., 1992. Turnip crinkle virus genes required for RNA replication and virus movement. Virology 186, 1–8.

Ham, B.K., Brandom, J.L., Xoconostle-Cázares, B., Ringgold, V., Lough, T.J., Lucas, W.J., 2009. A polypyrimidine tract binding protein, pumpkin RBP50, forms the basis of a phloem-mobile ribonucleoprotein complex. Plant Cell 21, 197–215.

Ham, B.K., Li, G., Jia, W., Leary, J.A., Lucas, W.J., 2014. Systemic delivery of siRNA in pumpkin by a plant phloem small RNA-binding protein 1–ribonucleoprotein complex. Plant J. 80, 683–694.

Hapiak, M., Li, Y.Z., Agama, K., Swade, S., Okenka, G., Falk, J., Khandekar, S., Raikhy, G., Anderson, A., Pollock, J., Zellner, W., Schoelz, J., Leisner, S.M., 2008. Cauliflower mosaic virus gene VI product N-terminus contains regions involved in resistance-breakage, self-association and interactions with movement protein. Virus Res. 138, 119–129.

Harries, P., Ding, B., 2011. Cellular factors in plant virus movement: at the leading edge of macromolecular trafficking in plants. Virology 411, 237–243.

Harries, P.A., Palanichelvam, K., Yu, W.C., Schoelz, J.E., Nelson, R.S., 2009. The Cauliflower mosaic virus protein P6 forms motile inclusions that traffic along actin microfilaments and stabilize microtubules. Plant Physiol. 149, 1005–1016.

Haupt, S., Cowan, G.H., Ziegler, A., Roberts, A.G., Oparka, K.J., Torrance, L., 2005. Two plant-viral movement proteins traffic in the endocytic recycling pathway. Plant Cell 17, 164–181.

Heinlein, M., 2015. Plant virus replication and movement. Virology 479-480, 657–671.

Heinlein, M., Epel, B.L., Padgett, H.S., Beachy, R.N., 1995. Interaction of tobamovirus movement proteins with the plant cytoskeleton. Science 270, 1983–1985.

Helenius, A., 2018. Virus entry: looking back and moving forward. J. Mol. Biol. 430, 1853–1862.

Herranz, M.C., Sanchez-Navarro, J.A., Saurí, A., Mingarro, I., Pallás, V., 2005. Mutational analysis of the RNA-binding domain of the Prunus necrotic ringspot virus (PNRSV) movement protein reveals its requirement for cell-to-cell movement. Virology 339, 31–41.

Herranz, M.C., Pallas, V., Aparicio, F., 2012. Multifunctional roles for the N-terminal basic motif of Alfalfa mosaic virus coat protein, nucleolar/cytoplasmic shuttling, modulation of RNA-binding activity and virion formation. Mol. Plant Microbe Interact. 258, 1093–1103.

Hipper, C., Brault, V., Ziegler-Graff, V., Revers, F., 2013. Viral and cellular factors involved in phloem transport of plant viruses. Front. Plant Sci. 4, a154.

Hofmann, C., Niehl, A., Sambade, A., Steinmetz, A., Heinlein, M., 2009. Inhibition of tobacco mosaic virus movement by expression of an actin-binding protein. Plant Physiol. 149, 1810–1823.

Howard, A.R., Heppler, M.L., Ju, H.J., Krishnamurthy, K., Payton, M.E., Verchot-Lubicz, J., 2004. Potato virus X TGBp1 induces plasmodesmata gating and moves between cells in several host species whereas CP moves only in N. benthamiana leaves. Virology 328, 185–197.

Hsu, H.T., Tseng, Y.H., Chou, Y.L., Su, S.H., Hsu, Y.H., Chang, B.Y., 2009. Characterization of the RNA-binding properties of the triple-gene-block protein 2 of Bamboo mosaic virus. Virol. J. 6, 50.

Huang, Z., Han, Y., Howell, S.H., 2000. Formation of surface tubules and fluorescent foci in Arabidopsis thaliana protoplasts expressing a fusion between the green fluorescent protein and the cauliflower mosaic virus movement protein. Virology 271, 58–64.

Hull, R., 2014. Chapter 12—plant to plant movement. In: Hull, R. (Ed.), Plant Virology, fifth ed. Academic Press, Boston, pp. 669–751.

Hyodo, K., Mine, A., Taniguchi, T., Kaido, M., Mise, K., Taniguchi, H., Okuno, T., 2013. ADP ribosylation factor 1 plays an essential role in the replication of a plant RNA virus. J. Virol. 87, 163–176.

Jackson, A.O., Lim, H.S., Bragg, J., Ganesan, U., Lee, M.Y., 2009. Hordeivirus replication, movement, and pathogenesis. Annu. Rev. Phytopathol. 47, 385–422.

Johansen, I.E., Dougherty, W.G., Keller, K.E., Wang, D., Hampton, R.O., 1996. Multiple viral determinants affect seed transmission of pea seedborne mosaic virus in Pisum sativum. J. Gen. Virol. 77, 3149–3154.

Jones, R.A.C., 2018. Plant and insect viruses in managed and natural environments: novel and neglected transmission pathways. Adv. Virus Res. 101, 149–187.

Ju, H.J., Samuels, T.D., Wang, Y.S., Blancaflor, E., Payton, M., Mitra, R., Krishnamurthy, K., Nelson, R.S., Verchot-Lubicz, J., 2005. The potato virus X TGBp2 movement protein associates with endoplasmic reticulum-derived vesicles during virus infection. Plant Physiol. 138, 1877–1895.

Ju, H.J., Brown, J.E., Ye, C.M., Verchot-Lubicz, J., 2007. Mutations in the central domain of potato virus X TGBp2 eliminate granular vesicles and virus cell-to-cell trafficking. J. Virol. 81, 1899–1911.

Kaido, M., Tsuno, Y., Mise, K., Okuno, T., 2009. Endoplasmic reticulum targeting of the Red clover necrotic mosaic virus movement protein is associated with the replication of viral RNA1 but not that of RNA2. Virology 395, 232–242.

Kaido, M., Abe, K., Mine, A., Hyodo, K., Taniguchi, T., Taniguchi, H., Mise, K., Okuno, T., 2014. GAPDH—a recruits a plant virus movement protein to cortical virus replication complexes to facilitate viral cell-to-cell movement. PLoS Pathog. 10, e1004505.

Kaplan, I.B., Zhang, L., Palukaitis, P., 1998. Characterization of cucumber mosaic virus—V. Cell-to-cell movement requires capsid protein but not virions. Virology 246, 221–231.

Karpova, O.V., Zayakina, O.V., Arkhipenko, M.V., Sheval, E.V., Kiselyova, O.I., Poljakov, V.Y., Yaminsky, I.V., Rodionova, N.P., Atabekov, J.G., 2006. Potato virus X RNA-mediated assembly of single-tailed ternary 'coat protein-RNA-movement protein' complexes. J. Gen. Virol. 87, 2731–2740.

Kasteel, D.T.J., Perbal, M.C., Boyer, J.C., Wellink, J., Goldbach, R.W., Maule, A.J., van Lent, J.W.M., 1996. The movement proteins of cowpea mosaic virus and cauliflower mosaic virus induce tubular structures in plant and insect cells. J. Gen. Virol. 77, 2857–2864.

Kawakami, S., Watanabe, Y., Beachy, R.N., 2004. Tobacco mosaic virus infection spreads cell to cell as intact replication complexes. Proc. Natl. Acad. Sci. U. S. A. 101, 6291–6296.

Kehr, J., Kragler, F., 2018. Long distance RNA movement. New Phytol. 218, 29–40.

Killiny, N., Hijaz, F., Harper, S.J., Dawson, W.O., 2017. Effects of Citrus tristeza closterovirus infection on phloem sap and released volatile organic compounds in Citrus macrophylla. Physiol. Mol. Plant Pathol. 98, 25–36.

Kim, S.H., Kalinina, N.O., Andreev, I., Ryabov, E.V., Fitzgerald, A.G., Taliansky, M.E., Palukaitis, P., 2004. The C-terminal 33 amino acids of the cucumber mosaic virus 3a protein affect virus movement, RNA binding and inhibition of infection and translation. J. Gen. Virol. 85, 221–230.

Kim, I., Cho, E., Crawford, K., Hempel, F.D., Zambryski, P.C., 2005a. Cell-to-cell movement of GFP during embryogenesis and early seedling development in Arabidopsis. Proc. Natl. Acad. Sci. U. S. A. 102, 2227–2231. https://doi.org/10.1073/pnas.0409193102.

Kim, I., Kobayashi, K., Cho, E., Zambryski, P.C., 2005b. Subdomains for transport via plasmodesmata corresponding to the apical-basal axis are established during Arabidopsis embryogenesis. Proc. Natl. Acad. Sci. U. S. A. 102, 11945–11950.

Kim, G., LeBlanc, M.L., Wafula, E.K., dePamphilis, C.W., Westwood, J.H., 2014. Genomic-scale exchange of mRNA between a parasitic plant and its hosts. Science 345, 808–811.

Kragler, F., 2010. RNA in the phloem. A crisis or a return on investment? Plant Sci. 178, 99–104.

Kragler, F., Curin, M., Trutnyeva, K., Gansch, A., Waigmann, E., 2003. MPB2C, a microtubule-associated plant protein binds to and interferes with cell-to-cell transport of tobacco mosaic virus movement protein. Plant Physiol. 132, 1870–1883.

Kusumanegara, K., Mine, A., Hyodo, K., Kaido, M., Mise, K., Okuno, T., 2012. Identification of domains in p27 auxiliary replicase protein essential for its association with the endoplasmic reticulum membranes in Red clover necrotic mosaic virus. Virology 433, 131–141.

Laporte, C., Vetter, G., Loudes, A.M., Robinson, D.G., Hillmer, S., Stussi-Garaud, C., Ritzenthaler, C., 2003. Involvement of the secretory pathway and the cytoskeleton in intracellular targeting and tubule assembly of Grapevine fanleaf virus movement protein in tobacco BY-2 cells. Plant Cell 15, 2058–2075.

Lartey, R.T., Ghoshroy, S., Citovsky, V., 1998. Identification of an Arabidopsis thaliana mutation vsm1 that restricts systemic movement of tobamoviruses. Mol. Plant Microbe Interact. 11, 706–709.

Leastro, M.O., Pallás, V., Resende, R.O., Sánchez-Navarro, J.A., 2015. The movement proteins (NSm) of distinct tospoviruses peripherally associate with cellular membranes and interact with homologous and heterologous NSm and nucleocapsid proteins. Virology 478, 39–49.

Lee, J.Y., Frank, M., 2018. Plasmodesmata in phloem, different gateways for different cargoes. Curr. Opin. Plant Biol. 43, 119–124.

Lee, J.Y., Lucas, W.J., 2001. Phosphorylation of viral movement proteins—regulation of cell-to-cell trafficking. Trends Microbiol. 9, 5–8.

Lee, S.C., Wu, C.H., Wang, C.W., 2010. Traffic of a viral movement protein complex to the highly curved tubules of the cortical endoplasmic reticulum. Traffic 11, 912–930.

Leshchiner, A.D., Minina, E.A., Rakitina, D.V., Vishnichenko, V.K., Solovyev, A.G., Morozov, S.Y., Kalinina, N.O., 2008. Oligomerization of the potato virus X 25-kD movement protein. Biochemistry 73, 50–55.

Levy, A., Zheng, J.Y., Lazarowitz, S.G., 2013. The tobamovirus turnip vein clearing virus 30-kiloDalton movement protein localizes to novel nuclear filaments to enhance virus infection. J. Virol. 87, 6428–6440.

Levy, A., Zheng, J.Y., Lazarowitz, S.G., 2015. Synaptotagmin SYTA forms ER-plasma membrane junctions that are recruited to plasmodesmata for plant virus movement. Curr. Biol. 25, 2018–2025.

Lewis, J.D., Lazarowitz, S.G., 2010. Arabidopsis synaptotagmin SYTA regulates endocytosis and virus movement protein cell-to-cell transport. Proc. Natl. Acad. Sci. U. S. A. 107, 2491–2496.

Li, W.Z., Qu, F., Morris, T.J., 1998. Cell-to-cell movement of turnip crinkle virus is controlled by two small open reading frames that function in trans. Virology 244, 405–416.

Li, Y., Wu, M.Y., Song, H.H., Hu, X., Qiu, B.S., 2005. Identification of a tobacco protein interacting with Tomato mosaic virus coat protein and facilitating long distance movement of virus. Arch. Virol. 150, 1993–2008.

Li, Y., Cui, H., Cui, X., Wang, A., 2016. The altered photosynthetic machinery during compatible virus infection. Curr. Opin. Virol. 17, 19–24.

Li, Z., Zhang, Y., Jiang, Z., Jin, X., Zhang, K., Wang, X., Han, C., Yu, J., Li, D., 2018. Hijacking of the nucleolar protein fibrillarin by TGB1 is required for cell-to-cell movement of Barley stripe mosaic virus. Mol. Plant Pathol. 195, 1222–1237.

Lim, H.S., Bragg, J.N., Ganesan, U., Lawrence, D.M., Yu, J., Isogai, M., Hammond, J., Jackson, A.O., 2008. Triple gene block protein interactions involved in movement of Barley stripe mosaic virus. J. Virol. 82, 4991–5006.

Lim, H.S., Bragg, J.N., Ganesan, U., Ruzin, S., Schichnes, D., Lee, M.Y., Vaira, A.M., Ryu, K.H., Hammond, J., Jackson, A.O., 2009. Subcellular localization of the barley stripe mosaic virus triple gene block proteins. J. Virol. 83, 9432–9448.

Lim, H.S., Vaira, A.M., Bae, H., Bragg, J.N., Ruzin, S.E., Bauchan, G.R., Dienelt, M.M., Owens, R.A., Hammond, J., 2010. Mutation of a chloroplast-targeting signal in alternanthera mosaic virus TGB3 impairs cell-to-cell movement and eliminates long-distance virus movement. J. Gen. Virol. 91, 2102–2115.

Lim, H.S., Lee, M.Y., Moon, J.S., Moon, J.K., Yu, Y.M., Cho, I.S., Bae, H., deBoer, M., Ju, H., Hammond, J., Jackson, A.O., 2013. Actin cytoskeleton and golgi involvement in Barley stripe mosaic virus movement and cell wall localization of triple gene block proteins. Plant Pathol. J. 29, 17–30.

Lin, M.K., Lee, Y.J., Lough, T.J., Phinney, B.S., Lucas, W.J., 2009. Analysis of the pumpkin phloem proteome provides insights into angiosperm sieve tube function. Mol. Cell. Proteomics. 8, 343–356.

Linnik, O., Liesche, J., Tilsner, J., Oparka, K.J., 2013. Unraveling the structure of viral replication complexes at super-resolution. Front. Plant Sci. 4, 6.

Liou, M.R., Hu, C.C., Chou, Y.L., Chang, B.Y., Lin, N.S., Hsu, Y.H., 2015. Viral elements and host cellular proteins in intercellular movement of Bamboo mosaic virus. Curr. Opin. Virol. 12, 99–108.

Liu, C., Nelson, R.S., 2013. The cell biology of tobacco mosaic virus replication and movement. Front. Plant Sci. 4, 12.

Liu, Y., Burch-Smith, T., Schiff, M., Feng, S., Dinesh-Kumar, S.P., 2004. Molecular chaperone Hsp90 associates with resistance protein N and its signaling proteins SGT1 and Rar1 to modulate an innate immune response in plants. J. Biol. Chem. 279, 2101–2108.

Liu, C., Ye, L., Lang, G., Zhang, C., Hong, J., Zhou, X., 2011. The VP37 protein of broad bean wilt virus 2 induces tubule-like structures in both plant and insect cells. Virus Res. 155, 42–47.

Lucas, W.J., Ding, B., Vanderschoot, C., 1993. Plasmodesmata and the supracellular nature of plants. New Phytologist 125, 435–476.

Lucas, W.J., Yoo, B.C., Kragler, F., 2001. RNA as a long-distance information macromolecule in plants. Nat. Rev. Mol. Cell Biol. 2, 849–857.

Lucas, W.J., Groover, A., Lichtenberger, R., Furuta, K., Yadav, S.R., Helariutta, Y., He, X.Q., Fukuda, H., Kang, J., Brady, S.M., Patrick, J.W., Sperry, J., Yoshida, A., Lopez-Millan, A.F., Grusak, M.A., Kachroo, P., 2013. The plant vascular system, evolution, development and functions. J. Integr. Plant Biol. 55, 294–388.

Lukhovitskaya, N.I., Cowan, G.H., Vetukuri, R.R., Tilsner, J., Torrance, L., Savenkov, E.I., 2015. Importin-a-mediated nucleolar localization of potato mop-top virus triple gene block1 TGB1 protein facilitates virus systemic movement, whereas TGB1 self-interaction is required for cell-to-cell movement in Nicotiana benthamiana. Plant Physiol. 167, 738–752.

Luo, K.R., Huang, N.C., Yu, T.S., 2018. Selective targeting of mobile mRNAs to plasmodesmata for cell-to-cell movement. Plant Physiol. 177, 604–614.

Malter, D., Wolf, S., 2011. Melon phloem-sap proteome, developmental control and response to viral infection. Protoplasma 248, 217–224.

Marcos, J.F., Vilar, M., Perez-Paya, E., Pallas, V., 1999. In vivo detection, RNA-binding properties and characterization of the RNA-binding domain of the p7 putative movement protein from carnation mottle carmovirus (CarMV). Virology 255, 354–365.

Martinez-Gil, L., Sauri, A., Vilar, M., Pallas, V., Mingarro, I., 2007. Membrane insertion and topology of the p7B movement protein of Melon Necrotic Spot Virus (MNSV). Virology 367, 348–357.

Martinez-Gil, L., Johnson, A.E., Mingarro, I., 2010. Membrane insertion and biogenesis of the Turnip crinkle virus p9 movement protein. J. Virol. 84, 5520–5527.
Martinez-Turino, S., Hernandez, C., 2011. A membrane-associated movement protein of Pelargonium flower break virus shows RNA-binding activity and contains a biologically relevant leucine zipper-like motif. Virology 413, 310–319.
Martín-Hernández, A.M., Baulcombe, D.C., 2008. Tobacco rattle virus 16-kilodalton protein encodes a suppressor of RNA silencing that allows transient viral entry in meristems. J. Virol. 82, 4064–4071.
Mas, P., Beachy, R.N., 1999. Replication of tobacco mosaic virus on endoplasmic reticulum and role of the cytoskeleton and virus movement protein in intracellular distribution of viral RNA. J. Cell Biol. 147, 945–958.
Mas, P., Beachy, R.N., 2000. Role of microtubules in the intracellular distribution of tobacco mosaic virus movement protein. Proc. Natl. Acad. Sci. U. S. A. 97, 12345–12349.
Medina, V., Peremyslov, V.V., Hagiwara, Y., Dolja, V.V., 1999. Subcellular localization of the HSP70-homolog encoded by beet yellows closterovirus. Virology 260, 173–181.
Melcher, U., 2000. The "30K" superfamily of viral movement proteins. J. Gen. Virol. 81, 257–266.
Milne, R.J., Grof, C.P.L., Patrick, J.W., 2018. Mechanisms of phloem unloading, shaped by cellular pathways, their conductances and sink function. Curr. Opin. Plant Biol. 43, 8–15.
Mine, A., Hyodo, K., Tajima, Y., Kusumanegara, K., Taniguchi, T., Kaido, M., Mise, K., Taniguchi, H., Okuno, T., 2012. Differential roles of Hsp70 and Hsp90 in the assembly of the replicase complex of a positive-strand RNA plant virus. J. Virol. 86, 12091–12104.
Moreno, I.M., Thompson, J.R., García-Arenal, F., 2004. Analysis of the systemic colonization of cucumber plants by cucumber green mottle mosaic virus. J. Gen. Virol. 85, 749–759.
Morozov, S.Y., Solovyev, A.G., 2003. Triple gene block: modular design of a multifunctional machine for plant virus movement. J. Gen. Virol. 84, 1351–1366.
Moussu, S., Doll, N.M., Chamot, S., Brocard, L., Creff, A., Fourquin, C., Widiez, T., Nimchuk, Z.L., Ingram, G., 2017. ZHOUPI and KERBEROS mediate embryo/endosperm separation by promoting the formation of an extracuticular sheath at the embryo surface. Plant Cell 29, 1642–1656.
Movahed, N., Patarroyo, C., Sun, J., Vali, H., Laliberte, J.F., Zheng, H., 2017. Cylindrical inclusion protein of turnip mosaic virus serves as a docking point for the intercellular movement of viral replication vesicles. Plant Physiol. 175, 1732–1744.
Movahed, N., Sun, J., Vali, H., Laliberte, J.F., Zheng, H., 2019. A host ER fusogen is recruited by Turnip mosaic virus for maturation of viral replication vesicles. Plant Physiol. 179, 507–518.
Mushegian, A.R., Koonin, E.V., 1993. Cell-to-cell movement of plant viruses. Insights from amino acid sequence comparisons of movement proteins and from analogies with cellular transport systems. Arch. Virol. 133, 239–257.
Nagano, H., Mise, K., Okuno, T., Furusawa, I., 1999. The cognate coat protein is required for cell-to-cell movement of a chimeric brome mosaic virus mediated by the cucumber mosaic virus movement protein. Virology 265, 226–234.
Nagano, H., Mise, K., Furusawa, I., Okuno, T., 2001. Conversion in the requirement of coat protein in cell-to-cell movement mediated by the cucumber mosaic virus movement protein. J. Virol. 75, 8045–8053.
Nagy, P.D., 2016. Tombusvirus-host interactions, co-opted evolutionarily conserved host factors take center court. Annu. Rev. Virol. 3, 491–515.
Napuli, A.J., Falk, B.W., Dolja, V.V., 2000. Interaction between HSP70 homolog and filamentous virions of the beet yellows virus. Virology 274, 232–239.
Napuli, A.J., Alzhanova, D.V., Doneanu, C.E., Barofsky, D.F., Koonin, E.V., Dolja, V.V., 2003. The 64-kiloDalton capsid protein homolog of beet yellows virus is required for assembly of virion tails. J. Virol. 77, 2377–2384.

Navarro, J.A., Genoves, A., Climent, J., Sauri, A., Martinez-Gil, L., Mingarro, I., Pallas, V., 2006. RNA-binding properties and membrane insertion of Melon necrotic spot virus (MNSV) double gene block movement proteins. Virology 356, 57–67.

Nelson, R.S., van Bel, A.J.E., 1998. The mystery of virus trafficking into, through and out of vascular tissue. Prog. Bot. 59, 476–533.

Newbigin, E., Bacic, A., Read, S., 2009. Chapter 4.4.3—Callose and its role in pollen and embryo sac development in flowering plants. In: Bacic, A., Fincher, G.B., Stone, B.A. (Eds.), Chemistry, Biochemistry, and Biology of 1-3 Beta Glucans and Related Polysaccharides. Academic Press, San Diego, pp. 465–498.

Ng, J.C.K., Falk, B.W., 2006a. Bemisia tabaci transmission of specific lettuce infectious yellows virus genotypes derived from in vitro synthesized transcript-inoculated protoplasts. Virology 352, 209–215.

Ng, J.C.K., Falk, B.W., 2006b. Virus-vector interactions mediating nonpersistent and semipersistent transmission of plant viruses. Annu. Rev. Phytopathol. 44, 183–212.

Niehl, A., Heinlein, M., 2011. Cellular pathways for viral transport through plasmodesmata. Protoplasma 248, 75–99.

Niehl, A., Amari, K., Gereige, D., Brandner, K., Mely, Y., Heinlein, M., 2012. Control of tobacco mosaic virus movement protein fate by cell-division-cycle protein48. Plant Physiol. 160, 2093–2108.

Niehl, A., Pena, E.J., Amari, K., Heinlein, M., 2013. Microtubules in viral replication and transport. Plant J. 75, 290–308.

Ohki, T., Akita, F., Mochizuki, T., Kanda, A., Sasaya, T., Tsuda, S., 2010. The protruding domain of the coat protein of Melon necrotic spot virus is involved in compatibility with and transmission by the fungal vector Olpidium bornovanus. Virology 402, 129–134.

Opalka, N., Brugidou, C., Bonneau, C., Nicole, M., Beachy, R.N., Yeager, M., Fauquet, C., 1998. Movement of rice yellow mottle virus between xylem cells through pit membranes. Proc. Natl. Acad. Sci. U. S. A. 95, 3323–3328.

Oparka, K.J., Turgeon, R., 1999. Sieve elements and companion cells—traffic control centers of the phloem. Plant Cell 11, 739–750.

Ouko, M.O., Sambade, A., Brandner, K., Niehl, A., Pena, E., Ahad, A., Heinlein, M., Nick, P., 2010. Tobacco mutants with reduced microtubule dynamics are less susceptible to TMV. Plant J. 62, 829–839.

Overall, R.L., Blackman, L.M., 1996. A model of the macromolecular structure of plasmodesmata. Trends Plant Sci. 1, 307–311.

Pallas, V., Gomez, G., 2013. Phloem RNA-binding proteins as potential components of the long-distance RNA transport system. Front. Plant Sci. 4, 1–5.

Pallas, V., Genovés, A., Sanchez-Pina, M.A., Navarro, J.A., 2011. Systemic movement of viruses via the plant phloem. In: Caranta, C., Aranda, M.A., Tepfer, M., Lopez-Moya, J.J. (Eds.), Recent Advances in Plant Virology. Caister Academic Press, ISBN: 978-1-904455-75-2, p. 470.

Pallas, V., Aparicio, F., Herranz, M.C., Sanchez-Navarro, J.A., Scott, S.W., 2012. The molecular biology of ilarviruses. Adv. Virus Res. 87, 139–181.

Park, S.H., Sit, T.L., Kim, K.H., Lommel, S.A., 2013. The red clover necrotic mosaic virus capsid protein N-terminal amino acids possess specific RNA binding activity and are required for stable virion assembly. Virus Res. 176, 107–118.

Park, M.-R., Jeong, R.-D., Kim, K.-H., 2014. Understanding the intracellular trafficking and intercellular transport of potex viruses in their host plants. Front. Plant Sci. 5, a60.

Peiro, A., Cañizares, M.C., Rubio, L., Lopez, C., Moriones, E., Aramburu, J., Sanchez-Navarro, J.A., 2014a. The movement protein (NSm) of Tomato spotted wilt virus is the avirulence determinant in the tomato Sw-5 gene-based resistance. Mol. Plant Pathol. 15, 802–813.

Peiro, A., Martinez-Gil, L., Tamborero, S., Pallas, V., Sanchez-Navarro, J.A., Mingarro, I., 2014b. The tobacco mosaic virus movement protein associates with but does not integrate into biological membranes. J. Virol. 88, 3016–3026.
Peña, E.J., Heinlein, M., 2012. RNA transport during TMV cell-to-cell movement. Front. Plant Sci. 3, 193.
Peña, E.J., Heinlein, M., 2013. Cortical microtubule-associated ER sites: organization centers of cell polarity and communication. Curr. Opin. Plant Biol. 16, 764–773.
Peña, E., Niehl, A., Heinlein, M., 2012. Viral studies point the way, mechanisms of intercellular transport. In: Kragler, F., Hülskamp, M. (Eds.), Short and Long Distance Signaling, Advances in Plant Biology 3. Springer Science+Business Media, LLC 2012.
Peremyslov, V.V., Hagiwara, Y., Dolja, V.V., 1999. HSP70 homolog functions in cell-to-cell movement of a plant virus. Proc. Natl. Acad. Sci. U. S. A. 96, 14771–14776.
Peremyslov, V.V., Pan, Y.W., Dolja, V.V., 2004. Movement protein of a closterovirus is a type III integral transmembrane protein localized to the endoplasmic reticulum. J. Virol. 78, 3704–3709.
Pitzalis, N., Heinlein, M., 2017. The roles of membranes and associated cytoskeleton in plant virus replication and cell-to-cell movement. J. Exp. Bot. 69, 117–132.
Pouwels, J., van der Krogt, G.N.M., Van Lent, J., Bisseling, T., Wellink, J., 2002. The cytoskeleton and the secretory pathway are not involved in targeting the cowpea mosaic virus movement protein to the cell periphery. Virology 297, 48–56.
Pouwels, J., van der Velden, T., Willemse, J., Borst, J.W., van Lent, J., Bisseling, T., Wellink, J., 2004. Studies on the origin and structure of tubules made by the movement protein of Cowpea mosaic virus. J. Gen. Virol. 85, 3787–3796.
Prokhnevsky, A.I., Peremyslov, V.V., Dolja, V.V., 2005. Actin cytoskeleton is involved in targeting of a viral Hsp70 homolog to the cell periphery. J. Virol. 79, 14421–14428.
Qian, X., Xiang, Q., Yang, T., Ma, H., Ding, X.S., Tao, X., 2018. Molecular co-chaperone SGT1 Is critical for cell-to-cell movement and systemic infection of tomato spotted wild virus in Nicotiana benthamiana. Viruses 10, a647.
Raffaele, S., Bayer, E., Lafarge, D., Cluzet, S., German, R.S., Boubekeur, T., Leborgne-Castel, N., Carde, J.P., Lherminier, J., Noirot, E., Satiat-Jeunemaitre, B., Laroche-Traineau, J., Moreau, P., Ott, T., Maule, A.J., Reymond, P., Simon-Plas, F., Farmer, E.E., Bessoule, J.J., Mongrand, S., 2009. Remorin, a solanaceae protein resident in membrane rafts and plasmodesmata, impairs potato virus X movement. Plant Cell 21, 1541–1555.
Reichel, C., Beachy, R.N., 1998. Tobacco mosaic virus infection induces severe morphological changes of the endoplasmic reticulum. Proc. Natl. Acad. Sci. U. S. A. 95, 11169–11174.
Ritzenthaler, C., Hofmann, C., 2007. Tubule-guided movement of plant viruses. In: Viral Transport in Plants. Springer Berlin Heidelberg, Berlin, Heidelberg, pp. 63–83.
Roberts, A.G., Oparka, K.J., 2003. Plasmodesmata and the control of symplastic transport. Plant Cell Environ. 26, 103–124.
Roberts, I.M., Wang, D., Findlay, K., Maule, A.J., 1998. Ultrastructural and temporal observations of the potyvirus cylindrical inclusions (Cls) show that the Cl protein acts transiently in aiding virus movement. Virology 245, 173–181.
Roberts, I.M., Wang, D., Thomas, C.L., Maule, A.J., 2003. Pea seed-borne mosaic virus seed transmission exploits novel symplastic pathways to infect the pea embryo and is, in part, dependent upon chance. Protoplasma 222, 31–43.
Robles Luna, G., Peña, E.J., Borniego, M.B., Heinlein, M., García, M.L., 2018. Citrus psorosis virus movement protein contains an aspartic protease required for autocleavage and the formation of tubule-like structures at plasmodesmata. J. Virol. 92. e00355-18.

Rochon, D., Kakani, K., Robbins, M., Reade, R., 2004. Molecular aspects of plant virus transmission by olpidium and plasmodiophorid vectors. Annu. Rev. Phytopathol. 42, 211–241.

Rodionova, N.P., Karpova, O.V., Kozlovsky, S.V., Zayakina, O.V., Arkhipenko, M.V., Atabekov, J.G., 2003. Linear remodeling of helical virus by movement protein binding. J. Mol. Biol. 333, 565–572.

Roossinck, M.J., 2010. Lifestyles of plant viruses. Philos. Trans. R. Soc. Lond. B Biol. Sci. 365, 1899–1905.

Ruiz-Medrano, R., Hinojosa-Moya, J., Xoconostle-Cázares, B., Lucas, W.J., 2007. Influence of cucumber mosaic virus infection on the mRNA population present in the phloem translocation stream of pumpkin plants. Funct. Plant Biol. 34 (4), 292–301.

Saito, T., Hosokawa, D., Meshi, T., Okada, Y., 1987. Immunocytochemical localization of the 130K and 180K proteins (putative replicase components) of tobacco mosaic virus. Virology 160, 477–481.

Salanki, K., Gellert, A., Huppert, E., Naray-Szabo, G., Balazs, E., 2004. Compatibility of the movement protein and the coat protein of cucumoviruses is required for cell-to-cell movement. J. Gen. Virol. 85, 1039–1048.

Sambade, A., Brandner, K., Hofmann, C., Seemanpillai, M., Mutterer, J., Heinlein, M., 2008. Transport of TMV movement protein particles associated with the targeting of RNA to plasmodesmata. Traffic 9, 2073–2088.

Samuels, T.D., Ju, H.J., Ye, C.M., Motes, C.M., Blancaflor, E.B., Verchot-Lubicz, J., 2007. Subcellular targeting and interactions among the potato virus X TGB proteins. Virology 367, 375–389.

Sanchez-Navarro, J.A., Bol, J.F.F., 2001. Role of the Alfalfa mosaic virus movement protein and coat protein in virus transport. Mol. Plant Microbe Interact. 14, 1051–1062.

Sanchez-Navarro, J.A., Reusken, C.B., Bol, J.F., Pallas, V., 1997. Replication of alfalfa mosaic virus RNA 3 with movement and coat protein genes replaced by corresponding genes of Prunus necrotic ringspot ilarvirus. J. Gen. Virol. 78 (12), 3171–3176.

Sanchez-Navarro, J.A., Herranz, M.C., Pallas, V., 2006. Cell-to-cell movement of Alfalfa mosaic virus can be mediated by the movement proteins of Ilar-, bromo-, cucumo-, tobamo- and comoviruses and does not require virion formation. Virology 346, 66–73.

Sánchez-Navarro, J., Fajardo, T., Zicca, S., Pallás, V., Stavolone, L., 2010. Caulimoviridae tubule-guided transport is dictated by movement protein properties. J. Virol. 84, 4109–4112.

Sastry, K.S., 2013. Mechanism of seed transmission. In: Seed-Borne Plant Virus Diseases. Springer India, India, pp. 85–100.

Sauri, A., Saksena, S., Salgado, J., Johnson, A.E., Mingarro, I., 2005. Double-spanning plant viral movement protein integration into the endoplasmic reticulum membrane is signal recognition particle-dependent, translocon-mediated, and concerted. J. Biol. Chem. 280, 25907–25912.

Schepetilnikov, M.V., Solovyev, A.G., Gorshkova, E.N., Schiemann, J., Prokhnevsky, A.I., Dolja, V.V., Morozov, S.Y., 2008. Intracellular targeting of a hordeiviral membrane-spanning movement protein: sequence requirements and involvement of an unconventional mechanism. J. Virol. 82, 1284–1293.

Scholthof, H.B., 2005. Plant virus transport, motions of functional equivalence. Trends Plant Sci. 10, 376–382.

Schwach, F., Vaistij, F.E., Jones, L., Baulcombe, D.C., 2005. An RNA-dependent RNA polymerase prevents meristem invasion by potato virus X and is required for the activity but not the production of a systemic silencing signal. Plant Physiol. 138, 1842–1852.

Seemanpillai, M., Elamawi, R., Ritzenthaler, C., Heinlein, M., 2006. Challenging the role of microtubules in tobacco mosaic virus movement by drug treatments is disputable. J. Virol. 80, 6712–6715.

Senshu, H., Yamaji, Y., Minato, N., Shiraishi, T., Maejima, K., Hashimoto, M., Miura, C., Neriya, Y., Namba, S., 2011. A dual strategy for the suppression of host antiviral silencing: two distinct suppressors for viral replication and viral movement encoded by potato virus M. J. Virol. 85, 10269–10278.

Serra-Soriano, M., Pallas, V., Navarro, J.A., 2014. A model for transport of a viral membrane protein through the early secretory pathway: minimal sequence and endoplasmic reticulum lateral mobility requirements. Plant J. 77, 863–879.

Serra-Soriano, M., Navarro, J.A., Genoves, A., Pallas, V., 2015. Comparative proteomic analysis of melon phloem exudates in response to viral infection. J. Proteomics 124, 11–24.

Shemyakina, E.A., Erokhina, T.N., Gorshkova, E.N., Schiemann, J., Solovyev, A.G., Morozov, S.Y., 2011a. Formation of protein complexes containing plant virus movement protein TGBp3 is necessary for its intracellular trafficking. Biochimie 93, 742–748.

Shemyakina, E.A., Solovyev, A.G., Leonova, O.G., Popenko, V.I., Schiemann, J., Morozov, S.Y., 2011b. The role of microtubule association in plasmodesmal targeting of potato mop-top virus movement protein TGBp1. Open Virol. J. 5, 1–11.

Shen, J., Chen, X., Chen, J., Sun, L., 2016. A phloem-limited fijivirus induces the formation of neoplastic phloem tissues that house virus multiplication in the host plant. Sci. Rep. 6, 29848.

Sicard, A., Pirolles, E., Gallet, R., Vernerey, M.S., Yvon, M., Urbino, C., Peterschmitt, M., Gutierrez, S., Michalakis, Y., Blanc, S., 2019. A multicellular way of life for a multipartite virus. eLife 8, e43599. https://doi.org/10.7554/eLife.43599.

Silva, M.S., Wellink, J., Goldbach, R.W., van Lent, J.W.M., 2002. Phloem loading and unloading of Cowpea mosaic virus in Vigna unguiculata. J. Gen. Virol. 83, 1493–1504.

Silva, C., Tereso, S., Nolasco, G., Oliveira, M.M., 2003. Cellular location of Prune dwarf virus in almond sections by in situ reverse transcription-polymerase chain reaction. Phytopathology 93, 278–285.

Simon-Buela, L., Garcia-Arenal, F., 1999. Virus particles of cucumber green mottle mosaic tobamovirus move systemically in the phloem of infected cucumber plants. Mol. Plant Microbe Interact. 12, 112–118.

Simpson, C., Thomas, C., Findlay, K., Bayer, E., Maule, A.J., 2009. An Arabidopsis GPI-anchor plasmodesmal neck protein with callose binding activity and potential to regulate cell-to-cell trafficking. Plant Cell 21, 581–594.

Smirnova, E., Firth, A.E., Miller, W.A., Scheidecker, D., Brault, V., Reinbold, C., Rakotondrafara, A.M., Chung, B.Y.-W., Ziegler-Graffet, V., 2015. Discovery of a small non-AUG-initiated ORF in poleroviruses and luteoviruses that is required for long-distance movement. PLoS Pathog. 115, e1004868.

Sofer, L., Garcia-Cabanillas, D., Gayral, M., Teplier, R., Pouzoulet, J., Ducousso, M., Dufin, L., Brehelin, C., Ziegler-Graff, V., Brault, V., Revers, F., 2017. Identification of host factors potentially involved in RTM-mediated resistance during potyvirus long distance movement. Arch. Virol. 162, 1855–1865.

Solovyev, A.G., Savenkov, E.I., 2014. Factors involved in the systemic transport of plant RNA viruses, the emerging role of the nucleus. J. Exp. Bot. 65, 1689–1697.

Solovyev, A.G., Kalinina, N.O., Morozov, S.Y., 2012. Recent advances in research of plant virus movement mediated by triple gene block. Front. Plant Sci. 3, 276.

Spiechowicz, M., Zylicz, A., Bieganowski, P., Kuznicki, J., Filipek, A., 2007. Hsp70 is a new target of Sgt1—an interaction modulated by S100A6. Biochem. Biophys. Res. Commun. 357, 1148–1153.

Spiegelman, Z., Golan, G., Wolf, S., 2013. Don't kill the messenger, Long-distance trafficking of mRNA molecules. Plant Sci. 213, 1–8.

Stavolone, L., Lionetti, V., 2017. Extracellular matrix in plants and animals, hooks and locks for viruses. Front. Microbiol. 8, 1760.

Stavolone, L., Villani, M.E., Leclerc, D., Hohn, T., 2005. A coiled-coil interaction mediates cauliflower mosaic virus cell-to-cell movement. Proc. Natl. Acad. Sci. U. S. A. 102, 6219–6224.

Storms, M.M.H., Kormelink, R., Peters, D., van Lent, J.W., Goldbach, R.W., vanLent, J.W.M., Goldbach, R.W., 1995. The nonstructural NSm protein of tomato spotted wilt virus induces tubular structures in plant and insect cells. Virology 214, 485–493.

Sudarshana, M.R., Wang, H.L., Lucas, W.J., Gilbertson, R.L., 1998. Dynamics of bean dwarf mosaic geminivirus cell-to-cell and long-distance movement in Phaseolus vulgaris revealed, using the green fluorescent protein. Mol. Plant Microbe Interact. 11, 277–291.

Szécsi, J., Ding, X.S., Lim, C.O., Bendahmane, M., Cho, M.J., Nelson, R.S., Beachy, R.N., 1999. Development of tobacco mosaic virus infection sites in nicotiana benthamiana. Mol. Plant Microbe Interact. 12, 143–152.

Tagami, Y., Watanabe, Y., 2007. Effects of brefeldin A on the localization of Tobamovirus movement protein and cell-to-cell movement of the virus. Virology 361, 133–140.

Taliansky, M.E., Brown, J.W.S., Rajamaki, M.L., Valkonen, J.P.T., Kalinina, N.O., 2010. Involvement of the plant nucleolus in virus and viroid infections, parallels with animal pathosystems. Adv. Virus Res. 77, 119–158.

Tamai, A., Meshi, T., 2001. Cell-to-cell movement of potato virus X: the role of p12 and p8 encoded by the second and third open reading frames of the triple gene block. Mol. Plant Microbe Interact. 14, 1158–1167.

Thieme, C.J., Rojas-Triana, M., Stecyk, E., Schudoma, C., Zhang, W., Yang, L., Minambres, M., Walther, D., Schulze, W.X., Paz-Ares, J., Scheible, W.R., Kragler, F., 2015. Endogenous Arabidopsis messenger RNAs transported to distant tissues. Nat. Plants 1, 15025.

Thomas, C.L., Maule, A.J., 1999. Identification of inhibitory mutants of cauliflower mosaic virus movement protein function after expression in insect cells. J. Virol. 73, 7886–7890.

Thomas, C.L., Bayer, E.M., Ritzenthaler, C., Fernandez-Calvino, L., Maule, A.J., 2008. Specific targeting of a plasmodesmal protein affecting cell-to-cell communication. PLoS Biol. 6, e7.

Thompson, J.R., Garcia-Arenal, F., 1998. The bundle sheath-phloem interface of Cucumis sativus is a boundary to systemic infection by tomato aspermy virus. Mol. Plant Microbe Interact. 11, 109–114.

Tilsner, J., Linnik, O., Christensen, N.M., Bell, K., Roberts, I.M., Lacomme, C., Oparka, K.J., 2009. Live-cell imaging of viral RNA genomes using a Pumilio-based reporter. Plant J. 57, 758–770.

Tilsner, J., Cowan, G.H., Roberts, A.G., Chapman, S.N., Ziegler, A., Savenkov, E., Torrance, L., 2010. Plasmodesmal targeting and intercellular movement of potato mop-top pomovirus is mediated by a membrane anchored tyrosine-based motif on the lumenal side of the endoplasmic reticulum and the C-terminal transmembrane domain in the TGB3 movement protein. Virology 402, 41–51.

Tilsner, J., Linnik, O., Louveaux, M., Roberts, I.M., Chapman, S.N., Oparka, K.J., 2013. Replication and trafficking of a plant virus are coupled at the entrances of plasmodesmata. J. Cell Biol. 201, 981–995.

Tilsner, J., Nicolas, W., Rosado, A., Bayer, E.M., 2016. Staying Tight: plasmodesmal membrane contact sites and the control of cell-to-cell connectivity in plants. Annu. Rev. Plant Biol. 67, 337–364.

Tomenius, K., Clapham, D., Meshi, T., 1987. Localization by immunogold cytochemistry of the virus-coded 30K protein in plasmodesmata of leaves infected with tobacco mosaic virus. Virology 160, 363–371.

Torrance, L., Wright, K.M., Crutzen, F., Cowan, G.H., Lukhovitskaya, N.I., Bragard, C., Savenkov, E.I., 2011. Unusual features of pomoviral RNA movement. Front. Microbiol. 2, 259.

Ueki, S., Citovsky, V., 2007. Spread throughout the plant, systemic transport of viruses. In: Heinlein, M., Waigmann, E. (Eds.), Viral Transport in Plants, Plant Cell Monographs. Springer-Verlag, pp. 86–117.
Underwood, W., 2012. The plant cell wall, a dynamic barrier against pathogen invasion. Front. Plant Sci. 3, 385.
Vaewhongs, A.A., Lommel, S.A., 1995. Virion formation is required for the long-distance movement of Red-clover necrotic mosaic virus in movement protein transgenic plants. Virology 212, 607–613.
Vaira, A.M., Lim, H.S., Bauchan, G., Gulbronson, C.J., Miozzi, L., Vinals, N., Natilla, A., Hammond, J., 2018. The interaction of lolium latent virus major coat protein with ankyrin repeat protein NbANKr redirects it to chloroplasts and modulates virus infection. J. Gen. Virol. 99, 730–742.
van der Vossen, E.A., Neeleman, L., Bol, J.F., 1994. Early and late functions of alfalfa mosaic virus coat protein can be mutated separately. Virology 202, 891–903.
Verchot-Lubicz, J., Torrance, L., Solovyev, A.G., Morozov, S.Y., Jackson, A.O., Gilmer, D., 2010. Varied movement strategies employed by triple gene block-encoding viruses. Mol. Plant Microbe Interact. 23, 1231–1247.
Vijayapalani, P., Maeshima, M., Nagasaki-Takekuchi, N., Miller, W.A., 2012. Interaction of the trans-frame potyvirus protein P3N-PIPO with host protein PCaP1 facilitates potyvirus movement. PLoS Pathog. 8, e1002639.
Vilar, M., Esteve, V., Pallas, V., Marcos, J.F., Perez-Paya, E., 2001. Structural properties of carnation mottle virus p7 movement protein and its RNA-binding domain. J. Biol. Chem. 276, 18122–18129.
Vilar, M., Sauri, A., Monne, M., Marcos, J.F., von Heijne, G., Perez-Paya, E., Mingarro, I., 2002. Insertion and topology of a plant viral movement protein in the endoplasmic reticulum membrane. J. Biol. Chem. 277, 23447–23452.
Vilar, M., Sauri, A., Marcos, J.F., Mingarro, I., Perez-Paya, E., 2005. Transient structural ordering of the RNA-binding domain of carnation mottle virus p7 movement protein modulates nucleic acid binding. Chembiochem 6, 1391–1396.
Voinnet, O., Lederer, C., Baulcombe, D.C., 2000. A viral movement protein prevents spread of the gene silencing signal in Nicotiana benthamiana. Cell 103, 157–167.
Vuorinen, A.L., Kelloniemia, J., Valkonen, J.P.T., 2011. Why do viruses need phloem for systemic invasion of plants? Plant Sci. 181, 355–363.
Waigmann, E., Ueki, S., Trutnyeva, K., Citovsky, V., 2004. The ins and out of non-destructive cell-to-cell and systemic movement of plant viruses. Crit. Rev. Plant Sci. 23, 195–250.
Waigmann, E., Curin, M., Heinlein, M., 2007. Tobacco mosaic virus—a model for macromolecular cell-to-cell spread. In: Waigmann, E., Heinlein, M. (Eds.), Viral Transport in Plants. Springer-Verlag, Berlin Heidelberg, pp. 29–62.
Wan, J., Garcia-Cabanillas, D., Zheng, H., Laliberté, J.F., 2015. Turnip mosaic virus moves systemically through both phloem and xylem as membrane-associated complexes. Plant Phys. 167, 1374–1388.
Wang, A., 2015. Dissecting the molecular network of virus-plant interactions, the complex roles of host factors. Annu. Rev. Phytopathol. 53, 45–66.
Wang, D., Maule, A.J., 1994. A model for seed transmission of a plant virus: genetic and structural analyses of pea embryo invasion by pea seed-borne mosaic virus. Plant Cell 6, 777–787.
Wang, D., Maule, A.J., 1997. Contrasting patterns in the spread of two seed-borne viruses in pea embryos. Plant J. 11, 1333–1340.
Wang, D., MacFarlane, S.A., Maule, A.J., 1997. Viral determinants of pea early browning virus seed transmission in pea. Virology 234, 112–117.

Wang, X., Sager, R., Cui, W., Zhang, C., Lu, H., Lee, J.Y., 2013. Salicylic acid regulates plasmodesmata closure during innate immune responses in Arabidopsis. Plant Cell 25, 2315–2329.
Wei, T., Huang, T.S., McNeil, J., Laliberte, J.F., Hong, J., Nelson, R.S., Wang, A., 2010a. Sequential recruitment of the endoplasmic reticulum and chloroplasts for plant potyvirus replication. J. Virol. 84, 799–809.
Wei, T., Zhang, C., Hong, J., Xiong, R., Kasschau, K.D., Zhou, X., Carrington, J.C., Wang, A., 2010b. Formation of complexes at plasmodesmata for potyvirus intercellular movement is mediated by the viral protein P3N-PIPO. PLoS Pathog. 6, e1000962.
Wellink, J., van Lent, J.W., Verver, J., Sijen, T., Goldbach, R.W., van Kammen, A., 1993. The cowpea mosaic virus M RNA-encoded 48-kiloDalton protein is responsible for induction of tubular structures in protoplasts. J. Virol. 67, 3660–3664.
White, R.G., Barton, D.A., 2011. The cytoskeleton in plasmodesmata: a role in intercellular transport? J. Exp. Bot. 62, 5249–5266.
Wintermantel, W.M., Banerjee, N., Oliver, J.C., Paolillo, D.J., Zaitlin, M., 1997. Cucumber mosaic virus is restricted from entering minor veins in transgenic tobacco exhibiting replicase-mediated resistance. Virology 231, 248–257.
Wolf, S., Deom, C.M., Beachy, R.N., Lucas, W.J., 1989. Movement protein of tobacco mosaic virus modifies plasmodesmatal size exclusion limit. Science 246, 377–379.
Wright, K.M., Chapman, S., Roberts, A.G., 2007. Plasmodesmal targeting and accumulation of TMV movement protein. Plant Signal. Behav. 2, 180–181.
Wu, C.H., Lee, S.C., Wang, C.W., 2011. Viral protein targeting to the cortical endoplasmic reticulum is required for cell-cell spreading in plants. J. Cell Biol. 193, 521–535.
Xie, L., Shang, W., Liu, C., Zhang, Q., Sunter, G., Hong, J., Zhou, X., 2016. Mutual association of Broad bean wilt virus 2 VP37-derived tubules and plasmodesmata obtained from cytological observation. Sci. Rep. 6, 21552.
Xiong, Z., Kim, K.H., Giesman-Cookmeyer, D., Lommel, S.A., 1993. The roles of the Red-clover necrotic mosaic virus capsid and cell to cell movement proteins in systemic infection. Virology 192, 27–32.
Xoconostle-Cazares, B., Xiang, Y., Ruiz-Medrano, R., Wang, H.L., Monzer, J., Yoo, B.C., McFarland, K.C., Franceschi, V.R., Lucas, W.J., 1999. Plant paralog to viral movement protein that potentiates transport of mRNA into the phloem. Science 283, 94–98.
Xu, J., Xu, Q., Peng, L., Wei, T., Yang, Z., Dong, M., Cheng, G., 2017. Dissecting the molecular mechanism of the subcellular localization and cell-to-cell movement of the sugarcane mosaic virus P3N-PIPO. Sci. Rep. 7, 1–11.
Yamaji, Y., Kobayashi, T., Hamada, K., Sakurai, K., Yoshii, A., Suzuki, M., Namba, S., Hibi, T., 2006. In vivo interaction between Tobacco mosaic virus RNA-dependent RNA polymerase and host translation elongation factor 1A. Virology 347, 100–108.
Yelina, N.E., Savenkov, E.I., Solovyev, A.G., Morozov, S.Y., Valkonen, J.P., 2002. Long-distance movement, virulence, and RNA silencing suppression controlled by a single protein in hordei- and potyviruses: complementary functions between virus families. J. Virol. 76, 12981–12991.
Yuan, C., Lazarowitz, S.G., Citovsky, V., 2016. Identification of a functional plasmodesmal localization signal in a plant viral cell-to-cell-movement protein. mBio 7, e02052-15.
Yuan, C., Lazarowitz, S.G., Citovsky, V., 2018. The plasmodesmal localization signal of TMV MP is recognized by plant synaptotagmin SYTA. mBio 9, e01314–e01318. https://doi.org/10.1128/MBIO.01314-18.
Zamyatnin Jr., A.A., Solovyev, A.G., Savenkov, E.I., Germundsson, A., Sandgren, M., Valkonen, J.P., Morozov, S.Y., 2004. Transient coexpression of individual genes encoded by the triple gene block of potato mop-top virus reveals requirements for TGBp1 trafficking. Mol. Plant Microbe Interact. 17, 921–930.

Zamyatnin Jr., A.A., Solovyev, A.G., Bozhkov, P.V., Valkonen, J.P., Morozov, S.Y., Savenkov, E.I., 2006. Assessment of the integral membrane protein topology in living cells. Plant J. 46, 145–154.
Zayakina, O., Arkhipenko, M., Kozlovsky, S., Nikitin, N., Smirnov, A., Susi, P., Rodionova, N., Karpova, O., Atabekov, J., 2008. Mutagenic analysis of potato virus X movement protein (TGBp1) and the coat protein (CP): in vitro TGBp1-CP binding and viral RNA translation activation. Mol. Plant Pathol. 9, 37–44.
Zeenko, V.V., Ryabova, L.A., Spirin, A.S., Rothnie, H.M., Hess, D., Browning, K.S., Hohn, T., 2002. Eukaryotic elongation factor 1A interacts with the upstream pseudoknot domain in the 3' untranslated region of tobacco mosaic virus RNA. J. Virol. 76, 5678–5691.
Zhang, C., Turgeon, R., 2018. Mechanisms of phloem loading. Curr. Opin. Plant Biol. 43, 71–75.
Zhang, C.Z., Liu, Y.Y., Sun, X.C., Qian, W.D., Zhang, D.D., Qiu, B.S., 2008. Characterization of a specific interaction between IP-L, a tobacco protein localized in the thylakoid membranes, and tomato mosaic virus coat protein. Biochem. Biophys. Res. Commun. 374, 253–257.
Zhang, W., Thieme, C.J., Kollwig, G., Apelt, F., Yang, L., Winter, N., Adresen, N., Walther, D., Kragler, F., 2016. tRNA-related sequences trigger systemic mRNA transport in plants. Plant Cell 28, 1237–1249.
Zhang, Y., Fan, J., Sun, J., Francis, F., Chen, J., 2017. Transcriptome analysis of the salivary glands of the grain aphid, Sitobion avenae. Sci. Rep. 7, 15911.
Zhao, J., Liu, Q., Zhang, H., Jia, Q., Hong, Y., Liu, Y., 2013. The rubisco small subunit is involved in Tobamovirus movement and Tm-22-mediated extreme resistance. Plant Physiol. 161, 374–383.
Zhao, J., Zhang, X., Hong, Y., Liu, Y., 2016. Chloroplast in plant-virus interaction. Front. Microbiol. 7, 1565.
Zheng, L., Du, Z., Lin, C., Mao, Q., Wu, K., Wu, J., Wei, T., Wu, Z., Xie, L., 2015. Rice stripe tenuivirus p2 may recruit or manipulate nucleolar functions through an interaction with fibrillarin to promote virus systemic movement. Mol. Plant Pathol. 16, 921–930.

Further reading

Chen, M.H., Tian, G.W., Gafni, Y., Citovsky, V., 2005. Effects of calreticulin on viral cell-to-cell movement. Plant Physiol. 138, 1866–1876.
Deom, C.M., Oliver, M.J., Beachy, R.N., 1987. The 30-kiloDalton gene product of tobacco mosaic virus potentiates virus movement. Science 237, 389–394.
Gutierrez, C., Martinez-Salas, E., 2001. DNA plant and animal virus replication. In: Encyclopedia of Life Sciences. John Wiley & Sons, Ltd, Chichester, UK.
Harries, P.A., Park, J.W., Sasaki, N., Ballard, K.D., Maule, A.J., Nelson, R.S., 2009b. Differing requirements for actin and myosin by plant viruses for sustained intercellular movement. Proc. Nati. Acad. Sci. U. S. A. 106, 17594–17599.
Ibrahim, A., Hutchens, H.M., Howard Berg, R., Sue Loesch-Fries, L., 2012. Alfalfa mosaic virus replicase proteins, P1 and P2, localize to the tonoplast in the presence of virus RNA. Virology 433, 449–461.
Ishikawa, K., Hashimoto, M., Yusa, A., Koinuma, H., Kitazawa, Y., Netsu, O., Yamaji, Y., Namba, S., 2017. Dual targeting of a virus movement protein to ER and plasma membrane subdomains is essential for plasmodesmata localization. PLoS Pathog. 13, e1006463.
Kehr, J., 2006. Phloem sap proteins, their identities and potential roles in the interaction between plants and phloem-feeding insects. J. Exp. Bot. 57, 767–774.

Kim, S.H., Macfarlane, S., Kalinina, N.O., Rakitina, D.V., Ryabov, E.V., Gillespie, T., Haupt, S., Brown, J.W.S., Taliansky, M., 2007a. Interaction of a plant virus-encoded protein with the major nucleolar protein fibrillarin is required for systemic virus infection. Proc. Natl. Acad. Sci. U. S. A. 104, 11115–11120.

Kim, S.H., Ryabov, E.V., Kalinina, N.O., Rakitina, D.V., Gillespie, T., MacFarlane, S., Haupt, S., Brown, J.W.S., Taliansky, M., 2007b. Cajal bodies and the nucleolus are required for a plant virus systemic infection. EMBO J. 26, 2169–2179.

Lekkerkerker, A., Wellink, J., Yuan, P., van Lent, J., Goldbach, R., van Kammen, A.V., 1996. Distinct functional domains in the Cowpea mosaic virus movement protein. J. Virol. 70, 5658–5661.

Maule, A., Faulkner, C., Benitez-Alfonso, Y., 2012. Plasmodesmata "in communicado". Front. Plant Sci. 3, 30.

Navarro, J.A., Pallas, V., 2017. An update on the intracellular and intercellular trafficking of carmoviruses. Front. Plant Sci. 8, 1801.

Oparka, K.J., Prior, D.A.M., Crawford, J.W., 1994. Behaviour of plasma membrane, cortical ER and plasmodesmata during plasmolysis of onion epidermal cells. Plant Cell Environ. 17, 163–171.

Ritzenthaler, C., Laporte, C., Gaire, F., Dunoyer, P., Schmitt, C., Duval, S., Piéquet, A., Loudes, A.M., Rohfritsch, O., Stussi-Garaud, C., Pfeiffer, P., 2002. Grapevine fanleaf virus replication occurs on endoplasmic reticulum-derived membranes. J. Virol. 76, 8808–8819.

Sasaki, N., Takashima, E., Nyunoya, H., 2018. Altered subcellular localization of a tobacco membrane raft-associated remorin protein by tobamovirus infection and transient expression of viral replication and movement proteins. Front. Plant Sci. 9, 619.

Schaad, M.C., Jensen, P.E., Carrington, J.C., 1997. Formation of plant RNA virus replication complexes on membranes: role of an endoplasmic reticulum-targeted viral protein. EMBO J. 16, 4049–4059.

Semashko, M.A., González, I., Shaw, J., Leonova, O.G., Popenko, V.I., Taliansky, M.E., Canto, T., Kalinina, N.O., 2012. The extreme N-terminal domain of a hordeivirus TGB1 movement protein mediates its localization to the nucleolus and interaction with fibrillarin. Biochimie 94, 1180–1188.

Zhang, B.C., Tolstikov, V., Turnbull, C., Hicks, L.M., Fiehn, O., 2010. Divergent metabolome and proteome suggest functional independence of dual phloem transport systems in cucurbits. Proc. Natl. Acad. Sci. U. S. A. 107, 13532–13537.

CHAPTER TWO

Entry of bunyaviruses into plants and vectors

Yuting Chen[a], Moshe Dessau[b], Dorith Rotenberg[a], David A. Rasmussen[a], Anna E. Whitfield[a,*]
[a]Department of Entomology and Plant Pathology, North Carolina State University, Raleigh, NC, United States
[b]Azrieli Faculty of Medicine, Bar Ilan University, Safed, Israel
*Corresponding author: e-mail address: awhitfi@ncsu.edu

Contents

Abstract

The majority of plant-infecting viruses are transmitted by arthropod vectors that deliver them directly into a living plant cell. There are diverse mechanisms of transmission ranging from direct binding to the insect stylet (non-persistent transmission) to persistent-propagative transmission in which the virus replicates in the insect vector. Despite this diversity in interactions, most arthropods that serve as efficient vectors have feeding strategies that enable them to deliver the virus into the plant cell without extensive damage to the plant and thus effectively inoculate the plant. As such, the primary virus entry mechanism for plant viruses is mediated by the biological vector. Remarkably, viruses that are transmitted in a propagative manner (bunyaviruses, rhabdoviruses, and reoviruses) have developed an ability to replicate in hosts from two kingdoms. Viruses in the order *Bunyavirales* are of emerging importance and with the advent of new sequencing technologies, we are getting unprecedented glimpses into the diversity of these viruses. Plant-infecting bunyaviruses are transmitted in a persistent, propagative manner must enter two unique types of host cells, plant and insect.

Advances in Virus Research, Volume 104
ISSN 0065-3527
https://doi.org/10.1016/bs.aivir.2019.07.001

In the insect phase of the virus life cycle, the propagative viruses likely use typical cellular entry strategies to traverse cell membranes. In this review, we highlight the transmission and entry strategies of three genera of plant-infecting bunyaviruses: orthotospoviruses, tenuiviruses, and emaraviruses.

1. Overview of the plant virus transmission cycle and mechanisms

1.1 The plant virus transmission cycle

The plant virus transmission cycle involves a complex series of interactions among the vector, virus and plant host. Different plant-associated organisms are capable of serving as biological vectors of viruses including insects, arachnids (mites), nematodes, fungi, and plasmodiophorids. The vast majority of known plant viruses (~76%) are transmitted by vectors and of those ~55% are transmitted by hemipterans (e.g., aphids, leafhoppers, mealybugs, planthoppers, and whiteflies) (Hogenhout et al., 2008). The plant surface is covered by a layer of cuticle and plant cells are surrounded by rigid walls made of cellulose, hemicellulose, and pectin that are a significant barrier to virus entry. Vectors of plant viruses are able to breach these structures and deliver virus into the cytoplasm. Viruses may have distinct tissue tropisms in the plant, with some infecting diverse tissues and other being more specialized, for example the phloem-limited viruses. In each case, there are distinct interactions with vectors that enable viruses to be delivered to the appropriate tissue for inoculation and subsequent replication to occur. During this process, the vector delivers the virus into the cytoplasm of the cell and uncoating of viral particles may occur through destabilization of the virion due to physiological conditions in the cells and subsequent disassembly driven by various viral replication steps (reviewed in Culver, 2002; Mandahar, 2006). In the plant cell, the replication cycle is initiated, and progeny are then able to move to a new cell via the plasmodesmata, channels in the cell wall that connect cells and allow molecules and substances to move back and forth as needed (reviewed in Heinlein, 2015). Because plants are connected by plasmodesmata for short distance transport and the plant vascular system, phloem and xylem, for long distance movement, once a virus is delivered into a susceptible plant host, the cell wall and plasma membrane are no longer barriers to entry and exit of virions.

Because plant virus entry is largely dependent on the biological vector delivering the virion into a living plant cell, hemipteran insects are the most common plant virus vectors. These insects have specialized structures (stylets) for plant-feeding that enable them to successfully feed with very little damage to the plant (reviewed in Fereres, 2016; Ng and Walker, 2016). In addition to anatomical features, vector behavior and ecology of the vector-virus-host interaction also drives the success of vector transmission of plant viruses (Carr et al., 2018; Fereres, 2016; Ng and Walker, 2016). Plants are sensed by hemipteran vectors using long- distance visual cues and short-distance olfactory cues. In a generalized transmission cycle, the vector must first find the infected plant host, feed from the plant, acquire the virus, retain the virus, and finally deliver virus to a new host plant through another feeding period. Each step in the transmission process can be modulated by external and internal factors to increase or decrease transmission efficiency (for thorough coverage of this topic please refer to reviews by Carr et al., 2018; Mauck et al., 2018).

1.2 Plant virus transmission mechanisms

Plant virus transmission by vectors is divided using two classification strategies (1) timing and duration of the interaction (i.e., nonpersistent, semi-persistent, and persistent) and (2) virus retention sites in the vector (i.e., circulative vs. non-circulative, Fig. 1) (reviewed in Whitfield et al., 2015). Prior to the modern era of molecular virology, scientists realized that plant viruses had varying times of retention in their insect vectors and these qualities became an important determinant of virus classification and provided essential information for devising control strategies for the diseases caused by viruses. Non-persistent viruses are characterized by short acquisition and inoculation times (seconds-minutes) and semi-persistent viruses have longer periods of association with the vector (hours-days) (Ng and Falk, 2006). Viruses transmitted in a persistent manner generally have a life-long association with the vector. As researchers learned more about the relationships between plant viruses and vectors, they began to understand that the non- and semi-persistent viruses attach to the feeding structures of the vector body (stylet or foregut) and were thus termed non-circulative (reviewed in Blanc et al., 2014). Non-circulative viruses do not circulate in the body of the insect and are lost during insect molts because they attach to the exoskeleton. A differentiating characteristic between non-circulative and circulative viruses is transstadial transmission. Because circulative viruses enter the insect body (typically through the gut tissues) and traverse several membrane barriers to eventually reach

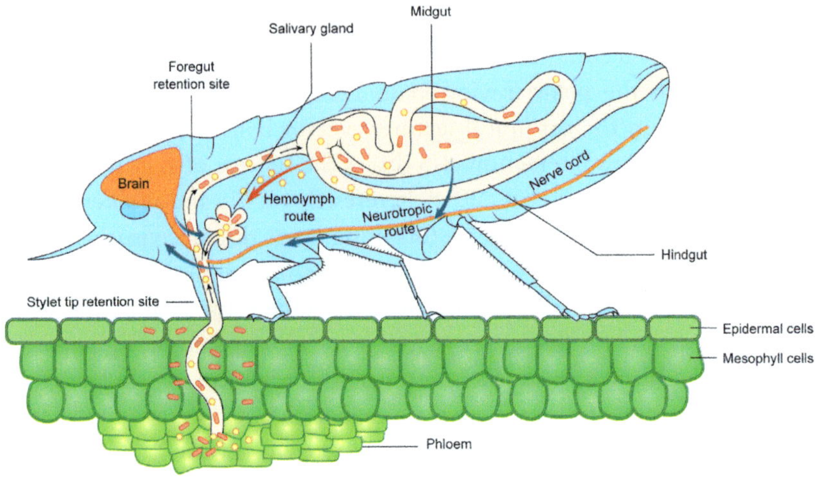

Fig. 1 Viruses localize to different sites in the plant-feeding insect vector depending on their modes of transmission. Non-circulative viruses bind to the insect stylet or foregut. Non-propagative circulative (yellow circles) viruses are generally phloem limited and move through the insect body via the midgut or hindgut. Circulative viruses enter the hemolymph and then enter the salivary glands. By contrast, circulative propagative viruses (red ovals) tend to enter the insect through the midgut and replicate in insect tissues. Some propagative viruses are phloem limited while others are widely distributed in plant tissues. The salivary glands are the final destination for circulative transmission, and viruses reach the salivary glands via the hemolymph or other routes such as the nervous tissue (neurotropic route) or through connective tissues. *Used with permission from Whitfield, A.E., Rotenberg, D. 2015. Disruption of insect transmission of plant viruses. Curr. Opin. Insect Sci., 8, 79–87.*

the salivary glands, they are passaged through the different insect life stages. Circulative, persistent viruses can be further divided based on whether or not they replicate in the vector and viruses that do replicate are called propagative. Transmission of viruses by beetles deviates from this process and was recently reviewed by Tolin et al. (2016).

The common characteristic among the different types of transmission is that the plant cell is inoculated when the vector feeds on the plant. The pathway of entry into the plant is with the insect saliva and/or regurgitant. For the non-circulative viruses that attach to the tip of the insect stylet, the virus binds to the acrostyle, a region of the stylet in the common canal, where the food canal and salivary canal are fused (Uzest et al., 2010). This binding site enables virus to be released from the stylet during initial test probes on a new plant (Fig. 1). The first aphid receptor for a stylet-borne virus, cauliflower mosaic virus, was found to be a cuticle protein that is located in the acrostyle (Webster et al., 2018) and this finding is consistent with previous work implicating cuticle proteins in potyvirus transmission by aphids (Dombrovsky et al., 2007). Other non-circulative viruses attach to the foregut and the whitefly-borne criniviruses are the best-studied vector-borne viruses in the semi-persistent, non-circulative transmission mechanism category (reviewed in Ng and Walker, 2016). Virus particles bind to sites further into the alimentary tract (e.g., the cibarium or anterior foregut) and virus inoculation is hypothesized to occur during the egestion process (Chen et al., 2011). Ng and Walker (2016) described the egestion process as a time when the specific cibarial muscles relax and the precibarial valve muscles contract thus creating an opportunity for the flow of ingested fluid (phloem sap) down the maxillary food canal and back into the plant. In this way, virions that attached to the foregut region of the vector can be released back into the plant for inoculation to occur.

The circulative, persistently-transmitted viruses have a longer association with their vectors and circulate through the insect body (Hogenhout et al., 2008). Virus is acquired when the insect feeds on the infected plant and virus particles enter the alimentary canal of the vector. Diverse regions of the gut can serve as the entry site for circulative viruses. For example, many of the well-characterized luteovirids enter through the aphid hindgut while tospoviruses enter through the anterior region of the thrips midgut (Gildow, 1992; Montero-Astua et al., 2016b). Plant viruses are thought to enter the vector gut cells via receptor-mediated endocytosis. For the non-propagative viruses (e.g., nanoviruses, luteoviruses, and geminiviruses), virions are transcytosed across the gut cell layer and then delivered into the hemolymph side of the insect gut. Transmission electron microscopy combined with virus transmission assays has documented the gut and salivary gland barriers that viruses traverse (Gildow, 1992; Gray and Gildow, 2003). Once past the epithelial cells of the gut, the virus encounters the basal lamina, an extracellular matrix, that can be a barrier to virus dissemination. Beyond the basal lamina are the circular and longitudinal muscle cells and then the insect hemocoel. Within the insect hemocoel, some viruses can be found associated with hemocytes and others with bacterial endoymbionts (Granados et al., 1968; Su et al., 2013; Van Den Heuvel et al., 1994; Yao et al., 2019). The other transmission relevant site of virus entry is the salivary gland. Virions face a similar series of barriers in this insect organ with the basal lamina on the hemolymph side and a basal membrane to bind and enter. Once inside the salivary gland viruses enter the secretory pathway and can be delivered with saliva into the plant during feeding. Non-propagative viruses simply transcytose through the salivary gland cells, but the propagative viruses (reoviruses, bunyaviruses, and rhabdoviruses) replicate in the gut and salivary gland cells (Ammar et al., 2009; Czosnek et al., 2017; Gray and Gildow, 2003; Montero-Astua et al., 2016b; Wei and Li, 2016). Some propagative viruses use tubule structures made of virally-encoded proteins to disseminate in their vectors (Jia et al., 2014; Liao et al., 2017). The dissemination routes for orthotospoviruses and tenuiviruses are presented in Sections 3.1 and 3.2. The tenuiviruses are also transovarially transmitted to offspring and the mechanism of entry using vitellogenin has been described (Huo et al., 2014). Other transovarially-transmitted propagative viruses hijack endosymbiont transport pathways to enter eggs (Jia et al., 2017; Wu et al., 2019).

2. The order *Bunyavirales*

The order *Bunyavirales* includes a rapidly-expanding list of viruses that infect plant, invertebrate, and vertebrate hosts that use arthropods as well as mammals (e.g., viral particles are present in aerosolized excretions and secretions from small rodents and transmitted to a new host through this route) as their transmitting vectors (Abudurexiti et al., 2019). The classical virus features that once grouped the viruses together were genome polarity (negative-sense ssRNA), genome segment number, and virion morphology. The primary classification criteria used currently is sequence comparisons and phylogenetic analyses. Further characterization of plant-infecting viruses in this order has revealed a diverse number of genome segments between genera, variations in virion structure, and distinct arthropod vectors. In this review, we present the virus entry/transmission mechanisms for the longest-standing genus in the *Bunyavirales*, the *Orthotospoviruses*, and the exciting new developments with *Tenuiviruses*, members of the *Phenuiviridae*. The most recent additions to this group of viruses are the *Emaraviruses* that are mite transmitted and very little is known about their vector interactions. A newly described citrus-infecting virus, citrus concave gum-associated virus (CCGaV), is the sole member of the new genus *Coguvirus* that is currently not taxonomically placed within a family.

The diversity of the described bunyaviruses has greatly increased in recent years thanks to more intensive sampling and de novo virus discovery through metagenomic sequencing (Junglen, 2016; Li et al., 2015). We reconstructed a maximum likelihood (ML) phylogeny of the bunyaviruses, including representatives of each family, the type species of many genera and several recently described species. Our phylogenetic analysis supports earlier reports (Junglen, 2016; Marklewitz et al., 2015) that plant-infecting bunyaviruses evolved multiple times independently from invertebrate-infecting viruses. The emaraviruses, orthotospoviruses and tenuiviruses form distinct clades that are more closely related to other invertebrate viruses than to one another (Fig. 2; Table S1 in the online version at https://doi.org/10.1016/bs.aivir.2019.07.001). Interestingly, newly described plant-infected viruses like citrus concave gum-associated virus also appear to have evolved independently from invertebrate viruses (Navarro et al., 2018a). New virus discoveries and phylogenetic analysis support the hypothesis that bunyaviruses are arthropod viruses that evolved to infect other hosts. It is therefore tempting to speculate that additional evolutionary transitions to plant-infecting lifestyles will be discovered as new bunyaviruses are described.

Bunyaviruses vary in their morphology that ranges from high symmetry icosahedral particles as seen in phleboviruses (Freiberg et al., 2008; Halldorsson et al., 2018) to spherical and pleomorphic particles as seen in orthobunyaviruses (Obijeski et al., 1976) and hantaviruses (Huiskonen et al., 2010). A classic bunyavirus would have a tripartite negative-sense ssRNA genome enveloped in a lipid bilayer in which the viral glycoproteins N and C (termed Gn and Gc) are embedded. For the majority of animal-infecting viruses and the orthotospoviruses the genome is comprised of three RNA segments designated large (L), medium (M), and small (S). Each of the genome segments is encapsidated by multiple copies of nucleoprotein (N). In addition, each of the segments is bound to the RNA-dependent RNA polymerase (RdRp, commonly called the L protein) that is coded by the L segment. In all viruses characterized to-date, the L RNA codes for a single gene. Interestingly, the tenuiviruses and emaraviruses have additional genome segments whereas the coguvirus appears to have only two genome segments. The numbers of genes and expression strategies for the M and S segments varies between virus genera and sometimes within a genus. For the plant-infecting orthotospoviruses, the M segment encodes the two glycoproteins, Gn and Gc, and the movement protein, NSm. The S segment codes for N protein and the silencing suppressor, NSs, an important viral anti-defense protein in the plant and insect vector (Margaria et al., 2014; Takeda et al., 2002).

2.1 Entry of Bunyaviruses

All arthropod-borne bunyaviruses are transmitted in a persistent, circulative manner and replicate in their arthropod vector/hosts. In the arthropod

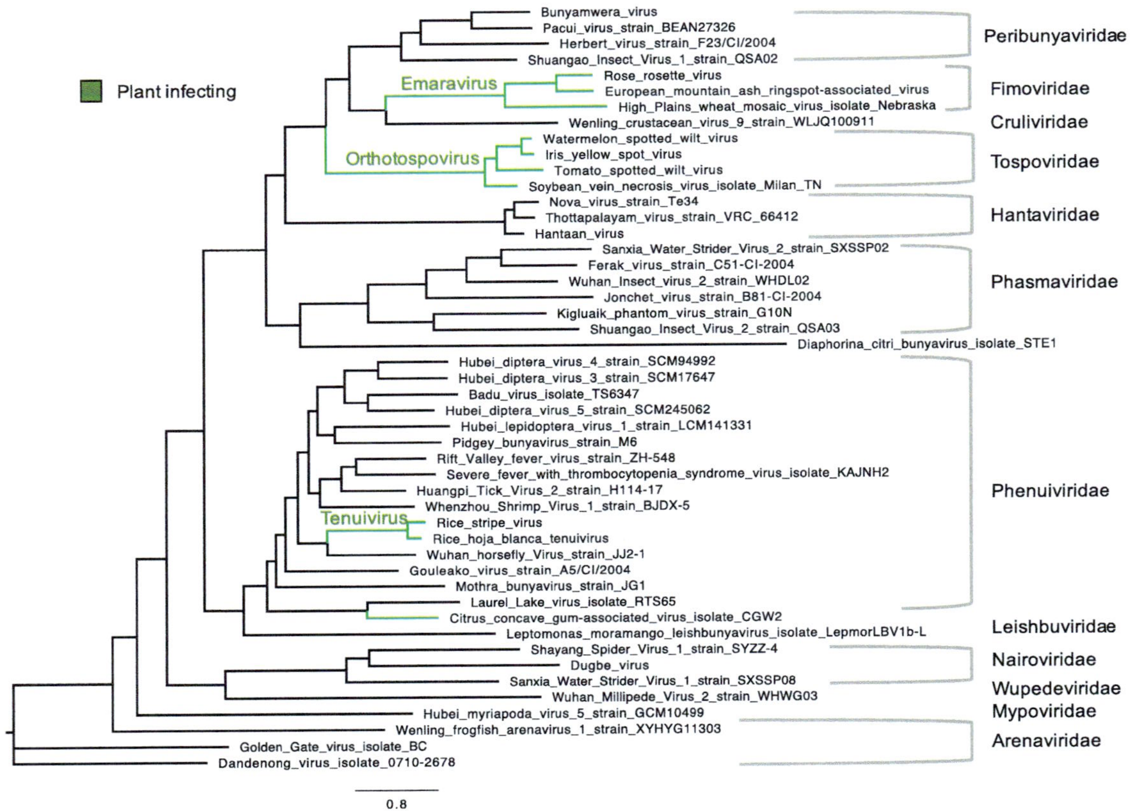

Fig. 2 See legend on opposite page.

vector, virions must traverse membrane and structural barriers in the guts and salivary glands. The majority of research on virus entry for bunyaviruses has focused on animal-infecting viruses and the use of cells cultures and infectious clones has enabled straightforward experiments with synchronized infection thus enabling the steps of virus entry process and host factors to be characterized (reviewed in Albornoz et al., 2016). Virus entry is mediated by the two glycoproteins, Gn and Gc, that interact with cell surface receptors. Bunyaviruses enter diverse host cell types via receptor-mediated endocytosis. The precise endocytic pathway varies among bunyaviruses and host cell types, but in general, receptor binding activates cellular signaling pathways and the virus particle is endocytosed. Once in the endocytic pathway the particles move from early endosomes to late endosomes and may eventually reach the lysosomes. As the virus particles progress through the endocytic pathway, the vesicle pH decreases from ~6.5 to ~5.5–5.0 (Albornoz et al., 2016; Lozach et al., 2011). The decrease in pH is the trigger for Gc conformational changes and rearrangements that mediate the fusion of the viral membrane with the endosomal membrane resulting in the release of the virus' genetic cargo into the cytoplasm.

Receptors have been identified for a few bunyaviruses and common cell surface receptors for viruses can be proteins, lipids, or glycans (Albornoz et al., 2016). Polysaccharides on glycoproteins and glycolipids are present on the surface of many animal cells (Albornoz et al., 2016). Heparan sulfate, a glycosaminoglycan, was shown to be the first docking site for two phleboviruses, Rift Valley fever virus and Toscana virus, and the interaction occurs through low affinity electrostatic interactions (De Boer et al., 2012; Pietrantoni et al., 2015; Riblett et al., 2016). The receptor for two phleboviruses, Rift Valley

Fig. 2 Maximum likelihood phylogenetic tree of the bunyaviruses. The tree was reconstructed from the RdRp protein sequences of representative viral species. Viruses known to infect plants are highlighted in green. We reconstructed a molecular phylogeny of representative *Bunyavirales* species based on the RdRp protein sequences. A full length sequence of the L segment encoding RdRp was downloaded from NCBI GenBank for each species. Accession numbers for each species are provided in Table S1. We then extracted the RdRp protein encoding region from each sequence and created an amino acid alignment using Muscle (Edgar, 2004). Poorly aligned regions with extensive gaps or amino acid divergence were removed using TrimAl under the "automated1" setting (Capella-Gutierrez et al., 2009). A maximum likelihood phylogeny was then reconstructed from the trimmed RdRp alignment using RAxML (Stamatakis, 2014) under the "PROTGAMMALGF" model, which assumes a LG substitution model (Le and Gascuel, 2008) with gamma distributed rate heterogeneity across sites. Phylogenetic uncertainty was assessed using bootstrap support values.

fever virus and Uukuniemi virus, has been identified and the viruses enter dendritic cells via the C-type lectin dendritic cell-specific intercellular adhesion molecule-3-grabbing non-integrin (DC-SIGN) (Lozach et al., 2011). The initial identification of DC-SIGN as a receptor led to other studies documenting its role as a receptor for other bunyaviruses including orthobunyaviruses, nairoviruses, and mammarenaviruses (Goncalves et al., 2013; Hofmann et al., 2013; Suda et al., 2016). For the orthohantaviruses, several integrins (β1, β2, and β3) have been identified as receptors (Gavrilovskaya et al., 1998, 1999; Matthys et al., 2010; Mou et al., 2006; Raftery et al., 2014). Recently it was shown that the surface glycoproteins from new-world hantaviruses directly interact with procadherin-1 and this interaction is essential to their entry. No receptor has been identified for the arthropod vector of bunyaviruses, but Gn-interacting proteins have been identified for thrips vectors of tomato spotted wilt virus (TSWV) (Badillo-Vargas et al., 2018).

Determination of glycoprotein structures and electron cryomicroscopy have greatly increased our understanding of bunyavirus particle morphology and glycoprotein function (reviewed in Albornoz et al., 2016; Guardado-Calvo and Rey, 2017). A consistent trend for the important mammalian bunyaviruses that have been well characterized is that the Gn-Gc proteins form heteromultimers on the surface of the virion that form diverse ultrastructure assemblies (Bowden et al., 2013). The phleboviruses have been shown to have a $T=12$ icosahedral symmetry while orthobunyaviruses have tripod GP arrangement and orthohantaviruses a tetrameric square arrangement (Battisti et al., 2011; Bowden et al., 2013; Freiberg et al., 2008; Huiskonen et al., 2010, 2009; Overby et al., 2008; Sherman et al., 2009). The cryo-EM structure of the plant viruses has not been resolved. With the uniqueness of these individual families and genera described above, it would not be surprising to find that orthotospovirus GP structural arrangements in the virus particle differ. Based on sequence similarity of the GPs, we can hypothesize that the TSWV structure would be most similar to viruses in the *Peribunyaviridae*, including the orthobunyaviruses. The tenuiviruses do not appear to form enveloped virions and their structure is radically different from other members of the order. The emaraviruses differ from the classical bunyaviruses by having larger virions that encapsulate additional genome segments, and we expect their GP structure to be quite different as well. Viruses in the new genus *Coguvirus* do not appear to encode glycoproteins (Navarro et al., 2018b).

The three-dimensional structure of the Gc protein of several bunyaviruses has been resolved and the proteins characterized thus far are class II fusion proteins (Dessau and Modis, 2013; Guardado-Calvo et al., 2016; Hellert et al., 2019; Willensky et al., 2016). To mediate fusion of the viral and host membranes the viral fusion proteins change their conformation as a response to the acidic pH and rearrange as trimers. These conformational changes result in the exposure of a hydrophobic loop, termed the fusion loop, that serves as an anchor in the host's target membrane. Once anchored to the endosomal membrane the class II membrane fusion proteins fold back upon themselves, bringing the viral and host membranes to proximity that promote membrane fusion (reviewed in Modis, 2013). Cryo EM tomography studies by Halldersson et al. (2018) showed that in the Rift Valley fever virus mature virion (genus: *Phlebovirus*, family: *Phenuiviridae*) Gn protects Gc's fusion loop from premature membrane insertion and an un-shielding mechanism exposes Gc, allowing its extension and insertion of its fusion loops in the outer leaflet of the target membrane. It was recently shown that Gn-Gc spikes in hantaviruses exhibit temperature dependent dynamics between "closed" and "open" forms, and in the open form, the Gc fusion loop is exposed. However, this "open" state resulted in inefficient membrane fusion activity (Bignon et al., 2019).

Further comparisons of the Gc structures and sequences by Guardado-Calvo and Rey (2017) showed that the peribunyaviruses, nairoviruses, hantanviruses, and tospoviruses all share a common motif suggesting that the domain II of these fusion proteins is conserved. Interestingly, the phenuiviruses (including the tenuiviruses) and emaraviruses do not contain this motif. It is possible that the plant-infecting bunyaviruses may use different entry and dissemination strategies in their arthropod vectors based on the diversity of fusion protein biology.

3. Plant-infecting members of the Bunyavirales

3.1 Orthotospoviruses

Orthotospoviruses are an emerging group of plant viruses with 18 ICTV-approved species and an additional 11 described viruses that have not been taxonomically validated, Table 1 (Abudurexiti et al., 2019; Rotenberg and Whitfield, 2018). Orthotospoviruses have the typical virion structure for bunyaviruses (Fig. 3) and are transmitted to plants by thrips vectors. The virus transmission cycle is unique and outlined in Fig. 4. Generally, thrips are attracted to virus-infected plants and when a female thrips oviposits into infected plant tissue, the resulting progeny thrips have direct access to the plant and the virus contained within. This attraction to virus-infected plants

Table 1 List of described viruses in the genus *Orthotospovirus* and their thrips vectors.

***Orthotospovirus* species name**[a]	**Acronym**	**Vectors**[b]
Alstroemeria necrotic streak virus	ANSV	*Frankliniella occidentalis*
Alstroemeria yellow spot virus[c]	AYSV	*Thrips tabaci*[c]
Bean necrotic mosaic orthotospovirus	BeNMV	Unknown
Calla lily chlorotic spot orthotospovirus	CCSV	*Thrips palmi*
Capsicum chlorosis orthotospovirus	CaCV	*Ceratothripoides claratris*
		T. palmi
		F. schultzei
Chrysanthemum stem necrosis orthotospovirus	CSNV	*F. intonosa*
		F. occidentalis
		F. schultzei
Groundnut bud necrosis orthotospovirus	GBNV	*F. schultzei*
		Scirtothrips dorsalis
		T. palmi
Groundnut chlorotic fan-spot virus	GCFSV	*S. dorsalis*
Groundnut ringspot orthotospovirus	GRSV	*F. gemina*
		F. occidentalis
		F. schultzei
Groundnut yellow spot orthotospovirus	GYSV	*S. dorsalis*
Hippeastrum chlorotic ringspot virus	HCRV	Unknown
Impatiens necrotic spot orthotospovirus	INSV	*F. fusca*
		F. intonsa
		F. occidentalis
Iris yellow spot orthotospovirus	IYSV	*F. fusca*
		T. tabaci
Lisianthus necrotic ringspot virus	LNRV	Unknown
Melon severe mosaic orthotospovirus	MSMV	Unknown
Melon yellow spot orthotospovirus	MYSV	*T. palmi*
Mulberry vein banding associated virus	MVBaV	Unknown
Pepper chlorotic spot virus	PCSV	unknown

Table 1 List of described viruses in the genus *Orthotospovirus* and their thrips vectors.—cont'd

***Orthotospovirus* species name**[a]	**Acronym**	**Vectors**[b]
Pepper necrotic spot virus	PNSV	Unknown
Polygonum ringspot orthotospovirus	PolRSV	*Dictyothrips betae*
Soybean vein necrosis orthotospovirus	SVNV	*F. fusca*
		F. fusca
		Neohydatothrips variabilis
Tomato chlorotic spot orthotospovirus	TCSV	*F. intonsa*
		F. occidentalis
		F. schultzei
Tomato necrotic ringspot virus [Tomato necrotic ring virus]	TNRV	*C. claratris*
		T. palmi
Tomato necrotic spot virus	TNSV	Unknown
Tomato spotted wilt orthotospovirus	TSWV	*Frankliniella bispinosa*
		Frankliniella cephalica
		Frankliniella fusca
		Frankliniella gemina
		Frankliniella intonsa
		Frankliniella occidentalis
		Frankliniella schultzei
		Thrips setosus
		Thrips tabaci
Tomato yellow ring virus [Tomato yellow fruit ring virus]	TYRV	*T. tabaci*
Tomato zonate spot virus	TZSV	*F. occidentalis*
Watermelon bud necrosis orthotospovirus	WBNV	*T. palmi*
Watermelon silver mottle orthotospovirus	WSMoV	*T. palmi*
Zucchini lethal chlorosis orthotospovirus	ZLCV	*Frankliniella zucchini*

[a]Viruses with italicized names are officially recognized by the ICTV.
[b]List updated from the following reviews (Oliver and Whitfield, 2016; Riley et al., 2011; Rotenberg and Whitfield, 2018; Rotenberg et al., 2015).
[c]New tospovirus infecting Alstroemeria (Hassani-Mehraban et al., 2019).

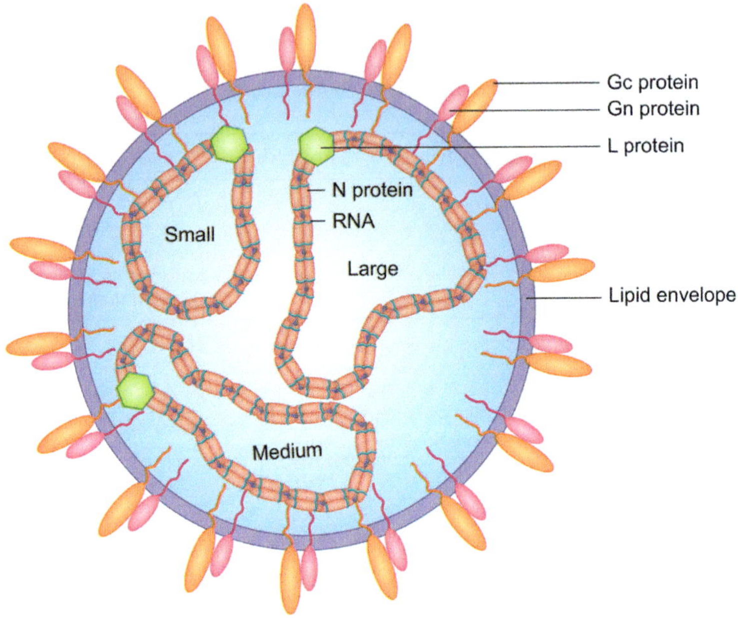

Fig. 3 Schematic representation of the tospovirus virion. Tospovirus genomes are comprised of three RNA segments (small, medium, and large). The negative sense viral RNA is encapsidated by the nucleocapsid protein (N) to form the ribonucleoprotein (RNP) complexes. The RNPs associate with the RNA-dependent RNA polymerase (RdRp), and the RNPs are packaged in the lipid envelope. The two glycoproteins, Gn and Gc, decorate the surface of the virion and span the membrane. The glycoproteins are drawn as heterodimers but higher order oligomeric structures may be a feature of the mature virion. The glycoproteins span the lipid bilayer and the cytoplasmic tails have been shown to interact with the N protein. *Used with permission from Rotenberg, D., Jacobson, A. L., Schneweis, D. J. Whitfield, A. E. 2015. Thrips transmission of tospoviruses. Curr. Opin. Virol., 15, 80–89.*

is important for the transmission cycle because only the young larval thrips can effectively acquire the virus. Early stages of the thrips vector are the most susceptible to virus infection and virus acquisition and dissemination are tightly linked to thrips development (Montero-Astua et al., 2016a; Rotenberg and Whitfield, 2018; Schneweis et al., 2017). It is well documented that adult thrips that feed on virus do not become virus transmitters. Once inside the insect, the virus is transstadially passaged. The pupal stages are non-feeding and do not acquire or inoculate virus. After the quiescent pupal stages, the adult thrips emerge and this life stage is the most effective inoculator of tospoviruses. The adult thrips are winged and highly mobile thus effectively moving virus from plant to plant during feeding. In he adult

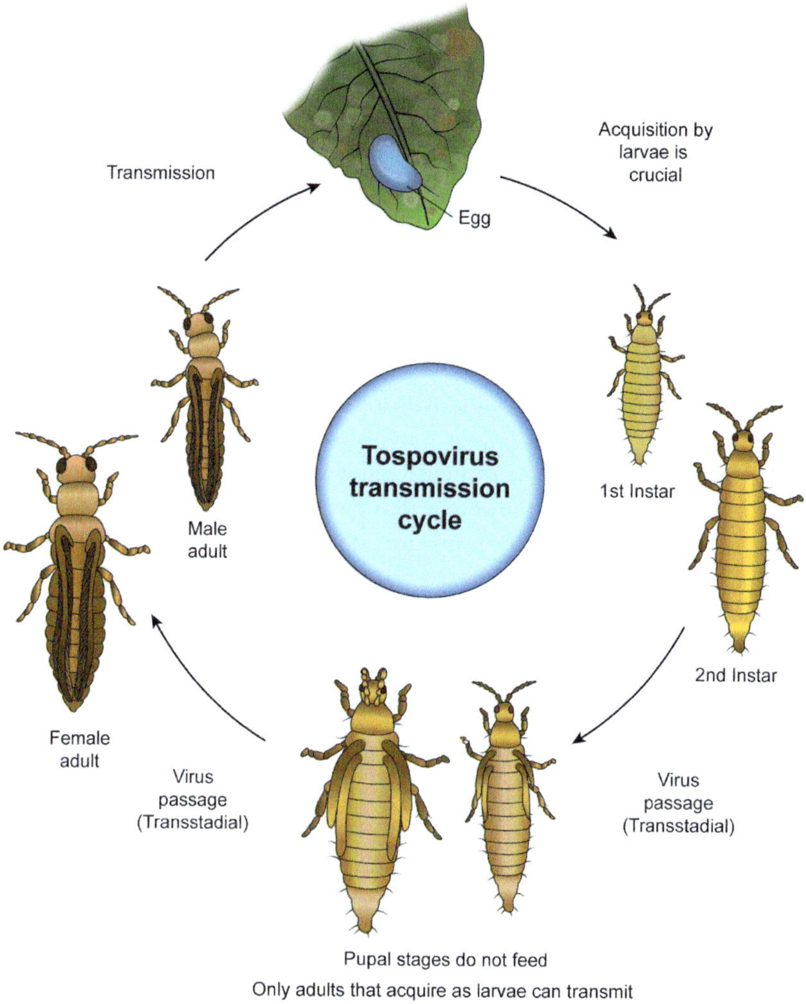

Fig. 4 Schematic representation of the thrips life cycle and the tospovirus transmission cycle. Thrips eggs are oviposited into plant tissue and within a few days the first instar larvae emerge. Virus acquisition occurs during the larval stages after which the virus persists in the vector and is passed transstadially to the adult. The pupal insects may occur in the soil or leaf tissue and these stages are nonfeeding. Adults emerge and may disperse widely. Only adult thrips (male and female) that acquired the virus during their larval stages can transmit tospoviruses. *Used with permission from Rotenberg, D., Jacobson, A. L., Schneweis, D. J. Whitfield, A. E. 2015. Thrips transmission of tospoviruses. Curr. Opin. Virol., 15, 80–89.*

lifestage, the virus has reached the salivary glands and they are the primary site of virus infection where the virus can be delivered with insect saliva through the thrips stylet during feeding on epidermal or mesophyll cells (Montero-Astua et al., 2016a,b).

The TSWV dissemination route in thrips has been documented for the highly efficient vector, *Frankliniella occidentalis*. Virus first enters the epithelial cells at anterior region of midgut 1 and from there spreads to other gut cells (Montero-Astua et al., 2016b). After this initial infection of the epithelial cells the virus particles have two possible routes for reaching the salivary glands. One model proposes that virions traverse the basement membrane, basal lamina, and muscle cells and directly infect the principal salivary glands during the larval stage of development (Moritz et al., 2004). Alternatively, and perhaps not exclusively, the virus may use the tubular salivary glands that connect the gut tissues to the principal salivary glands as a non-hemolymph route for virus transport in the vector (Kritzman et al., 2002; Montero-Astua et al., 2016b). Confocal and transmission electron microscopy showed virus accumulating in the tubular salivary glands of thrips. An open question in orthotosopovirus research is the mechanism for cell-to-cell spread of virus in the vector and the potential role of intercellular junctions in this process.

It is hypothesized that virus entry into thrips' midguts, the initial site of virus infection, is mediated by a specific cellular receptor. While there are alternative hypotheses, there is evidence to support the existence of a gut receptor. The TSWV Gn protein binds directly to insect midgut cells indicating that Gn is a viral attachment protein (Whitfield et al., 2004). Additionally, exogenous Gn protein fed to larval thrips prior to or concomitant with virus acquisition results in reduced virus accumulation in the insect gut and body (Whitfield et al., 2004, 2008). This reduction in virus entry was further substantiated using tomato plants that express a soluble form of Gn in transmission-blocking experiments (Montero-Astua et al., 2014). The transgenic tomato plants were inoculated with TSWV and used as acquisition hosts for larval thrips. The insects that fed on Gn transgene-expressing TSWV-infected plants showed a significant reduction in virus titer that persisted over insect development. This reduction in virus abundance was observed in larval and adult thrips and resulted in a decrease in virus transmission efficiency (Montero-Astua et al., 2014). The direct binding of Gn to thrips' midguts and the ability to saturate potential binding sites with a specific viral protein is suggestive of a specific virus protein-host receptor interaction. These findings have implications for the basic biology of the thrips-tospovirus interaction and control strategies. First, the ability to block virus acquisition at early life stages translates into reduced virus abundance throughout the insect life span and a reduction in transmission by adults. By decreasing the virus infection early on it is possible that virus is unable to reach sufficient titer to overcome insect defenses and/or disseminate to the salivary glands. Although the thrips vector supports virus replication, an initial reduction in virus entry in larval thrips effectively reduces transmission. The ability to disrupt the virus entry process during the acquisition phase of the transmission cycle means that this control strategy may be effective under field and/or greenhouse conditions. A strong positive correlation between virus titer in larval and adult thrips and transmission efficiency has been demonstrated (Montero-Astua et al., 2014; Rotenberg et al., 2009).

The identification of the thrips receptor has remained elusive. While it is hypothesized that tospoviruses enter thrips cells via receptor-mediated endocytosis alternative hypotheses cannot be ruled out at this time. The small size of the thrips vector and the complex transmission cycle have been confounding factors in the study of virus-thrips interactions. There are specific interactions between thrips vectors and the viruses that they transmit. For example, there are more than 7000 species of thrips that have been described, but only 16 transmit tospoviruses. It is likely that there are specific genetic determinants of vector competence and behavioral components as well. The development of -omics technologies and their application to non-model organisms have ushered in a new era of understanding for thrips-orthotospovirus interactions. For the model system of *F. occidentalis* and TSWV, the proteome, transcriptomes, genome and even tissue specific transcriptomes have been characterized (Badillo-Vargas et al., 2012; Schneweis et al., 2017; Stafford-Banks et al., 2014). Work with *F.occidentalis* has paved the way for new transcriptome-wide studies with other important vectors including *F. fusca* (Shrestha et al., 2017) and *Thrips palmi* (Widana Gamage et al., 2018).

Although the identification of the vector determinants of virus transmission have not been defined, there has been progress towards identifying thrips proteins that bind to orthotospoviruses. Bandla et al. (1998) and Kikkert et al. (1998) used gel overlay assays to identify thrips proteins that bound to purified TSWV particles. They saw discreet bands of proteins that bound to the virus but they were not able to characterize the proteins due to a lack of genome sequence data and the limited proteomics technologies of the time. Recent work by Badillo-Vargas et al. (2018) has identified the first thrips proteins that bind to TSWV. Six unique proteins bound to purified virus or soluble Gn protein and they were identified from interacting spots using electrospray ionization mass spectrometry. The sequences were annotated as cuticle proteins (three of the six), enolase, cyclophilin, and mATPase. It is intriguing that thrips cuticle proteins bound to TSWV because these proteins have been shown to be receptors and/or interactors for diverse viruses such as the non-circulative, stylet-borne virus cauliflower mosaic virus and the related tenuivirus, rice stripe virus (Liu et al., 2015; Webster et al., 2018). Another interacting protein, cyclophilin, was shown to be important for two circulative viruses, geminiviruses and luteoviruses in their respective vectors, whiteflies and aphids (Kanakala and Ghanim, 2016; Tamborindeguy et al., 2013). Further work examining the function of these TSWV-interacting proteins of thrips will be of interest to the vector biology community.

3.2 Tenuiviruses

The genus *Tenuivirus* has seven species (Table 2) and *Rice stripe tenuivirus* is the type member (Abudurexiti et al., 2019). Tenuiviruses have four to six single stranded negative-sense RNAs with both negative and ambisense coding strategies (Falk and Tsai, 1998; Liu et al., 2018). The 5′ and 3′ termini (~20 bases) of each segmented RNA are reverse complementary. Two structural proteins, RdRp and nucleocapsid protein, as well as other non-structural proteins are translated from different RNAs and subgenomic RNAs (Ramirez and Haenni, 1994). Tenuiviruses have filamentous virions that are also referred to as ribonucleoprotein particles (RNPs). Viral RNA, nucleocapsid protein and RdRp are present in the purified RNPs of tenuiviruses but viral GPs are absent in what is thought to be the virion (Falk and Tsai, 1998; Ramirez and Haenni, 1994; Toriyama, 1986).

Table 2 Tenuiviruses and their insect vectors.

Tenuivirus species	Acronym	Insect vector
Echinochloa hoja blanca tenuivirus	EHBV	*Tagosodes cubanus*
Iranian wheat stripe tenuivirus	IWSV	*Unkanodes tanasijevici*
Maize stripe tenuivirus	MSpV	*Peregrinus maidis*
Rice grassy stunt tenuivirus	RGSV	*Nilaparvata lugens*
Rice hoja blanca tenuivirus	RHBV	*Tagosodes orizicolus*
Urochloa hoja blanca tenuivirus	UHBV	*Caenodelphax teapae*
Rice stripe tenuivirus	RSV	*Laodelphax striatellus*

Modified from Falk, B. W. Tsai, J. H. 1998. Biology and molecular biology of viruses in the genus Tenuivirus. Annu. Rev. Phytopathol., 36, 139–163.

Tenuiviruses are horizontally transmitted by delphacid planthoppers in a persistent propagative manner, and vertically transmitted by the viruliferous females to their offspring (Falk and Tsai, 1998; Heydarnejad et al., 2007; Huo et al., 2014). The infection route of rice grassy stunt virus in the brown planthopper vector, *Nilaparvata lugens*, as well as rice stripe virus in the small brown planthopper, *Laodelphax striatellus*, were studied using immunofluorescence microscopy. Similar to other persistently transmitted bunyaviruses, tenuiviruses first infected the midgut epithelium, spread into visceral muscle tissues, disseminated into the hemolymph or other organs, and finally moved into salivary glands (Wu et al., 2014; Zheng et al., 2014).

Further study of the molecular transmission mechanism by insect vectors has focused on the type species, rice stripe virus and its vector, *L. striatellus*. However, unlike tospoviruses or other bunyaviruses, purified preparations of tenuiviruses are composed of RNPs that do not have an envelope, and the glycoproteins are not present in the purified extracts (Falk and Tsai, 1998; Ramirez and Haenni, 1994). Therefore, the nucleocapsid protein (pc3) was hypothesized to be the best candidate as the viral determinant for transmission (Liu et al., 2015). Five vector proteins (atlasin, novel cuticle protein (CPR1), jagunal, nascent polypeptide-associated complex (NAC) domain protein, and vitellogenin) interacted with pc3, and a novel cuticular protein (CPR1) colocalized with pc3 in the insect hemolymph and stabilized the virus (Liu et al., 2015). In addition, pc3 was also identified to interact with the 60S ribosomal protein L18 as well as G protein pathway suppressor 2 to promote viral replication and translation in *L. striatellus* (Li et al., 2018; Wang et al., 2017). The interaction between pc3 and the non-structural movement protein NS4 played a critical role in viral RNP accumulation in *L. striatellus* (Wu et al., 2014). Interestingly, the interaction between pc3 and hemocyte-produced vitellogenin of *L. striatellus* plays a critical role in RSV transovarial transmission, specifically in entry and spread in the insect ovariole. RSV RNPs (comprised mainly of pc3) bind hemocyte-produced vitellogenin and are then able to enter the insect eggs using this protein-protein interaction. RNP-vitellogenin binding enables the virus to invade the germarium, the anterior side of ovariole, and then spread to oocytes through nutritive cords, which is the same route as vitellogenin transportation (He et al., 2019; Huo et al., 2014, 2018, 2019). This intriguing mechanism of vertical transmission enables these viruses to overwinter in eggs and to persist in the field in the absence of a susceptible plant host.

Although the two tenuivirus-encoded GPs do not appear to be part of the viral particle, they do behave similarly to typical bunyavirus GPs in cellular localization experiments and insect binding experiments. In addition to pc3, the viral non-structural glycoprotein, NSvc2-N (analogous to Gn) was found localized to the Golgi apparatus in *Nicotiana benthamiana* cells, where the assembly and maturation of enveloped particles occurs for enveloped bunyaviruses. Additionally, NSvc2-N could redirect NSvc2-C (analogous to Gc) from the endoplasmic reticulum (ER) to the Golgi bodies, hence the NSvc2-N could also play a key role in insect vector transmission (Yao et al., 2014). Purified rice stripe RNPs could not enter *L. striatellus* gut cells without the addition of the glycoproteins, thus suggesting that NSvc2-N and NSvc2-C are required for rice stripe virus entry into vector midgut cells (Lu et al., 2019). A recent study (Lu et al., 2019) elucidated the helper function of the GPs in the rice stripe virus midgut entry mechanism. They found that both NSvc2-N and NSvc2-C could interact with RNPs. The NSvc2N-NSvc2C-rice stripe virion complex was recognized by *L. striatellus* midgut microvilli via the direct binding of NSvc2-N to the insect midgut. Then the cell membrane formed vesicles to compartmentalize the viral complex through endocytosis. These vesicles further developed into early and late endosomes, followed by the membrane fusion triggered by NSvc2-C under acidic conditions. After membrane fusion, RNPs are released from endosomes to the cytosol for replication and spread among midgut epithelial cells. The identification of the receptor(s) localized on the surface of midgut cells that recognize NSvc2-N is the next mystery to be solved. Despite the lack of a virion membrane, the tenuivirus entry pathway into animal cells is very similar to other bunyaviruses.

3.3 Emaraviruses

The genus *Emaravirus* includes nine species (Table 3), and *European mountain ash ringspot-associated emaravirus* (EMARaV) is the type member (Abudurexiti et al., 2019). Emaraviruses have multipartite single-stranded genomes consisting of 4–8 negative-sense RNAs (Elbeaino et al., 2009a,b, 2015; Di Bello et al., 2015, 2016; Ishikawa et al., 2012; Jensen et al., 1996; Jones et al., 2004; Kumar et al., 2003, 2017; Laney et al., 2011; Lebas et al., 2005; McGavin et al., 2012; Mielke-Ehret and Muhlbach, 2012; Mielke and Muehlbach, 2007; Mielke et al., 2008; Tatineni et al., 2014). RNAs 1 to 4 encode the viral RdRp, glycoprotein precursor, nucleocapsid proteins and movement protein respectively (Mielke-Ehret and Muhlbach, 2012). With the development of deep sequencing techniques, more genomic RNAs were discovered for viruses in this group (Elbeaino et al., 2014, 2015;

Table 3 Transmission features of nine viral species belonging to the *Emaravirus* genus.

Emaravirus species	Acronym	Arthropod vector	Mechanical inoculation	Grafting	References
Actinidia chlorotic ringspot-associated emaravirus	AcCRaV	Not confirmed	Yes	Not confirmed	Zheng et al. (2017)
European mountain ash ringspot-associated emaravirus	EMARaV	*Phytoptus pyri* (putative)	Not confirmed	Yes	Grimová et al. (2015) and Mielke-Ehret et al. (2010)
Fig mosaic emaravirus	FMV	*Aceria ficus*	Not confirmed	Yes	Caglayan et al. (2012) and Flock and Wallace (1955)
High Plains wheat mosaic emaravirus	HPWMoV	*Aceria tosichella*	Yes (vascular puncture inoculation)	Not confirmed	Louie et al. (2006), Seifers et al. (1997), and Wosula et al. (2016)
Pigeonpea sterility mosaic emaravirus 1	PPSMV 1	*Aceria cajani*	Yes	Yes	Kulkarni et al. (2002) and Kumar et al. (2002, 2003)
Pigeonpea sterility mosaic emaravirus 2	PPSMV 2	*Aceria cajani*	Not confirmed	Not confirmed	Elbeaino et al. (2015)
Raspberry leaf blotch emaravirus	RLBV	*Phyllocoptes gracilis*	Yes	Not confirmed	McGavin et al. (2012)
Redbud yellow ringspot-associated emaravirus	RYRSaV	Not confirmed	Not confirmed	Yes	Di Bello et al. (2016)
Rose rosette emaravirus	RRV	*Phyllocoptes fructiphilus*	Not confirmed	Yes	Allington et al. (1968) and Epstein and Hill (1995)

Modified from Mielke-Ehret, N. Muhlbach, H. P. 2012. Emaravirus: a novel genus of multipartite, negative strand RNA plant viruses. Viruses, 4, 1515–1536; Patil, B.L., Kumar, P.L., 2015. Pigeonpea sterility mosaic virus: a legume-infecting Emaravirus from South Asia. Mol. Plant Pathol., 16, 775–786.

Tatineni et al., 2014; Zheng et al., 2017), however, although phylogenetic analysis clustered RdRp, glycoprotein precursor and nucleocapsid proteins of different emaraviruses within the same clade, there was no homologous protein sequences available for other newly discovered ORFs, therefore, the functions of those proteins are still unknown (Elbeaino et al., 2009b; Di Bello et al., 2016; Hassan et al., 2017; Ishikawa et al., 2012; Kumar et al., 2017; Lu et al., 2015; McGavin et al., 2012; Tatineni et al., 2014; Zheng et al., 2017).

The double-membrane-bound bodies (DMBs) that were identified from several emaravirus-infected plant tissues associated with typical disease symptoms were considered to be virions (Ahn et al., 1998; Kumar et al., 2003, 2002; Martelli et al., 1993; Silvestro and Chapman, 2004; Skare et al., 2006; Zheng et al., 2017). These DMBs are 80–200 nm in diameter and localized at the ER and Golgi (Ahn et al., 1998; Kumar et al., 2002; Silvestro and Chapman, 2004; Zheng et al., 2017), indicating the viral membranes could be derived from these intracellular host membranes (Mielke-Ehret and Muhlbach, 2012; Zheng et al., 2017), which is similar to tospovirus virion morphogenesis (Whitfield et al., 2005).

Emaraviruses are mainly transmitted by eriophyid mites, although several emaravirues can also be transmitted by grafting (European mountain ash ringspot-associated virus, fig mosaic virus, pigeonpea sterility mosaic virus 1, rose rosette virus and redbud yellow ringspot-associated virus) and mechanical inoculation (actinidia chlorotic ringspot-associated virus, High Plains wheat mosaic virus, pigeonpea sterility mosaic virus 1 and raspberry leaf blotch virus). The transmission mechanisms of emaraviruses are not well studied (Mielke-Ehret and Muhlbach, 2012; Mielke-Ehret et al., 2010). Table 3 shows the arthropod vector and transmission features for each viral species (Allington et al., 1968; Caglayan et al., 2012; Di Bello et al., 2016; Elbeaino et al., 2015; Epstein and Hill, 1995; Flock and Wallace, 1955; Grimová et al., 2015; Kulkarni et al., 2002; Kumar et al., 2002, 2003; Louie et al., 2006; McGavin et al., 2012; Mielke-Ehret et al., 2010; Seifers et al., 1997; Wosula et al., 2016; Zheng et al., 2017). Although EMARaV is the type member of this genus, the arthropod vector of this virus is suspected to be *Phytoptus pyri* but is still not definitively confirmed (Mielke-Ehret et al., 2010). This huge knowledge gap could be due to the difficulty of establishment of mite colonies (Skare et al., 2003), which is essential for the study of vector-virus interactions. The eriophyid mites are extremely small (~0.2 mm in length) and not visible with the naked eye making any experiments with the vector challenging. In addition, unlike persistent propagative transmitted tenuiviruses and tospoviruses, emaraviruses, (e.g., *Aceria ajani* transmitted PPSMV 1) are hypothesized to be transmitted in a semi-persistent manner (Kulkarni et al., 2002; Kumar et al., 2003). With the short stylet and host specificity of eriophyid mites, as well as the short acquisition and retention period of the viruses, studying the transmission mechanisms of emaraviruses by arthropod vectors has many challenges. The transmission of other plant viruses by eriophyid mites defies convention and further developments in mite-virus interactions are needed to fully characterize the biology of these emerging mite-borne viruses (Tatineni and Hein, 2018).

4. Concluding remarks

Plant viruses have diverse transmission strategies for successful plant-to-plant spread, and transmission biology has been a sustained area of study for both applied and basic vector biologists. New and expanding areas of research will fuel the next phase of vector biology advances and enable the development of new antiviral strategies using genome editing tools for controlling vector populations, modifying vector competence, and creating resistant plant hosts. Creative strategies for disrupting the transmission cycle in field settings will be derived from new knowledge of the chemical ecology of vector-virus-host interactions and. The study of molecular interactions can lead to new chemically- and biologically-based control strategies with reduced off-target effects.

The bunyavirus research discussed in this review identifies these viruses as emerging threats and suggests that there are many viruses yet to be described. The discovery of large numbers of arthropod-infecting viruses may increase our understanding of relationships between bunyaviruses and possibly inform predictions of virus emergence. Greater sampling of virus diversity in herbivorous arthropods could provide additional branches in the bunyavirus phylogenetic tree (Fig. 2). The currently described plant bunyaviruses provide a spectrum of possible interactions with vectors and plant hosts. For example, the orthotospoviruses have an envelope and associated GPs, Gn and Gc, that likely mediate entry into the vector, tenuiviruses form non-enveloped particles with GP-associates that are required for transmission, and coguviruses are non-enveloped and do not encode GPs. The enveloped emaraviruses are mite-transmitted and the mechanism of transmission for viruses in this genus is still unresolved. In-depth analysis of the vector phase of the bunyavirus infection cycle is urgently needed for animal- and plant-infecting viruses. To date, no receptor has been identified for an arthropod vector of bunyaviruses despite the availability of cell cultures and infectious clones for some animal-infecting bunyaviruses. It may be that uncovering these tissue-specific interactions will require in vivo experimentation. Despite the recent flood of structural information on human infecting bunyaviruses, the assembly and entry of plant-infecting bunyaviruses is still poorly characterized from a structural perspective. Based on the virus-host and structural analyses performed to date, it is likely that we will discover a myriad of mechanisms of virus entry and transmission for this diverse virus order.

Acknowledgments

AEW was supported by the following grants: USDA-NIFA 2016-67013-27492 and USDA-FNRI 208 6034-22000-039-06S. DR was supported by the following grants USDA-NIFA 2016-67013-27492 and 2018-67013-28495. AEW, DR, and DAR were supported by North Carolina State University startup funds as part of the Global Food Security and Emerging Plant Disease Cluster that is sponsored by the Chancellor's Faculty Excellence Program.

References

Abudurexiti, A., Adkins, S., Alioto, D., Alkhovsky, S.V., Avsic-Zupanc, T., Ballinger, M.J., Bente, D.A., Beer, M., Bergeron, E., Blair, C.D., Briese, T., Buchmeier, M.J., Burt, F.J., Calisher, C.H., Chang, C., Charrel, R.N., Choi, I.R., Clegg, J.C.S., De La Torre, J.C., De Lamballerie, X., Deng, F., Di Serio, F., Digiaro, M., Drebot, M.A., Duan, X., Ebihara, H., Elbeaino, T., Ergunay, K., Fulhorst, C.F., Garrison, A.R., Gao, G.F., Gonzalez, J.J., Groschup, M.H., Gunther, S., Haenni, A.L., Hall, R.A., Hepojoki, J., Hewson, R., Hu, Z., Hughes, H.R., Jonson, M.G., Junglen, S., Klempa, B., Klingstrom, J., Kou, C., Laenen, L., Lambert, A.J., Langevin, S.A., Liu, D., Lukashevich, I.S., Luo, T., Lu, C., Maes, P., De Souza, W.M., Marklewitz, M., Martelli, G.P., Matsuno, K., Mielke-Ehret, N., Minutolo, M., Mirazimi, A., Moming, A., Muhlbach, H.P., Naidu, R., Navarro, B., Nunes, M.R.T., Palacios, G., Papa, A., Pauvolid-Correa, A., Paweska, J.T., Qiao, J., Radoshitzky, S.R., Resende, R.O., Romanowski, V., Sall, A.A., Salvato, M.S., Sasaya, T., Shen, S., Shi, X., Shirako, Y., Simmonds, P., Sironi, M., Song, J.W., Spengler, J.R., Stenglein, M.D., Su, Z., Sun, S., Tang, S., Turina, M., Wang, B., Wang, C., Wang, H., Wang, J., Wei, T., Whitfield, A.E., Zerbini, F.M., Zhang, J., Zhang, L., Zhang, Y., Zhang, Y.Z., Zhang, Y., et al., 2019. Taxonomy of the order *Bunyavirales*: update 2019. Arch. Virol. 164, 1949–1965.

Ahn, K.K., Kim, K.S., Gergerich, R.C., Jensen, S.G., 1998. High plains disease of corn and wheat: ultrastructural and serological aspects. J. Submicrosc. Cytol. Pathol. 30, 563–571.

Albornoz, A., Hoffmann, A.B., Lozach, P.Y., Tischler, N.D., 2016. Early bunyavirus-host cell interactions. Viruses 8, 143.

Allington, W.B., Staples, R., Viehmeyer, G., 1968. Transmission of rose rosette virus by the eriophyid mite *Phyllocoptes fructiphilus*. J. Econ. Entomol. 61, 1137–1140.

Ammar, E.D., Tsai, C.W., Whitfield, A.E., Redinbaugh, M.G., Hogenhout, S.A., 2009. Cellular and molecular aspects of rhabdovirus interactions with insect and plant hosts. Annu. Rev. Entomol. 54, 447–468.

Badillo-Vargas, I.E., Rotenberg, D., Schneweis, D.J., Hiromasa, Y., Tomich, J.M., Whitfield, A.E., 2012. Proteomic analysis of *Frankliniella occidentalis* and differentially expressed proteins in response to tomato spotted wilt virus infection. J. Virol. 86, 8793–8809.

Badillo-Vargas, I.E., Chen, Y., Martin, K.M., Rotenberg, D., Whitfield, A.E., 2018. Discovery of novel thrips vector proteins that bind to the plant bunyavirus, tomato spotted wilt virus. bioRxiv, p. 416560.

Bandla, M.D., Campbell, L.R., Ullman, D.E., Sherwood, J.L., 1998. Interaction of tomato spotted wilt tospovirus (TSWV) glycoproteins with a thrips midgut protein, a potential cellular receptor for TSWV. Phytopathology 88, 98–104.

Battisti, A.J., Chu, Y.K., Chipman, P.R., Kaufmann, B., Jonsson, C.B., Rossmann, M.G., 2011. Structural studies of Hantaan virus. J. Virol. 85, 835–841.

Bignon, E.A., Albornoz, A., Guardado-Calvo, P., Rey, F.A., Tischler, N.D., 2019. Molecular organization and dynamics of the fusion protein Gc at the hantavirus surface. elife 8 e46028.

Blanc, S., Drucker, M., Uzest, M., 2014. Localizing viruses in their insect vectors. Annu. Rev. Phytopathol. 52, 403–425.

Bowden, T.A., Bitto, D., Mclees, A., Yeromonahos, C., Elliott, R.M., Huiskonen, J.T., 2013. Orthobunyavirus ultrastructure and the curious tripodal glycoprotein spike. PLoS Pathog. 9 e1003374.

Caglayan, K., Elci, E., Ulubas Serce, C., Kaya, K., Gazel, M., Medina, V., 2012. Detection of fig mosaic virus in viruliferous eriophyid mite *Aceria ficus*. J. Plant Pathol. 94, 629–634.

Capella-Gutierrez, S., Silla-Martinez, J.M., Gabaldon, T., 2009. trimAl: a tool for automated alignment trimming in large-scale phylogenetic analyses. Bioinformatics 25, 1972–1973.

Carr, J.P., Donnelly, R., Tungadi, T., Murphy, A.M., Jiang, S., Bravo-Cazar, A., Yoon, J.Y., Cunniffe, N.J., Glover, B.J., Gilligan, C.A., 2018. Viral manipulation of plant stress responses and host interactions with insects. Adv. Virus Res. 102, 177–197.

Chen, A.Y., Walker, G.P., Carter, D., Ng, J.C., 2011. A virus capsid component mediates virion retention and transmission by its insect vector. Proc. Natl. Acad. Sci. U. S. A. 108, 16777–16782.

Culver, J.N., 2002. Tobacco mosaic virus assembly and disassembly: determinants in pathogenicity and resistance. Annu. Rev. Phytopathol. 40, 287–308.

Czosnek, H., Hariton-Shalev, A., Sobol, I., Gorovits, R., Ghanim, M., 2017. The incredible journey of begomoviruses in their whitefly vector. Viruses 9, 273.

De Boer, S.M., Kortekaas, J., De Haan, C.A., Rottier, P.J., Moormann, R.J., Bosch, B.J., 2012. Heparan sulfate facilitates Rift Valley fever virus entry into the cell. J. Virol. 86, 13767–13771.

Dessau, M., Modis, Y., 2013. Crystal structure of glycoprotein C from Rift Valley fever virus. Proc. Natl. Acad. Sci. U. S. A. 110, 1696–1701.

Di Bello, P.L., Ho, T., Tzanetakis, I.E., 2015. The evolution of emaraviruses is becoming more complex: seven segments identified in the causal agent of rose rosette disease. Virus Res. 210, 241–244.

Di Bello, P.L., Laney, A.G., Druciarek, T., Ho, T., Gergerich, R.C., Keller, K.E., Martin, R.R., Tzanetakis, I.E., 2016. A novel emaravirus is associated with redbud yellow ringspot disease. Virus Res. 222, 41–47.

Dombrovsky, A., Gollop, N., Chen, S., Chejanovsky, N., Raccah, B., 2007. In vitro association between the helper component–proteinase of zucchini yellow mosaic virus and cuticle proteins of *Myzus persicae*. J. Gen. Virol. 88, 1602–1610.

Edgar, R.C., 2004. MUSCLE: multiple sequence alignment with high accuracy and high throughput. Nucleic Acids Res. 32, 1792–1797.
Elbeaino, T., Digiaro, M., Alabdullah, A., De Stradis, A., Minafra, A., Mielke, N., Castellano, M.A., Martelli, G.P., 2009a. A multipartite single-stranded negative-sense RNA virus is the putative agent of fig mosaic disease. J. Gen. Virol. 90, 1281–1288.
Elbeaino, T., Digiaro, M., Martelli, G.P., 2009b. Complete nucleotide sequence of four RNA segments of fig mosaic virus. Arch. Virol. 154, 1719–1727.
Elbeaino, T., Digiaro, M., Uppala, M., Sudini, H., 2014. Deep sequencing of pigeonpea sterility mosaic virus discloses five RNA segments related to emaraviruses. Virus Res. 188, 27–31.
Elbeaino, T., Digiaro, M., Uppala, M., Sudini, H., 2015. Deep sequencing of dsRNAs recovered from mosaic-diseased pigeonpea reveals the presence of a novel emaravirus: pigeonpea sterility mosaic virus 2. Arch. Virol. 160, 2019–2029.
Epstein, A.H., Hill, J.H., 1995. The biology of rose rosette disease: a mite-associated disease of uncertain aetiology. J. Phytopathol. 143, 353–360.
Falk, B.W., Tsai, J.H., 1998. Biology and molecular biology of viruses in the genus *Tenuivirus*. Annu. Rev. Phytopathol. 36, 139–163.
Fereres, A., 2016. Aphid behavior and the transmission of noncirculative viruses. In: Brown, J.K. (Ed.), Vector-Mediated Transmission of Plant Pathogens. The American Phytopathological Society, pp. 31–45.
Flock, R.A., Wallace, J.M., 1955. Transmission of fig mosaic by the eriophyid mite *Aceria ficus*. Phytopathology 45, 52–54.
Freiberg, A.N., Sherman, M.B., Morais, M.C., Holbrook, M.R., Watowich, S.J., 2008. Three-dimensional organization of Rift Valley fever virus revealed by cryoelectron tomography. J. Virol. 82, 10341–10348.
Gavrilovskaya, I.N., Shepley, M., Shaw, R., Ginsberg, M.H., Mackow, E.R., 1998. β_3 integrins mediate the cellular entry of hantaviruses that cause respiratory failure. Proc. Natl. Acad. Sci. U. S. A. 95, 7074.
Gavrilovskaya, I.N., Brown, E.J., Ginsberg, M.H., Mackow, E.R., 1999. Cellular entry of hantaviruses which cause hemorrhagic fever with renal syndrome is mediated by β_3 integrins. J. Virol. 73, 3951.
Gildow, F.E., 1992. Evidence for receptor-mediated endocytosis regulating luteovirus acquisition by aphids. Phytopathology 83, 270–277.
Goncalves, A.R., Moraz, M.L., Pasquato, A., Helenius, A., Lozach, P.Y., Kunz, S., 2013. Role of DC-SIGN in Lassa virus entry into human dendritic cells. J. Virol. 87, 11504.
Granados, R.R., Ward, L.S., Maramorosch, K., 1968. Insect viremia caused by a plant-pathogenic virus: electron microscopy of vector hemocytes. Virology 34, 790–796.
Gray, S., Gildow, F.E., 2003. Luteovirus-aphid interactions. Annu. Rev. Phytopathol. 41, 539–566.
Grimová, L., Marek, M., Konrady, M., Ryšánek, P., 2015. Newly identified host range of European mountain ash ringspot-associated virus (EMARaV) and its distribution in the Czech Republic. For. Pathol. 45, 177–189.
Guardado-Calvo, P., Rey, F.A., 2017. The envelope proteins of the *Bunyavirales*. Adv. Virus Res. 98, 83–118.
Guardado-Calvo, P., Bignon, E.A., Stettner, E., Jeffers, S.A., Perez-Vargas, J., Pehau-Arnaudet, G., Tortorici, M.A., Jestin, J.L., England, P., Tischler, N.D., Rey, F.A., 2016. Mechanistic insight into bunyavirus-induced membrane fusion from structure-function analyses of the hantavirus envelope glycoprotein Gc. PLoS Pathog. 12, e1005813.
Halldorsson, S., Li, S., Li, M., Harlos, K., Bowden, T.A., Huiskonen, J.T., 2018. Shielding and activation of a viral membrane fusion protein. Nat. Commun. 9, 349.

Hassan, M., Di Bello, P.L., Keller, K.E., Martin, R.R., Sabanadzovic, S., Tzanetakis, I.E., 2017. A new, widespread emaravirus discovered in blackberry. Virus Res. 235, 1–5.

Hassani-Mehraban, A., Dullemans, A.M., Verhoeven, J.T.J., Roenhorst, J.W., Peters, D., Van Der Vlugt, R.A.A., Kormelink, R., 2019. Alstroemeria yellow spot virus (AYSV): a new orthotospovirus species within a growing Eurasian clade. Arch. Virol. 164, 117–126.

He, K., Lin, K., Ding, S., Wang, G., Li, F., 2019. The vitellogenin receptor has an essential role in vertical transmission of rice stripe virus during oogenesis in the small brown plant hopper. Pest Manag. Sci. 75, 1370–1382.

Heinlein, M., 2015. Plant virus replication and movement. Virology 479–480, 657–671.

Hellert, J., Aebischer, A., Wernike, K., Haouz, A., Brocchi, E., Reiche, S., Guardado-Calvo, P., Beer, M., Rey, F.A., 2019. Orthobunyavirus spike architecture and recognition by neutralizing antibodies. Nat. Commun. 10, 879.

Heydarnejad, J., Izadpanah, K., Hunter, F.R., Gooding, M.J., 2007. Transmission properties of Iranian wheat stripe virus. Australas. Plant Pathol. 36, 354–357.

Hofmann, H., Li, X., Zhang, X., Liu, W., Kuhl, A., Kaup, F., Soldan, S.S., Gonzalez-Scarano, F., Weber, F., He, Y., Pohlmann, S., 2013. Severe fever with thrombocytopenia virus glycoproteins are targeted by neutralizing antibodies and can use DC-SIGN as a receptor for pH-dependent entry into human and animal cell lines. J. Virol. 87, 4384–4394.

Hogenhout, S.A., Ammar El, D., Whitfield, A.E., Redinbaugh, M.G., 2008. Insect vector interactions with persistently transmitted viruses. Annu. Rev. Phytopathol. 46, 327–359.

Huiskonen, J.T., Overby, A.K., Weber, F., Grunewald, K., 2009. Electron cryo-microscopy and single-particle averaging of Rift Valley fever virus: evidence for G_N-G_C glycoprotein heterodimers. J. Virol. 83, 3762–3769.

Huiskonen, J.T., Hepojoki, J., Laurinmaki, P., Vaheri, A., Lankinen, H., Butcher, S.J., Grunewald, K., 2010. Electron cryotomography of Tula Hantavirus suggests a unique assembly paradigm for enveloped viruses. J. Virol. 84, 4889–4897.

Huo, Y., Liu, W., Zhang, F., Chen, X., Li, L., Liu, Q., Zhou, Y., Wei, T., Fang, R., Wang, X., 2014. Transovarial transmission of a plant virus is mediated by vitellogenin of its insect vector. PLoS Pathog. 10, e1003949.

Huo, Y., Yu, Y., Chen, L., Li, Q., Zhang, M., Song, Z., Chen, X., Fang, R., Zhang, L., 2018. Insect tissue-specific vitellogenin facilitates transmission of plant virus. PLoS Pathog. 14 e1006909.

Huo, Y., Yu, Y., Liu, Q., Liu, D., Zhang, M., Liang, J., Chen, X., Zhang, L., Fang, R., 2019. Rice stripe virus hitchhikes the vector insect vitellogenin ligand-receptor pathway for ovary entry. Philos. Trans. R. Soc. Lond. B Biol. Sci. 374, 20180312.

Ishikawa, K., Maejima, K., Komatsu, K., Kitazawa, Y., Hashimoto, M., Takata, D., Yamaji, Y., Namba, S., 2012. Identification and characterization of two novel genomic RNA segments of fig mosaic virus, RNA5 and RNA6. J. Gen. Virol. 93, 1612–1619.

Jensen, S.G., Lane, L.C., Seifers, D.L., 1996. A new disease of maize and wheat in the high plains. Plant Dis. 80, 1387–1390.

Jia, D., Mao, Q., Chen, H., Wang, A., Liu, Y., Wang, H., Xie, L., Wei, T., 2014. Virus-induced tubule: a vehicle for rapid spread of virions through basal lamina from midgut epithelium in the insect vector. J. Virol. 88, 10488–10500.

Jia, D., Mao, Q., Chen, Y., Liu, Y., Chen, Q., Wu, W., Zhang, X., Chen, H., Li, Y., Wei, T., 2017. Insect symbiotic bacteria harbour viral pathogens for transovarial transmission. Nat. Microbiol. 2, 17025.

Jones, A.T., Kumar, P.L., Saxena, K.B., Kulkarni, N.K., Muniyappa, V., Waliyar, F., 2004. Sterility mosaic disease-the "green plague" of pigeonpea: advances in understanding the etiology, transmission and control of a major virus disease. Plant Dis. 88, 436–445.

Junglen, S., 2016. Evolutionary origin of pathogenic arthropod-borne viruses-a case study in the family *Bunyaviridae*. Curr. Opin. Insect Sci. 16, 81–86.
Kanakala, S., Ghanim, M., 2016. Implication of the whitefly *Bemisia tabaci* cyclophilin B protein in the transmission of tomato yellow leaf curl virus. Front. Plant Sci. 7, 1702.
Kikkert, M., Meurs, C., Van De Wetering, F., Dorfmuller, S., Peters, D., Kormelink, R., Goldbach, R., 1998. Binding of tomato spotted wilt virus to a 94-kDa thrips protein. Phytop athology 88, 63–69.
Kritzman, A., Gera, A., Raccah, B., Van Lent, J.W., Peters, D., 2002. The route of tomato spotted wilt virus inside the thrips body in relation to transmission efficiency. Arch. Virol. 147, 2143–2156.
Kulkarni, N.K., Kumar, P.L., Muniyappa, V., Jones, A.T., Reddy, D.V.R., 2002. Transmission of pigeon pea sterility mosaic virus by the eriophyid mite, *Aceria cajani* (Acari: Arthropoda). Plant Dis. 86, 1297–1302.
Kumar, P.L., Duncan, G.H., Roberts, I.M., Teifion, J., Reddy, A., D. V. R., 2002. Cytopathology of plgeonpea sterility mosaic virus in pigeonpea and *Nicotiana benthamiana*: similarities with those of eriophyid mite-borne agents of undefined aetiology. Ann. Appl. Biol. 140, 87–96.
Kumar, P.L., Jones, A.T., Reddy, D.V., 2003. A novel mite-transmitted virus with a divided RNA genome closely associated with pigeonpea sterility mosaic disease. Phytopathology 93, 71–81.
Kumar, S., Subbarao, B.L., Hallan, V., 2017. Molecular characterization of emaraviruses associated with pigeonpea sterility mosaic disease. Sci. Rep. 7, 11831.
Laney, A.G., Keller, K.E., Martin, R.R., Tzanetakis, I.E., 2011. A discovery 70 years in the making: characterization of the rose rosette virus. J. Gen. Virol. 92, 1727–1732.
Le, S.Q., Gascuel, O., 2008. An improved general amino acid replacement matrix. Mol. Biol. Evol. 25, 1307–1320.
Lebas, B.S.M., Ochoa-Corona, F.M., Elliott, D.R., Tang, Z., Alexander, B.J.R., 2005. Development of an RT-PCR for high plains virus indexing scheme in New Zealand post-entry quarantine. Plant Dis. 89, 1103–1108.
Li, C.X., Shi, M., Tian, J.H., Lin, X.D., Kang, Y.J., Chen, L.J., Qin, X.C., Xu, J., Holmes, E.C., Zhang, Y.Z., 2015. Unprecedented genomic diversity of RNA viruses in arthropods reveals the ancestry of negative-sense RNA viruses. Elife 4, e05378.
Li, S., Li, X., Zhou, Y., 2018. Ribosomal protein L18 is an essential factor that promote rice stripe virus accumulation in small brown planthopper. Virus Res. 247, 15–20.
Liao, Z., Mao, Q., Li, J., Lu, C., Wu, W., Chen, H., Chen, Q., Jia, D., Wei, T., 2017. Virus-induced tubules: a vehicle for spread of virions into ovary oocyte cells of an insect vector. Front. Microbiol. 8, 475.
Liu, W., Gray, S., Huo, Y., Li, L., Wei, T., Wang, X., 2015. Proteomic analysis of interaction between a plant virus and its vector insect reveals new functions of hemipteran cuticular protein. Mol. Cell. Proteomics 14, 2229–2242.
Liu, W., Hajano, J.U., Wang, X., 2018. New insights on the transmission mechanism of tenuiviruses by their vector insects. Curr. Opin. Virol. 33, 13–17.
Louie, R., Seifers, D.L., Bradfute, O.E., 2006. Isolation, transmission and purification of the high plains virus. J. Virol. Methods 135, 214–222.
Lozach, P.Y., Huotari, J., Helenius, A., 2011. Late-penetrating viruses. Curr. Opin. Virol. 1, 35–43.
Lu, Y., McGavin, W., Cock, P.J., Schnettler, E., Yan, F., Chen, J., Macfarlane, S., 2015. Newly identified RNAs of raspberry leaf blotch virus encoding a related group of proteins. J. Gen. Virol. 96, 3432–3439.
Lu, G., Li, S., Zhou, C., Qian, X., Xiang, Q., Yang, T., Wu, J., Zhou, X., Zhou, Y., Ding, X.S., Tao, X., 2019. Tenuivirus utilizes its glycoprotein as a helper component to overcome insect midgut barriers for its circulative and propagative transmission. PLoS Pathog. 15, e1007655.

Mandahar, C.L., 2006. Infection by and uncoating of virus particles. In: Multiplication of RNA Plant Viruses. Springer, Dordrecht, pp. 71–85.

Margaria, P., Bosco, L., Vallino, M., Ciuffo, M., Mautino, G.C., Tavella, L., Turina, M., 2014. The NSs protein of tomato spotted wilt virus is required for persistent infection and transmission by *Frankliniella occidentalis*. J. Virol. 88, 5788–5802.

Marklewitz, M., Zirkel, F., Kurth, A., Drosten, C., Junglen, S., 2015. Evolutionary and phenotypic analysis of live virus isolates suggests arthropod origin of a pathogenic RNA virus family. Proc. Natl. Acad. Sci. U. S. A. 112, 7536–7541.

Martelli, G.P., Castellano, M.A., Lafortezza, R., 1993. An ultrastructural study of fig mosaic. Phytopathol. Mediterr. 32, 33–43.

Matthys, V.S., Gorbunova, E.E., Gavrilovskaya, I.N., Mackow, E.R., 2010. Andes virus recognition of human and Syrian hamster β_3 integrins is determined by an L33P substitution in the PSI domain. J. Virol. 84, 352–360.

Mauck, K.E., Chesnais, Q., Shapiro, L.R., 2018. Evolutionary determinants of host and vector manipulation by plant viruses. Adv. Virus Res. 101, 189–250.

McGavin, W.J., Mitchell, C., Cock, P.J., Wright, K.M., Macfarlane, S.A., 2012. Raspberry leaf blotch virus, a putative new member of the genus *Emaravirus*, encodes a novel genomic RNA. J. Gen. Virol. 93, 430–437.

Mielke, N., Muehlbach, H.P., 2007. A novel, multipartite, negative-strand RNA virus is associated with the ringspot disease of European mountain ash (*Sorbus aucuparia* L.). J. Gen. Virol. 88, 1337–1346.

Mielke, N., Weber, M., Khan, S., Muehlbach, H.P., 2008. Detection of European mountain ash ringspot-associated virus (EMARaV) in *Sorbus aucuparia* L. by a specific antiserum and reverse transcription-PCR. For. Pathol. 38, 371–380.

Mielke-Ehret, N., Muhlbach, H.P., 2012. *Emaravirus*: a novel genus of multipartite, negative strand RNA plant viruses. Viruses 4, 1515–1536.

Mielke-Ehret, N., Thoma, J., Schlatermund, N., Muhlbach, H.P., 2010. Detection of European mountain ash ringspot-associated virus-specific RNA and protein P3 in the pear leaf blister mite *Phytoptus pyri* (Eriophyidae). Arch. Virol. 155, 987–991.

Modis, Y., 2013. Class II fusion proteins. In: Pöhlmann, S., Simmons, G. (Eds.), Viral Entry into Host Cells. Advances in Experimental Medicine and Biology. Springer, New York, NY, pp. 150–166.

Montero-Astua, M., Rotenberg, D., Leach-Kieffaber, A., Schneweis, B.A., Park, S., Park, J.K., German, T.L., Whitfield, A.E., 2014. Disruption of vector transmission by a plant-expressed viral glycoprotein. Mol. Plant-Microbe Interact. 27, 296–304.

Montero-Astua, M., Stafford-Banks, C.A., Badillo-Vargas, I.E., Rotenberg, D., Ullman, D.E., Whitfield, A.E., 2016a. Tospovirus–thrips biology in vector-mediated transmission of plant pathogens. In: Brown, J.K. (Ed.), Vector-Mediated Transmission of Plant Pathogens. The American Phytopathological society, pp. 289–308.

Montero-Astua, M., Ullman, D.E., Whitfield, A.E., 2016b. Salivary gland morphology, tissue tropism and the progression of tospovirus infection in *Frankliniella occidentalis*. Virology 493, 39–51.

Moritz, G., Kumm, S., Mound, L., 2004. Tospovirus transmission depends on thrips ontogeny. Virus Res. 100, 143–149.

Mou, D.L., Wang, Y.P., Huang, C.X., Li, G.Y., Pan, L., Yang, W.S., Bai, X.F., 2006. Cellular entry of Hantaan virus A9 strain: specific interactions with β_3 integrins and a novel 70kDa protein. Biochem. Biophys. Res. Commun. 339, 611–617.

Navarro, B., Minutolo, M., De Stradis, A., Palmisano, F., Alioto, D., Di Serio, F., 2018a. The first phlebo-like virus infecting plants: a case study on the adaptation of negative-stranded RNA viruses to new hosts. Mol. Plant Pathol. 19, 1075–1089.

Navarro, B., Zicca, S., Minutolo, M., Saponari, M., Alioto, D., Di Serio, F., 2018b. A negative-stranded RNA virus infecting citrus trees: the second member of a new genus within the order *Bunyavirales*. Front. Microbiol. 9, 2340.

Ng, J.C., Falk, B.W., 2006. Virus-vector interactions mediating nonpersistent and semipersistent transmission of plant viruses. Annu. Rev. Phytopathol. 44, 183–212.
Ng, J., Walker, G.P., 2016. Whitefly feeding behavior and transmission of noncirculative plant viruses. In: Brown, J.K. (Ed.), Vector-Mediated Transmission of Plant Pathogens. The American Phytopathological society, pp. 47–58.
Obijeski, J.F., Bishop, D.H., Murphy, F.A., Palmer, E.L., 1976. Structural proteins of La Crosse virus. J. Virol. 19, 985–997.
Oliver, J.E., Whitfield, A.E., 2016. The genus *Tospovirus*: emerging bunyaviruses that threaten food security. Annu. Rev. Virol. 3, 101–124.
Overby, A.K., Pettersson, R.F., Grunewald, K., Huiskonen, J.T., 2008. Insights into bunyavirus architecture from electron cryotomography of Uukuniemi virus. Proc. Natl. Acad. Sci. U. S. A. 105, 2375–2379.
Pietrantoni, A., Fortuna, C., Remoli, M.E., Ciufolini, M.G., Superti, F., 2015. Bovine lactoferrin inhibits Toscana virus infection by binding to heparan sulphate. Viruses 7, 480–495.
Raftery, M.J., Lalwani, P., Krautkrmer, E., Peters, T., Scharffetter-Kochanek, K., Kruger, R., Hofmann, J., Seeger, K., Kruger, D.H., Schonrich, G., 2014. β_2 integrin mediates hantavirus-induced release of neutrophil extracellular traps. J. Exp. Med. 211, 1485–1497.
Ramirez, B.C., Haenni, A.L., 1994. Molecular biology of tenuiviruses, a remarkable group of plant viruses. J. Gen. Virol. 75, 467–475.
Riblett, A.M., Blomen, V.A., Jae, L.T., Altamura, L.A., Doms, R.W., Brummelkamp, T.R., Wojcechowskyj, J.A., 2016. A haploid genetic screen identifies heparan sulfate proteoglycans supporting Rift Valley fever virus infection. J. Virol. 90, 1414–1423.
Riley, D.G., Joseph, S.V., Srinivasan, R., Diffie, S., 2011. Thrips vectors of tospoviruses. J. Integr. Pest Manag. 2, 1–10.
Rotenberg, D., Whitfield, A.E., 2018. Molecular interactions between tospoviruses and thrips vectors. Curr. Opin. Virol. 33, 191–197.
Rotenberg, D., Krishna Kumar, N.K., Ullman, D.E., Montero-Astua, M., Willis, D.K., German, T.L., Whitfield, A.E., 2009. Variation in tomato spotted wilt virus titer in *Frankliniella occidentalis* and its association with frequency of transmission. Phytopathology 99, 404–410.
Rotenberg, D., Jacobson, A.L., Schneweis, D.J., Whitfield, A.E., 2015. Thrips transmission of tospoviruses. Curr. Opin. Virol. 15, 80–89.
Schneweis, D.J., Whitfield, A.E., Rotenberg, D., 2017. Thrips developmental stage-specific transcriptome response to tomato spotted wilt virus during the virus infection cycle in *Frankliniella occidentalis*, the primary vector. Virology 500, 226–237.
Seifers, D.L., Harvey, T.L., Martin, T.J., Jensen, S.G., 1997. Identification of the wheat curl mite as the vector of the high plains virus of corn and wheat. Plant Dis. 81, 1161–1166.
Sherman, M.B., Freiberg, A.N., Holbrook, M.R., Watowich, S.J., 2009. Single-particle cryo-electron microscopy of Rift Valley fever virus. Virology 387, 11–15.
Shrestha, A., Champagne, D.E., Culbreath, A.K., Rotenberg, D., Whitfield, A.E., Srinivasan, R., 2017. Transcriptome changes associated with tomato spotted wilt virus infection in various life stages of its thrips vector, *Frankliniella fusca* (Hinds). J. Gen. Virol. 98, 2156–2170.
Silvestro, S.R., Chapman, G.B., 2004. A transmission electron microscope study of "New Dawn" climber rose (*Rosa wichuraiana* x *safrano*) exhibiting rose rosette disease. Plant Cell Rep. 23, 345–351.
Skare, J.M., Wijkamp, I., Rezende, J., Michels, G., Rush, C., Scholthof, K.B., Scholthof, H.B., 2003. Colony establishment and maintenance of the eriophyid wheat curl mite *Aceria tosichella* for controlled transmission studies on a new virus-like pathogen. J. Virol. Methods 108, 133–137.

Skare, J.M., Wijkamp, I., Denham, I., Rezende, J.A., Kitajima, E.W., Park, J.W., Desvoyes, B., Rush, C.M., Michels, G., Scholthof, K.B., Scholthof, H.B., 2006. A new eriophyid mite-borne membrane-enveloped virus-like complex isolated from plants. Virology 347, 343–353.

Stafford-Banks, C.A., Rotenberg, D., Johnson, B.R., Whitfield, A.E., Ullman, D.E., 2014. Analysis of the salivary gland transcriptome of *Frankliniella occidentalis*. PLoS One 9, e94447.

Stamatakis, A., 2014. RAxML version 8: a tool for phylogenetic analysis and post-analysis of large phylogenies. Bioinformatics 30, 1312–1313.

Su, Q., Pan, H., Liu, B., Chu, D., Xie, W., Wu, Q., Wang, S., Xu, B., Zhang, Y., 2013. Insect symbiont facilitates vector acquisition, retention, and transmission of plant virus. Sci. Rep. 3, 1367.

Suda, Y., Fukushi, S., Tani, H., Murakami, S., Saijo, M., Horimoto, T., Shimojima, M., 2016. Analysis of the entry mechanism of Crimean-Congo hemorrhagic fever virus, using a vesicular stomatitis virus pseudotyping system. Arch. Virol. 161, 1447–1454.

Takeda, A., Sugiyama, K., Nagano, H., Mori, M., Kaido, M., Mise, K., Tsuda, S., Okuno, T., 2002. Identification of a novel RNA silencing suppressor, NSs protein of tomato spotted wilt virus. FEBS Lett. 532, 75–79.

Tamborindeguy, C., Bereman, M.S., Deblasio, S., Igwe, D., Smith, D.M., White, F., Maccoss, M.J., Gray, S.M., Cilia, M., 2013. Genomic and proteomic analysis of *Schizaphis graminum* reveals cyclophilin proteins are involved in the transmission of cereal yellow dwarf virus. PLoS One 8 e71620.

Tatineni, S., Hein, G.L., 2018. Genetics and mechanisms underlying transmission of Wheat streak mosaic virus by the wheat curl mite. Curr. Opin. Virol. 33, 47–54.

Tatineni, S., Mcmechan, A.J., Wosula, E.N., Wegulo, S.N., Graybosch, R.A., French, R., Hein, G.L., 2014. An eriophyid mite-transmitted plant virus contains eight genomic RNA segments with unusual heterogeneity in the nucleocapsid protein. J. Virol. 88, 11834–11845.

Tolin, S.A., Langham, M.A.C., Gergerich, R.C., 2016. Beetle transmission: a unique alliance of virus, vector, and host. In: Brown, J.K. (Ed.), Vector-Mediated Transmission of Plant Pathogens. The American Phytopathological Society, pp. 131–146.

Toriyama, S., 1986. An RNA-dependent RNA polymerase associated with the filamentous nucleoproteins of Rice stripe virus. J. Gen. Virol. 67, 1247–1255.

Uzest, M., Gargani, D., Dombrovsky, A., Cazevieille, C., Cot, D., Blanc, S., 2010. The "acrostyle": a newly described anatomical structure in aphid stylets. Arthropod Struct. Dev. 39, 221–229.

Van Den Heuvel, J.F., Verbeek, M., Van Der Wilk, F., 1994. Endosymbiotic bacteria associated with circulative transmission of potato leafroll virus by *Myzus persicae*. J. Gen. Virol. 75, 2559–2565.

Wang, W., Zhao, W., Li, J., Luo, L., Kang, L., Cui, F., 2017. The c-Jun N-terminal kinase pathway of a vector insect is activated by virus capsid protein and promotes viral replication. elife 6, e26591.

Webster, C.G., Pichon, E., Van Munster, M., Monsion, B., Deshoux, M., Gargani, D., Calevro, F., Jimenez, J., Moreno, A., Krenz, B., Thompson, J.R., Perry, K.L., Fereres, A., Blanc, S., Uzest, M., 2018. Identification of plant virus receptor candidates in the stylets of their aphid vectors. J. Virol. 92, e00432-18.

Wei, T., Li, Y., 2016. Rice reoviruses in insect vectors. Annu. Rev. Phytopathol. 54, 99–120.

Whitfield, A.E., Ullman, D.E., German, T.L., 2004. Expression and characterization of a soluble form of tomato spotted wilt virus glycoprotein G_N. J. Virol. 78, 13197–13206.

Whitfield, A.E., Ullman, D.E., German, T.L., 2005. Tospovirus-thrips interactions. Annu. Rev. Phytopathol. 43, 459–489.

Whitfield, A.E., Kumar, N.K., Rotenberg, D., Ullman, D.E., Wyman, E.A., Zietlow, C., Willis, D.K., German, T.L., 2008. A soluble form of the *tomato spotted wilt virus* (TSWV) glycoprotein G_N (G_N-S) inhibits transmission of TSWV by *Frankliniella occidentalis*. Phytopathology 98, 45–50.

Whitfield, A.E., Falk, B.W., Rotenberg, D., 2015. Insect vector-mediated transmission of plant viruses. Virology 479–480, 278–289.

Widana Gamage, S.M.K., Rotenberg, D., Schneweis, D.J., Tsai, C.W., Dietzgen, R.G., 2018. Transcriptome-wide responses of adult melon thrips (*Thrips palmi*) associated with capsicum chlorosis virus infection. PLoS One 13, e0208538.

Willensky, S., Bar-Rogovsky, H., Bignon, E.A., Tischler, N.D., Modis, Y., Dessau, M., 2016. Crystal structure of glycoprotein C from a hantavirus in the post-fusion conformation. PLoS Pathog. 12, e1005948.

Wosula, E.N., Mcmechan, A.J., Oliveira-Hofman, C., Wegulo, S.N., Hein, G.L., 2016. Differential transmission of two isolates of wheat streak mosaic virus by five wheat curl mite populations. Plant Dis. 100, 154–158.

Wu, W., Zheng, L., Chen, H., Jia, D., Li, F., Wei, T., 2014. Nonstructural protein NS4 of rice stripe virus plays a critical role in viral spread in the body of vector insects. PLoS One 9, e88636.

Wu, W., Huang, L., Mao, Q., Wei, J., Li, J., Zhao, Y., Zhang, Q., Jia, D., Wei, T., 2019. Interaction of viral pathogen with porin channels on the outer membrane of insect bacterial symbionts mediates their joint transovarial transmission. Philos. Trans. R. Soc. Lond. B Biol. Sci. 374, 20180320.

Yao, M., Liu, X., Li, S., Xu, Y., Zhou, Y., Zhou, X., Tao, X., 2014. Rice stripe tenuivirus NSvc2 glycoproteins targeted to the golgi body by the N-terminal transmembrane domain and adjacent cytosolic 24 amino acids via the COP I- and COP II-dependent secretion pathway. J. Virol. 88, 3223–3234.

Yao, J., Rotenberg, D., Whitfield, A.E., 2019. Delivery of maize mosaic virus to planthopper vectors by microinjection increases infection efficiency and facilitates functional genomics experiments in the vector. J. Virol. Methods 270, 153–162.

Zheng, L., Mao, Q., Xie, L., Wei, T., 2014. Infection route of rice grassy stunt virus, a tenuivirus, in the body of its brown planthopper vector, *Nilaparvata lugens* (Hemiptera: Delphacidae) after ingestion of virus. Virus Res. 188, 170–173.

Zheng, Y., Navarro, B., Wang, G., Wang, Y., Yang, Z., Xu, W., Zhu, C., Wang, L., Serio, F.D., Hong, N., 2017. Actinidia chlorotic ringspot-associated virus: a novel emaravirus infecting kiwifruit plants. Mol. Plant Pathol. 18, 569–581.

Further reading

Jangra, R.K., Herbert, A.S., Li, R., Jae, L.T., Kleinfelter, L.M., Slough, M.M., Barker, S.L., Guardado-Calvo, P., Román-Sosa, G., Dieterle, M.E., Kuehne, A.I., Muena, N.A., Wirchnianski, A.S., Nyakatura, E.K., Fels, J.M., Ng, M., Mittler, E., Pan, J., Bharrhan, S., Wec, A.Z., Lai, J.R., Sidhu, S.S., Tischler, N.D., Rey, F.A., Moffat, J., Brummelkamp, T.R., Wang, Z., Dye, J.M., Chandran, K., 2018. Protocadherin-1 is essential for cell entry by New World Hantaviruses. Nature 563 (7732), 559–563.

Mandahar, C.L., 2006. Infection by and uncoating of virus particles. In: Multiplication of RNA Plant Viruses. Springer, Dordrecht, pp. 71–85.

Navarro J.A., Sanchez-Navarro J.A. and Pallas V., Key checkpoints in the movement of plant viruses through the host, in press. Corrected Proof, Available online 18 July 2019 and the forthcoming chapter on Hantaan virus entry.

Whitfield, A.E., Rotenberg, D., 2015. Disruption of insect transmission of plant viruses. Curr. Opin. Insect Sci. 8, 79–87.

CHAPTER THREE

How non-enveloped viruses hijack host machineries to cause infection

Chelsey C. Spriggs[a,†], Mara C. Harwood[a,b,†], Billy Tsai[a,b,*]

[a]Department of Cell and Developmental Biology, University of Michigan Medical School, Ann Arbor, MI, United States

[b]Cellular and Molecular Biology Program, University of Michigan Medical School, Ann Arbor, MI, United States

*Corresponding author: e-mail address: btsai@umich.edu

Contents

Abstract

Viruses must navigate the complex endomembranous network of the host cell to cause infection. In the case of a non-enveloped virus that lacks a surrounding lipid bilayer, endocytic uptake from the plasma membrane is not sufficient to cause infection. Instead, the virus must travel within organelle membranes to reach a specific cellular destination that supports exposure or arrival of the virus to the cytosol. This is achieved by viral penetration across a host endomembrane, ultimately enabling entry of the virus

† Indicates co-first authors.

Advances in Virus Research, Volume 104
ISSN 0065-3527
https://doi.org/10.1016/bs.aivir.2019.05.002

into the nucleus to initiate infection. In this review, we discuss the entry mechanisms of three distinct non-enveloped DNA viruses—adenovirus (AdV), human papillomavirus (HPV), and polyomavirus (PyV)—highlighting how each exploit different intracellular transport machineries and membrane penetration apparatus associated with the endosome, Golgi, and endoplasmic reticulum (ER) membrane systems to infect a host cell. These processes not only illuminate a highly-coordinated interplay between non-enveloped viruses and their host, but may provide new strategies to combat non-enveloped virus-induced diseases.

1. Introduction

To cause infection, a virus typically binds to a receptor displayed on the surface of the host cell (Helenius, 2018; Mercer et al., 2010). This engagement triggers receptor-mediated endocytosis, initiating a complex journey for the viral particle through the highly inter-connected cellular endomembrane network (Staring et al., 2018). Transport via this network, which includes the endosome, lysosome, Golgi, and endoplasmic reticulum (ER), can lead to either non-productive or productive infection. In a non-productive infection, the virus is delivered to an organelle where it is either trapped or becomes degraded by a protease. By contrast, for productive infection to take place, the virus instead must be directed down a pathway that enables it to penetrate a membrane barrier so that it can reach the cytosol or the nucleus, which is essential for successful infection (Greber, 2016). The strategies devised by viruses as they exploit the myriad of host intracellular transport machineries and membrane penetration apparatus to promote infection, which are unique and vary widely, are often dependent on the physical properties of the viral particle.

One critical property of a virus that governs its productive entry route is the presence or absence of a lipid bilayer surrounding the viral particle. A virus encased within a lipid bilayer is called an enveloped virus, while a virus lacking this bilayer is called a non-enveloped virus. An enveloped virus, such as the human immunodeficiency virus (HIV) (Melikyan, 2014) or influenza virus (Smrt and Lorieau, 2017), penetrates at either the plasma membrane or the endosome membrane of the host. Membrane penetration is accomplished when the viral and host membrane fuse, thereby delivering the core viral particle into the cytosol to promote infection (Skehel and Wiley, 2000; Weissenhorn et al., 1999).

By contrast, as a lipid bilayer does not surround a non-enveloped virus, its membrane penetration mechanism must be fundamentally distinct from an enveloped virus. Although the precise molecular basis by which a non-enveloped virus penetrates a host membrane remains enigmatic, a more coherent understanding of this process is emerging (Kumar et al., 2018; Moyer and Nemerow, 2011; Tsai, 2007). The first step is delivery of the viral particle to the membrane penetration site. For a virus that penetrates the plasma membrane, no intracellular transport is required as virus recruitment to the host cell surface from the extracellular milieu is sufficient. However, for a virus that penetrates an intracellular membrane such as the endosome, Golgi, or ER membrane, it must co-opt intracellular transport machineries to reach these compartments. Next, upon reaching the specific membrane penetration site, cellular cues such as low pH, receptors, proteases, or chaperones, confer critical conformational changes to the virus (Yamauchi and Greber, 2016). This results in formation of a hydrophobic viral intermediate capable of penetrating the limiting membrane. Alternatively, the host-induced structural alterations may trigger the release of a viral lytic factor previously hidden in the native virion. This lytic factor in turn binds to and compromises the integrity of the host membrane, thereby allowing the remaining core viral particle to undergo membrane penetration. In some instances, these host-induced conformational changes may be initiated prior to the virus arriving to the membrane penetration site. In the third and final step, depending on the specific penetration site, cytosolic or nuclear components are recruited to extract the virus into the cytosol or nucleus to complete the penetration process.

We focus this review on clarifying the molecular entry mechanisms of three different non-enveloped DNA viruses—adenovirus (AdV), human papillomavirus (HPV), and polyomavirus (PyV). These viruses are highlighted because they illustrate how distinct intracellular transport machineries and membrane penetration apparatus linked to the endosome, Golgi, and endoplasmic reticulum (ER) membranes can be exploited to promote infection.

Specifically, post-endocytosis, AdV traffics to the endosome and relies on cues largely within this compartment to penetrate the endosome membrane in order to reach the cytosol (Fig. 1, green arrows); from the cytosol, the virus is delivered to the nucleus to cause infection. By contrast, HPV membrane penetration occurs in two steps (Fig. 1, orange arrows). Following entry to the endosome, HPV first initiates membrane penetration by partially inserting across the endosomal membrane. With a portion of

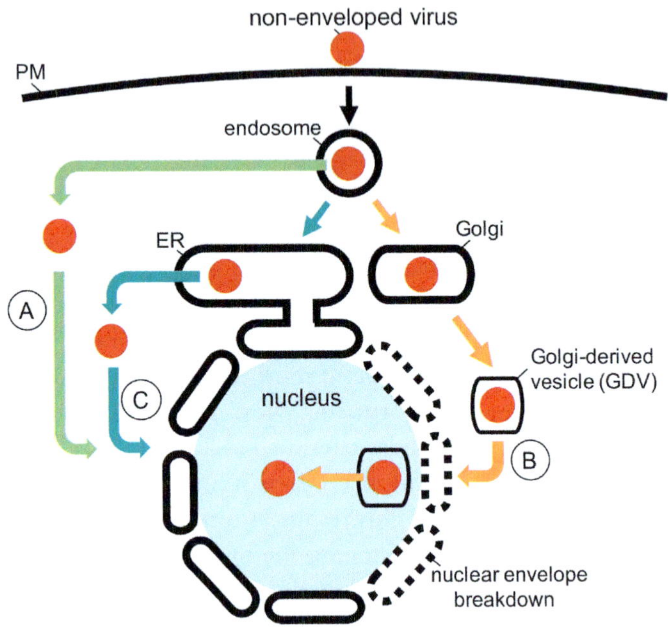

Fig. 1 Cellular entry of AdV, HPV, and PyV. (A) To infect cells, AdV traffics from the plasma membrane (PM) to the endosome where it penetrates the endosome membrane to reach the cytosol (green arrows); from the cytosol, the virus mobilizes to the nucleus to promote infection. (B) HPV infects cells by reaching the endosome where it initiates partial membrane penetration, exposing a segment of its viral protein to the cytosol. The membrane-inserted virus then sorts to the Golgi (orange arrow). Golgi membrane fragmentation during mitosis forms Golgi-derived vesicles (GDVs) containing HPV, which enter the nucleus when the nuclear envelope breaks down at the onset of mitosis. Upon complete translocation of the virus across the GDV membrane, HPV is deposited into the nucleoplasm to initiate infection. (C) After reaching the endosome from the PM, PyV is targeted to the ER from where it penetrates the ER membrane to gain entry into the cytosol (blue arrows). The virus is further transported to the nucleus to trigger infection.

the virus exposed to the cytosol, HPV is further targeted to the Golgi apparatus. During mitosis, Golgi membrane fragmentation generates Golgi-derived vesicles (GDVs) harboring HPV. These GDVs in turn gain access to the nucleus due to nuclear envelope breakdown at the onset of mitosis. Within the nucleus, HPV completes membrane penetration across the GDV membrane, fully depositing the viral particle into the nucleoplasm to enable infection. Yet a different journey is experienced by PyV. After reaching the endosome from the cell surface, PyV targets directly to the

ER, bypassing the Golgi (Fig. 1, blue arrows). From the ER lumen, PyV crosses the ER membrane to reach the cytosol where it mobilizes further to the nucleus to trigger infection. By emphasizing cell entry of AdV, HPV, and PyV, this review reveals the diversity of mechanisms used by non-enveloped viruses during infectious entry.

2. Co-opting the endosome membrane system during cell entry of AdV

Although human AdV is responsible for mild respiratory and gastrointestinal infections, it can also cause life-threatening diseases in immunocompromised individuals (Lion, 2014). Paradoxically, modified forms of AdV are actively being explored as oncolytic agents, as well as vectors in gene transfer and vaccine delivery experiments, in combating different human diseases (Pascual-Pasto et al., 2019; Sayedahmed et al., 2018). Thus, clarifying AdV cell entry not only directly impacts our understanding of AdV-induced diseases, but has potential therapeutic implications for other diseases.

2.1 Receptor-mediated entry to the endosome

To date, there are at least 60 types of AdV. Structurally, AdV measures 90 nm in diameter and harbors a 36-kilobase-pair double-stranded DNA genome. Thirteen structural proteins are encoded by its genome. When properly assembled, the viral particle displays 12 defined vertices, with each vertex composed of 2 distinct viral proteins—fiber, which binds to the host attachment receptor (van Raaij et al., 1999), and penton base, which interacts with the host internalization receptor (Zubieta et al., 2005). The primary host attachment receptor is the Coxsackie and Adenovirus Receptor (CAR) (Bergelson et al., 1997), although for some subtypes of AdV, the transmembrane proteins desmosomal desmoglein-2 (Wang et al., 2011) and CD46 of the complement system (Gaggar et al., 2003) can act in this capacity. Atomic-resolution of the protein-protein contact sites between fiber and CAR (Seiradake et al., 2006), or with CD46 (Cupelli et al., 2010), have been identified. Functionally, these attachment receptors serve to recruit the viral particles from the extracellular milieu to the plasma membrane of the host cell (Fig. 2, step 1).

Post-recruitment, AdV engages $\alpha v\beta 3$ or $\alpha v\beta 5$ integrin, which act as internalization receptors via interaction with the penton base of the virus (Wickham et al., 1993). Not surprisingly, given their roles as internalization

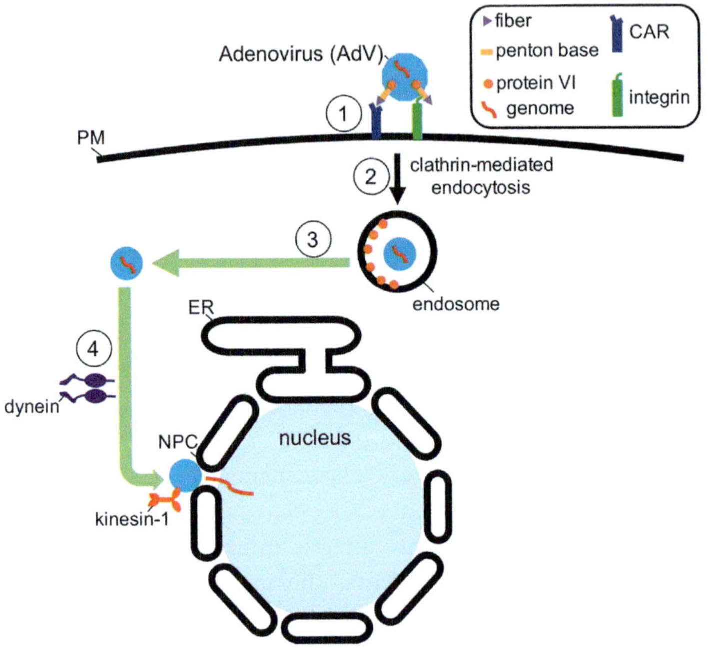

Fig. 2 Co-opting the endosome membrane system during cell entry of AdV. AdV entry is initiated when it binds to the CAR attachment receptor via the viral protein called fiber (step 1), followed by interaction with the integrin internalization receptor through the penton base viral protein. This engagement promotes clathrin-mediated endocytosis (step 2), delivering the virus to the endosome. In the endosome, protein VI is released from the virus. Protein VI then binds to and ruptures the endosome membrane, depositing the core viral particle in the cytosol (step 3). Cytosol-localized virus is targeted to the NPC by the dynein motor; here, the kinesin-1 motor pulls on the virus and disrupts the integrity of the viral structure and NPC, enabling the released viral genome to enter the nucleus.

receptors, integrins promote endocytosis of AdV. However, prior to virus internalization, coincident engagement of a single AdV particle with both CAR and an integrin receptor at the cell surface can initiate viral disassembly. This is thought to occur because integrins remain stationary within the lipid bilayer of the plasma membrane, while CAR experiences local drifting (Burckhardt et al., 2011). Hence, mechanical tension is generated and imparted on the virus when it binds simultaneously to these receptors (Burckhardt et al., 2011). This force begins the virus disassembly process and primes the virus for additional uncoating when it reaches the endosome, a step essential in virus penetration across the endosome membrane.

AdV internalization occurs through the classic receptor-mediated, clathrin-dependent pathway, delivering the virus to the endosome (Fig. 2, step 2) (Wang et al., 1998). However, there is also evidence that caveolae-mediated endocytosis (Yousuf et al., 2013) and non-selective macropinocytosis (Amstutz et al., 2008) can serve as alternative routes for AdV internalization. Cell-type differences between these experiments likely account for the varied endocytic pathways. Indeed, the ability to exploit different endocytic machineries is a common theme observed during entry of numerous non-enveloped viruses, including reovirus (Schulz et al., 2012), rotavirus (Gutierrez et al., 2010), and parvovirus (Boisvert et al., 2010).

2.2 Membrane penetration

Upon reaching the endosome, AdV is thought to gain access to the cytosol by penetrating the endosome membrane (Fig. 2, step 3). This idea is based on early negative-stain electron microscopy (EM) and fluorescence imaging studies (Greber et al., 1993; Morgan et al., 1969). In addition to these descriptive reports, functional analyses using inhibitors of endosomal acidification supported this idea (FitzGerald et al., 1983). However, one caveat is that although perturbing endosomal pH by a chemical inhibitor can disrupt appearance of AdV in the cytosol (Seth et al., 1984), this effect might in fact result from a block in the initial internalization of the virus before it reaches the endosome (Perez and Carrasco, 1994).

A more definitive understanding regarding the role of endosome membrane penetration during AdV entry was revealed when the viral lytic factor responsible for compromising the endosome membrane was identified. Specifically, in vitro biochemical and cell-based analyses demonstrated that the AdV protein VI, via its N-terminal amphipathic helix, is responsible for disruption of the endosome membrane (Maier et al., 2010; Wiethoff et al., 2005). High-resolution structural studies suggested that protein VI is normally hidden inside the virus where it acts to stabilize the peripentonal hexon and hexon interactions important for maintaining the overall virus architecture (Reddy and Nemerow, 2014). Hence, to release protein VI from AdV requires significant viral disassembly within the endosome (Nguyen et al., 2010), a process that may be triggered by the low pH of this compartment. Mechanistically, upon release from the virus, protein VI is believed to fragment the endosome membrane by binding to and inducing positive curvature to the lipid bilayer (Maier and Wiethoff, 2010), resulting in membrane rupture. In fact, AdV-induced membrane rupture can allow a 70kDa soluble marker to gain access into the cytosol (Prchla et al., 1995).

While other non-enveloped viruses also penetrate the endosome membrane, the AdV membrane penetration mechanism is fundamentally different to the one experienced by parvovirus. In this case, the virus uses the lipolytic enzyme phospholipase A_2 (located at the N-terminus of its VP1 coat protein) in order to modify the endosome membrane (Farr et al., 2005). This triggers transient changes to the physical property of the lipid bilayer without imposing permanent damage so that a subviral complex can successfully breach the membrane (Suikkanen et al., 2003b). Similarly, reovirus releases a lytic factor (called μ1 N) that interacts with the endosome membrane (Chandran et al., 2002) to induce specific size-selective pores in the membrane without inflicting significant damage. This allows the virus to penetrate this membrane and access the cytosol (Agosto et al., 2006; Ivanovic et al., 2008). Thus, although the endosome is a common site for membrane penetration of many non-enveloped viruses, the precise mechanism deployed to transport across this membrane barrier is vastly different.

2.3 Nuclear import

After endosome membrane fragmentation, the core AdV particle is deposited in the cytosol. Cytosol-localized AdV in turn engages the microtubule minus-end directed cytoplasmic dynein motor (hereafter called dynein) to transport to the nuclear pore complex (NPC) before entering the nucleus (Fig. 2, step 4) (Bremner et al., 2009). This directed movement can be countered by an unidentified motor (Suomalainen et al., 1999). Regardless, when AdV reaches the NPC, it docks to the NPC via interaction with the cytoplasmic filament Nup214 of the NPC (Trotman et al., 2001). The microtubule positive-end directed motor kinesin-1 is then recruited to the virus, pulling it away from the nucleus. This movement generates a mechanical force that simultaneously disassembles the virus to promote genome release and compromises the integrity of the NPC. These events enable the released viral genome to gain entry into the nucleus. Thus, generation of mechanical forces is important at both "ends" of productive AdV infection: prior to endocytosis and before nuclear entry. As dynein is also implicated in the delivery of cytosol-localized parvovirus (Suikkanen et al., 2003a) and likely reovirus (Mainou et al., 2013) into the nucleus post endosome escape, co-opting the cargo-transport function of dynein may be a shared strategy used by non-enveloped viruses to complete the last phase of productive infection.

3. Hijacking the dual endosome-Golgi membrane systems during HPV cell entry

Although HPV also exploits elements of the endosome during entry, a striking distinction is that this non-enveloped virus couples machineries associated with both the endosome and Golgi membrane systems to promote infection. HPV is the most common sexually transmitted infection in the United States (Satterwhite et al., 2013) and high-risk HPV types are the primary cause of cervical, anogenital, and oropharyngeal cancers (Centers for Disease Control and Prevention, 2018). Prophylactic vaccines against HPV are available; however, vaccine uptake is low (Walker et al., 2018) and the annual number of HPV-associated oropharyngeal cancers continues to rise (Van Dyne et al., 2018). There is no cure for HPV infection and despite its significant impact on human health, there is limited understanding of its cellular entry mechanism.

3.1 Cell-surface binding and entry to the endosome

To date, there are over 200 types of HPV (Muhr et al., 2018). Structurally, HPV is composed of the L1 and L2 capsid proteins. The L1 major capsid protein forms an icosahedral outer shell of 72 pentamers. Harbored within the L1 pentamers is the L2 minor coat protein. These viral proteins encapsulate an 8-kilobase-pair, double-stranded DNA genome (Buck et al., 2008). The viral genome encodes eight genes, some of which are necessary for cellular transformation (Doorbar, 2005).

Through miniscule wounds or abrasions, HPV infects the basal epithelial cells of cutaneous and mucosal tissues (Doorbar, 2005). To initiate infection, L1 directly binds to heparin sulfate proteoglycans (HSPGs) expressed at the surface of the host cell (Giroglou et al., 2001; Richards et al., 2013). Following HSPG binding, L1 is cleaved by the serine protease KLK8, and L2 interacts with the peptidylprolyl isomerase cyclophilin B (Bienkowska-Haba et al., 2009; Cerqueira et al., 2015). These steps result in exposure of L2 to the capsid surface, revealing a furin cleavage site at its N-terminus (Richards et al., 2006). While extracellular furin cleavage of L2 does not appear to be immediately essential for cell-surface events, it is required to engage factors important at later steps of viral entry (Richards et al., 2006). Ultimately, the conformational changes of the viral capsid allow for transfer to an unknown entry receptor at the plasma membrane

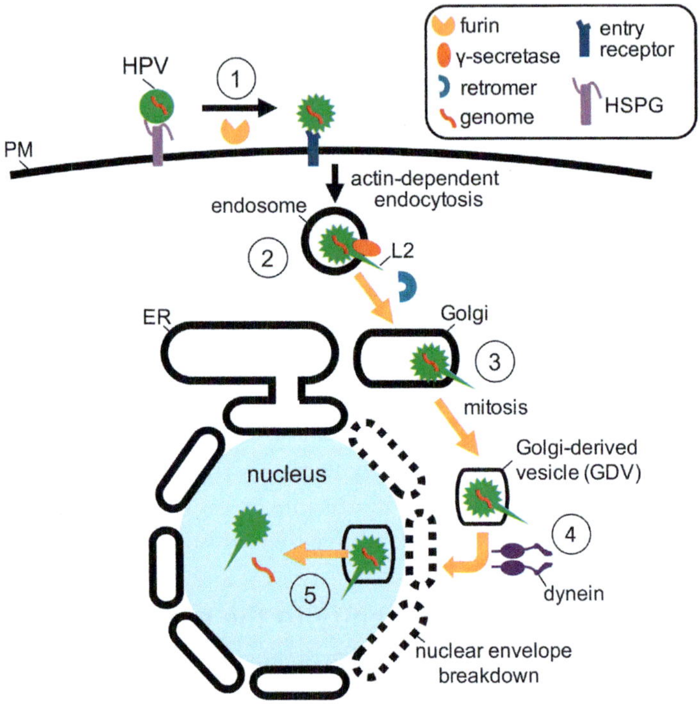

Fig. 3 Hijacking the dual endosome-Golgi membrane systems during HPV cell entry. To initiate infection, HPV attaches to HSPGs on the cell surface, resulting in conformational changes to the virus that enable furin-cleavage of L2 and subsequent transfer of the virus to an unknown entry receptor (step 1). The virus is then endocytosed by an actin-dependent mechanism to reach the endosome. As the endosome matures, the viral capsid proteins dissociate in a pH-dependent manner, and L2 engages γ-secretase. The C-terminus of L2 is then inserted across the endosomal membrane by γ-secretase, with the N-terminus remaining complexed to the viral DNA within the endosomal lumen (step 2). From here, the cytosolic portion of L2 engages retromer, and the virus is trafficked to the Golgi apparatus (step 3). Golgi fragmentation during mitosis generates HPV-harboring vesicles, which are thought to undergo dynein-mediated transport toward the nucleus (step 4). With the concomitant breakdown of the nuclear envelope during mitosis, the HPV-harboring vesicles enter the nucleus, where they remain intact until completion of mitosis and nuclear envelope reformation. Through an unknown mechanism, the viral genome is exposed to the nucleoplasm, and viral replication ensues (step 5).

(Fig. 3, step 1) (Day et al., 2008; Selinka et al., 2007). The identity of the entry receptor and mode of endocytosis remains a point of contention. Numerous putative factors are thought to support cellular uptake of HPV, including integrins α6 and β4, as well as tetraspanins CD63 and

CD151 (Aksoy et al., 2014; Evander et al., 1997; Scheffer et al., 2013; Spoden et al., 2008). Annexin A2 and epidermal growth factor receptors have also been implicated in viral entry, though their contributions are less clear (Dziduszko and Ozbun, 2013; Surviladze et al., 2012; Taylor et al., 2018).

Further complicating the process, emerging evidence suggests that endocytosis of HPV is independent of clathrin-, caveolin-, cholesterol-, or dynamin-mediated endocytic mechanisms (Schelhaas et al., 2012; Spoden et al., 2013). Rather, HPV internalization appears to be guided by an actin-dependent mechanism related to macropinocytosis. In vitro, endocytosis of HPV occurs in a highly asynchronous fashion, with virions continually internalized from 2 to 20 h after addition of virus to cell culture (Schelhaas et al., 2012). This asynchronicity may be a result of the multiple conformational changes and extracellular processing events required for HPV entry. In addition, this asynchronicity suggests the presence of a secondary entry receptor as a limiting factor of viral uptake.

3.2 Membrane penetration

Upon internalization, HPV is delivered to the endosome. As the endosome matures and acidifies, L1 and L2 begin to dissociate in a pH-dependent manner (Smith et al., 2008). Proper dissociation and early endocytic trafficking of the virus also appears to involve cyclophilin B (Bienkowska-Haba et al., 2012), tetraspanin CD63, syntenin-1, and the ESCRT protein ALIX (Grassel et al., 2016), as well as interaction with sorting nexin 17 (Bergant Marusic et al., 2012). The extent to which L1-L2 disassembles at the endosome is unclear, as emerging evidence suggests that a subset of L1 remains associated with L2 and the viral genome through later trafficking steps (DiGiuseppe et al., 2017).

Similar to AdV, initial membrane penetration of HPV to the cytosol occurs at the endosomal membrane. However, HPV differs in that while it exposes the C-terminal portion of its L2 capsid protein to the cytosol, its N-terminal portion remains complexed to the viral DNA within the endosomal lumen (Fig. 3, step 2) (DiGiuseppe et al., 2015; Kamper et al., 2006). To achieve this topology, L2 must physically insert into the endosomal membrane. Although this topology is in principle possible because L2 contains a membrane-penetrating peptide and a transmembrane-like domain, how L2 is inserted into the endosomal membrane remains largely mysterious (Bronnimann et al., 2013; Zhang et al., 2018).

A clue to this enigma was recently revealed when the transmembrane protease γ-secretase was shown to act as a chaperone, promoting insertion of L2 into the endosomal membrane in a pH-dependent manner (Inoue et al., 2018). Proper membrane insertion of L2 by γ-secretase is a critical step, as mutation of L2's transmembrane-like domain, or of γ-secretase itself, terminates productive infection (Bronnimann et al., 2013; Inoue et al., 2018). The molecular basis by which γ-secretase promotes membrane penetration of L2 remains unclear. Despite the well-established proteolytic activity of this transmembrane protease, cleavage of L2 by γ-secretase is not required to promote HPV infection—only γ-secretase-induced membrane insertion of L2 is essential (Inoue et al., 2018). Thus, HPV appears to exploit a novel chaperone function of γ-secretase.

γ-secretase-dependent insertion of L2 is a decisive step of HPV infection, as it diverts the virus away from the non-productive lysosomal pathway and instead targets HPV along the productive route (Inoue et al., 2018; Zhang et al., 2014). Specifically, with a portion of L2 now exposed to the cytosol, it engages the cytosolic retromer sorting complex (Lipovsky et al., 2013; Popa et al., 2015). Retromer normally functions to sort a cellular cargo from the endosome to the Golgi apparatus. HPV hijacks this retrograde transport function of the retromer, allowing for the endosome-localized virus to be delivered to the Golgi apparatus (Fig. 3, step 3) (Day et al., 2013; Popa et al., 2015; Zhang et al., 2014). Upon reaching the Golgi, the viral DNA remains shielded from the cytosol within the Golgi lumen, with only L2 protruding into the cytosolic space.

3.3 Nuclear import

HPV remains in the Golgi until the onset of mitosis, when this organelle fragments to form Golgi-derived vesicles (GDVs) (Aydin et al., 2014; Champion et al., 2017). HPV takes advantage of Golgi fragmentation by budding off into GDVs, thus escaping the Golgi while still remaining protected within a membrane (Fig. 3, step 4) (DiGiuseppe et al., 2016). As mitosis progresses, the HPV-harboring vesicles are transported along the astral microtubules to the nuclear region, where the virus appears to colocalize with the centrosome (Calton et al., 2017; DiGiuseppe et al., 2016). This trafficking step is thought to involve dynein-mediated transport, though additional evidence is required to support this observation (Florin et al., 2006; Schneider et al., 2011).

With the concomitant nuclear envelope breakdown (NEB) during mitosis (Champion et al., 2017), the HPV-harboring vesicles gain entry to the nucleus (Aydin et al., 2014; Calton et al., 2017). In this manner, HPV accesses the nucleus without engaging the NPC, in contrast to AdV and PyV (see below). The virus-harboring vesicles associate with the condensed chromosomes of the host cell (Calton et al., 2017; DiGiuseppe et al., 2016). During this highly dynamic process, it is unclear how HPV anchors itself in the nucleus. One possibility is that the virus directly binds to the mitotic chromosomes through a chromatin binding domain on L2 (Aydin et al., 2017; Calton et al., 2017). Alternatively, L2 might actively target the viral DNA to promyelocytic leukemia nuclear bodies of the host cell (Day et al., 2004). As a third option, an unidentified nuclear host factor may help target and retain the viral complex at the nucleus during specific stages of mitosis. Though L2 appears to contain nuclear localization signals (NLSs) (Mamoor et al., 2012), the functional significance of these signals during infection remains to be addressed.

Upon completion of mitosis, when the nuclear envelope reforms, the HPV-harboring vesicle is effectively trapped within the nucleus (Calton et al., 2017, DiGiuseppe et al., 2016). How L2 and the viral genome fully translocate across the protective vesicular membrane into the nucleoplasm is entirely unknown (Fig. 3, step 5). In fact, the presence of cargo-transport vesicles within the nucleus is itself a unique phenomena, with no previously described examples.

What mechanism might translocate and expose the viral genome to the nucleoplasm? There is minimal evidence that nuclear phospholipases could play a role in degrading the vesicular membrane (Aydin et al., 2014; Lipovsky et al., 2013). Furthermore, whether L2 executes a function in exposing the viral DNA is not known. Coordinating nuclear entry and viral DNA translocation with mitotic NEB appears to be unique to HPV among non-enveloped viruses, though some enveloped viruses such as the murine-leukemia virus are known to rely on NEB as well (Cohen et al., 2011). Regardless, upon nuclear exposure of HPV's genome, host cell machinery initiates viral transcription and replication, thereby promoting infection.

4. Exploiting the ER during cell entry of PyV

In contrast to AdV and HPV, by far the most unique feature of PyV cell entry is its reliance on ER-associated machineries. PyV infection is relatively common among the general population with a seroprevalence of up to 90% in some cases (DeCaprio and Garcea, 2013). PyVs cause debilitating disease in humans, particularly in immunocompromised individuals. These include neuropathy by BK PyV and progressive multifocal leukoencephalopathy (PML) by JC PyV. With the recent discovery of 11 new human PyVs, including Merkel cell PyV, which causes the often-fatal Merkel cell skin carcinoma (Arora et al., 2012; DeCaprio and Garcea, 2013), the study of these small DNA tumor viruses is as essential as ever.

4.1 Receptor-mediated entry to the endosome

Historically, much of our knowledge on PyV was gained through the study of two animal viruses—the murine PyV and the simian virus 40 (SV40), the archetype PyV. In this chapter, we will discuss mainly the entry of SV40, which shares structural and genomic organization with human PyVs as well as the same infectious life cycle (Howley and Livingston, 2009). SV40 has a 5-kilobase-pair double-stranded DNA genome that is encased within an icosahedral capsid shell. The viral capsid is composed of 72 pentamers of the VP1 major capsid protein. Each pentamer is associated with one copy of either the VP2 or VP3 minor capsid proteins, which associate with the viral genome and contain an NLS (Chen et al., 1998; Liddington et al., 1991). When fully assembled, the virus is approximately 45 nm in diameter.

To initiate viral entry, PyV attaches to a ganglioside glycolipid receptor on the host cell surface through its VP1 protein (Fig. 4, step 1). The exact receptor varies between PyVs, with SV40 associating with GM1, murine PyV binding GD1a and BK, and JC and Merkel cell PyVs interacting instead with either GD1b or GT1b (Erickson et al., 2009; Komagome et al., 2002; Low et al., 2006; Tsai et al., 2003). Following receptor binding, the virus is internalized by a caveolae/lipid-raft dependent mechanism and subsequently transported to the early endosome (Fig. 4, step 2) (Anderson et al., 1996). Endosomal maturation is essential for PyV entry as blocking maturation or acidification of the late endosome disrupts downstream steps of viral trafficking during entry (Qian et al., 2009).

4.2 Membrane penetration of the ER membrane

Unlike AdV and HPV, the initial site of PyV membrane penetration is not at the endosomal membrane. Instead, virus-containing endosomes are sorted to the ER, at least in part, through the guidance of the ganglioside glycolipid receptor and in a microtubule-dependent fashion (Fig. 4, step 3)

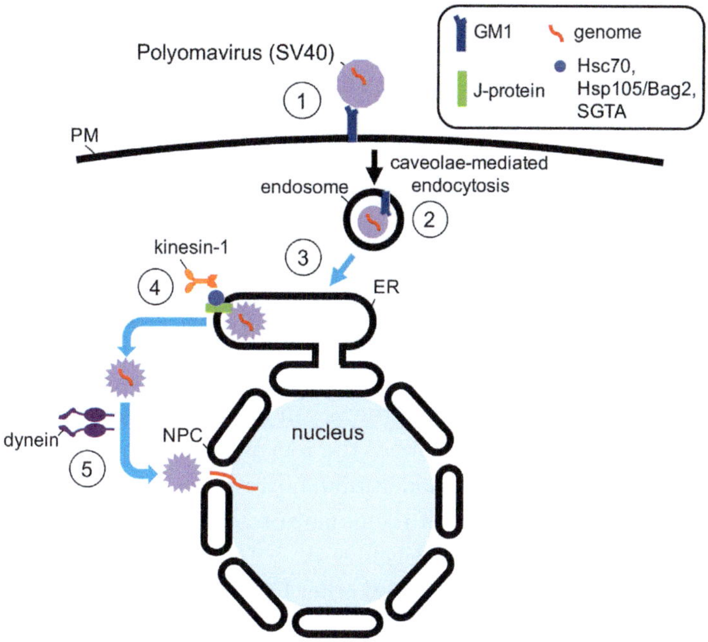

Fig. 4 Unique viral transport pathway through the ER during PyV entry. To infect cells, SV40 binds to the ganglioside GM1 receptor on the cell surface (step 1) and is endocytosed by a caveolae/lipid raft-dependent mechanism to reach the endosome (step 2). From here, the virus is targeted directly to the ER (step 3) where it undergoes conformational changes to penetrate the ER membrane and reach the cytosol (step 4). This requires the concerted effort of both ER resident proteins, a cytosolic chaperone complex, and the kinesin-1 motor. Once in the cytosol, the dynein motor transports SV40 to the nucleus where the viral genome enters through the NPC (step 5).

(Norkin et al., 2002; Pelkmans et al., 2001; Qian et al., 2009). The discovery that endocytosed SV40 bypasses the Golgi and is targeted directly from the endosome to the ER was the first evidence of this unique cellular transport route (Kartenbeck et al., 1989). At present, PyV is the only virus explicitly shown to utilize this pathway, although one study suggests that a subset of HPV may also transport through the ER during entry (Zhang et al., 2014).

PyV was initially thought to disguise itself as a misfolded substrate that hijacks the ER-associated degradation (ERAD) pathway to reach the cytosol. This cellular quality control pathway ejects misfolded proteins from the ER into the cytosol through a Hrd1 E3 ubiquitin ligase-containing channel for degradation by the proteasome (Tsai et al., 2002). Entry of SV40, however, is Hrd1-independent and avoids ubiquitination (Geiger et al., 2011). Further, because SV40 crosses the ER membrane as a relatively large particle

(45 nm) that is unlikely to move through the pore of any known channel (Inoue and Tsai, 2011), a new model has emerged in which the virus reorganizes the ER membrane to create a membrane penetration site. In support of this model, SV40 infection induces mobilization of the BAP31 ER membrane protein as well as the transmembrane DNA J-proteins B12, B14, and C18 into discrete puncta called "foci" (Bagchi et al., 2015; Geiger et al., 2011; Goodwin et al., 2011; Walczak et al., 2014). This virus-induced transmembrane complex recruits a cytosolic chaperone complex composed of Hsc70, Hsp105, Bag2, and SGTA to the foci, which then extracts the virus into the cytosol (Fig. 4, step 4) (Dupzyk and Tsai, 2018; Dupzyk et al., 2017; Ravindran et al., 2015; Walczak et al., 2014). A more detailed review of this extraction process can be found in Dupzyk and Tsai (2016). The architectural structure of the foci as well as how SV40 induces its formation is not completely understood; however, recent studies show that this process relies on both the ER membrane protein complex 1 (EMC1) and the kinesin-1 molecular motor (Bagchi et al., 2016; Ravindran et al., 2017). It is likely that future research will reveal other ER-associated proteins involved in coordinating this virus-specific phenomenon.

In addition to being its site of membrane penetration, transport through the ER is essential to PyV infection as ER luminal proteins impart conformational changes on the virus that are essential for subsequent steps of the membrane transport pathway. Several members of the protein disulfide isomerase (PDI) family of ER luminal proteins induce structural rearrangement of the virus by reducing VP1 inter-pentameric disulfide bonds in the capsid shell (Schelhaas et al., 2007; Walczak and Tsai, 2011). This reduction is required for productive infection as it exposes the underlying hydrophobic VP2 and VP3 proteins to generate a viral particle that is primed and competent for membrane penetration (Magnuson et al., 2005; Norkin et al., 2002). This initial exposure of VP2 and VP3 may also be the first step in disassembly of the viral capsid in the cytosol.

4.3 Nuclear import

Upon reaching the cytosol, two major steps still stand between PyV and successful infection—physical transport to the nucleus and capsid disassembly to expose the viral genome (Inoue and Tsai, 2011). Similar to other non-enveloped viruses, SV40 engages the dynein motor when the virus arrives in the cytosol, which is essential for nuclear entry of the virus (Fig. 4, step 5) (Ravindran et al., 2018). This interaction is likely important for transporting the viral particle to the nucleus, as inhibiting the motor's ATPase motor domain significantly impaired infection. How the virus is handed off from kinesin-1 to dynein after ER membrane penetration has not been established, and is a point of interest in the field. In addition to its role in transport, inhibiting dynein activity also impairs viral disassembly in the cytosol. We envision that, similar to AdV, dynein movement generates a mechanical force that physically disassembles the capsid shell to expose the viral genome. Alternatively, dynein may instead transport PyV to a secondary site where disassembly occurs by an unknown mechanism. In either case, viral disassembly is a rate-limiting step in PyV entry.

While PyV nuclear entry has not been fully characterized, evidence indicates that it also gains access to the nucleus through the NPC. Cytoplasmic injection of mAB414, an anti-nucleoporin antibody that blocks entry of signal-mediated nuclear proteins, inhibits the nuclear expression of the viral large T-antigen (TAg), a marker of successful nuclear entry (Clever et al., 1991; Yamada and Kasamatsu, 1993). Moreover, SV40 VP3 was reported to bind to components of the nuclear import machinery, including importin α/β (Nakanishi et al., 2002), which transport cargos with an NLS through the NPC (Lange et al., 2007). Interestingly, the first NLS was discovered in SV40 (Kalderon et al., 1984), and as VP2 and VP3 each contain this signal, it is possible that these minor capsid proteins remain associated with viral DNA after disassembly in order to direct the genome into the nucleus.

Finally, a proteomics approach using an immunoprecipitation-coupled mass spectrometry strategy identified several NPC components as interacting partners of cytosol-localized SV40, including the cytoplasmic filament Nup358/RanBP2 (Ravindran et al., 2018). Intriguingly, RanBP2, which associates with Nup214 during AdV nuclear import, has also been implicated in the nuclear import of enveloped viruses such as vaccinia virus and HIV (Hutten et al., 2009; Khuperkar et al., 2017). Therefore, while the entry pathways of different DNA viruses can take them along unique journeys through the cell, many of these routes appear to converge at their final destination for nuclear entry.

5. Conclusion and future directions

In order to cause infection, viruses must navigate the complex intracellular environment of their host cells to reach the cytosol or nucleus where replication occurs. In the case of non-enveloped DNA viruses, such as AdV, HPV and PyV, this infectious route can vary greatly. Prior to reaching the nucleus, their ultimate destination during entry, these viruses must first successfully penetrate a host membrane. Whether this occurs at the endosome, as seen with AdV and HPV, or at the ER membrane, as seen with PyV, membrane penetration remains a decisive step in productive infection.

The study of how non-enveloped viruses co-opt cellular membrane penetration apparatus and transport machineries has revealed much about how these viruses infect cells, but several questions remain. For instance, what precise structural changes are imparted on AdV that enables release of protein VI, and how might low pH be involved in this process? Regarding HPV, identifying the mysterious viral entry receptor is of foremost importance. Pinpointing this essential host factor will clarify the mode of HPV endocytosis and reveal a novel target for therapeutic development. In addition, resolving how this virus hijacks γ-secretase activity to promote penetration of the endosome membrane requires further research, as probing this phenomena will provide critical insights into the trafficking functions of both HPV and γ-secretase. Unlike other non-enveloped DNA viruses, HPV's genome remains in a protective vesicle throughout nuclear entry. How the viral genome egresses from the transport vesicles into the nucleoplasm is perhaps the most intriguing question of HPV entry, yet remains largely unexplored. In the case of PyV, how this virus induces the construction of its own membrane penetration site is still a key question in the field. More specifically, how PyV signals to cellular factors, including kinesin-1, to mobilize into a virus-induced foci and whether this coordinates with membrane penetration are widely unknown. Additionally, identifying other host proteins involved in membrane penetration and the subsequent trafficking of PyV to the nucleus could provide new targets for combating infection.

The leveraging of both new and traditional techniques will be instrumental in addressing these outstanding questions. For instance, with increasing advances in high-resolution microscopy techniques, these state-of-the-art imaging approaches could be used to resolve the structure and formation of the PyV membrane penetration site. Along the same lines, these sophisticated methods may illuminate how γ-secretase mediates HPV insertion, as well as how the viral DNA translocates across the limiting membrane upon arrival in the nucleus. Because identifying additional host factors involved in viral entry is a criticial objective in clarifying the infection mechanisms of these non-enveloped viruses, ongoing proteomic advances may aid in this endeavor. For example, biotin-proximity labeling (BioID) can be used to screen for relevant protein-protein interactions, and if coupled with mass-spectrometry, may reveal even the most transient of associations (Kim et al., 2014, 2016).

In addition to modern methods, however, the use of classical biochemical techniques is still very much needed. In particular, developing in vitro assays to reconstitute viral entry events will clarify the precise molecular mechanisms of these processes, which is imperative for answering these questions.

Even still, perhaps the greatest and most deliberated question remains. Why have these, and other non-enveloped viruses devised such different entry strategies to reach the nucleus to cause infection? It is even more puzzling when considering the variance in entry routes between types of the same virus. It is possible that while seemingly similar to us, the unique characteristics of each viral particle necessitate a selective path during entry: acidification in the endosome to reveal a hidden viral protein required for membrane penetration; ER luminal proteins to remodel the viral capsid for subsequent disassembly in the cytosol; or perhaps a Golgi-dependent action that has yet to be uncovered. While we can only speculate on the reasons, one thing that is certain is that after many years of viral evolution, these distinctions provide us the best clues to understanding and preventing virus infection and are only waiting to be discovered.

Acknowledgments

The authors would like to thank Alex Kukreja (University of Michigan) for critically reviewing this article. C.C.S. is supported by the National Institutes of Health (NIH) (T32-AI007528). M.C.H. is supported by NIH (T32-GM007315). B.T. is funded by the NIH (RO1-AI064296).

References

Agosto, M.A., Ivanovic, T., Nibert, M.L., 2006. Mammalian reovirus, a nonfusogenic non-enveloped virus, forms size-selective pores in a model membrane. Proc. Natl. Acad. Sci. U. S. A. 103, 16496–16501.

Aksoy, P., Abban, C.Y., Kiyashka, E., Qiang, W., Meneses, P.I., 2014. HPV16 infection of HaCaTs is dependent on beta4 integrin, and alpha6 integrin processing. Virology 449, 45–52.

Amstutz, B., Gastaldelli, M., Kalin, S., Imelli, N., Boucke, K., Wandeler, E., Mercer, J., Hemmi, S., Greber, U.F., 2008. Subversion of CtBP1-controlled macropinocytosis by human adenovirus serotype 3. EMBO J. 27, 956–969.

Anderson, H.A., Chen, Y., Norkin, L.C., 1996. Bound simian virus 40 translocates to caveolin-enriched membrane domains, and its entry is inhibited by drugs that selectively disrupt caveolae. Mol. Biol. Cell 7, 1825–1834.

Arora, R., Chang, Y., Moore, P.S., 2012. MCV and Merkel cell carcinoma: a molecular success story. Curr. Opin. Virol. 2, 489–498.

Aydin, I., Weber, S., Snijder, B., Samperio Ventayol, P., Kuhbacher, A., Becker, M., Day, P.M., Schiller, J.T., Kann, M., Pelkmans, L., Helenius, A., Schelhaas, M., 2014. Large scale RNAi reveals the requirement of nuclear envelope breakdown for nuclear import of human papillomaviruses. PLoS Pathog. 10, e1004162.

Aydin, I., Villalonga-Planells, R., Greune, L., Bronnimann, M.P., Calton, C.M., Becker, M., Lai, K.Y., Campos, S.K., Schmidt, M.A., Schelhaas, M., 2017. A central region in the minor capsid protein of papillomaviruses facilitates viral genome tethering and membrane penetration for mitotic nuclear entry. PLoS Pathog. 13 ,e1006308.

Bagchi, P., Walczak, C.P., Tsai, B., 2015. The endoplasmic reticulum membrane J protein C18 executes a distinct role in promoting simian virus 40 membrane penetration. J. Virol. 89, 4058–4068.

Bagchi, P., Inoue, T., Tsai, B., 2016. EMC1-dependent stabilization drives membrane penetration of a partially destabilized non-enveloped virus. Elife 5, 1–23.

Bergant Marusic, M., Ozbun, M.A., Campos, S.K., Myers, M.P., Banks, L., 2012. Human papillomavirus L2 facilitates viral escape from late endosomes via sorting nexin 17. Traffic 13, 455–467.

Bergelson, J.M., Cunningham, J.A., Droguett, G., Kurt-Jones, E.A., Krithivas, A., Hong, J.S., Horwitz, M.S., Crowell, R.L., Finberg, R.W., 1997. Isolation of a common receptor for Coxsackie B viruses and adenoviruses 2 and 5. Science 275, 1320–1323.

Bienkowska-Haba, M., Patel, H.D., Sapp, M., 2009. Target cell cyclophilins facilitate human papillomavirus type 16 infection. PLoS Pathog. 5, e1000524.

Bienkowska-Haba, M., Williams, C., Kim, S.M., Garcea, R.L., Sapp, M., 2012. Cyclophilins facilitate dissociation of the human papillomavirus type 16 capsid protein L1 from the L2/DNA complex following virus entry. J. Virol. 86, 9875–9887.

Boisvert, M., Fernandes, S., Tijssen, P., 2010. Multiple pathways involved in porcine parvovirus cellular entry and trafficking toward the nucleus. J. Virol. 84, 7782–7792.

Bremner, K.H., Scherer, J., Yi, J., Vershinin, M., Gross, S.P., Vallee, R.B., 2009. Adenovirus transport via direct interaction of cytoplasmic dynein with the viral capsid hexon subunit. Cell Host Microbe 6, 523–535.

Bronnimann, M.P., Chapman, J.A., Park, C.K., Campos, S.K., 2013. A transmembrane domain and GxxxG motifs within L2 are essential for papillomavirus infection. J. Virol. 87, 464–473.

Buck, C.B., Cheng, N., Thompson, C.D., Lowy, D.R., Steven, A.C., Schiller, J.T., Trus, B.L., 2008. Arrangement of L2 within the papillomavirus capsid. J. Virol. 82, 5190–5197.

Burckhardt, C.J., Suomalainen, M., Schoenenberger, P., Boucke, K., Hemmi, S., Greber, U.F., 2011. Drifting motions of the adenovirus receptor CAR and immobile integrins initiate virus uncoating and membrane lytic protein exposure. Cell Host Microbe 10, 105–117.

Calton, C.M., Bronnimann, M.P., Manson, A.R., Li, S., Chapman, J.A., Suarez-Berumen, M., Williamson, T.R., Molugu, S.K., Bernal, R.A., Campos, S.K., 2017. Translocation of the papillomavirus L2/vDNA complex across the limiting membrane requires the onset of mitosis. PLoS Pathog. 13, e1006200.

Centers For Disease Control And Prevention, 2018. Cancers associated with human papillomavirus, United States—2011–2015. USCS data brief, no. 4. Atlanta, GA: Centers for Disease Control and Prevention.

Cerqueira, C., Samperio Ventayol, P., Vogeley, C., Schelhaas, M., 2015. Kallikrein-8 proteolytically processes human papillomaviruses in the extracellular space To facilitate entry into host cells. J. Virol. 89, 7038–7052.

Champion, L., Linder, M.I., Kutay, U., 2017. Cellular reorganization during mitotic entry. Trends Cell Biol. 27, 26–41.

Chandran, K., Farsetta, D.L., Nibert, M.L., 2002. Strategy for nonenveloped virus entry: a hydrophobic conformer of the reovirus membrane penetration protein micro 1 mediates membrane disruption. J. Virol. 76, 9920–9933.

Chen, X.S., Stehle, T., Harrison, S.C., 1998. Interaction of polyomavirus internal protein VP2 with the major capsid protein VP1 and implications for participation of VP2 in viral entry. EMBO J. 17, 3233–3240.

Clever, J., Yamada, M., Kasamatsu, H., 1991. Import of simian virus 40 virions through nuclear pore complexes. Proc. Natl. Acad. Sci. U. S. A. 88, 7333–7337.
Cohen, S., Au, S., Pante, N., 2011. How viruses access the nucleus. Biochim. Biophys. Acta 1813, 1634–1645.
Cupelli, K., Muller, S., Persson, B.D., Jost, M., Arnberg, N., Stehle, T., 2010. Structure of adenovirus type 21 knob in complex with CD46 reveals key differences in receptor contacts among species B adenoviruses. J. Virol. 84, 3189–3200.
Day, P.M., Baker, C.C., Lowy, D.R., Schiller, J.T., 2004. Establishment of papillomavirus infection is enhanced by promyelocytic leukemia protein (PML) expression. Proc. Natl. Acad. Sci. U. S. A. 101, 14252–14257.
Day, P.M., Lowy, D.R., Schiller, J.T., 2008. Heparan sulfate-independent cell binding and infection with furin-precleaved papillomavirus capsids. J. Virol. 82, 12565–12568.
Day, P.M., Thompson, C.D., Schowalter, R.M., Lowy, D.R., Schiller, J.T., 2013. Identification of a role for the trans-Golgi network in human papillomavirus 16 pseudovirus infection. J. Virol. 87, 3862–3870.
DeCaprio, J.A., Garcea, R.L., 2013. A cornucopia of human polyomaviruses. Nat. Rev. Microbiol. 11, 264–276.
DiGiuseppe, S., Keiffer, T.R., Bienkowska-Haba, M., Luszczek, W., Guion, L.G., Muller, M., Sapp, M., 2015. Topography of the human papillomavirus minor capsid protein L2 during vesicular trafficking of infectious entry. J. Virol. 89, 10442–10452.
DiGiuseppe, S., Luszczek, W., Keiffer, T.R., Bienkowska-Haba, M., Guion, L.G., Sapp, M.J., 2016. Incoming human papillomavirus type 16 genome resides in a vesicular compartment throughout mitosis. Proc. Natl. Acad. Sci. U. S. A. 113, 6289–6294.
DiGiuseppe, S., Bienkowska-Haba, M., Guion, L.G.M., Keiffer, T.R., Sapp, M., 2017. Human papillomavirus major capsid protein L1 remains associated with the incoming viral genome throughout the entry process. J. Virol. 91 (16), 1–17.
Doorbar, J., 2005. The papillomavirus life cycle. J. Clin. Virol. 32 (Suppl. 1), S7–15.
Dupzyk, A., Tsai, B., 2016. How polyomaviruses exploit the ERAD machinery to cause infection. Viruses 8, 1–14.
Dupzyk, A., Tsai, B., 2018. Bag2 is a component of a cytosolic extraction machinery that promotes membrane penetration of a nonenveloped virus. J. Virol. 92, 1–15.
Dupzyk, A., Williams, J.M., Bagchi, P., Inoue, T., Tsai, B., 2017. SGTA-dependent regulation of Hsc70 promotes cytosol entry of simian virus 40 from the endoplasmic reticulum. J. Virol. 91, 1–14.
Dziduszko, A., Ozbun, M.A., 2013. Annexin A2 and S100A10 regulate human papillomavirus type 16 entry and intracellular trafficking in human keratinocytes. J. Virol. 87, 7502–7515.
Erickson, K.D., Garcea, R.L., Tsai, B., 2009. Ganglioside GT1b is a putative host cell receptor for the Merkel cell polyomavirus. J. Virol. 83, 10275–10279.
Evander, M., Frazer, I.H., Payne, E., Qi, Y.M., Hengst, K., McMillan, N.A., 1997. Identification of the alpha6 integrin as a candidate receptor for papillomaviruses. J. Virol. 71, 2449–2456.
Farr, G.A., Zhang, L.G., Tattersall, P., 2005. Parvoviral virions deploy a capsid-tethered lipolytic enzyme to breach the endosomal membrane during cell entry. Proc. Natl. Acad. Sci. U. S. A. 102, 17148–17153.
FitzGerald, D.J., Padmanabhan, R., Pastan, I., Willingham, M.C., 1983. Adenovirus-induced release of epidermal growth factor and pseudomonas toxin into the cytosol of KB cells during receptor-mediated endocytosis. Cell 32, 607–617.
Florin, L., Becker, K.A., Lambert, C., Nowak, T., Sapp, C., Strand, D., Streeck, R.E., Sapp, M., 2006. Identification of a dynein interacting domain in the papillomavirus minor capsid protein l2. J. Virol. 80, 6691–6696.
Gaggar, A., Shayakhmetov, D.M., Lieber, A., 2003. CD46 is a cellular receptor for group B adenoviruses. Nat. Med. 9, 1408–1412.

Geiger, R., Andritschke, D., Friebe, S., Herzog, F., Luisoni, S., Heger, T., Helenius, A., 2011. BAP31 and BiP are essential for dislocation of SV40 from the endoplasmic reticulum to the cytosol. Nat. Cell Biol. 13, 1305–1314.

Giroglou, T., Florin, L., Schafer, F., Streeck, R.E., Sapp, M., 2001. Human papillomavirus infection requires cell surface heparan sulfate. J. Virol. 75, 1565–1570.

Goodwin, E.C., Lipovsky, A., Inoue, T., Magaldi, T.G., Edwards, A.P., Van Goor, K.E., Paton, A.W., Paton, J.C., Atwood, W.J., Tsai, B., Dimaio, D., 2011. BiP and multiple DNAJ molecular chaperones in the endoplasmic reticulum are required for efficient simian virus 40 infection. MBio 2, e00101–e00111.

Grassel, L., Fast, L.A., Scheffer, K.D., Boukhallouk, F., Spoden, G.A., Tenzer, S., Boller, K., Bago, R., Rajesh, S., Overduin, M., Berditchevski, F., Florin, L., 2016. The CD63-Syntenin-1 Complex Controls Post-Endocytic Trafficking of Oncogenic Human Papillomaviruses. Sci. Rep. 6, 32337.

Greber, U.F., 2016. Virus and host mechanics support membrane penetration and cell entry. J. Virol. 90, 3802–3805.

Greber, U.F., Willetts, M., Webster, P., Helenius, A., 1993. Stepwise dismantling of adenovirus 2 during entry into cells. Cell 75, 477–486.

Gutierrez, M., Isa, P., Sanchez-San Martin, C., Perez-Vargas, J., Espinosa, R., Arias, C.F., Lopez, S., 2010. Different rotavirus strains enter MA104 cells through different endocytic pathways: the role of clathrin-mediated endocytosis. J. Virol. 84, 9161–9169.

Helenius, A., 2018. Virus entry: looking back and moving forward. J. Mol. Biol. 430, 1853–1862.

Howley, P.M., Livingston, D.M., 2009. Small DNA tumor viruses: large contributors to biomedical sciences. Virology 384, 256–259.

Hutten, S., Walde, S., Spillner, C., Hauber, J., Kehlenbach, R.H., 2009. The nuclear pore component Nup358 promotes transportin-dependent nuclear import. J. Cell Sci. 122, 1100–1110.

Inoue, T., Tsai, B., 2011. A large and intact viral particle penetrates the endoplasmic reticulum membrane to reach the cytosol. PLoS Pathog. 7, e1002037.

Inoue, T., Zhang, P., Zhang, W., Goodner-Bingham, K., Dupzyk, A., Dimaio, D., Tsai, B., 2018. gamma-Secretase promotes membrane insertion of the human papillomavirus L2 capsid protein during virus infection. J. Cell Biol. 217, 3545–3559.

Ivanovic, T., Agosto, M.A., Zhang, L., Chandran, K., Harrison, S.C., Nibert, M.L., 2008. Peptides released from reovirus outer capsid form membrane pores that recruit virus particles. EMBO J. 27, 1289–1298.

Kalderon, D., Roberts, B.L., Richardson, W.D., Smith, A.E., 1984. A short amino acid sequence able to specify nuclear location. Cell 39, 499–509.

Kamper, N., Day, P.M., Nowak, T., Selinka, H.C., Florin, L., Bolscher, J., Hilbig, L., Schiller, J.T., Sapp, M., 2006. A membrane-destabilizing peptide in capsid protein L2 is required for egress of papillomavirus genomes from endosomes. J. Virol. 80, 759–768.

Kartenbeck, J., Stukenbrok, H., Helenius, A., 1989. Endocytosis of simian virus 40 into the endoplasmic reticulum. J. Cell Biol. 109, 2721–2729.

Khuperkar, D., Kamble, A., Singh, A., Ghate, A., Nawadkar, R., Sahu, A., Joseph, J., 2017. Selective recruitment of nucleoporins on vaccinia virus factories and the role of Nup358 in viral infection. Virology 512, 151–160.

Kim, D.I., Birendra, K.C., Zhu, W., Motamedchaboki, K., Doye, V., Roux, K.J., 2014. Probing nuclear pore complex architecture with proximity-dependent biotinylation. Proc. Natl. Acad. Sci. U. S. A. 111, E2453–E2461.

Kim, D.I., Jensen, S.C., Noble, K.A., Kc, B., Roux, K.H., Motamedchaboki, K., Roux, K.J., 2016. An improved smaller biotin ligase for BioID proximity labeling. Mol. Biol. Cell 27, 1188–1196.

Komagome, R., Sawa, H., Suzuki, T., Suzuki, Y., Tanaka, S., Atwood, W.J., Nagashima, K., 2002. Oligosaccharides as receptors for JC virus. J. Virol. 76, 12992–13000.
Kumar, C.S., Dey, D., Ghosh, S., Banerjee, M., 2018. Breach: host membrane penetration and entry by nonenveloped viruses. Trends Microbiol. 26, 525–537.
Lange, A., Mills, R.E., Lange, C.J., Stewart, M., Devine, S.E., Corbett, A.H., 2007. Classical nuclear localization signals: definition, function, and interaction with importin alpha. J. Biol. Chem. 282, 5101–5105.
Liddington, R.C., Yan, Y., Moulai, J., Sahli, R., Benjamin, T.L., Harrison, S.C., 1991. Structure of simian virus 40 at 3.8-A resolution. Nature 354, 278–284.
Lion, T., 2014. Adenovirus infections in immunocompetent and immunocompromised patients. Clin. Microbiol. Rev. 27, 441–462.
Lipovsky, A., Popa, A., Pimienta, G., Wyler, M., Bhan, A., Kuruvilla, L., Guie, M.A., Poffenberger, A.C., Nelson, C.D., Atwood, W.J., Dimaio, D., 2013. Genome-wide siRNA screen identifies the retromer as a cellular entry factor for human papillomavirus. Proc. Natl. Acad. Sci. U. S. A. 110, 7452–7457.
Low, J.A., Magnuson, B., Tsai, B., Imperiale, M.J., 2006. Identification of gangliosides GD1b and GT1b as receptors for BK virus. J. Virol. 80, 1361–1366.
Magnuson, B., Rainey, E.K., Benjamin, T., Baryshev, M., Mkrtchian, S., Tsai, B., 2005. ERp29 triggers a conformational change in polyomavirus to stimulate membrane binding. Mol. Cell 20, 289–300.
Maier, O., Wiethoff, C.M., 2010. N-terminal alpha-helix-independent membrane interactions facilitate adenovirus protein VI induction of membrane tubule formation. Virology 408, 31–38.
Maier, O., Galan, D.L., Wodrich, H., Wiethoff, C.M., 2010. An N-terminal domain of adenovirus protein VI fragments membranes by inducing positive membrane curvature. Virology 402, 11–19.
Mainou, B.A., Zamora, P.F., Ashbrook, A.W., Dorset, D.C., Kim, K.S., Dermody, T.S., 2013. Reovirus cell entry requires functional microtubules. MBio 4, 1–10.
Mamoor, S., Onder, Z., Karanam, B., Kwak, K., Bordeaux, J., Crosby, L., Roden, R.B., Moroianu, J., 2012. The high risk HPV16 L2 minor capsid protein has multiple transport signals that mediate its nucleocytoplasmic traffic. Virology 422, 413–424.
Melikyan, G.B., 2014. HIV entry: a game of hide-and-fuse? Curr. Opin. Virol. 4, 1–7.
Mercer, J., Schelhaas, M., Helenius, A., 2010. Virus entry by endocytosis. Annu. Rev. Biochem. 79, 803–833.
Morgan, C., Rosenkranz, H.S., Mednis, B., 1969. Structure and development of viruses as observed in the electron microscope. V. Entry and uncoating of adenovirus. J. Virol. 4, 777–796.
Moyer, C.L., Nemerow, G.R., 2011. Viral weapons of membrane destruction: variable modes of membrane penetration by non-enveloped viruses. Curr. Opin. Virol. 1, 44–49.
Muhr, L.S.A., Eklund, C., Dillner, J., 2018. Towards quality and order in human papillomavirus research. Virology 519, 74–76.
Nakanishi, A., Shum, D., Morioka, H., Otsuka, E., Kasamatsu, H., 2002. Interaction of the Vp3 nuclear localization signal with the importin alpha 2/beta heterodimer directs nuclear entry of infecting simian virus 40. J. Virol. 76, 9368–9377.
Nguyen, E.K., Nemerow, G.R., Smith, J.G., 2010. Direct evidence from single-cell analysis that human {alpha}-defensins block adenovirus uncoating to neutralize infection. J. Virol. 84, 4041–4049.
Norkin, L.C., Anderson, H.A., Wolfrom, S.A., Oppenheim, A., 2002. Caveolar endocytosis of simian virus 40 is followed by brefeldin A-sensitive transport to the endoplasmic reticulum, where the virus disassembles. J. Virol. 76, 5156–5166.

Pascual-Pasto, G., Bazan-Peregrino, M., Olaciregui, N.G., Restrepo-Perdomo, C.A., Mato-Berciano, A., Ottaviani, D., Weber, K., Correa, G., Paco, S., Vila-Ubach, M., Cuadrado-Vilanova, M., Castillo-Ecija, H., Botteri, G., Garcia-Gerique, L., Moreno-Gilabert, H., Gimenez-Alejandre, M., Alonso-Lopez, P., Farrera-Sal, M., Torres-Manjon, S., Ramos-Lozano, D., Moreno, R., Aerts, I., Doz, F., Cassoux, N., Chapeaublanc, E., Torrebadell, M., Roldan, M., Konig, A., Sunol, M., Claverol, J., Lavarino, C., Carmen De, T., Fu, L., Radvanyi, F., Munier, F.L., Catala-Mora, J., Mora, J., Alemany, R., Cascallo, M., Chantada, G.L., Carcaboso, A.M., 2019. Therapeutic targeting of the RB1 pathway in retinoblastoma with the oncolytic adenovirus VCN-01. Sci. Transl. Med. 11, 1–11.

Pelkmans, L., Kartenbeck, J., Helenius, A., 2001. Caveolar endocytosis of simian virus 40 reveals a new two-step vesicular-transport pathway to the ER. Nat. Cell Biol. 3, 473–483.

Perez, L., Carrasco, L., 1994. Involvement of the vacuolar H(+)-ATPase in animal virus entry. J. Gen. Virol. 75 (Pt. 10), 2595–2606.

Popa, A., Zhang, W., Harrison, M.S., Goodner, K., Kazakov, T., Goodwin, E.C., Lipovsky, A., Burd, C.G., Dimaio, D., 2015. Direct binding of retromer to human papillomavirus type 16 minor capsid protein L2 mediates endosome exit during viral infection. PLoS Pathog. 11, e1004699.

Prchla, E., Plank, C., Wagner, E., Blaas, D., Fuchs, R., 1995. Virus-mediated release of endosomal content in vitro: different behavior of adenovirus and rhinovirus serotype 2. J. Cell Biol. 131, 111–123.

Qian, M., Cai, D., Verhey, K.J., Tsai, B., 2009. A lipid receptor sorts polyomavirus from the endolysosome to the endoplasmic reticulum to cause infection. PLoS Pathog. 5, e1000465.

Ravindran, M.S., Bagchi, P., Inoue, T., Tsai, B., 2015. A Non-enveloped virus hijacks host disaggregation machinery to translocate across the endoplasmic reticulum membrane. PLoS Pathog. 11, e1005086.

Ravindran, M.S., Engelke, M.F., Verhey, K.J., Tsai, B., 2017. Exploiting the kinesin-1 molecular motor to generate a virus membrane penetration site. Nat. Commun. 8, 15496.

Ravindran, M.S., Spriggs, C.C., Verhey, K.J., Tsai, B., 2018. Dynein engages and disassembles cytosol-localized SV40 to promote infection. J. Virol. 92 (12), 1–13.

Reddy, V.S., Nemerow, G.R., 2014. Structures and organization of adenovirus cement proteins provide insights into the role of capsid maturation in virus entry and infection. Proc. Natl. Acad. Sci. U. S. A. 111, 11715–11720.

Richards, R.M., Lowy, D.R., Schiller, J.T., Day, P.M., 2006. Cleavage of the papillomavirus minor capsid protein, L2, at a furin consensus site is necessary for infection. Proc. Natl. Acad. Sci. U. S. A. 103, 1522–1527.

Richards, K.F., Bienkowska-Haba, M., Dasgupta, J., Chen, X.S., Sapp, M., 2013. Multiple heparan sulfate binding site engagements are required for the infectious entry of human papillomavirus type 16. J. Virol. 87, 11426–11437.

Satterwhite, C.L., Torrone, E., Meites, E., Dunne, E.F., Mahajan, R., Ocfemia, M.C., Su, J., Xu, F., Weinstock, H., 2013. Sexually transmitted infections among US women and men: prevalence and incidence estimates, 2008. Sex. Transm. Dis. 40, 187–193.

Sayedahmed, E.E., Kumari, R., Shukla, S., Hassan, A.O., Mohammed, S.I., York, I.A., Gangappa, S., Sambhara, S., Mittal, S.K., 2018. Longevity of adenovirus vector immunity in mice and its implications for vaccine efficacy. Vaccine 36, 6744–6751.

Scheffer, K.D., Gawlitza, A., Spoden, G.A., Zhang, X.A., Lambert, C., Berditchevski, F., Florin, L., 2013. Tetraspanin CD151 mediates papillomavirus type 16 endocytosis. J. Virol. 87, 3435–3446.

Schelhaas, M., Malmstrom, J., Pelkmans, L., Haugstetter, J., Ellgaard, L., Grunewald, K., Helenius, A., 2007. Simian virus 40 depends on ER protein folding and quality control factors for entry into host cells. Cell 131, 516–529.

Schelhaas, M., Shah, B., Holzer, M., Blattmann, P., Kuhling, L., Day, P.M., Schiller, J.T., Helenius, A., 2012. Entry of human papillomavirus type 16 by actin-dependent, clathrin- and lipid raft-independent endocytosis. PLoS Pathog. 8, e1002657.

Schneider, M.A., Spoden, G.A., Florin, L., Lambert, C., 2011. Identification of the dynein light chains required for human papillomavirus infection. Cell. Microbiol. 13, 32–46.

Schulz, W.L., Haj, A.K., Schiff, L.A., 2012. Reovirus uses multiple endocytic pathways for cell entry. J. Virol. 86, 12665–12675.

Seiradake, E., Lortat-Jacob, H., Billet, O., Kremer, E.J., Cusack, S., 2006. Structural and mutational analysis of human Ad37 and canine adenovirus 2 fiber heads in complex with the D1 domain of coxsackie and adenovirus receptor. J. Biol. Chem. 281, 33704–33716.

Selinka, H.C., Florin, L., Patel, H.D., Freitag, K., Schmidtke, M., Makarov, V.A., Sapp, M., 2007. Inhibition of transfer to secondary receptors by heparan sulfate-binding drug or antibody induces noninfectious uptake of human papillomavirus. J. Virol. 81, 10970–10980.

Seth, P., Willingham, M.C., Pastan, I., 1984. Adenovirus-dependent release of 51Cr from KB cells at an acidic pH. J. Biol. Chem. 259, 14350–14353.

Skehel, J.J., Wiley, D.C., 2000. Receptor binding and membrane fusion in virus entry: the influenza hemagglutinin. Annu. Rev. Biochem. 69, 531–569.

Smith, J.L., Campos, S.K., Wandinger-Ness, A., Ozbun, M.A., 2008. Caveolin-1-dependent infectious entry of human papillomavirus type 31 in human keratinocytes proceeds to the endosomal pathway for pH-dependent uncoating. J. Virol. 82, 9505–9512.

Smrt, S.T., Lorieau, J.L., 2017. Membrane fusion and infection of the influenza hemagglutinin. Adv. Exp. Med. Biol. 966, 37–54.

Spoden, G., Freitag, K., Husmann, M., Boller, K., Sapp, M., Lambert, C., Florin, L., 2008. Clathrin- and caveolin-independent entry of human papillomavirus type 16—involvement of tetraspanin-enriched microdomains (TEMs). PLoS One 3, e3313.

Spoden, G., Kuhling, L., Cordes, N., Frenzel, B., Sapp, M., Boller, K., Florin, L., Schelhaas, M., 2013. Human papillomavirus types 16, 18, and 31 share similar endocytic requirements for entry. J. Virol. 87, 7765–7773.

Staring, J., Raaben, M., Brummelkamp, T.R., 2018. Viral escape from endosomes and host detection at a glance. J. Cell Sci. 131, 1–8.

Suikkanen, S., Aaltonen, T., Nevalainen, M., Valilehto, O., Lindholm, L., Vuento, M., Vihinen-Ranta, M., 2003a. Exploitation of microtubule cytoskeleton and dynein during parvoviral traffic toward the nucleus. J. Virol. 77, 10270–10279.

Suikkanen, S., Antila, M., Jaatinen, A., Vihinen-Ranta, M., Vuento, M., 2003b. Release of canine parvovirus from endocytic vesicles. Virology 316, 267–280.

Suomalainen, M., Nakano, M.Y., Keller, S., Boucke, K., Stidwill, R.P., Greber, U.F., 1999. Microtubule-dependent plus- and minus end-directed motilities are competing processes for nuclear targeting of adenovirus. J. Cell Biol. 144, 657–672.

Surviladze, Z., Dziduszko, A., Ozbun, M.A., 2012. Essential roles for soluble virion-associated heparan sulfonated proteoglycans and growth factors in human papillomavirus infections. PLoS Pathog. 8, e1002519.

Taylor, J.R., Fernandez, D.J., Thornton, S.M., Skeate, J.G., Luhen, K.P., Da Silva, D.M., Langen, R., Kast, W.M., 2018. Heterotetrameric annexin A2/S100A10 (A2t) is essential for oncogenic human papillomavirus trafficking and capsid disassembly, and protects virions from lysosomal degradation. Sci. Rep. 8, 11642.

Trotman, L.C., Mosberger, N., Fornerod, M., Stidwill, R.P., Greber, U.F., 2001. Import of adenovirus DNA involves the nuclear pore complex receptor CAN/Nup214 and histone H1. Nat. Cell Biol. 3, 1092–1100.

Tsai, B., 2007. Penetration of nonenveloped viruses into the cytoplasm. Annu. Rev. Cell Dev. Biol. 23, 23–43.

Tsai, B., Ye, Y., Rapoport, T.A., 2002. Retro-translocation of proteins from the endoplasmic reticulum into the cytosol. Nat. Rev. Mol. Cell Biol. 3, 246–255.
Tsai, B., Gilbert, J.M., Stehle, T., Lencer, W., Benjamin, T.L., Rapoport, T.A., 2003. Gangliosides are receptors for murine polyoma virus and SV40. EMBO J. 22, 4346–4355.
Van Dyne, E.A., Henley, S.J., Saraiya, M., Thomas, C.C., Markowitz, L.E., Benard, V.B., 2018. Trends in Human Papillomavirus-Associated Cancers—United States, 1999-2015. MMWR Morb. Mortal. Wkly Rep. 67, 918–924.
van Raaij, M.J., Mitraki, A., Lavigne, G., Cusack, S., 1999. A triple beta-spiral in the adenovirus fibre shaft reveals a new structural motif for a fibrous protein. Nature 401, 935–938.
Walczak, C.P., Tsai, B., 2011. A PDI family network acts distinctly and coordinately with ERp29 to facilitate polyomavirus infection. J. Virol. 85, 2386–2396.
Walczak, C.P., Ravindran, M.S., Inoue, T., Tsai, B., 2014. A cytosolic chaperone complexes with dynamic membrane J-proteins and mobilizes a nonenveloped virus out of the endoplasmic reticulum. PLoS Pathog. 10, e1004007.
Walker, T.Y., Elam-Evans, L.D., Yankey, D., Markowitz, L.E., Williams, C.L., Mbaeyi, S.A., Fredua, B., Stokley, S., 2018. National, regional, state, and selected local area vaccination coverage among adolescents aged 13-17 years—United States, 2017. MMWR Morb. Mortal. Wkly Rep. 67, 909–917.
Wang, K., Huang, S., Kapoor-Munshi, A., Nemerow, G., 1998. Adenovirus internalization and infection require dynamin. J. Virol. 72, 3455–3458.
Wang, H., Li, Z.Y., Liu, Y., Persson, J., Beyer, I., Moller, T., Koyuncu, D., Drescher, M.R., Strauss, R., Zhang, X.B., Wahl 3rd, J.K., Urban, N., Drescher, C., Hemminki, A., Fender, P., Lieber, A., 2011. Desmoglein 2 is a receptor for adenovirus serotypes 3, 7, 11 and 14. Nat. Med. 17, 96–104.
Weissenhorn, W., Dessen, A., Calder, L.J., Harrison, S.C., Skehel, J.J., Wiley, D.C., 1999. Structural basis for membrane fusion by enveloped viruses. Mol. Membr. Biol. 16, 3–9.
Wickham, T.J., Mathias, P., Cheresh, D.A., Nemerow, G.R., 1993. Integrins alpha v beta 3 and alpha v beta 5 promote adenovirus internalization but not virus attachment. Cell 73, 309–319.
Wiethoff, C.M., Wodrich, H., Gerace, L., Nemerow, G.R., 2005. Adenovirus protein VI mediates membrane disruption following capsid disassembly. J. Virol. 79, 1992–2000.
Yamada, M., Kasamatsu, H., 1993. Role of nuclear pore complex in simian virus 40 nuclear targeting. J. Virol. 67, 119–130.
Yamauchi, Y., Greber, U.F., 2016. Principles of virus uncoating: cues and the snooker ball. Traffic 17, 569–592.
Yousuf, M.A., Zhou, X., Mukherjee, S., Chintakuntlawar, A.V., Lee, J.Y., Ramke, M., Chodosh, J., Rajaiya, J., 2013. Caveolin-1 associated adenovirus entry into human corneal cells. PLoS One 8, e77462.
Zhang, W., Kazakov, T., Popa, A., Dimaio, D., 2014. Vesicular trafficking of incoming human papillomavirus 16 to the Golgi apparatus and endoplasmic reticulum requires gamma-secretase activity. MBio 5, e01777-14.
Zhang, P., Monteiro Da Silva, G., Deatherage, C., Burd, C., Dimaio, D., 2018. Cell-penetrating peptide mediates intracellular membrane passage of human papillomavirus L2 protein to trigger retrograde trafficking. Cell 174. 1465-1476.e13.
Zubieta, C., Schoehn, G., Chroboczek, J., Cusack, S., 2005. The structure of the human adenovirus 2 penton. Mol. Cell 17, 121–135.

CHAPTER FOUR

Developments in single-molecule and single-particle fluorescence-based approaches for studying viral envelope glycoprotein dynamics and membrane fusion

Angela R. Howard, James B. Munro*

Department of Molecular Biology and Microbiology, Tufts University School of Medicine, Boston, MA, United States

*Corresponding author: e-mail address: james.munro@tufts.edu

Contents

Advances in Virus Research, Volume 104
ISSN 0065-3527
https://doi.org/10.1016/bs.aivir.2019.05.004

Abstract

Fusion of viral and cellular membranes is an essential step in the entry pathway of all enveloped viruses. This is a dynamic and multistep process, which has been extensively studied, resulting in the endpoints of the reaction being firmly established, and many essential cellular factors identified. What remains is to elucidate the dynamic events that underlie this process, including the order and timing of glycoprotein conformational changes, receptor-binding events, and movement of the glycoprotein on the surface of the virion. Due to the inherently asynchronous nature of these dynamics, there has been an increased focus on the study of single virions and single molecules. These techniques provide researchers the high precision and resolution necessary to bridge the gaps in our understanding of viral membrane fusion. This review highlights the advancement of single-molecule and single-particle fluorescence-based techniques, with a specific focus on how these techniques have been used to study the dynamic nature of the viral fusion pathway.

1. Introduction

Membrane fusion is a required step during infection for all enveloped viruses. This process is mediated by viral envelope glycoproteins found on the surface of virions. Envelope glycoproteins facilitate attachment to cellular membranes by way of interaction with receptors. Following receptor engagement envelope glycoproteins promote fusion of the viral and cellular membranes through a stepwise process that has been reviewed extensively and is believed to be highly-conserved in its essential features across diverse species of viruses (Fig. 1) (Harrison, 2015; Kielian, 2014). In all cases, during promotion of membrane fusion the envelope glycoproteins transition from a metastable pre-fusion conformation to a highly stable post-fusion conformation. As a result of this release of energy, transition to the post-fusion conformation is irreversible (Melikyan et al., 2000; Skehel et al., 1982). Therefore, the triggering of these envelope glycoproteins is tightly regulated, since premature transition to the post-fusion conformation renders the virus non-infectious.

Class I viral fusion proteins are the most extensively studied envelope glycoproteins (Harrison, 2015). Members of this group are translated as single polypeptides, but require cleavage by host proteases to form the metastable pre-fusion trimer of heterodimers (Bosch et al., 1981; Colman and Lawrence, 2003; Volchkov et al., 1998). The heterodimer consists of a domain that mediates binding to a host receptor, and a helical domain that promotes membrane fusion. The fusion domain contains a hydrophobic

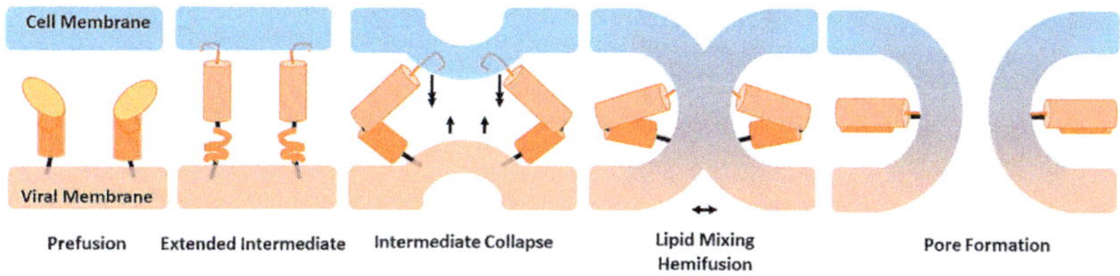

Fig. 1 A simple mechanistic model of viral fusion. The fusion of viral and cellular membranes is mediated by the fusion domains of envelope glycoprotein trimers, which undergo a dramatic refolding in response to a trigger. For clarity, only pair of monomers are shown. The glycoprotein binds to a specific receptor in the host cell membrane. Conformational changes in the fusion domain move the hydrophobic fusion peptide away from the virion surface, allowing it to insert into the host membrane forming an extended intermediate. The intermediate collapses into a helical bundle, allowing the outer leaflets of the two membranes to fuse (hemifusion). Finally, the inner leaflets fuse, forming a pore, which allows transfer of the viral genome into the host.

region, located at or near the N terminus that inserts into the target membrane during the fusion reaction (Fig. 1) (Kielian, 2014). These heterodimers form functional trimers on the surface of the virus. Upon triggering and fusion, the fusion domains undergo dramatic conformational rearrangement to a six-helical bundle structure during the formation of the fusion pore; these characteristics are believed to be universally conserved among members of the class (Harrison, 2015).

The class I fusion group features many viral envelope glycoproteins of importance, such as those of retroviruses, filoviruses (Chan et al., 2000), arenaviruses (Torriani et al., 2017), orthopneumoviruses (Rima et al., 2017), and ortho- and paramyxoviruses (Baker et al., 1999; Chen et al., 2001; Dutch, 2010), including the canonical representative of class I fusion, influenza hemagglutinin (Skehel and Wiley, 2000). These proteins share similar structural characteristics. But there are key differences between them, particularly with respect to the mechanisms of triggering. These triggers vary from simple stimuli such as the lowering of pH, which occurs in the endocytic compartment of the cell during virus entry, or interaction with a cellular ligand, to mechanisms that remain incompletely understood (Kielian, 2014; Skehel et al., 1982; White et al., 2008). In all cases, however, methods poised to elucidate the dynamic events on timescales relevant to triggering and fusion have been lacking.

What has become well established is that upon triggering, class I fusion glycoproteins undergo a significant conformational change from the metastable pre-fusion conformation to the highly stable post-fusion helical

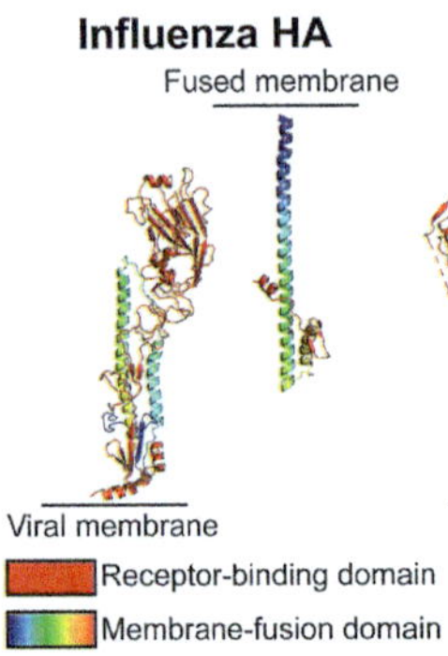

Fig. 2 Conserved structural features of class-I viral envelope glycoproteins. All class-I viral envelope glycoproteins are trimers of heterodimers on the surface of the virion. Each heterodimer consists of a receptor-binding domain and a membrane fusion domain. Shown here are individual heterodimers from the influenza, Ebola, HIV-1, and Lassa viruses in pre-fusion and post-fusion conformations shown. All undergo dramatic refolding events during transition from the pre- to postfusion conformations. The coloring of the receptor-binding domains (red) and fusion domains (rainbow) demonstrates the conserved structural elements typical of the class I family of viral fusion glycoproteins. The viral membrane and the fused membranes are indicated. PDB accession codes: 2FK0 (pre-fusion influenza HA), 1HTM (post-fusion influenza HA), 5JQ3 (pre-fusion Ebola GP), 1EBO (post-fusion Ebola GP), 5FUU (pre-fusion HIV-1 Env), 1ENV (post-fusion HIV-1 Env), 5VK2 (pre-fusion Lassa GP), and 5OMI (post-fusion Lassa GP).

bundle, which facilitates fusion (Fig. 2). For many class I glycoproteins, pre-fusion (Kwon et al., 2012; McLellan et al., 2013; Stevens et al., 2006; Wilson et al., 1981; Yin et al., 2006b; Zhao et al., 2016) and post-fusion (Bullough et al., 1994; McLellan et al., 2011; Swanson et al., 2010; Weissenhorn et al., 1997, 1998) crystal structures have been described at atomic resolution. The apparent differences between the pre-fusion and post-fusion structures imply that significant rearrangement of the protein is required for viral entry and fusion (Fig. 2). But the sequence of events that connect these two states is only just coming into view.

In addition to the dramatic change in conformation between the pre- and postfusion structures, recent applications of single-molecule imaging have directly demonstrated that class I fusion glycoproteins are intrinsically dynamic in their pre-fusion state, and appear to sample intermediate conformations that are on-pathway to membrane fusion (Das et al., 2018; Munro et al., 2014). Modulation of these intrinsic dynamics may therefore be critical to regulating the timing of envelope glycoprotein triggering. The existence of intermediate conformations has been long hypothesized, but remains incompletely characterized (Garcia et al., 2015; Lee et al., 2011; Stegmann et al., 1990). Static crystal structures, while deeply informative, cannot capture the transient or unstable intermediate conformations that have been hypothesized. This structural data must be supplemented by methods capable of providing time-resolved information on glycoprotein dynamics. In this regard, single-molecule methods show great promise in bridging the gaps between structure, dynamics, and function. In addition to its putative role in membrane fusion, this dynamicity provides a means of obscuring functional centers from attack by antibodies, which is a key strategy for immune evasion likely employed by many viruses (Kwong et al., 2002). In this review, we focus on current developments in single-particle and single-molecule techniques that have advanced the study of class I fusion glycoprotein-mediated viral entry (these developments have been similarly reviewed for the class II glycoprotein of flaviviruses; Sharma et al., 2018).

2. Förster resonance energy transfer (FRET)

2.1 Principal applications of FRET in viral fusion research

FRET is used widely in virology to assess protein and nucleic acid association, movement, and dynamics. FRET involves the energy transfer of directly excited donor fluorophores to proximal acceptor fluorophores, which requires that the emission spectrum of the donor and excitation spectrum of the acceptor are overlapping (Ishikawa-Ankerhold et al., 2012; Stryer, 1978). The efficiency of this energy transfer, and resultant change in donor and acceptor fluorescence, is a function of the distance between the two fluorophores, according to.

$$FRET = 1/\left(1 + (R/R_0)^6\right)$$

where R is the distance between the two fluorophores, and R_0 is the Förster distance, which determines the scale on which FRET is a sensitive measure of inter-fluorophore distance (30–100 Å for most commonly used fluorophore pairs). For this reason, FRET is known as a "spectroscopic ruler." That is, the efficiency of energy transfer provides a readout for fluorophore proximity. Many well-known virology techniques, such as the β-lactamase (BlaM) assay for virus entry (Cavrois et al., 2002; Dale et al., 2011; Jones and Padilla-Parra, 2016; Marin et al., 2015), and some viral fusion and lipid mixing assays (Struck et al., 1981), use the fundamental principles of FRET to measure viral infectivity and fusion. This can be done by detecting direct association or dissociation of fluorescently labeled proteins (Banning et al., 2010; Jones and Padilla-Parra, 2015, 2016) or changes in concentration of fluorescent components that occur upon some stimulus or change in condition (Engel et al., 2010; Struck et al., 1981). The exceptional flexibility and versatility of FRET as a reporter of molecular distance has promoted its continued use and development.

2.2 Single-molecule FRET (smFRET) imaging of biomolecular conformational and compositional dynamics

The principal of FRET has been used to address many fundamental questions in virology, and the development of single-molecule FRET (smFRET) lends the technique more specificity and precision. Single-molecule methods in general enable researchers to probe biomolecules on an individualized basis, circumventing the averaging of ensembles of molecules, which masks asynchronous events and non-accumulating intermediates. smFRET, a technique born in biophysical investigations of nucleic acid conformational and compositional dynamics (Ha, 2001), has been a crossover success in addressing exciting questions in virology. smFRET allows the targeted measurement of the relative movement of a single fluorophore pair site-specifically attached to the protein or nucleic acid of interest (Fig. 3) (Lerner et al., 2018; van den Wildenberg et al., 2011).

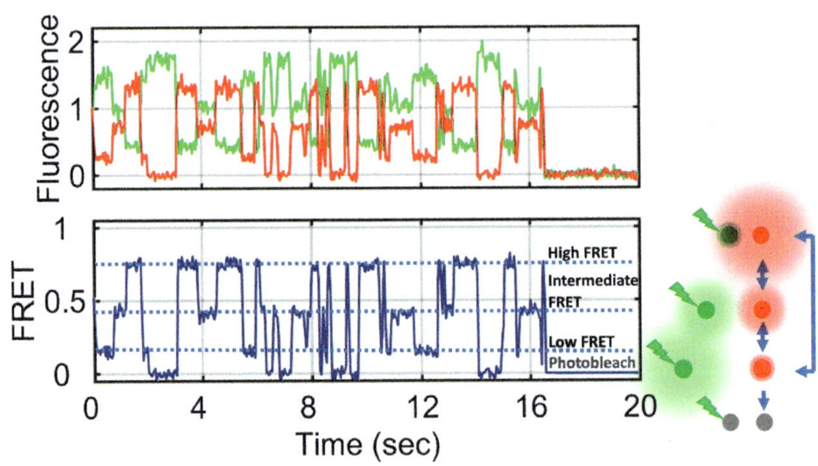

Fig. 3 Example trajectory indicating changes in FRET efficiency resulting from movement of the donor and acceptor fluorophores. (Top) Example donor (green) and acceptor (red) fluorescence traces showing anti-correlated fluctuations in intensity, which give rise to (bottom) changes in FRET efficiency. The FRET trajectory provides a time-resolved report on the relative positions of the two fluorophores. Intrinsic instabilities of the fluorophores lead to photobleaching. Hypothetical fluorophore pair positions are shown at right.

By attaching donor and acceptor fluorophores to a single biomolecule smFRET can be used to measure conformational dynamics within that molecule. Alternatively, the fluorophores can be attached to separate molecules to monitor their association or dissociation. Immobilization of fluorescently labeled molecules on surfaces prevents them from diffusing outside of the focal plane or detection volume, thus permitting observation of individual molecules for extended periods of time (~minutes). In this case, surface-based smFRET provides trajectories of biomolecular dynamics, which gives information about both the thermodynamics and the kinetics of the underlying process.

In addition to the advantages of focusing on individual molecules over extended periods, surface-based smFRET imaging has several practical strengths as well. Implementation using wide-field total internal reflection fluorescence (TIRF) microscopy allows one to image hundreds to thousands of individual molecules simultaneously at high signal-to-noise ratio, thus providing relatively rapid and high-throughput data collection. The technique is also forgiving to incomplete or inefficient fluorescent labeling, since a FRET signal requires the presence of both a donor and acceptor fluorophore. Therefore, molecules that lack either a donor or an acceptor fluorophore are not visualized and can be easily ignored. This is especially helpful in applications aimed at observing the dynamics of viral glycoproteins on the surface of virions (see below) in which purification of labeled molecules away from unlabeled molecules is not currently feasible.

Several technical challenges complicate the application of smFRET to new biomolecular systems. A major bottleneck in the application of smFRET imaging is addressing the need to site-specifically attach donor and acceptor fluorophores to the protein or nucleic acid of interest. Conventionally, labeling proteins is accomplished through cysteine mutagenesis and subsequent fluorophore attachment with thiol-maleimide chemistry. However, application to more complex systems, such as viral envelope glycoproteins, which can contain numerous native cysteine residues, requires more sophisticated and specific techniques. To this end, enzymatic methods have been used where a target peptide, 6–12 amino acids in length, is inserted into the protein of interest. An enzyme then ligates a fluorophore-conjugate to the peptide (Lotze et al., 2016; Ma et al., 2018; Munro et al., 2014; Yin et al., 2006a). Although this is a highly specific and efficient technique, the insertion of the target peptide can interfere with the function of the protein of interest. Therefore, more minimalist modifications are advantageous. To address this limitation, researchers have utilized genetic code expansion technology whereby amber stop codons are introduced at the desired sites of fluorophore attachment. These stop codons are then used to code for non-canonical amino acids (ncAAs) with a suppressor tRNA and an engineered aminoacyl-tRNA synthetase, which has been optimized for expression in mammalian cells (Nikić et al., 2014, 2016). Fluorophores are then attached to the ncAAs by copper-free click chemistry. This approach has recently been applied to both smFRET imaging of hemagglutinin dynamics, as well as for super-resolved localization of HIV Env (Das et al., 2018; Sakin et al., 2017). Finally, fluorescence, and FRET in particular, is exquisitely sensitive to the environment of the fluorophore. Thus, changes in the environment of the fluorophores can give rise to photophysical effects that can make smFRET data less reliable and difficult to interpret as resulting from changes in conformation or bound state of the labeled protein or nucleic acid. As labeling strategies, fluorophores, and imaging modalities continue to improve, smFRET has the capacity to bring new understanding to the inter- and intramolecular dynamics involved in numerous biological processes.

2.3 smFRET imaging reveals novel conformations of class-I viral fusion machines

smFRET imaging has recently been applied to visualize the conformational changes in the viral envelope glycoproteins from influenza and HIV that underlie the mechanisms of receptor recognition, membrane fusion, and interactions with antibodies (Alsahafi et al., 2018, 2019; Das et al., 2018; Lu et al., 2019; Ma et al., 2018; Munro et al., 2014). Taken together, these studies provide direct observation of not only pre- and postfusion conformation states, but also transient intermediate states that may be important or necessary for viral entry and attack by antibodies.

Through the use of smFRET, Das and colleagues were able to address important, long-standing questions in influenza hemagglutinin (HA)-mediated membrane fusion: what are the characteristics of pre-fusion intermediate structures formed during HA-mediated fusion? and what is the nature of reversibility in the conformation of prefusion HA? (Das et al., 2018). By attaching both donor and acceptor fluorophores within the membrane fusion domain of HA (HA2, see Fig. 2) on a pseudotyped HIV core, this research was able to characterize the existence and reversibility of prefusion intermediates of HA (Fig. 4).

They found that HA exists in a dynamic equilibrium, sampling the canonical pre-fusion conformation as well as intermediate states in which

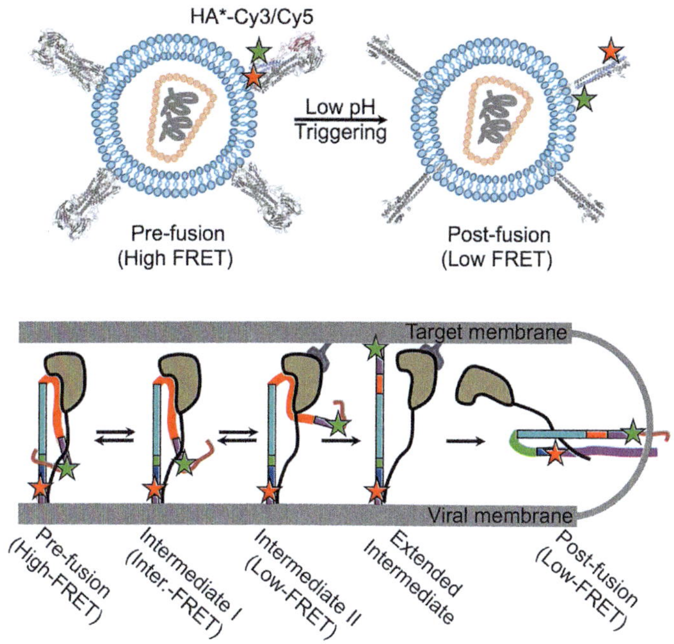

Fig. 4 A revised model of HA-mediated membrane fusion generated through smFRET imaging. Das et al. used smFRET to visualize the conformational changes in individual HA molecules on the surface of virions. (Top) Virions were formed with a single HA site-specifically labeled with donor (green star) and acceptor (red star) fluorophores. FRET efficiency then reported on the conformational changes that HA undergoes in response to triggering by acidic pH. (Bottom) These experiments led to a revised model of HA-mediated membrane fusion where the fusion peptide (brown) can adopt multiple positions prior to insertion into the target membrane.

the fusion peptide at the N-terminus of HA2 readily transitions out of a hydrophobic pocket. Acidification of pH shifted the equilibrium in favor of an intermediate conformation. These dynamics were previously predicted by hydrogen-deuterium exchange coupled to mass spectrometry (HDX-MS), which demonstrated exposure of the fusion peptide under acidic conditions (Garcia et al., 2015). After transient exposures to acidic pH, followed by reneutralization of the pH, the equilibrium returned to the predominant pre-fusion conformation, as previously predicted by bulk experiments (Leikina et al., 2002). However, upon longer exposure to acidic pH HA began to transition to the putative coiled-coil conformation indicative of the post-fusion state, which was irreversible. When liposomes containing the cellular ligand sialic acid were present, HA more rapidly adopted the coiled-coil conformation. Further analyses indicated that binding sialic acid initiated the release of the fusion peptide, whereas interaction with the target membrane stimulated transition to the irreversible coiled-coil conformation. The difference in stability of the observed pre-fusion intermediate states with and without sialic acid alludes to a functional role of the receptor during *in vivo* infection to prevent against premature triggering prior to the virus arriving at a permissive cellular membrane. The time-resolved nature of these smFRET measurements allowed the characterization and visualization of the order and timing of HA conformational changes. Furthermore, this work extends the canonical model of class-I viral fusion that describes HA triggering and fusion loop release as irreversible steps (Harrison, 2015; Kielian, 2014). The work of Das and colleagues demonstrates that long-standing scientific questions can be tackled using real-time observation of the dynamics of the individual molecules.

With a similar approach Munro and colleagues demonstrated that HIV Env samples multiple pre-fusion conformations (Munro et al., 2014). In the predominant pre-fusion "closed" conformation the receptor-binding sites are obscured, protecting HIV from attack by most antibodies (state 1). Env then transitions to an asymmetric intermediate state in which a single CD4 receptor molecule is capable of binding (state 2) (Ma et al., 2018). From here, Env can adopt a symmetric intermediate competent for interaction with three CD4 molecules and antibodies that target the co-receptor binding site (state 3); from this state viral fusion and entry putatively proceed (Herschhorn et al., 2016). Around the same time, HDX-MS data provided a more holistic structural picture, though not time resolved, of the Env dynamics induced by CD4 binding (Guttman et al., 2014). Neutralizing antibodies that target Env and make up a small minority of the total pool of antibodies in infected individuals appeared to recognize the closed state 1 and prevent subsequent transition to the observed intermediate states (Munro et al., 2014). Here again, HDX-MS data supported this conclusion (Guttman et al., 2015). This modulation of Env conformation likely relates to the mechanisms of inhibition of the neutralizing antibodies investigated.

Also recently, Alsahafi and colleagues conducted a study of a group of non-neutralizing antibodies, which target epitopes at the Env trimer interface obscured in state 1 (Alsahafi et al., 2019). These antibodies were of particular interest as they possess potent antibody-mediated cellular cytotoxicity (ADCC) activity. When combined with a small-molecule CD4 mimetic and a co-receptor-binding site antibody, the authors observed a fourth distinct Env conformation (state 2A). Antibodies with ADCC activity that were capable of inducing this conformation were also detected in

HIV-infected patient sera. Finally, aided by extensive virological assays, Alsahafi and colleagues were able to identify conditions to stabilize the novel Env conformation, which permitted structure determination by cryo-electron microscopy.

3. Super-resolution fluorescence microscopy (SRFM) in virus entry

3.1 Stochastic optical reconstruction microscopy (STORM) delineates individual molecular events during HIV entry

Light microscopy is an indispensable tool in viral fusion research (Witte et al., 2018). Indeed, all of the techniques addressed in this review are reliant on some form of light microscopy. However, the diffraction limit of light, which is minimally about 200 nm, is larger than many individual virions (Huang et al., 2009). As a result, observation of sub-virion structure and organization using light microscopy has conventionally not been possible. The development of SRFM has overcome this practical barrier, and as a result, has become widely used in virology over the last decade.

One approach to breaking the diffraction limit is reflected in two similar techniques, stochastic optical reconstruction microscopy (STORM) (Rust et al., 2006), and photoactivated localization microscopy (PALM) (Betzig et al., 2006) or similarly, fluorescence photoactivation localization microscopy (FPALM) (Hess et al., 2006). These methods rely on the fact that individual fluorophores can be localized with nanometer accuracy, far smaller than the diffraction limit of light. By imaging sparse fields of blinking fluorophores, and localizing each fluorophore individually, researchers can reconstruct images at near molecular resolution. The trade-off of this exceptional resolution is the increased time, on the scale of minutes, required to develop an image (Hanne et al., 2016). That is, the protocol requires a large number of captured images, as well as post-processing to develop a final image (Zhuang, 2009).

In a notable example of direct STORM microscopy, Iliopolou and colleagues visualized intermolecular dynamics and stoichiometry of Env:receptor binding during HIV entry into live cells (Iliopoulou et al., 2018). The authors labeled both the HIV virion and the cellular receptors, allowing them to colocalize multiple components and deduce conformational changes of Env during binding in real time (Fig. 5). They proposed a model describing the stepwise process of HIV binding CD4 and coreceptor

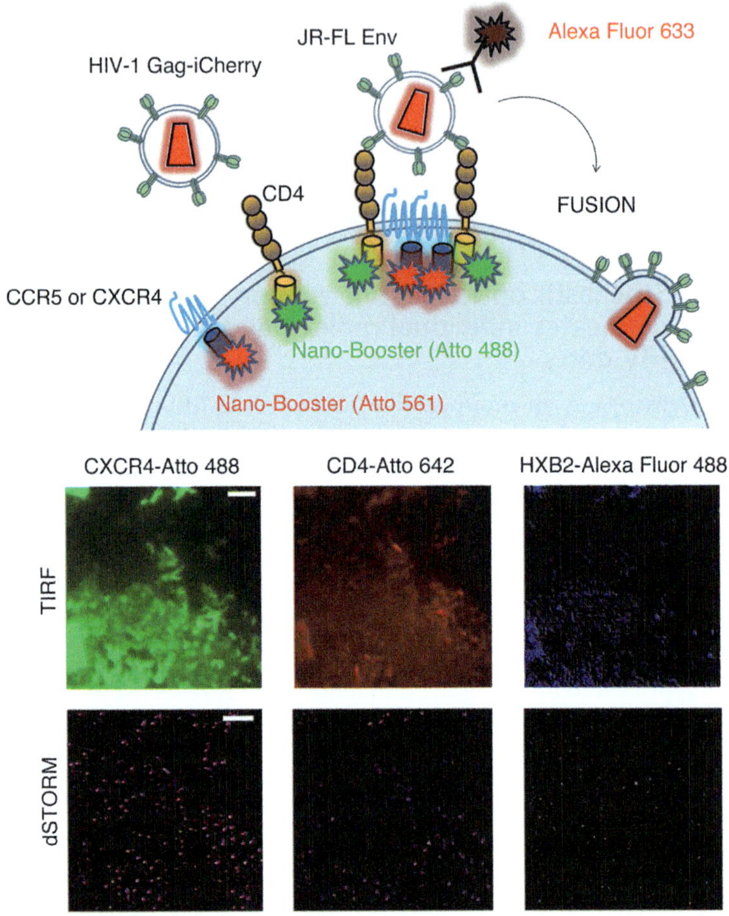

Fig. 5 Super-resolution microscopy for tracking individual HIV virions in real time during entry into cells. Iliopolou et al. used a multi-fluorophore labeling system (top) and dSTORM (bottom) to colocalize single HIV-1 virions, the receptor CD4, and the co-receptors CXCR4 or CCR5 at high resolution in real time. Either CXCR4-tropic (HXB2) or CCR5-tropic (JR-FL) HIV-1 virions were labeled with Gag-iCherry, CD4 was labeled with mRFP, and the relevant co-receptor was labeled with mYFP. Virus entry into cells was visualized using TIRF microscopy. At the same time, dSTORM imaging was enabled through labeling HIV-1 Env with antibodies bound to Alexa Fluor 633, and the receptors were labeled with anti-fluorescent protein nanobodies bound to Atto 488 or 642. The multiple fluorescent signals were colocalized to dissect the individual receptor-binding events during HIV-1 entry.

molecules sequentially during activation of Env for fusion, which includes a symmetric intermediate Env conformation bound to a single CD4 molecule, as had been suggested by smFRET experiments (Ma et al., 2018; Munro et al., 2014). They further suggest that the details of this mechanism differ slightly among laboratory-adapted and clinical HIV isolates, suggestive of altered conformational dynamics, as had also been demonstrated in smFRET experiments (Munro et al., 2014). This research was also able to provide insight into the number of Env trimers on the surface required for promoting fusion in CCR5-tropic and CXCR4-tropic HIV; there is some suggestion that the number of Env trimers required might differ for triggering by these two coreceptors. The use of SRFM in these experiments was vital to determining the localization of sub-diffraction limit sized viral structures and protein-protein interactions at a resolution that would be otherwise impossible.

3.2 Stimulated emission depletion (STED) microscopy for visualizing HIV Env organization and dynamics

Sub-diffraction limit resolution makes SRFM techniques particularly enticing in virology due to the small size of viruses, and STED has been especially useful in probing intra-virion organization and dynamics (Chojnacki and Eggeling, 2018). STED microscopy uses a cylindrical beam (the "STED beam") that is dark on the optical axis to deplete fluorescence in the outer regions of the excitation point-spread function (Willig et al., 2006). A separate beam excites the sample, while the STED beam depletes fluorophores surrounding the center excitation point (Chojnacki and Eggeling, 2018; Vicidomini et al., 2018). This selective depletion of surrounding fluorophores creates a sub-diffraction limit central excitation volume, which is then scanned across the desired field of view as in normal scanning confocal microscopy. STED does not require post-processing of images, unlike the other forms of SRFM (Chojnacki and Eggeling, 2018). Though STED is generally not as high resolution as the other SRFM methods, it still averages around 20–40 nm resolution with a temporal resolution many times faster than the other methods (Chojnacki and Eggeling, 2018; Hanne et al., 2016). The speed of STED makes is particularly suited to study viral infection of live cells (Chojnacki and Eggeling, 2018; Chojnacki et al., 2017). However, the powerful lasers required for STED may induce deleterious phototoxic effects. But successful live-cell virology experiments have been demonstrated in the literature (Chojnacki et al., 2017; Sakin et al., 2017).

After newly assembled HIV virions bud from a host cell, they must undergo maturation before they are competent to enter a new cell. Maturation occurs when the viral protease becomes activated and cleaves the viral Gag and Gag-Pol polyproteins into their constituent parts. This leads to a dramatic reorganization of the viral capsid (Fig. 6). Chojnacki and colleagues

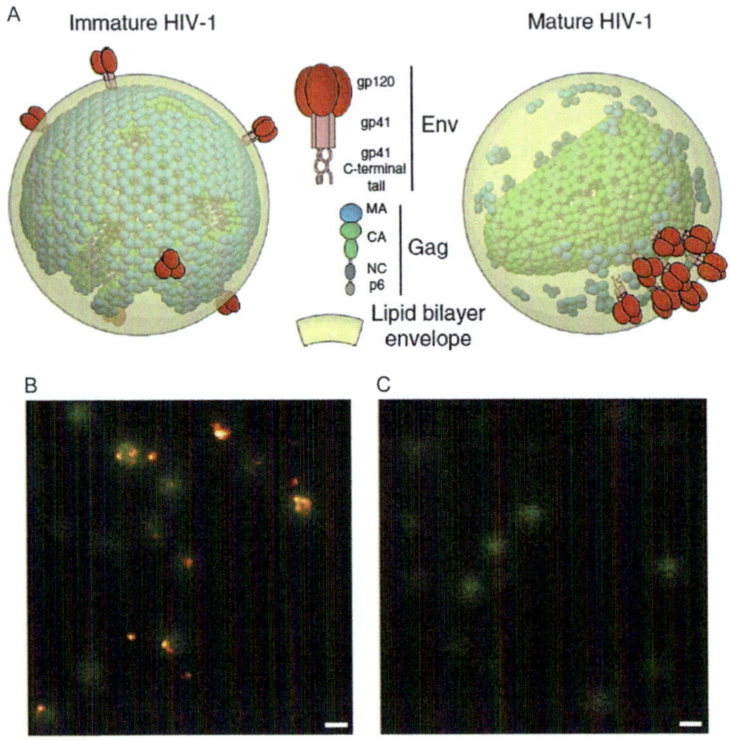

Fig. 6 STED microscopy for tracking Env motility on the surface of HIV-1 virions. Chojnacki et al. used fluorescence correlation spectroscopy on a STED microscope to detect the diffusion of individual Env trimers on the surface of immature and mature HIV-1 virions. (A) Cartoon representation of immature and mature HIV-1 virions with viral components indicated. (B) The interior of HIV-1 virions was labeled with eGFP-Vpr fusion (green) and colocalized with Env trimers stained with antibodies labeled with organic fluorophores (orange). (C) Virions lacking Env showed no immunostaining. This approach permitted detection of differential motilities of Env on the surface of mature and immature virions. From this data, the authors proposed that virus maturation allows Env to diffuse on the surface of the virion, forming clusters that enable efficient virus fusion with the cell membrane.

explored the differential distribution and mobility of Env on the immature and mature viral surface prior to entry (Chojnacki et al., 2012, 2017). First, the authors sought to demonstrate whether the distribution of Env on the virion changed after virus maturation, and whether this could explain the infectivity phenotype of immature virions. Using STED microscopy, they found that Env on mature virions clustered at the point of virus-cell contact, although the Envs had been dispersed across the virion surface prior to attachment. This clearly implies that Env is dynamic on the surface of the virion. In contrast, on immature virions Env failed to cluster at the point of attachment. This suggested that Env molecules are not motile prior to maturation, which could explain why maturation is necessary for Env-mediated membrane fusion. These results corroborated earlier studies with cryo-electron microscopy (Liu et al., 2008; Sougrat et al., 2007).

In a second study, Chojnacki and colleagues directly compare the mobility of Env on mature and immature virions using a combination of STED and scanning fluorescence correlation spectroscopy (Fig. 6). As suggested by the earlier study they demonstrate that Env motility is indeed significantly lower on the surface of immature virions—almost as attenuated as Env motility on the surface of fixed cells. Since glycoprotein clustering is widely believed to be necessary for Env-mediated fusion, this lack of Env motility on the surface of immature HIV virions likely contributes to the lack of infectivity. Though this study does not directly address conformational dynamics during entry or fusion, these experiments may predict that super-resolved observation of the viral entry pathway may soon be possible.

4. Viral membrane fusion

4.1 Fluorescence-based detection of membrane fusion in bulk

Bulk fluorescence assays have long been used to study lipid mixing that occurs during viral fusion. Such assays utilize lipophilic dyes that spontaneously incorporate into membranes. In some cases, the fluorophore can be incorporated at sufficient concentration to self-quench the fluorescence. Upon fusion with an unlabeled membrane the fluorescence increases due to dequenching (Hoekstra et al., 1984; Pal et al., 1988). Alternatively, both membranes can be labeled at sub-quenching concentrations such that FRET occurs due to mixing of the two fluorophores in the fused membrane (Domanska et al., 2013; Struck et al., 1981). Bulk measurements of fluorescence are still widely used due to their ease and robustness, but they face the same shortfalls as any ensemble measurement. Specifically, the membrane fusion process consists of several steps, and bulk assays cannot easily distinguish transient intermediates during asynchronous fusion events. Furthermore, large scale syncing of fusion over a population in bulk is effectively impossible. Furthermore, lipid mixing assays as described here cannot differentiate between hemifusion and pore formation (Floyd et al., 2008; Wessels et al., 2007).

4.2 Visualizing fusion of individual virions to target membranes

To address the shortcomings in bulk measurements of membrane fusion, single virion fusion assays have been developed (Wessels et al., 2007). As with single-molecule fluorescence assays, single-virion fusion is free from ensemble averaging, which masks transient intermediates. Compared to the other single-particle and single-molecule techniques covered in this review, single-particle fusion is well-established in the virology field. Single-particle fusion has been used to study membrane interactions and fusion of influenza to live-cells, and *in vitro* to immobilized liposomes or planar bilayers supported by a glass surface (Floyd et al., 2008; Lakadamyali et al., 2004; Padilla-Parra et al., 2012; Rawle et al., 2016; Wessels et al., 2007). Elegant extensions on these assays utilized proton uncaging by a photoisomerizable molecule to synchronize fusion events, and planar bilayers formed from cellular membranes that contain protein receptors, which permits investigations of more complex fusion reactions (Costello et al., 2012, 2013). Furthermore, single virion fusion assays can also incorporate additional fluorescence signals that report on viral content release, which made possible distinguishing between hemifusion and fusion pore formation, as well as directly correlating fusion events with endosomal pH in live cells (de la Vega et al., 2011; Floyd et al., 2008; Jones and Padilla-Parra, 2015; Miyauchi et al., 2009; Padilla-Parra et al., 2012). For the first time single-particle fusion assays permitted detailed analysis of the kinetics of fusion, which provides information on the number of rate-limiting steps and short-lived intermediates (Floyd et al., 2008; Ivanovic et al., 2013).

In a particularly insightful application of single-virion fusion, Ivanovic and colleagues monitored hemifusion of individual influenza virions to supported planar bilayers (Ivanovic et al., 2013). They observed that the virions diffused on the planar bilayer for an interval of time before arresting. Hemifusion always occurred following virion arrest. Mutational analysis supported the notion that arrest of the virion resulted from insertion of the fusion peptide into the planar bilayer. Kinetic analyses, supported by stochastic simulations, led to the conclusion that hemifusion required the cooperative action of 3–4 HA trimers that transition through an intermediate conformation in which the fusion peptide in embedded in the target membrane. This study provides a prime example of how quantitative analysis of single-particle fusion data can result in deep mechanistic insights unattainable in bulk measurements.

5. Live-cell single-virus tracking

5.1 Quantum dots (QD) and single-particle tracking (SPT)

QDs are fluorescent inorganic nanoparticles that have become a prominent tool in SPT assays. Their small 10–20 nm size, adjustable emission spectra, and broad excitation spectra have made them a versatile component of single- and multifluorophore imaging experiments. When the potential of QD for live-cell biological application was first heralded, their strengths, namely, signal intensity, photostability, nontoxicity, and wide spectral range were equally balanced by limitations like large size, relative to organic fluorophores and fluorescence proteins, and limited conjugation methods (Jaiswal and Simon, 2004). It was also unclear whether single-QD tracking would be possible. Since then, single-QD tracking has been demonstrated *in vitro* (Li et al., 2008), and many examples of single-virion tracking using QD with a variety of labeling and incorporation methods have been shown in the literature (Joo et al., 2008; Li et al., 2017; Liu et al., 2012; Qin et al., 2019; Zhang et al., 2013). Initially, QD labeling was primarily restricted to viral surface proteins (Joo et al., 2008; Liu et al., 2012). But it is now possible to label nucleic acids or nucleoproteins with QDs, or encapsulate QD-protein conjugates into virions and track virus entry and uncoating (Li et al., 2017; Qin et al., 2019; Zhang et al., 2013). This significantly increases the versatility of this SPT method.

5.2 A novel use of QD viral tracking in recent literature

In a recent novel application of QD tracking, Qin and colleagues mapped influenza virus infection, beginning with membrane fusion, through uncoating and disassembly of the capsid, and viral genome diffusion at high time resolution (Qin et al., 2019). The authors genetically fused a biotinylation sequence to the viral PA protein. They then transfected infected cells with BirA biotin ligase to facilitate biotinylation of PA. They then transfected streptavidin-coated QDs into the infected cells leading to their conjugation to the biotinylated PA protein. In the end, PA-coated QDs were encapsidated into newly formed virions. Virions containing PA-QD conjugates were then purified from the culture supernatant. This elegant labeling strategy permitted direct visualization of virion entry, uncoating, and genomic diffusion in a live-cell context for the first time, further characterizing the influenza entry pathway. They observed that influenza virus commonly fuses in perinuclear late endosomes, at which point the viral RNPs transit to the nucleus individually and display diverse dynamics once inside the nucleus. This novel approach will undoubtably find great utility in further visualization studies of virus entry and uncoating.

6. Conclusions: Future advances in single-particle techniques in viral fusion

In this review, we have covered many recent scientific developments in class I viral fusion that have utilized developments in single-molecule and single-particle technology to advance the field. Of note are the fluorescence-based techniques used to address these cutting-edge questions in virology (Fig. 7). At the center of these sophisticated fluorescence-based approaches are advances in labeling strategies of viral and cellular proteins that allow further exploration of viral fusion. Though the size of fluorophores and labels has decreased in recent history, the size and strategy of fluorescent labeling remains a potential limitation for (Fig. 7). The small size of viral proteins inherently limits the resolution and efficacy of these labeling strategies,

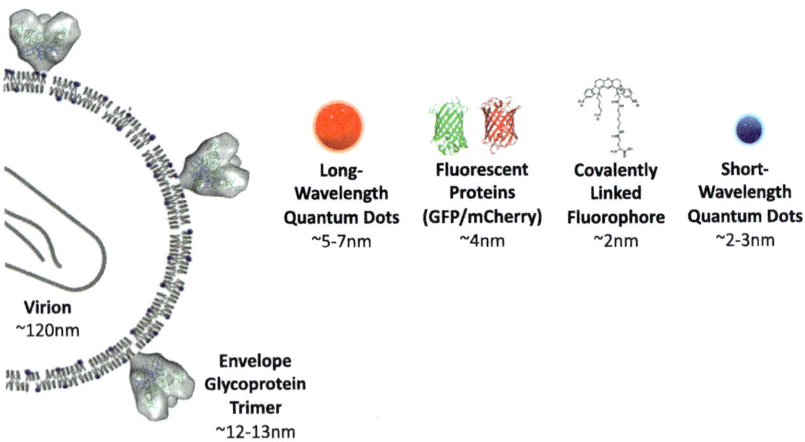

Fig. 7 Comparative size of fluorescent labeling strategies. The relative size (in nm) of the fluorescent probes used in various popular labeling strategies. Measurements were estimated using PyMOL; objects are not to scale. Though the labels most commonly used in single-virion entry experiments are smaller than an Env trimer, most are comparable in size of to a HIV Env monomer. Fluorophore sizes and labeling strategies remains a challenge in the application of single-molecule methods to studies of virus entry. PDB accession codes: GFP, 4KW4; mCherry, 2H5Q.

particularly in maintaining function of these labeled viral proteins (Sakin et al., 2016). As labeling methods become more sophisticated, we will be able to address more complex questions in viral entry dynamics.

These developments discussed here in single-molecule and single-virion technology have allowed the characterization of dynamic events during virus entry that were impossible to examine before these advances. These events include pre-fusion conformational dynamics of viral glycoproteins and co-receptors (Das et al., 2018; Iliopoulou et al., 2018; Ma et al., 2018; Munro et al., 2014), high-resolution single-virion tracking (Qin et al., 2019), and real-time viral protein dynamics on the viral capsid (Chojnacki et al., 2012, 2017). These developments have promoted a fuller understanding of the entry process for class I viral proteins as a whole, and will undoubtably continue to elucidate new biology in the years to come.

References

Alsahafi, N., Anand, S.P., Castillo-Menendez, L., Verly, M.M., Medjahed, H., Prévost, J., Herschhorn, A., Richard, J., Schön, A., Melillo, B., Freire, E., Smith, A.B., Sodroski, J., Finzi, A., 2018. SOSIP changes affect human immunodeficiency virus (HIV-1) envelope glycoprotein conformation and CD4 engagement. J. Virol. 92. e01080-18.

Alsahafi, N., Bakouche, N., Kazemi, M., Richard, J., Ding, S., Bhattacharyya, S., Das, D., Anand, S.P., Prévost, J., Tolbert, W.D., Lu, H., Medjahed, H., Gendron-Lepage, G., Ortega Delgado, G.G., Kirk, S., Melillo, B., Mothes, W., Sodroski, J., Smith, A.B., Kaufmann, D.E., Wu, X., Pazgier, M., Rouiller, I., Finzi, A., Munro, J.B., 2019. An asymmetric opening of HIV-1 envelope mediates antibody-dependent cellular cytotoxicity. Cell Host Microbe 25. 578–587.e5.

Baker, K.A., Dutch, R.E., Lamb, R.A., Jardetzky, T.S., 1999. Structural basis for paramyxovirus-mediated membrane fusion. Mol. Cell 3, 309–319.

Banning, C., Votteler, J., Hoffmann, D., Koppensteiner, H., Warmer, M., Reimer, R., Kirchhoff, F., Schubert, U., Hauber, J., Schindler, M., 2010. A flow cytometry-based FRET assay to identify and analyse protein-protein interactions in living cells. PLoS ONE 5, e9344.

Betzig, E., Patterson, G.H., Sougrat, R., Lindwasser, O.W., Olenych, S., Bonifacino, J.S., Davidson, M.W., Lippincott-Schwartz, J., Hess, H.F., 2006. Imaging intracellular fluorescent proteins at nanometer resolution. Science 313, 1642–1645.

Bosch, F.X., Garten, W., Klenk, H.-D., Rott, R., 1981. Proteolytic cleavage of influenza virus hemagglutinins: primary structure of the connecting peptide between HA1 and HA2 determines proteolytic cleavability and pathogenicity of avian influenza viruses. Virology 113, 725–735.

Bullough, P.A., Hughson, F.M., Skehel, J.J., Wiley, D.C., 1994. Structure of influenza haemagglutinin at the pH of membrane fusion. Nature 371, 37.

Cavrois, M., De Noronha, C., Greene, W.C., 2002. A sensitive and specific enzyme-based assay detecting HIV-1 virion fusion in primary T lymphocytes. Nat. Biotechnol. 20, 1151–1154.

Chan, S.Y., Speck, R.F., Ma, M.C., Goldsmith, M.A., 2000. Distinct mechanisms of entry by envelope glycoproteins of Marburg and Ebola (Zaire) viruses. J. Virol. 74, 4933–4937.

Chen, L., Gorman, J.J., McKimm-Breschkin, J., Lawrence, L.J., Tulloch, P.A., Smith, B.J., Colman, P.M., Lawrence, M.C., 2001. The structure of the fusion glycoprotein of

newcastle disease virus suggests a novel paradigm for the molecular mechanism of membrane fusion. Structure 9, 255–266.
Chojnacki, J., Eggeling, C., 2018. Super-resolution fluorescence microscopy studies of human immunodeficiency virus. Retrovirology 15, 41.
Chojnacki, J., Staudt, T., Glass, B., Bingen, P., Engelhardt, J., Anders, M., Schneider, J., Müller, B., Hell, S.W., Kräusslich, H.-G., 2012. Maturation-dependent HIV-1 surface protein redistribution revealed by fluorescence nanoscopy. Science 338, 524–528.
Chojnacki, J., Waithe, D., Carravilla, P., Huarte, N., Galiani, S., Enderlein, J., Eggeling, C., 2017. Envelope glycoprotein mobility on HIV-1 particles depends on the virus maturation state. Nat. Commun. 8, 545.
Colman, P.M., Lawrence, M.C., 2003. The structural biology of type I viral membrane fusion. Nat. Rev. Mol. Cell Biol. 4, 309–319.
Costello, D.A., Lee, D.W., Drewes, J., Vasquez, K.A., Kisler, K., Wiesner, U., Pollack, L., Whittaker, G.R., Daniel, S., 2012. Influenza virus-membrane fusion triggered by proton uncaging for single particle studies of fusion kinetics. Anal. Chem. 84, 8480–8489.
Costello, D.A., Hsia, C.-Y., Millet, J.K., Porri, T., Daniel, S., 2013. Membrane fusion-competent virus-like proteoliposomes and proteinaceous supported bilayers made directly from cell plasma membranes. Langmuir 29, 6409–6419.
Dale, B.M., McNerney, G.P., Thompson, D.L., Hubner, W., de los Reyes, K., Chuang, F.Y.S., Huser, T., Chen, B.K., 2011. Cell-to-cell transfer of HIV-1 via virological synapses leads to endosomal virion maturation that activates viral membrane fusion. Cell Host Microbe 10, 551–562.
Das, D.K., Govindan, R., Nikić-Spiegel, I., Krammer, F., Lemke, E.A., Munro, J.B., 2018. Direct visualization of the conformational dynamics of single influenza hemagglutinin trimers. Cell 174.926–937.e12.
de la Vega, M., Marin, M., Kondo, N., Miyauchi, K., Kim, Y., Epand, R.F., Epand, R.M., Melikyan, G.B., 2011. Inhibition of HIV-1 endocytosis allows lipid mixing at the plasma membrane, but not complete fusion. Retrovirology 8, 99.
Domanska, M.K., Wrona, D., Kasson, P.M., 2013. Multiphasic effects of cholesterol on influenza fusion kinetics reflect multiple mechanistic roles. Biophys. J. 105, 1383–1387.
Dutch, R.E., 2010. Entry and fusion of emerging paramyxoviruses. PLoS Pathog. 6, e1000881.
Engel, S., Scolari, S., Thaa, B., Krebs, N., Korte, T., Herrmann, A., Veit, M., 2010. FLIM-FRET and FRAP reveal association of influenza virus haemagglutinin with membrane rafts. Biochem. J. 425, 567–573.
Floyd, D.L., Ragains, J.R., Skehel, J.J., Harrison, S.C., van Oijen, A.M., 2008. Single-particle kinetics of influenza virus membrane fusion. Proc. Natl. Acad. Sci. U. S. A. 105, 15382–15387.
Garcia, N.K., Guttman, M., Ebner, J.L., Lee, K.K., 2015. Dynamic changes during acid-induced activation of influenza hemagglutinin. Structure 23, 665–676.
Guttman, M., Garcia, N.K., Cupo, A., Matsui, T., Julien, J.-P., Sanders, R.W., Wilson, I.A., Moore, J.P., Lee, K.K., 2014. CD4-induced activation in a soluble HIV-1 Env trimer. Structure 1993 (22), 974–984.
Guttman, M., Cupo, A., Julien, J.-P., Sanders, R.W., Wilson, I.A., Moore, J.P., Lee, K.K., 2015. Antibody potency relates to the ability to recognize the closed, pre-fusion form of HIV Env. Nat. Commun. 6, 6144.
Ha, T., 2001. Single-molecule fluorescence methods for the study of nucleic acids. Curr. Opin. Struct. Biol. 11, 287–292.
Hanne, J., Zila, V., Heilemann, M., Müller, B., Kräusslich, H.-G., 2016. Super-resolved insights into human immunodeficiency virus biology. FEBS Lett. 590, 1858–1876.
Harrison, S.C., 2015. Viral membrane fusion. Virology 479–480, 498–507. 60th Anniversary Issue.

Herschhorn, A., Ma, X., Gu, C., Ventura, J.D., Castillo-Menendez, L., Melillo, B., Terry, D.S., Smith, A.B., Blanchard, S.C., Munro, J.B., Mothes, W., Finzi, A., Sodroski, J., 2016. Release of gp120 restraints leads to an entry-competent intermediate state of the HIV-1 envelope glycoproteins. MBio 7, e01598–e01616.

Hess, S.T., Girirajan, T.P.K., Mason, M.D., 2006. Ultra-high resolution imaging by fluorescence photoactivation localization microscopy. Biophys. J. 91, 4258–4272.

Hoekstra, D., De Boer, T., Klappe, K., Wilschut, J., 1984. Fluorescence method for measuring the kinetics of fusion between biological membranes. Biochemistry (Mosc.) 23, 5675–5681.

Huang, B., Bates, M., Zhuang, X., 2009. Super-resolution fluorescence microscopy. Annu. Rev. Biochem. 78, 993–1016.

Iliopoulou, M., Nolan, R., Alvarez, L., Watanabe, Y., Coomer, C.A., Jakobsdottir, G.M., Bowden, T.A., Padilla-Parra, S., 2018. A dynamic three-step mechanism drives the HIV-1 pre-fusion reaction. Nat. Struct. Mol. Biol. 25, 814.

Ishikawa-Ankerhold, H.C., Ankerhold, R., Drummen, G.P.C., 2012. Advanced fluorescence microscopy techniques—FRAP, FLIP, FLAP, FRET and FLIM. Molecules 17, 4047–4132.

Ivanovic, T., Choi, J.L., Whelan, S.P., van Oijen, A.M., Harrison, S.C., 2013. Influenza-virus membrane fusion by cooperative fold-back of stochastically induced hemagglutinin intermediates. Elife 2, e00333.

Jaiswal, J.K., Simon, S.M., 2004. Potentials and pitfalls of fluorescent quantum dots for biological imaging. Trends Cell Biol. 14, 497–504.

Jones, D.M., Padilla-Parra, S., 2015. Imaging real-time HIV-1 virion fusion with FRET-based biosensors. Sci. Rep. 5 13449.

Jones, D.M., Padilla-Parra, S., 2016. The β-lactamase assay: harnessing a FRET biosensor to analyse viral fusion mechanisms. Sensors 16, 950.

Joo, K.-I., Lei, Y., Lee, C.-L., Lo, J., Hamm-Alvarez, J.X.S.F., Wang, P., 2008. Site-specific labeling of enveloped viruses with quantum dots for single virus tracking. ACS Nano 2, 1553–1562.

Kielian, M., 2014. Mechanisms of virus membrane fusion proteins. Annu. Rev. Virol. 1, 171–189.

Kwon, Y.D., Finzi, A., Wu, X., Dogo-Isonagie, C., Lee, L.K., Moore, L.R., Schmidt, S.D., Stuckey, J., Yang, Y., Zhou, T., Zhu, J., Vicic, D.A., Debnath, A.K., Shapiro, L., Bewley, C.A., Mascola, J.R., Sodroski, J.G., Kwong, P.D., 2012. Unliganded HIV-1 gp120 core structures assume the CD4-bound conformation with regulation by quaternary interactions and variable loops. Proc. Natl. Acad. Sci. U. S. A. 109, 5663–5668.

Kwong, P.D., Doyle, M.L., Casper, D.J., Cicala, C., Leavitt, S.A., Majeed, S., Steenbeke, T.D., Venturi, M., Chaiken, I., Fung, M., Katinger, H., Parren, P.W.I.H., Robinson, J., Van Ryk, D., Wang, L., Burton, D.R., Freire, E., Wyatt, R., Sodroski, J., Hendrickson, W.A., Arthos, J., 2002. HIV-1 evades antibody-mediated neutralization through conformational masking of receptor-binding sites. Nature 420, 678–682.

Lakadamyali, M., Rust, M.J., Zhuang, X., 2004. Endocytosis of influenza viruses. Microbes Infect. 6, 929–936.

Lee, K.K., Pessi, A., Gui, L., Santoprete, A., Talekar, A., Moscona, A., Porotto, M., 2011. Capturing a fusion intermediate of influenza hemagglutinin with a cholesterol-conjugated peptide, a new antiviral strategy for influenza virus. J. Biol. Chem. 286, 42141–42149.

Leikina, E., Ramos, C., Markovic, I., Zimmerberg, J., Chernomordik, L.V., 2002. Reversible stages of the low-pH-triggered conformational change in influenza virus hemagglutinin. EMBO J. 21, 5701–5710.

Lerner, E., Cordes, T., Ingargiola, A., Alhadid, Y., Chung, S., Michalet, X., Weiss, S., 2018. Toward dynamic structural biology: two decades of single-molecule Förster resonance energy transfer. Science 359 eaan1133.

Li, Q., Han, R., Meng, X., Gai, H., Yeung, E.S., 2008. Tracking single quantum dot and its spectrum in free solution with controllable thermal diffusion suppression. Anal. Biochem. 377, 176–181.

Li, Q., Li, W., Yin, W., Guo, J., Zhang, Z.-P., Zeng, D., Zhang, X., Wu, Y., Zhang, X.-E., Cui, Z., 2017. Single-particle tracking of human immunodeficiency virus type 1 productive entry into human primary macrophages. ACS Nano 11, 3890–3903.

Liu, J., Bartesaghi, A., Borgnia, M.J., Sapiro, G., Subramaniam, S., 2008. Molecular architecture of native HIV-1 gp120 trimers. Nature 455, 109–113.

Liu, S.-L., Tian, Z.-Q., Zhang, Z.-L., Wu, Q.-M., Zhao, H.-S., Ren, B., Pang, D.-W., 2012. High-efficiency dual labeling of influenza virus for single-virus imaging. Biomaterials 33, 7828–7833.

Lotze, J., Reinhardt, U., Seitz, O., Beck-Sickinger, A.G., 2016. Peptide-tags for site-specific protein labelling in vitro and in vivo. Mol. Biosyst. 12, 1731–1745.

Lu, M., Ma, X., Castillo-Menendez, L.R., Gorman, J., Alsahafi, N., Ermel, U., Terry, D.S., Chambers, M., Peng, D., Zhang, B., Zhou, T., Reichard, N., Wang, K., Grover, J.R., Carman, B.P., Gardner, M.R., Nikić-Spiegel, I., Sugawara, A., Arthos, J., Lemke, E.A., Smith, A.B., Farzan, M., Abrams, C., Munro, J.B., McDermott, A.B., Finzi, A., Kwong, P.D., Blanchard, S.C., Sodroski, J.G., Mothes, W., 2019. Associating HIV-1 envelope glycoprotein structures with states on the virus observed by smFRET. Nature 568, 415.

Ma, X., Lu, M., Gorman, J., Terry, D.S., Hong, X., Zhou, Z., Zhao, H., Altman, R.B., Arthos, J., Blanchard, S.C., Kwong, P.D., Munro, J.B., Mothes, W., 2018. HIV-1 Env trimer opens through an asymmetric intermediate in which individual protomers adopt distinct conformations. Elife 7, e34271.

Marin, M., Du, Y., Giroud, C., Kim, J.H., Qui, M., Fu, H., Melikyan, G.B., 2015. High-throughput HIV–cell fusion assay for discovery of virus entry inhibitors. Assay Drug Dev. Technol. 13, 155–166.

McLellan, J.S., Yang, Y., Graham, B.S., Kwong, P.D., 2011. Structure of respiratory syncytial virus fusion glycoprotein in the postfusion conformation reveals preservation of neutralizing epitopes. J. Virol. 85, 7788–7796.

McLellan, J.S., Chen, M., Leung, S., Graepel, K.W., Du, X., Yang, Y., Zhou, T., Baxa, U., Yasuda, E., Beaumont, T., Kumar, A., Modjarrad, K., Zheng, Z., Zhao, M., Xia, N., Kwong, P.D., Graham, B.S., 2013. Structure of RSV fusion glycoprotein trimer bound to a prefusion-specific neutralizing antibody. Science 340, 1113–1117.

Melikyan, G.B., Markosyan, R.M., Hemmati, H., Delmedico, M.K., Lambert, D.M., Cohen, F.S., 2000. Evidence that the transition of HIV-1 Gp41 into a six-helix bundle, not the bundle configuration, induces membrane fusion. J. Cell Biol. 151, 413–424.

Miyauchi, K., Kim, Y., Latinovic, O., Morozov, V., Melikyan, G.B., 2009. HIV enters cells via endocytosis and dynamin-dependent fusion with endosomes. Cell 137, 433–444.

Munro, J.B., Gorman, J., Ma, X., Zhou, Z., Arthos, J., Burton, D.R., Koff, W.C., Courter, J.R., Smith, A.B., Kwong, P.D., Blanchard, S.C., Mothes, W., 2014. Conformational dynamics of single HIV-1 envelope trimers on the surface of native virions. Science 346, 759–763.

Nikić, I., Plass, T., Schraidt, O., Szymański, J., Briggs, J.A.G., Schultz, C., Lemke, E.A., 2014. Minimal tags for rapid dual-color live-cell labeling and super-resolution microscopy. Angew. Chem. Int. Ed. 53, 2245–2249.

Nikić, I., Estrada Girona, G., Kang, J.H., Paci, G., Mikhaleva, S., Koehler, C., Shymanska, N.V., Ventura Santos, C., Spitz, D., Lemke, E.A., 2016. Debugging eukaryotic genetic code expansion for site-specific click-PAINT super-resolution microscopy. Angew. Chem. Int. Ed. Engl. 55, 16172–16176.

Padilla-Parra, S., Marin, M., Kondo, N., Melikyan, G.B., 2012. Synchronized retrovirus fusion in cells expressing alternative receptor isoforms releases the viral Core into distinct sub-cellular compartments. PLoS Pathog. 8, e1002694.

Pal, R., Barenholz, Y., Wagner, R.R., 1988. Pyrene phospholipid as a biological fluorescent probe for studying fusion of virus membrane with liposomes. Biochemistry (Mosc.) 27, 30–36.

Qin, C., Li, W., Li, Q., Yin, W., Zhang, X., Zhang, Z., Zhang, X.-E., Cui, Z., 2019. Real-time dissection of dynamic uncoating of individual influenza viruses. Proc. Natl. Acad. Sci. U. S. A. 116, 2577–2582.

Rawle, R.J., Boxer, S.G., Kasson, P.M., 2016. Disentangling viral membrane fusion from receptor binding using synthetic DNA-lipid conjugates. Biophys. J. 111, 123–131.

Rima, B., Collins, P., Easton, A., Fouchier, R., Kurath, G., Lamb, R.A., Lee, B., Maisner, A., Rota, P., Wang, L., 2017. ICTV virus taxonomy profile: Pneumoviridae. J. Gen. Virol. 98, 2912–2913.

Rust, M.J., Bates, M., Zhuang, X., 2006. Sub-diffraction-limit imaging by stochastic optical reconstruction microscopy (STORM). Nat. Methods 3, 793–796.

Sakin, V., Paci, G., Lemke, E.A., Müller, B., 2016. Labeling of virus components for advanced, quantitative imaging analyses. FEBS Lett. 590, 1896–1914.

Sakin, V., Hanne, J., Dunder, J., Anders-Össwein, M., Laketa, V., Nikić, I., Kräusslich, H.-G., Lemke, E.A., Müller, B., 2017. A versatile tool for live-cell imaging and super-resolution nanoscopy studies of HIV-1 Env distribution and mobility. Cell Chem. Biol. 24. 635–645.e5.

Sharma, K.K., Marzinek, J.K., Tantirimudalige, S.N., Bond, P.J., Wohland, T., 2018. Single-molecule studies of flavivirus envelope dynamics: experiment and computation. Prog. Biophys. Mol. Biol. 143, 38–51.

Skehel, J.J., Wiley, D.C., 2000. Receptor binding and membrane fusion in virus entry: the influenza hemagglutinin. Annu. Rev. Biochem. 69, 531–569.

Skehel, J.J., Bayley, P.M., Brown, E.B., Martin, S.R., Waterfield, M.D., White, J.M., Wilson, I.A., Wiley, D.C., 1982. Changes in the conformation of influenza virus hemagglutinin at the pH optimum of virus-mediated membrane fusion. Proc. Natl. Acad. Sci. U. S. A. 79, 968–972.

Sougrat, R., Bartesaghi, A., Lifson, J.D., Bennett, A.E., Bess, J.W., Zabransky, D.J., Subramaniam, S., 2007. Electron tomography of the contact between T cells and SIV/HIV-1: implications for viral entry. PLoS Pathog. 3, e63.

Stegmann, T., White, J.M., Helenius, A., 1990. Intermediates in influenza induced membrane fusion. EMBO J. 9, 4231–4241.

Stevens, J., Blixt, O., Tumpey, T.M., Taubenberger, J.K., Paulson, J.C., Wilson, I.A., 2006. Structure and receptor specificity of the hemagglutinin from an H5N1 influenza virus. Science 312, 404–410.

Struck, D.K., Hoekstra, D., Pagano, R.E., 1981. Use of resonance energy transfer to monitor membrane fusion. Biochemistry (Mosc.) 20, 4093–4099.

Stryer, L., 1978. Fluorescence energy transfer as a spectroscopic ruler. Annu. Rev. Biochem. 47, 819–846.

Swanson, K., Wen, X., Leser, G.P., Paterson, R.G., Lamb, R.A., Jardetzky, T.S., 2010. Structure of the Newcastle disease virus F protein in the post-fusion conformation. Virology 402, 372–379.

Torriani, G., Galan-Navarro, C., Kunz, S., 2017. Lassa virus cell entry reveals new aspects of virus-host cell interaction. J. Virol. 91, e01902–e01916.

van den Wildenberg, S.M.J.L., Prevo, B., Peterman, E.J.G., 2011. A brief introduction to single-molecule fluorescence methods. In: Peterman, E.J.G., Wuite, G.J.L. (Eds.), Single

Molecule Analysis: Methods and Protocols, Methods in Molecular Biology. Humana Press, Totowa, NJ, pp. 81–99.
Vicidomini, G., Bianchini, P., Diaspro, A., 2018. STED super-resolved microscopy. Nat. Methods 15, 173.
Volchkov, V.E., Feldmann, H., Volchkova, V.A., Klenk, H.-D., 1998. Processing of the Ebola virus glycoprotein by the proprotein convertase furin. Proc. Natl. Acad. Sci. U. S. A. 95, 5762–5767.
Weissenhorn, W., Dessen, A., Harrison, S.C., Skehel, J.J., Wiley, D.C., 1997. Atomic structure of the ectodomain from HIV-1 gp41. Nature 387, 426.
Weissenhorn, W., Carfi, A., Lee, K.-H., Skehel, J.J., Wiley, D.C., 1998. Crystal structure of the Ebola virus membrane fusion subunit, GP2, from the envelope glycoprotein ectodomain. Mol. Cell 2, 605–616.
Wessels, L., Elting, M.W., Scimeca, D., Weninger, K., 2007. Rapid membrane fusion of individual virus particles with supported lipid bilayers. Biophys. J. 93, 526–538.
White, J.M., Delos, S.E., Brecher, M., Schornberg, K., 2008. Structures and mechanisms of viral membrane fusion proteins. Crit. Rev. Biochem. Mol. Biol. 43, 189–219.
Willig, K.I., Kellner, R.R., Medda, R., Hein, B., Jakobs, S., Hell, S.W., 2006. Nanoscale resolution in GFP-based microscopy. Nat. Methods 3, 721–723.
Wilson, I.A., Skehel, J.J., Wiley, D.C., 1981. Structure of the haemagglutinin membrane glycoprotein of influenza virus at 3 Å resolution. Nature 289, 366.
Witte, R., Andriasyan, V., Georgi, F., Yakimovich, A., Greber, U.F., 2018. Concepts in light microscopy of viruses. Viruses 10, 202.
Yin, J., Lin, A.J., Golan, D.E., Walsh, C.T., 2006a. Site-specific protein labeling by Sfp phosphopantetheinyl transferase. Nat. Protoc. 1, 280–285.
Yin, H.-S., Wen, X., Paterson, R.G., Lamb, R.A., Jardetzky, T.S., 2006b. Structure of the parainfluenza virus 5 F protein in its metastable, prefusion conformation. Nature 439, 38–44.
Zhang, Y., Ke, X., Zheng, Z., Zhang, C., Zhang, Z., Zhang, F., Hu, Q., He, Z., Wang, H., 2013. Encapsulating quantum dots into enveloped virus in living cells for tracking virus infection. ACS Nano 7, 3896–3904.
Zhao, Y., Ren, J., Harlos, K., Jones, D.M., Zeltina, A., Bowden, T.A., Padilla-Parra, S., Fry, E.E., Stuart, D.I., 2016. Toremifene interacts with and destabilizes the Ebola virus glycoprotein. Nature 535, 169–172.
Zhuang, X., 2009. Nano-imaging with STORM. Nat. Photonics 3, 365–367.

Further reading

Pancera, M., Zhou, T., Druz, A., Georgiev, I.S., Soto, C., Gorman, J., Huang, J., Acharya, P., Chuang, G.-Y., Ofek, G., Stewart-Jones, G.B.E., Stuckey, J., Bailer, R.T., Joyce, M.G., Louder, M.K., Tumba, N., Yang, Y., Zhang, B., Cohen, M.S., Haynes, B.F., Mascola, J.R., Morris, L., Munro, J.B., Blanchard, S.C., Mothes, W., Connors, M., Kwong, P.D., 2014. Structure and immune recognition of trimeric pre-fusion HIV-1 Env. Nature 514, 455–461.

CHAPTER FIVE

Structural and cellular biology of rhabdovirus entry

Laura Belot, Aurélie Albertini, Yves Gaudin[1]
Institute for Integrative Biology of the Cell (I2BC), CEA, CNRS, University Paris-Sud, Université Paris-Saclay, Gif-sur-Yvette, France
[1]Corresponding author: e-mail address: yves.gaudin@i2bc.paris-saclay.fr

Contents

Abstract

Rhabdoviruses are enveloped viruses with a negative-sense single strand RNA genome and are widespread among a great variety of organisms. In their membrane, they have a single glycoprotein (G) that mediates both virus attachment to cellular receptors and fusion between viral and endosomal membranes allowing viral genome release in the cytoplasm. We present structural and cellular aspects of Rhabdovirus entry into their host cell with a focus on vesicular stomatitis virus (VSV) and rabies virus (RABV) for which the early events of the viral cycle have been extensively studied. Recent data have shown that the only VSV receptors are the members of the LDL-R family. This is in contrast with RABV for which multiple receptors belonging to unrelated families have been identified. Despite having different receptors, after attachment, rhabdovirus internalization occurs through clathrin-mediated endocytosis (CME) in an actin-dependent

Advances in Virus Research, Volume 104
ISSN 0065-3527
https://doi.org/10.1016/bs.aivir.2019.05.003

manner. There are still debates about the exact endocytic pathway of VSV in the cell and on RABV transport in the neuronal axon. In any case, fusion is triggered in the endosomal vesicle via a low-pH induced structural rearrangement of G from its pre- to its postfusion conformation. Vesiculovirus G is one of the best characterized fusion glycoproteins as the previously reported crystal structures of the pre- and postfusion states have been recently completed by those of intermediates during the structural transition. Understanding the entry pathway of rhabdoviruses may have strong impact in biotechnologies as, for example, VSV G is used for pseudotyping lentiviruses to promote efficient transduction, and VSV is a promising oncolytic virus.

1. Introduction

1.1 The Rhabdoviridae family

The Rhabdoviridae family belongs to the *Mononegavirales* order. Almost all rhabdoviruses are enveloped and have an unsegmented genome made of a single-strand RNA molecule of negative polarity (i.e., complementary of mRNAs). The virion is usually bullet shaped (having a flat base and a round tip) or bacilliform (with both tips that are rounded).

The genome (ranging from 10.8 to 16.1 kb (Walker et al., 2015)) contains up to a dozen genes, among which only five are common to all members of the family. These common genes encode the nucleoprotein (N), the phosphoprotein (P), the matrix protein (M), the glycoprotein (G) and the viral polymerase (L). The genome is associated with N, L and P to form the ribonucleoprotein (RNP), which is condensed into a tightly coiled helical structure inside the viral particle. The matrix protein forms a layer between the RNP and the lipid bilayer which contains the viral glycoprotein G that constitutes the spikes that protrude at the viral surface.

Rhabdoviruses are widespread among a great diversity of organisms such as vertebrates (mammals, birds, reptiles, fishes), invertebrates (insects, arachnids, crustaceans) and plants. Several members of the Rhabdoviridae are associated with significant pathologies in humans and livestock (Dietzgen et al., 2017).

In 2018, the Rhabdoviridae included 18 genera and one unassigned species (Moussa virus) (Walker et al., 2018). This classification is largely based on phylogenetic analyses of L protein sequences. Inside a genus, viruses generally have the same genome organization, including the number and locations of accessory genes. The two most studied genera are *Lyssavirus* and *Vesiculovirus*. Lyssaviruses infect only mammals in whom they cause deadly encephalitis. The prototype of the genus is rabies virus (RABV). Vesiculoviruses (prototype: vesicular stomatitis virus -VSV-) and members of the genera *Ephemerovirus* (prototype: bovine ephemeral fever virus -BEFV-), *Tibrovirus*, *Hapavirus*, *Curiovirus*, *Sripuvirus* and *Ledantevirus* infect vertebrates but are transmitted by arthropods. Members of the genus *Tupavirus* have only been isolated from vertebrates as well as the members of the genera *Novirhabdovirus* (prototype: viral hemorrhagic septicemia virus -VHSV-), *Sprivivirus* (prototype: spring viraemia of carp virus -SVCV-) and *Perhabdovirus* which infect only fishes. Members of the genus *Almendravirus* replicate only in insects as well as the members of the genus *Sigmavirus* which infect only dipteran flies of a single species and are transmitted vertically. Finally, plant rhabdoviruses are assigned to the genera *Cytorhabdovirus*, *Nucleorhabdovirus*, *Dichorhavirus* (which have bi-segmented genomes) and *Varicosavirus* (which lack an envelope) and are transmitted by either arthropods or fungi (Walker et al., 2018).

The viral cycle of animal rhabdoviruses is initiated by the binding of their glycoprotein G to cellular receptors. Thereafter, the viral particle enters the cell via the endocytic pathway. Then, the acidic environment of the endosome induces a huge conformational change of G from a native (or prefusion) state toward a final (or postfusion) state that catalyzes fusion of the viral envelope with the endosomal membrane. This results in the release of the viral RNP in the cytoplasm for the subsequent steps of infection.

1.2 Rhabdovirus glycoprotein G

1.2.1 Basic biochemical properties

Glycoprotein G is a classic type I transmembrane protein and is synthesized on ribosomes bound to the endoplasmic reticulum (ER). After cleavage of the amino terminal signal peptide, the mature glycoprotein is usually about 500 aa long (495 for VSV and 505 for RABV). However, members of the *Ephemerovirus* genus encode mature glycoproteins which are longer, having up to 640 amino acids in the case of Obhodiang virus (Blasdell et al., 2012). It is worth noting that ephemeroviruses also encode a second nonstructural glycoprotein (G_{NS}) of unknown function, homologous to the structural one (Blasdell et al., 2012).

G is anchored in the membrane by a single α-helical transmembrane (TM) segment (Fig. 1A). The bulk of the mass of G is located outside the viral membrane and constitutes the amino-terminal ectodomain. The intraviral domain is involved in interactions with the viral matrix protein

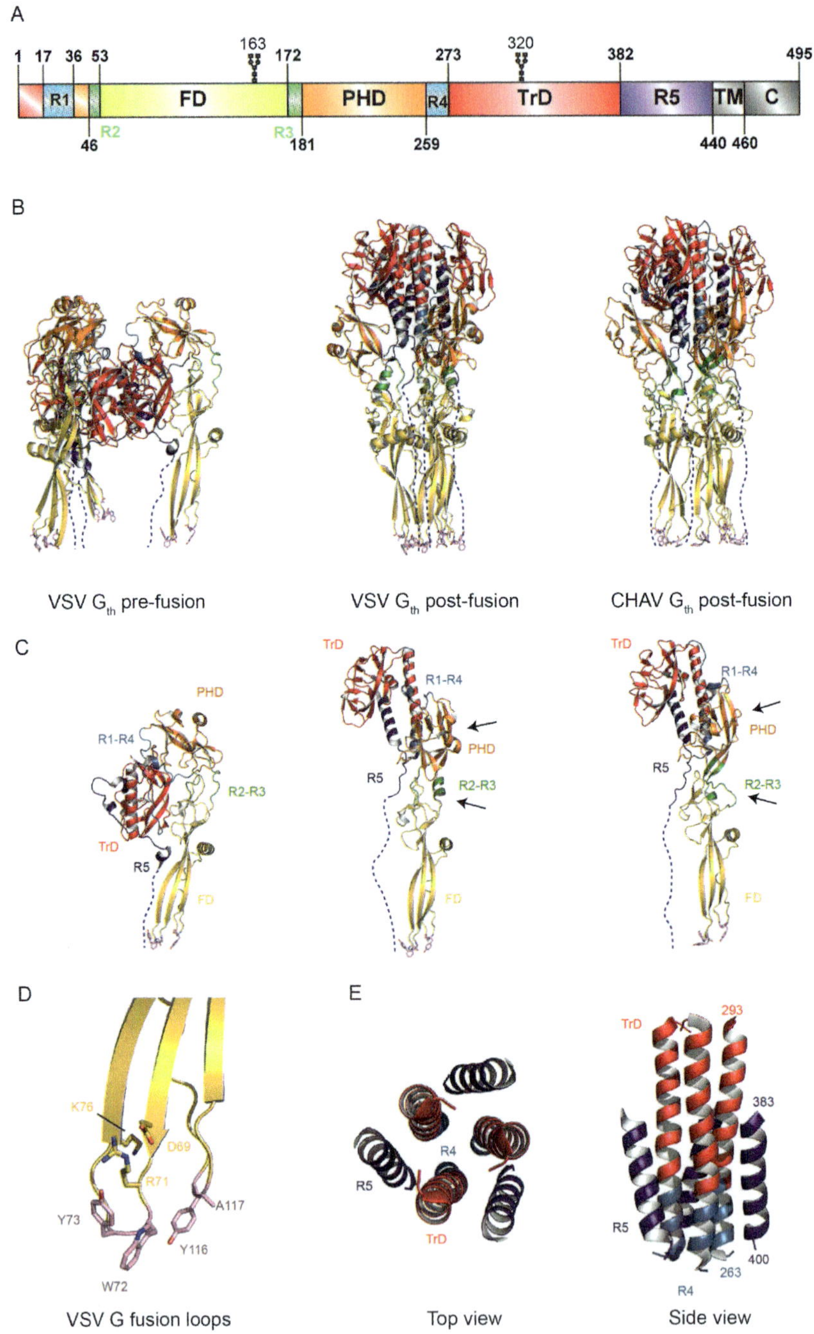

Fig. 1 See figure legend on opposite page.

(Ge et al., 2010; Lyles et al., 1992) but can also interfere with some cellular signaling pathways (Prehaud et al., 2010).

Rhabdovirus glycoproteins are N-glycosylated (Fig. 1A). The number of glycosylation sites varies from one virus to another. The oligosaccharide chains are required for correct folding of G (Gaudin, 1997; Hammond and Helenius, 1994; Mathieu et al., 1996). On the one hand, they increase the solubility of the folding intermediates in the ER (Gaudin, 1997; Mathieu et al., 1996). On the other, they favor the interaction of these intermediates with calnexin and calreticulin (Gaudin, 1997; Hammond and Helenius, 1994), which are chaperones with lectin properties.

1.2.2 Structure of G

VSV G is certainly the best characterized rhabdovirus glycoprotein (Abou-Hamdan et al., 2018; Roche et al., 2008). In the mid-2000s, the crystal structures of both the pre- and postfusion forms of the VSV G ectodomain, generated by limited proteolysis with thermolysin (G_{th}, aa residues 1–422), were determined (Roche et al., 2006, 2007). The structures of the two conformations of VSV G_{th} (Fig. 1B and C) reveal an organization distinct from those of other viral fusion proteins described so far belonging to class I (such as the influenza virus hemagglutinin -HA- and the paramyxovirus fusion protein -F-) or class II (such as the E protein of flaviviruses and E1 of alphaviruses) (Harrison, 2015). However, the comparison of the postfusion structure of VSV G_{th} with the structure of Herpes simplex virus 1 (HSV1) glycoprotein gB, which was determined at the same time (Heldwein et al., 2006), revealed that they share the same fold and,

Fig. 1 (A) Bar diagram showing the domain organization of VSV G. Glycosylation sites are indicated. Residue numbering do not take into account the signal peptide (i.e., residue 1 corresponds to the first residue of the ectodomain after cleavage of the signal peptide). (B) Ribbon diagrams of VSV G_{th} prefusion and postfusion trimers and CHAV G_{th} postfusion trimer (PDB codes: 5I2S, 5I2M and 4D6W). (C) Ribbon diagrams of VSV G_{th} prefusion and postfusion protomers and CHAV G_{th} postfusion protomer. Major structural differences between VSV G and CHAV G are indicated by black arrows. (D) Close-up view on VSV G fusion loops located at the tips of FD. Hydrophobic residues involved in membrane interaction are in pink sticks. Charged residues that limit penetration of FD inside the membrane are also represented in yellow sticks. (E) Central helix bundle of the VSV G trimer in its postfusion conformation. Color code: Trimerization domain (TrD) is in red, pleckstrin homology domain (PHD) is in orange, fusion domain (FD) is in yellow, segments R1 and R4 are in cyan, segments R2 and R3 are in green, the C-terminal segment (R5) is in purple. The dashed lines represent the missing C-terminal segment of the ectodomain that leads to the transmembrane (TM) segment.

therefore, that they have a common evolutionary origin. This fold defines the class III of viral fusion glycoproteins to which also belong the baculovirus gp64 (Kadlec et al., 2008), gB from other herpesviruses (Backovic et al., 2009; Burke and Heldwein, 2015; Chandramouli et al., 2015), and thogotovirus fusion glycoprotein (Peng et al., 2017). Amino acid sequence alignment shows that all rhabdoviruses glycoproteins have the same fold which is shared with the glycoprotein of bornaviruses (belonging to another family of *Mononegavirales*) (Garry and Garry, 2009).

The pre- and postfusion states are trimeric and the polypeptide chain of G ectodomain folds into three distinct domains (Fig. 1A–C) (Baquero et al., 2015a). The fusion domain (FD) is inserted in a loop of a pleckstrin homology domain (PHD) that is itself inserted in a trimerization domain (TrD) involved in the oligomerization of the glycoprotein in both states. Segments R1 and R4 connect TrD to PHD, while R2 and R3 connect PHD to FD. The TrD is connected to the C-terminal segment (R5) of the ectodomain. This segment itself connects to the transmembrane TM helix in the full-length glycoprotein (Baquero et al., 2015a).

The FD comprises an extended β-sheet structure terminating in a three-stranded sheet at the tip of which are located the two fusion loops which are hydrophobic motifs that interact with the target membrane during the fusion process (Fig. 1D) (Baquero et al., 2015a; Roche et al., 2006, 2007).

During the low pH-induced structural transition from pre- to postfusion state, FD, PHD and TrD retain their tertiary structure but undergo large rearrangements in their relative orientation (Fig. 1C). Globally, within the trimers, VSV G pre- and postfusion states of a protomer are related by flipping both FD and the C-terminal segment around TrD. This structural reorganization is driven by large refolding of segments R1–R5 (Baquero et al., 2015a; Roche et al., 2006, 2007).

At the end of the transition, in the postfusion state, the core of the structure is organized in a six-helix bundle, the trimeric core of which is made by the three TrD central helices (extended by the refolding of segment R4), whose grooves accommodate three lateral helices resulting in the refolding of R5 (Fig. 1E) (Baquero et al., 2015a; Roche et al., 2006, 2007).

The pre- and postfusions structures of VSV G have been observed at the viral surface by tomography either after negative staining (Libersou et al., 2010) and cryo-electron tomography (Si et al., 2018). The structures agree with the trimeric crystal structures in both conformations.

A particularity of rhabdovirus glycoprotein, and this is a major difference with other viral fusion glycoproteins, is that the low pH-induced conformational change is reversible. This was initially demonstrated for the full-length VSV G solubilized with detergents (Doms et al., 1987) and later, by electron microscopy, for RABV G at the surface of the viral particle (Gaudin et al., 1993). In fact, there is a pH-dependent equilibrium between the prefusion state, the postfusion state and monomeric conformations that are intermediates on the pathway of the structural transition (see below) (Albertini et al., 2012; Baquero et al., 2017). This equilibrium has been demonstrated for soluble forms of VSV G (Albertini et al., 2012; Lyles et al., 1990), for RABV G and VSV G at the surface of the virion (Albertini et al., 2012; Libersou et al., 2010; Roche and Gaudin, 2002), and in vivo for VSV G in the endoplasmic reticulum (Zagouras et al., 1991). Finally, even the low pH-induced insertion of the fusion domain of VSV G (Albertini et al., 2012; Kim et al., 2017; Pak et al., 1997) and Chandipura Virus (CHAV) G (Baquero et al., 2015b) in the target membrane has been demonstrated to be reversible.

1.2.3 G antigenicity and neutralizing antibodies

As rhabdovirus glycoprotein is the only viral protein with an external domain, it is the target of neutralizing antibodies. Its antigenicity has been extensively studied for both VSV and RABV.

In the case of RABV, several hundred monoclonal antibodies (Mabs) have been used to characterize the antigenic structure of G (Benmansour et al., 1991; Lafon et al., 1983; Prehaud et al., 1988; Raux et al., 1995; Seif et al., 1985). Two major antigenic sites (site II and III) have been identified. Antigenic site II is constituted by two segments of the protein, located between positions 34 and 42 and positions 198 and 200, which are clustered in the PHD of RABV G (Prehaud et al., 1988). Antigenic site III extends from amino acid 330 to 338 (Seif et al., 1985). Beside these major antigenic sites, one minor antigenic site and a few isolated epitopes have also been described (Benmansour et al., 1991; Dietzschold et al., 1990; Lafay et al., 1996; Lafon et al., 1983; Raux et al., 1995). RABV mutants resistant to MAbs neutralization often exhibit reduced pathogenicity in mice (Fallahi et al., 2016; Prehaud et al., 1988; Seif et al., 1985). This is particularly the case for mutants escaping MAbs directed against antigenic site III, of which some have an avirulent phenotype (Coulon et al., 1998; Seif et al., 1985; Tuffereau et al., 1989).

Less work has been performed on VSV G but the major epitopes have been also located in regions corresponding to the PHD and TrD (Luo et al., 1988; Munis et al., 2018; Vandepol et al., 1986). Interestingly, a comparison between VSV G and CHAV G domain structures (Fig. 1C) reveal that among the three domains of G, the PHD appears to be the most divergent and the largest differences are confined to its most exposed part. This is consistent with the idea that it is the major antigenic site of rhabdoviruses and that the main actor on G evolution is the circulating antibody mediated immune response (Baquero et al., 2015b).

The mechanism of rhabdovirus neutralization by MAbs has also been studied. Some MAbs interfere with receptor recognition, others clearly block membrane fusion (Flamand et al., 1993; Munis et al., 2018). However, in systematic studies performed on RABV, it was shown that neutralization is essentially dependent on the number of antibodies bound per virion (Flamand et al., 1993; Raux et al., 1995).

2. Rhabdovirus receptors

2.1 General

As any other viruses, in order to infect the host cell, rhabdoviruses must bind to molecules present at the cell surface. Depending on the consequences of the interaction, those molecules have been classified in two categories (Marsh and Helenius, 2006). The first category contains attachment factors. They are molecules (such as heparan sulfate or gangliosides) which interact weakly, in a nonspecific manner with the virions to concentrate them at the cell surface. The second category contains the viral receptors per se which actively trigger entry, particularly by promoting endocytic internalization.

2.2 Attachment factors of rhabdoviruses

Several rather ubiquitous molecules have long been considered to be rhabdovirus receptors. A reassessment of their role suggests that they are at best viral attachment factors.

As an example, phosphatidylserine (PS) has been proposed to be a VSV receptor (Schlegel et al., 1983). This was dubious as PS is only present at the surface of apoptotic cells and indeed it was demonstrated that it is not a receptor for VSV (Coil and Miller, 2004). Other older studies also proposed that gangliosides might play the role of the receptors in chicken embryo related (CER) cells (Sinibaldi et al., 1985). Taken together (and in view of the recent identification of real VSV receptors), those data suggest that negatively charged molecules probably concentrate VSV particles at the host cell surface.

In the case of RABV, host cell treatment with different phospholipases reduces binding, which suggests that some phospholipids can play a role at initial stages of infection (Superti et al., 1984b). Similarly, cell pretreatment with neuraminidase renders them nonsusceptible to RABV infection (Conti et al., 1986). After incorporation of exogenous gangliosides, cells recover their susceptibility to RABV infection (Superti et al., 1986). Here again, at best, those data suggest that negatively charged molecules concentrate RABV particles at the cell surface.

It is worth noting that all those attachment factors were identified with strains adapted to laboratory cell lines. In the case of RABV, except for the very beginning and end of the infectious process, non-lab-adapted isolates exclusively multiply and propagate in neurons, and in vitro, they can only infect established cell lines of neuronal origin. It is possible that the ability to bind negatively charged molecules at the host cell surface is a general consequence of adaptation and it is not clear whether the identified attachment factors are used by natural isolates during animal infection.

2.3 Rhabdovirus receptors

2.3.1 LDL receptor and the members of its family are VSV receptors

VSV has a wide host spectrum: it infects both vertebrate and insect cells. This broad tropism suggested that VSV enters cells through highly ubiquitous receptors. These receptors have been identified: they are the members of the low density lipoproteins receptor (LDL-R) family (Finkelshtein et al., 2013). The widespread expression of this receptor family, including in insects, accounts for the pantropism of VSV. It also explains why lentiviruses pseudotyped by VSV G do not allow efficient gene transfer into unstimulated T cells, B cells, and hematopoietic stem cells, as they have a very low expression level of LDL-R (Amirache et al., 2014).

The LDL-R is a type I transmembrane protein which regulates cholesterol homeostasis in mammalian cells (Brown and Goldstein, 1986). It removes cholesterol-carrying lipoproteins from plasma circulation. Ligands bound extracellularly by LDL-R at neutral pH are internalized. The intracellular release of the cargo, leading to its subsequent lysosomal degradation, is driven by a low pH-induced conformational change of LDL-R from an open to a more compact conformation (Beglova and Blacklow, 2005; Rudenko et al., 2002). The receptor then recycles back to the cell surface. LDL-R ectodomain is composed of a ligand-binding domain, an epidermal growth factor (EGF) precursor homology domain and a C-terminal domain enriched in O-linked oligosaccharides. The ligand binding domain is made of 7 cysteine-rich repeats (CR1–CR7). Each repeat is made of ~40 amino acids and contains 6 cysteine residues, engaged in 3 disulfide bridges, and an acidic residues cluster that forms an octahedral calcium cage (Rudenko et al., 2002) (Fig. 2A).

Both recombinant soluble CR2 and CR3 domains (but not the other ones) of LDL-R bind to VSV G with similar affinities (Kd ~ 5 μM) and neutralize viral infectivity when incubated with the viral inoculum prior infection (Nikolic et al., 2018).

G_{th} in complex with either CR2 or CR3 was crystallized. The binding site of CR domains on G is the same in both crystals (Fig. 2B and C). It is essentially constituted by two segments of the trimerization domain (TrD), one of the pleckstrin homology domain (PHD) and part of segment R2. The surface of interaction breaks apart upon G conformational change, which explains the inability of CR domains to bind G postfusion state (Nikolic et al., 2018).

Five residues (H8, K47, A51, Y209 and R354) of G were in direct contact with the CR domain in both complexes (Fig. 2B and C). Among those residues, K47 and R354 appear to be key for the interaction with both CR2 and CR3 as mutant glycoproteins in which they are replaced by an alanine or a glutamine are unable to bind CR domains. Remarkably, those mutant glycoproteins retain their ability to induce syncytia formation at low pH when expressed in cells by plasmid transfection. Therefore, in the case of VSV G, the fusion activity and receptor binding activity can be uncoupled (Nikolic et al., 2018).

Taking advantage of the properties of the receptor-associated protein (RAP), a common ligand of all LDL-R family members which blocks ligand binding to all LDL-R family members with the exception of LDL-R itself (Bu and Marzolo, 2000; Finkelshtein et al., 2013; Willnow, 1998), the role of other LDL-R family members as VSV receptors has been investigated. RAP was shown to significantly inhibit VSV infection in LDL-R deficient HAP cell lines but not in wild type (WT) HAP cells (Nikolic et al., 2018). Those results indicated that VSV can use other LDL-R family members as alternative receptors (Finkelshtein et al., 2013; Nikolic et al., 2018).

Finally, G mutants affected in CR binding cannot pseudotype a recombinant VSV (VSVΔG-GFP) in which the G gene has been replaced by the green fluorescent protein (GFP) gene. As the mutant glycoproteins are incorporated into the virions envelope as efficiently as WT G and as their fusion activity is unaffected, the loss of infectivity of pseudotypes bearing a mutant glycoprotein is due to their disability to recognize a cellular receptor (Nikolic et al., 2018).

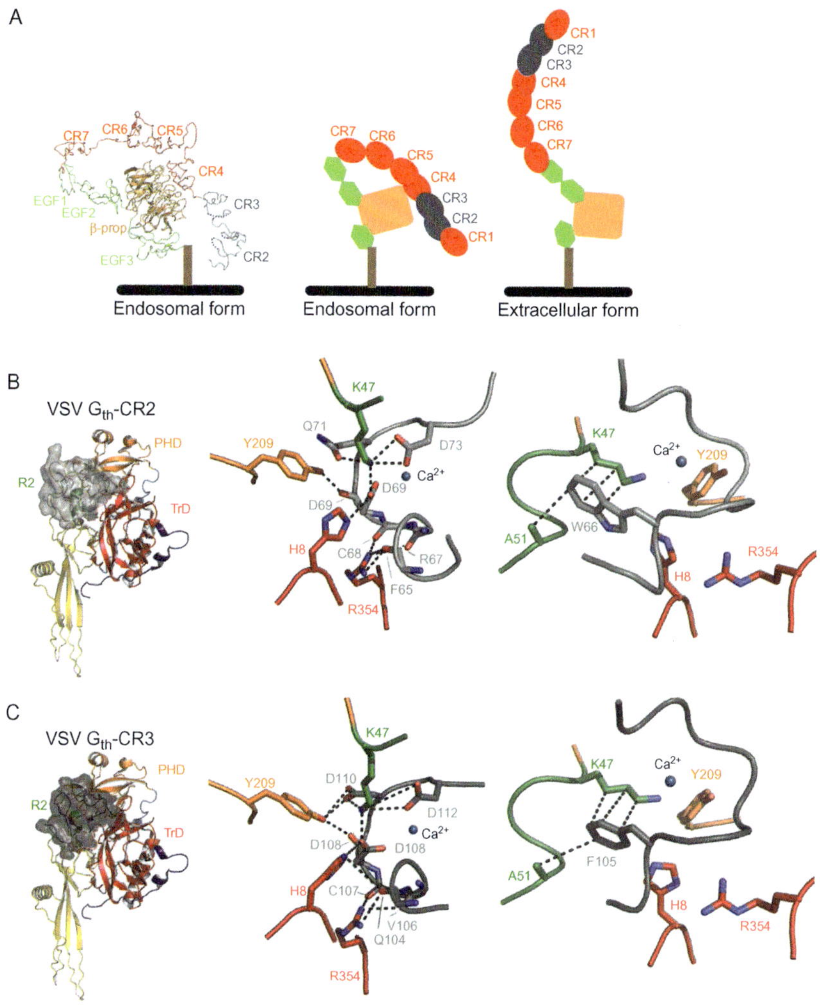

Fig. 2 (A) Ribbon diagram of the low pH structure of LDL-R (i.e., as in the endosome environment) (Rudenko et al., 2002) (PDB code: 1N7D) and cartoon representing the putative structure of LDL-R at the cell surface and inside the endosome after LDL release. (B) Overview of G_{th}-CR2 and (C) G_{th}-CR3 crystalline structures (PDB codes: 5OYL and 5OY9). In the first row, G is in ribbon representation and depicted by domains and CR domains are in full atom view. In both complexes the CR domain is nested in the same cavity of G. In the second row, close-up view on the G-CR interface showing the docking of G basic residues on the acidic patch of both CR2 and CR3. In both cases, the same G residues (H8, K47, Y209 and R354) are involved in the interaction. Third row, close-up view on the G-CR interface showing the hydrophobic interactions between the aromatic residue W66 of CR2 and F105 of CR3 and residues K47 and A51 of VSV G. Same color code than in Fig. 1, the calcium ion of the CR domains is depicted as a deep blue sphere. On each CR domain, residues involved in contacts with G are in gray and putative bonds are shown as black dashed lines. Oxygen atoms are in red and nitrogen atoms in blue.

This was observed in different cell lines (mammalian HEK, BSR, CHO and drosophila S2 cells). This strongly suggests that the only receptors of VSV, including in insect cells, are members of the LDL-R family or other transmembrane proteins containing similar CR domains. Therefore, VSV G has specifically evolved to interact with CR domains of the members of the LDL-R family (Nikolic et al., 2018).

Interestingly, it had been previously shown that the endoplasmic reticulum chaperone gp96 is essential for VSV infection (Bloor et al., 2010). Indeed, cells without gp96 or with catalytically inactive gp96 do not bind VSV G. Therefore, it had been proposed that gp96 is essential for the occurrence of functional VSV G receptors at the cell surface (Bloor et al., 2010). Indeed, it has been demonstrated that gp96-deficient cells present a marked decrease in cell surface expression of four members of the extended LDL receptor family (LDL-R, LRP6, Sorl1 and LRP8) (Weekes et al., 2012) in perfect agreement with the role of this receptor family in VSV entry.

2.3.2 Other VSV receptors?

Recently, it has been suggested that another cellular protein (Lgr4, leucine-rich repeat-containing G protein-coupled receptor 4) is essential for VSV and VSV G pseudotyped lentivirus to infect susceptible cells (Zhang et al., 2017). However, the effect of the deletion of the Lgr4 gene is particularly pronounced on VSV infection of peritoneal macrophages which are not the natural host cells of the virus.

Furthermore, the interpretation of several experiments is questionable as pointed out in a commentary of the article (Rubinstein, 2018). Particularly, to demonstrate the interaction between VSV G and Lgr4, the authors co-expressed the ectodomains of both proteins without their signal peptide (Zhang et al., 2017). Therefore, the proteins were not translocated into the ER and were probably misfolded and aggregated. Similarly, surface plasmon resonance experiments were difficult to interpret, as most of the binding is irreversible. Finally, the Lgr4 ectodomain was much less potent (by more than three orders of magnitude) than the LDL-R ectodomain in inhibition experiments (Rubinstein, 2018; Zhang et al., 2017).

Therefore, although a role of Lgr4 during VSV infection cannot be excluded, this molecule is probably not another viral receptor.

2.3.3 RABV receptors

Four proteins have been proposed to play the role of RABV receptors (Guo et al., 2019) (inset in Fig. 4).

First, since a segment of RABV G exhibits sequence similarity to the snake venom curaremimetic neurotoxins, which are potent ligands of nAChR, it has been proposed that the nicotinic acetylcholine receptor (nAChR) acts as a RABV receptor (Lentz et al., 1982, 1984). This hypothesis was reinforced by the demonstration that RABV interacts with purified Torpedo nAChR (Lentz et al., 1986), more precisely with its α subunit in an overlay assay (Gastka et al., 1996). Finally, a study in which mouse dorsal root ganglion (DRG) cells were treated with several nicotinic antagonists, showed that mecamylamine and d-tubocurarine reduce the percentage of infected neurons (Castellanos et al., 1997). However, there is no direct evidence in animal models that nAChR is a RABV receptor. Furthermore, RABV can infect neurons that do not express nAChR, and nAChR is located mainly on the muscle side of the neuromuscular junctions (McGehee and Role, 1995). Thus, although nAChR could account for the ability of RABV isolates to multiply locally in myotubes at the site of inoculation (Burrage et al., 1985), before subsequent penetration into neurons, other receptors are required for RABV entry into neurons.

The second protein that has been suggested to be a RABV receptor is the neural cell adhesion molecule (NCAM) (Thoulouze et al., 1998). This was based on the following observations. First, incubation of several cell lines with RABV decreased the surface expression of NCAM. Second, treatment of susceptible cells with heparan sulfate, a ligand for NCAM, or with NCAM antibodies reduced the rabies virus infection. Third, when RABV was preincubated with a soluble form of NCAM, it was unable to infect susceptible cells. Forth, the transfection of RABV-resistant L fibroblasts with NCAM-encoding gene induces RABV susceptibility. Finally, the infection of NCAM-deficient mice by RABV resulted in a slightly delayed mortality and a restricted brain invasion (Thoulouze et al., 1998).

At the same time, the murine low-affinity nerve-growth factor receptor, p75NTR was identified as a ligand of a soluble form of RABV G (Tuffereau et al., 1998). The ability of RABV G to bind p75NTR required the presence in G of a lysine and an arginine in positions 330 and 333, respectively, which were known to control virus penetration into the motor and sensory neurons of adult mice (Coulon et al., 1998; Seif et al., 1985; Tuffereau et al., 1989). Furthermore, expression of p75NTR in BSR cells (a clone of BHK21 cells) renders the cell line permissive for a non-adapted fox RABV isolate. Finally, G of lyssavirus genotype 6 (European bat lyssavirus type 2)—but not of genotype 3 (Mokola virus), 4 (Duvenhage virus) and 5 (European bat lyssavirus type 1)—was also shown to bind p75NTR (Tuffereau et al., 2001). Nevertheless, mice lacking all the extracellular part of p75NTR were still susceptible to infection (of which the rate and specificity were unchanged), indicating that the interaction between RABV G and p75NTR is not necessary for RABV infection of primary neurons (Tuffereau et al., 2007). However, recent data suggest that RABV may employ p75NTR-dependent transport as a fast pathway to facilitate transport to the central nervous system (CNS) (Gluska et al., 2014).

More recently, a global RNAi strategy developed to screen potential host factors for RABV infection in the human embryonic kidney cell line (HEK293) identified the metabotropic glutamate receptor subtype 2 (mGluR2) as essential for RABV infection (Wang et al., 2018). mGluR2 is a seven-transmembrane receptor and is conserved among mammals. It is abundant in the CNS and very rarely expressed in other tissues (Ohishi et al., 1993). By coimmunoprecipitation of tagged versions of RABV G and mGluR2, it was shown that both proteins interact together. Furthermore, mGluR2 antibodies block RABV infection in cells and a soluble version of the receptor neutralizes RABV infection. Finally, flow cytometry showed that RABV infection led to a decrease in cell surface expression of mGluR2 and that RABV and mGluR2 are internalized into cells and transported to early and late endosomes together. Therefore, mGluR2 has the characteristics to be a RABV receptor. However, in infected mice, the authors observed that some RABV-infected neuronal cells did not seem to express mGluR2 (Wang et al., 2018).

Although extensive studies have been performed to identify RABV receptors, no transmembrane protein has been definitively proved to have such a role. This complex situation may arise from the fact that RABV probably use several receptors. Indeed, experiments performed with an avirulent mutant have shown that some neuronal cells, permissive for the parental CVS strain, are not infected by the mutant. This indicates that the CVS strain is able to bind several distinct receptors to penetrate different neurons, while the avirulent mutant only recognizes some of them (Lafay et al., 1991).

Finally, African green monkey kidney cells (BS-C-1) which do not express p75NTR, nor NCAM, nor nAChR, nor most probably mGluR2 (as BS-C-1 have characteristics of epithelial cells) are susceptible to RABV infection (Piccinotti et al., 2013). Therefore, other factors remain to be identified. It is not excluded that these factors are specifically present in epithelial cells and are required for the final stages of disease, when virus spreads from the CNS to a variety of extraneural tissues.

2.3.4 *Receptors of other rhabdoviruses*

Much less is known concerning the receptors of other rhabdoviruses. The receptors or putative receptors that have been identified for VSV and RABV are not used by other rhabdoviruses even from the same genus. As mentioned above, lyssaviruses such as Mokola virus, Duvenhage virus or European bat lyssavirus type 1 do not bind to p75NTR (Tuffereau et al., 2001). Similarly, CHAV G does not bind CR domains from LDL-R (Nikolic et al., 2018).

In fact, hitherto, the only convincing data for another rhabdovirus receptor have been obtained for the viral hemorrhagic septicemia virus (VHSV), a salmonid novirhabdovirus, for which it has been shown that monoclonal antibodies (MAbs) directed against a fibronectin-containing complex protect cells from the infection. Since fibronectin, purified from rainbow trout, binds specifically to VHSV, fibronectin has been proposed to be a receptor for VHSV and some other fish rhabdoviruses (Bearzotti et al., 1999).

3. Pathway of rhabdoviruses entry into the host cell

3.1 Rhabdoviruses enter cells by clathrin mediated endocytosis

Binding of rhabdoviruses to their receptor triggers the endocytosis of the virion. All the rhabdoviruses studied so far, including RABV (Lewis and Lentz, 1998; Lewis et al., 1998; Piccinotti et al., 2013; Superti et al., 1984a), VSV (Cureton et al., 2009, 2010; Johannsdottir et al., 2009; Sun et al., 2005; Superti et al., 1987), BEFV (Cheng et al., 2012), SVCV (Shao et al., 2016) and infectious hematopoietic necrosis virus (IHNV, a Novirhabdovirus) (Liu et al., 2011) use clathrin-mediated endocytosis (CME) to enter the host cell.

RABV, VSV, BEFV and SVCV endocytosis was shown to be dependent on dynamin (Cheng et al., 2012; Johannsdottir et al., 2009; Piccinotti et al., 2013; Shao et al., 2016). However, for both RABV and VSV, it does not show a strict requirement of the adaptor protein AP-2 (Johannsdottir et al., 2009; Piccinotti et al., 2013). In the case of VSV, this is consistent with the CME of LDL-R which itself is not strictly dependent on AP-2 (Motley et al., 2003).

Endocytosis of VSV is a fast process with a half time-life of 2.5–3 min (Johannsdottir et al., 2009). After binding, virions do not laterally diffuse into preformed pits but are internalized by pits that form at or near viral particles (Cureton et al., 2009). Similar observations have been made with a recombinant VSV, in which the VSV G was replaced by RABV G, as a surrogate of RABV (Piccinotti et al., 2013).

3.2 Actin dependence of rhabdovirus entry

All rhabdoviruses are significantly larger than the dimensions of a typical clathrin-coated vesicle which has an average interior diameter of 60nm. Virion-containing vesicles are more elongated than empty vesicles arising from CME. On average, they contain more clathrin and clathrin adaptor molecules. However, this increase is insufficient to permit full coating of the vesicle (Cureton et al., 2009) and the electron-dense material (corresponding to the clathrin coat) is only visible on the vesicle surface next to the spherical end of the virion (Fig. 3).

The complete uptake of the clathrin structures that internalize VSV particles is dependent on actin polymerization (Fig. 3), and treatment of cells with either cytochalasin D or latrunculin B (which disrupt actin filaments) inhibits entry by blocking the transition of virus-containing pits to cytosolic vesicles (Cureton et al., 2009). The requirement of actin assembly for efficient VSV entry is due to the length of the viral particle as, in contrast, the entry of defective interfering particles that are shorter (75nm long vs 200nm for VSV), does not depend on actin polymerization (Cureton et al., 2010).

3.3 VSV transport in the endosomal pathway

Taking advantage of the fact that ammonium chloride (NH_4Cl) is a weak base that inhibits endosomal acidification and thus prevents the low pH-induced activation of VSV G, it was shown that fusion occurs rapidly after endocytosis, within 1–2min after internalization (Johannsdottir et al., 2009). Given the rapidity of the process, the low pH-induced structural transition of G, which catalyzes fusion, most likely occurs in early endosomes (Fig. 3, pathway (i)).

This is in agreement with experiments showing that dominant negative forms of RAB 5 that block the maturation of endocytic vesicles to early endosomes inhibit VSV infection whereas a dominant negative form of RAB 7 (that blocks the traffic from the early to the late endosome) or RAB 11 (that blocks transport of endocytosed cargoes to recycling endosomes) is without effect on VSV infection (Mire et al., 2010; Sieczkarski and Whittaker, 2003).

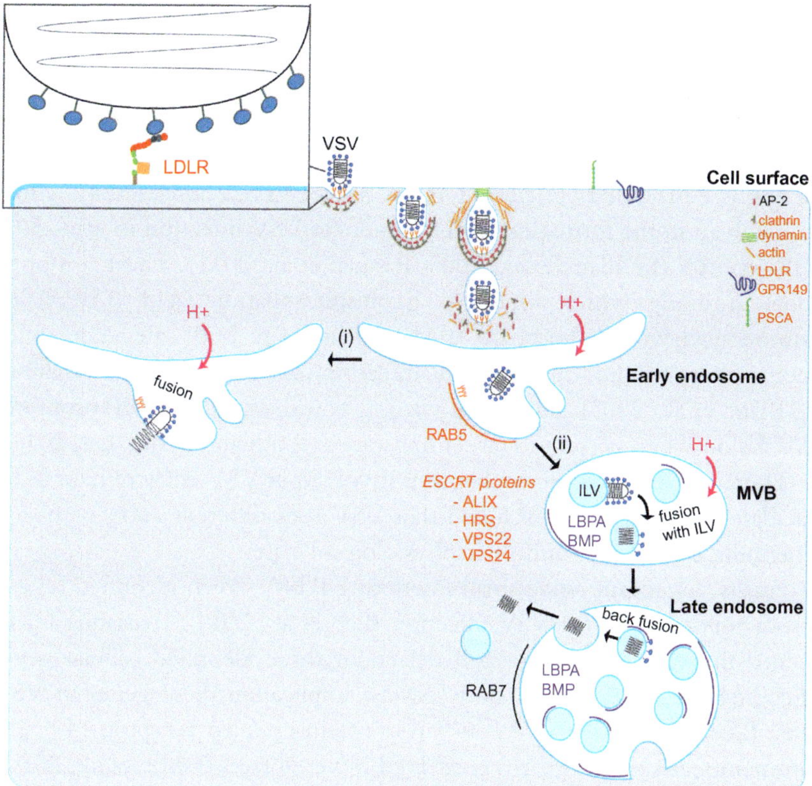

Fig. 3 Proposed pathways for VSV entry into the host cell. Binding of VSV G to the LDL-R triggers endocytosis of VSV particles. The endocytosis vesicle is only partially coated with clathrin. Efficient viral entry requires actin polymerization. Two pathways have been proposed. Pathway (i): fusion occurs rapidly in the early endosome and leads to the release of the RNP in the cytoplasm. Pathway (ii): inside multivesicular bodies, VSV fuses first with an intraluminal vesicle (ILV) in which the RNP is released. In the late endosome, these vesicles deliver the RNP to the cytoplasm through a back-fusion mechanism. Proteins that have been implicated in VSV entry have their name in red. Endosome-specific lipids which favor VSV fusion have their name in purple.

However, this classical view is challenged by experimental data which suggest that the acid-activated fusion of VSV and the release of the RNP into the cytosol occur in two distinct steps (Le Blanc et al., 2005; Luyet et al., 2008) (Fig. 3, pathway (ii)). The viral envelope would first fuse with an intraluminal vesicle (ILV) within the multivesicular bodies (MVB). A second fusion step (often called back-fusion, implicating the action of a cellular machinery) between the ILV and the limiting MVB membrane would then be required to release the RNP into the cytosol. This view is supported by several experimental data. First, fusion and RNA release into the cytosol can be uncoupled in presence of nocodazole which depolymerizes the microtubules. Indeed, nocodazole has no effect on the fusion of individual VSV particles while it significantly reduces RNA release (Le Blanc et al., 2005). Second, siRNA depletion of several subunits of the ESCRT machinery (including ALIX, HRS, VPS22 and VPS24), which are involved in the formation of ILVs, reduces VSV infection to 30%–50% of the control (Le Blanc et al., 2005; Pasqual et al., 2011). Third, a monoclonal antibody which binds to lysobisphosphatidic acid (LBPA) or bis(monoacylglycero)phosphate (BMP), which are both late endosome-specific phospholipids, inhibits VSV infection in a dose-dependent manner (Le Blanc et al., 2005). Finally, in vitro, it was shown that BMP promotes VSV fusion when present in target liposomes (Roth and Whittaker, 2011).

Therefore, the precise mechanisms involved in VSV entry remain to be elucidated. It cannot be excluded that VSV uses different entry pathways depending on the cells and their physiological state.

Finally, a genome-wide small interfering RNA screen identified several host factors required for VSV infection (Lee et al., 2014). Bypassing CME rescued the viral gene expression defect for three identified cellular genes (encoding GPR149, PSCA, and LSM5), implicating these genes in VSV entry (Lee et al., 2014). GPR149 (G-protein-coupled receptor 149) is a transmembrane protein that is conserved in vertebrates (Edson et al., 2010), while PSCA (prostate stem cell antigen) is a glycosylphosphatidylinositol-anchored membrane glycoprotein that is overexpressed in prostate cancers (Reiter et al., 1998). The cell membrane localization of GPR149 and PSCA is consistent with their assignment as potential entry factors (Fig. 3). In contrast, LSM5 (U6 snRNA-associated Sm-like protein) is mostly located in the nucleus and its role in VSV entry is probably indirect.

3.4 RABV transport in the axon

In the case of RABV, two different pathways were proposed for the transport through the axon to the soma where transcription and replication take place: either fusion occurs early and the RNP is transported alone or the whole viral particle is transported inside the endocytic vesicle. It is now clear that the predominant RABV entry pathway into neuron is the latter (Fig. 4).

First, it was demonstrated that lentiviruses pseudotyped with RABV G were transported in a retrograde way and reach the central nervous system

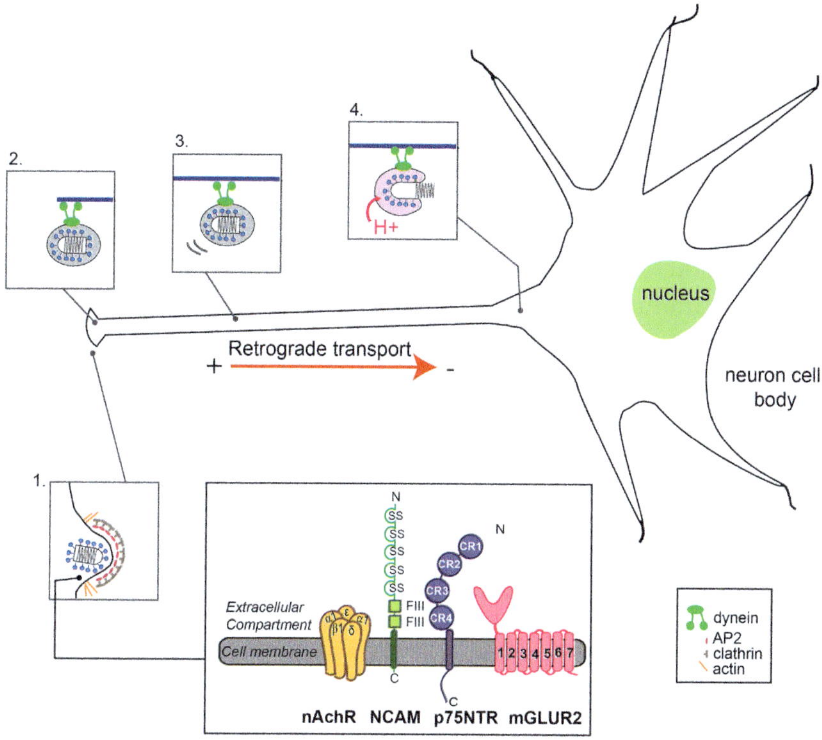

Fig. 4 Proposed mechanism for RABV entry and retrograde transport into neurons. After receptor recognition, the virion is endocytosed in partially clathrin-coated vesicles (1) and transported in an endocytic vesicle (2) along the axon via microtubules to the soma (3). Endosome acidification occurs at the soma leading to membrane fusion and RNP release (4).

after peripheral delivery (Mazarakis et al., 2001). This was not the case when VSV G-pseudotyped vectors were used. This indicated that RABV G but not VSV G allows long distance transport along axons and suggested that, in the case of RABV infection, the whole virion is transported inside a vesicle (Mazarakis et al., 2001).

This was demonstrated in murine neuroblastoma NS20Y cells using double fluorescently labeled RABV particles containing an eGFP fused to the RABV P protein and in whose membrane was inserted an engineered red fluorescent protein comprising RABV G C-terminal transmembrane segment and cytoplasmic domain. Such fluorescent particles allowed the discrimination of enveloped virus from free RNPs in live fluorescence microscopy experiments and indicated that in NS20Y cells enveloped RABV particles are transported as a cargo inside vesicles by typical retrograde axonal transport (Klingen et al., 2008). Similar observations were later made in sensory neurons from dorsal root ganglia and in motor neurons (Gluska et al., 2014; Piccinotti and Whelan, 2016).

Single particle experiments confirmed that, in primary neurons, CME is a major route of RABV endocytosis in both motor and sensory neurons. However, in motor neurons derived from the ventral spinal cords, a significant fraction of incoming particles did not colocalize with transferrin, or fluorescent dextran, suggesting that some virions enter independently of clathrin or macropinocytic uptake mechanisms (Piccinotti and Whelan, 2016).

There is still a debate about the acidification site of RABV containing vesicles. Some experiments with the SAD vaccine strain of RABV provided evidence for transport within acidified endosomes (Gluska et al., 2014) whereas another study performed with a recombinant VSV expressing CVS strain RABV G suggested that acidification of virion containing vesicle occurs primarily in the neuronal soma (Piccinotti and Whelan, 2016) (Fig. 4). The ability to tolerate mild acidification during axonal transport might be related to intrinsic properties of G from different strains as the kinetics of the low-pH induced structural transition vary significantly from one strain to another (Gaudin et al., 1995).

Finally, it is worth noting that some neurons seem to be resistant to retrograde infection by RABV. This is the case of two classes of unmyelinated sensory neurons (nonpeptidergic -NP- and C-fiber low-threshold mechanoreceptor -TH- neurons), which account for almost half of the total dorsal root ganglion neuron population, are largely resistant to rabies virus infection from the spinal cord (Albisetti et al., 2017). None of the expression patterns of the identified RABV receptors could explain the resistance of those neurons to rabies infection (Albisetti et al., 2017). It is thus possible that some yet unknown structural properties of NP and TH neurons or of their axons impede virus progression to the soma.

4. Rhabdovirus-induced membrane fusion

4.1 Fusion properties

Fusion of RABV and VSV with artificial liposomes (Gaudin et al., 1991, 1993, 1995, 1996; Roche and Gaudin, 2004; Roth and Whittaker, 2011; Yamada and Ohnishi, 1986) and with cells (Clague et al., 1990; Puri et al., 1988, 1992) have been largely characterized. From these studies, it appears that neither RABV (Gaudin et al., 1991) nor VSV (Herrmann et al., 1990; Yamada and Ohnishi, 1986) have an absolute specific lipid requirement for fusion. However, as mentioned above, some late endosome specific lipids such as BMP enhance VSV fusion when present in the target membrane (Roth and Whittaker, 2011).

The pH-dependence of fusion has been characterized for several rhabdoviruses and is very similar from one rhabdovirus to another being optimal around pH 6 (Clague et al., 1990; Gaudin et al., 1999; Roche and Gaudin, 2004).

As for other viruses that fuse at low pH, preincubation of the rhabdoviruses at low pH in the absence of a target membrane leads to the inactivation of the viral fusion properties (Clague et al., 1990; Gaudin et al., 1991). The inactivation process is more rapid at higher temperature and lower pH and is due to G structural transition toward its postfusion conformation. However, as G structural transition is reversible, re-incubation of the acidified virions at pH above pH 7 for 1 or 2 h leads to the recovery of their initial fusion activity (Gaudin et al., 1991, 1993).

The reversibility of the conformational change is required to allow newly synthesized G to be transported through the acidic compartments of the Golgi apparatus and to recover its native prefusion state when reaching the cell surface to be incorporated in newly synthesized virions (Gaudin et al., 1995). Other viruses, for which the low-pH induced structural transition is irreversible, have evolved different strategies to protect their fusion proteins in the acidic compartments of the Golgi apparatus (Gaudin, 2000b).

4.2 Rhabdovirus fusion pathway

The existence of a pH-dependent equilibrium between the different states of rhabdovirus G indicates that, differently from other viral fusion glycoproteins, the native conformation is not metastable. In general, it is proposed that the high amount of energy released during the conformational change from the metastable prefusion state to the final stable postfusion state is used to form the high energy lipidic intermediates during the fusion reaction (Carr et al., 1997; Harrison, 2015). In the case of rhabdoviruses, the amount of free energy released during the structural transition of a single G trimer is small compared to that released by fusion glycoproteins having a metastable native state such as HA and compared to the energy barrier of the fusion reaction (the activation energy of the fusion process has been estimated to be in the range of 40 kcal/mol (Kuzmin et al., 2001; Lee and Lentz, 1998)). Therefore, a concerted action of several G trimers is required to overcome this energy barrier. For RABV, the number of spikes required for the fusion reaction has been estimated to be approximately 15 (Roche and Gaudin, 2002).

Despite this major difference between rhabdoviral G and other viral fusion glycoprotein, experimental data, mainly obtained using lipids with various dynamic molecular shapes, suggest that the membrane fusion pathway is the same for rhabdoviruses as for other enveloped viruses (Chernomordik et al., 1998; Gaudin, 2000a; Zaitseva et al., 2005). Fusion proceeds via the formation of an intermediate stalk that forms a local lipid connection between the outer leaflets of the fusing membranes. Radial expansion of the stalk then induces the formation of a transient hemifusion diaphragm (i.e., a local bilayer composed of the two initial inner leaflets). The last step, the most energetically expensive (Cohen and Melikyan, 2004), is the formation and enlargement of a fusion pore within the hemifusion diaphragm, which results in complete fusion (Chernomordik and Kozlov, 2005; Chernomordik et al., 1995). In the case of RABV, the fusion process can be arrested at pH 6.4 and 4 °C and an intermediate can be trapped which is probably at the stage of initial pore formation (Gaudin, 2000a).

In the case of VSV, electron microscopy revealed that fusion of viral particles with liposomes occurs preferentially at the flat base of the virion (Libersou et al., 2010). In this experimental context, the VSV fusion pathway seems to involve two distinct stages. The first stage occurs at the flat base and involves local membrane deformations, leading to the formation of one or several stalks, and the formation of one or several initial fusion pores. From simulations of hemifusion kinetic data, the critical number of G trimers required at this stage was estimated to be 3–5 (Kim et al., 2017). Then, the formation of a helical glycoprotein network outside the contact zone, on the cylindrical part of the virion, leads to the enlargement of these initial pores, completing the fusion of the membranes (Libersou et al., 2010). Network formation involves favorable lateral interactions between glycoproteins, in their low pH conformation and probably allows overcoming the high energy barrier required for pore enlargement.

4.3 Working of rhabdovirus fusion machinery

4.3.1 Fusion loops interaction with the target membrane

The FD structural organization is well conserved among class III glycoproteins (Baquero et al., 2015a). As mentioned above, its core is a three-stranded β-sheet at the tip of which are located two hydrophobic loops that interact with the target membrane (and maybe also with the viral one) (Fig. 1D). An alignment of the amino-acid sequences of several rhabdovirus glycoproteins reveals that these loops always contain polar aromatic residues (tryptophans and tyrosines) (Roche et al., 2008). That these loops are indeed an essential part of the membrane-interacting motif is consistent with mutagenesis work performed on VSV (Stanifer et al., 2011; Sun et al., 2008). The presence of charged residues in the vicinity of the fusion loops precludes any deep penetration inside the membrane (Fig. 1D). Most probably, tryptophan and tyrosines position themselves at the interface between the fatty acid chains and head group layers of lipids (Wimley and White, 1992) and perturb the outer leaflet of the target membrane. This interfacial interaction facilitates the formation of point-like lipidic protrusions that have been suggested to be stalk precursors (Efrat et al., 2007).

4.3.2 pH-sensitive switches

Amino acids playing an important role during the structural transition have been identified. In the case of vesiculovirus G, a cluster of four histidine residues (in position 60, 162, 407 and 409) in the prefusion structure, which are conserved among the rhabdovirus family, has been proposed to play the role of pH-sensitive switch (Fig. 5A). Indeed, the histidine protonation might trigger the low pH-induced conformational change (Roche et al., 2007). Protonation of these histidine residues leads to a cluster of positive charges that might create a repulsive force between the FD and the C-terminal part of G ectodomain (segment R5) and initiate the glycoprotein refolding. This is consistent with the inhibition of VSV fusion by diethylpyrocarbonate, a reagent specifically modifying histidine residues (Carneiro et al., 2003).

Clusters of acidic residues are observed in the six-helix bundle constituting the core of the VSV G postfusion state (Fig. 5B). It had been proposed that their deprotonation at high pH destabilizes this central domain and, consequently, allows G refolding back toward its prefusion conformation. Indeed, a mutagenesis analysis demonstrated that the cluster of the three Asp 268 residues, buried in the center of the hydrophobic environment of the postfusion trimeric interface, constitutes a major pH-sensitive molecular switch while Asp274, Asp395, and Asp393 have additional contributions (Ferlin et al., 2014).

Interestingly, this VSV G switch is not found in CHAV G. In fact, the CHAV G switch triggering the conformational change back to the prefusion state, involves other acidic residues of which the deprotonation

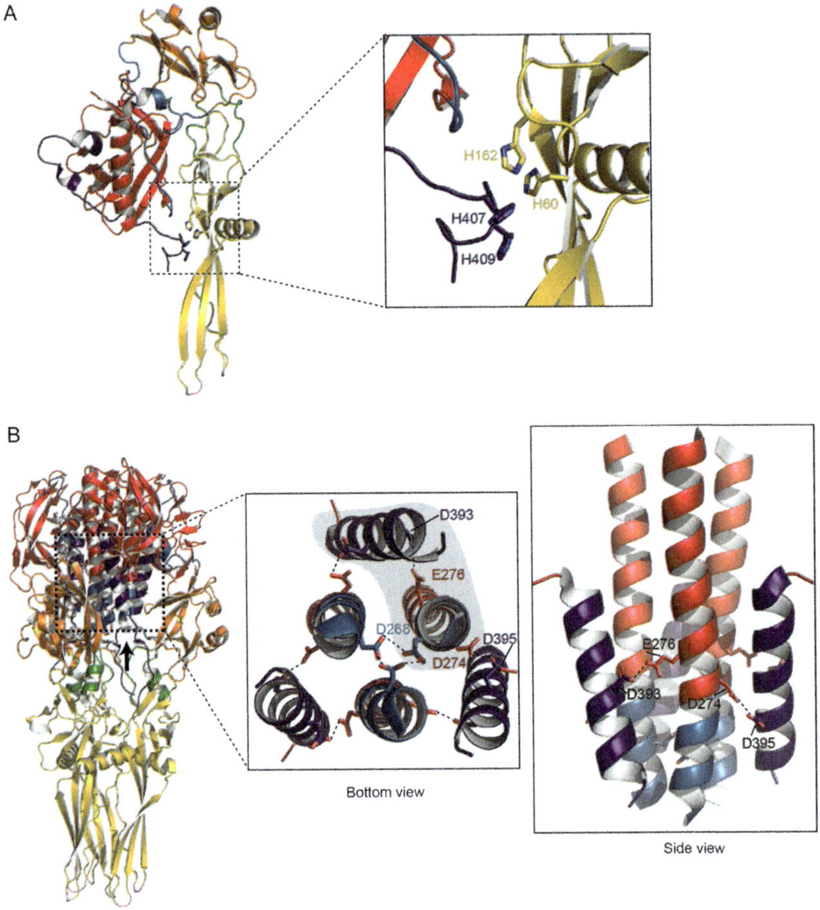

Fig. 5 (A) Close-up on the histidines cluster (represented in sticks) in the prefusion VSV G protomer. (B) Close-up on the acidic amino-acids buried in the six-helix bundle of the postfusion conformation. Hydrogen bonds between lateral chains in their protonated form are indicated. The black arrow on the whole trimeric structure indicates the position of the observer for the bottom view. In the bottom view, helices belonging to the same protomer are surrounded by a gray background. Oxygen atoms are in red and nitrogen atoms in blue.

at high pH also destabilizes the postfusion protomer of CHAV G (Baquero et al., 2015b).

It is worth noting that the regions corresponding to segment R4 and R5 in other Rhabdovirus glycoproteins are often enriched in acidic amino acid residues which can also play a similar role of pH-sensitive molecular switches.

4.3.3 G structural transition occurs through monomeric intermediates

The low pH-induced structural transition of G_{th} (the soluble ectodomain form that was previously crystallized) has been characterized in solution using several biophysical techniques. While G_{th} postfusion trimer is the major species detected at pH below 6.7, the prefusion trimer is never observed in solution (Albertini et al., 2012). This in perfect agreement with previous studies performed on a soluble form obtained after treatment of the virion with cathepsin D (Crimmins et al., 1983) but also on full length detergent solubilized G (Doms et al., 1987). In fact, above pH 7, G_{th} appears to be a flexible monomer that populates a manifold of conformations in a pH-dependent manner adopting more elongated conformations at lower pH (Albertini et al., 2012).

Analyses of vesiculovirus G structure at the viral surface using electron microscopy and tomography give very similar results. Indeed, below pH 6, the only structure observed in the viral membrane is the trimeric postfusion state which tends to reorganize into helical arrays (Libersou et al., 2010). Above pH 7, although some prefusion trimers are observed (Libersou et al., 2010), the majority of glycoproteins are flexible monomers (Albertini et al., 2012). At intermediate pH (between 6.5 and 7), the shape of the spikes is not homogeneous, and several forms of G are observed (Albertini et al., 2012). In some regions, G keeps its high pH organization while some spikes have already adopted the typical postfusion shape. Some elongated monomeric rod-shaped structures can also be observed (Albertini et al., 2012; Baquero et al., 2017). The sequential appearance of these species after lowering the pH suggested that VSV G structural transition occurs through monomeric intermediates.

Recently, a crystalline form of CHAV G_{th} has been obtained which contained four molecules in the asymmetric unit adopting two different conformations that were both distinct from the prefusion and postfusion protomer conformations (Baquero et al., 2017) (Figs. 6 and 7). The first conformation resembles VSV G prefusion protomer with one major difference: the R5 segment has already left the hydrophobic grooves it occupies in the prefusion conformation. Therefore, this conformation of CHAV G_{th} corresponds to an early intermediate (EI) on the transition pathway. In contrast, the second conformation of CHAV G_{th} in this crystal corresponds to a late intermediate (LI). It is already in an elongated hairpin although the refolding of segments R1, R4 and R5 is still incomplete. Those two structures suggest a plausible pathway for the conformational change which is detailed in Fig. 7.

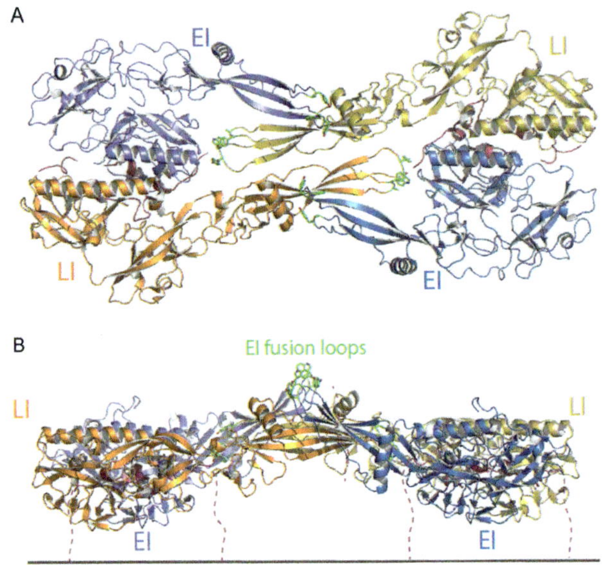

Fig. 6 Top (A) and lateral (B) views of the tetrameric assembly in the CHAV G_{th} crystal (PDB code: 5MDM). Protomers in the EI conformation are in blue shades. Protomers in the LI conformation are in yellowish shades. The residues in the fusion loops at the tip of the FD are in green sticks. In (B), the C-terminal segments which are linked to the transmembrane domain are depicted in dashed lines and the position of the membrane is indicated by the black line.

4.3.4 A role for dimers (or even tetramers) in vesiculovirus fusion?

In the crystal structure, two molecules of CHAV G_{th} EI and two molecules of G_{th} LI are assembled in a compact tetramer which forms a flat structure (Fig. 6) from which protrude the fusion loops of the two EI conformations (Baquero et al., 2017) (Fig. 6B). If such a structure would form at the viral surface, its orientation would optimally place those two loops to attack the target membrane while the two LI fusion loops would point toward the viral lipid bilayer.

In this crystalline tetrameric arrangement, the FDs are associated in an antiparallel manner which is stabilized by the formation of an intermolecular β-sheet between two EI and LI FDs, involving residues 75–80 in the EI strand (downstream the first G fusion loop) and 120–125 in the LI strand (downstream the second fusion loop) (Fig. 8). Remarkably, mutations of residues Asp121 and Glu123 were previously shown to affect the fusion properties of both VSV G (Fredericksen and Whitt, 1996) and VHSV G (Gaudin et al., 1999). Furthermore, for VSV G, a mutation in position 76 was shown to compensate for a deleterious mutation A117F in the fusion

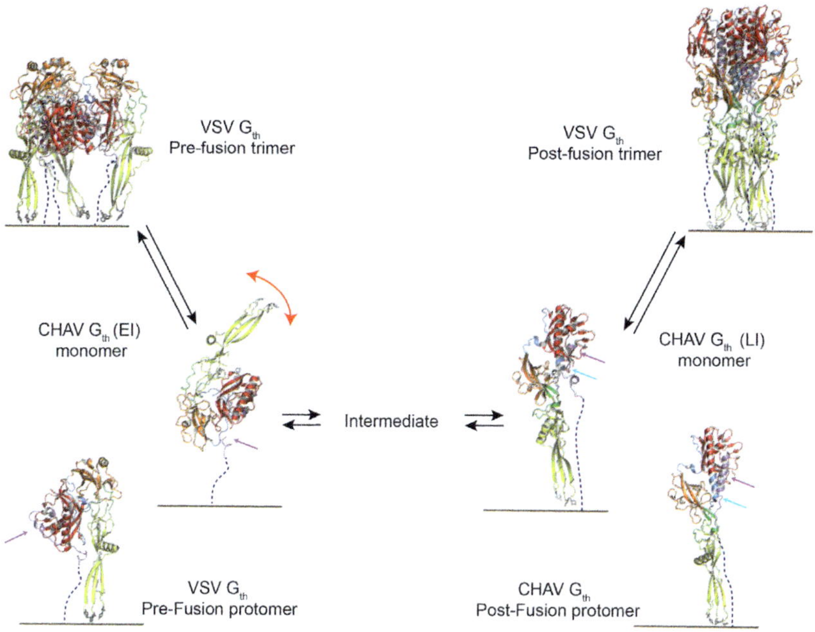

Fig. 7 Plausible pathway of vesiculovirus G conformational change. VSV G prefusion trimers dissociate into monomers. The R5 segment (indicated by purple arrows) leaves the hydrophobic grooves it occupied in the prefusion conformation to form the early intermediate state (EI). This confers orientational mobility to the rest of the molecule, even in the absence of any other conformational changes. Refolding of R2 and R3 and partial refolding of R1, R4 and R5 lead to the formation of the late intermediate state (LI) which resembles the postfusion protomer and is already in an elongated hairpin conformation. Final refolding of R1, R4 and R5 is required for retrimerization in the postfusion conformation. G crystalline structures are represented as ribbon diagrams with the same color code than in Fig. 1.

loop (Stanifer et al., 2011). Those phenotypes, which could not be explained by the crystal structures of the pre- and postconformation, were much more easily interpreted in the context of this novel structure in which residues in positions 121 and 123 on the LI side of the intermolecular β-sheet are facing the conserved His 80 on the EI side. Indeed, the H80A mutation and the E123L/D121L double mutation, which both abolish VSV G fusion properties, were rescued by the same compensatory Q112P mutation. In a single FD, residue 112, being on the opposite side of the three-stranded β-sheet, is far from residues 121 and 123. However, in the tetrameric CHAV G_{th} crystal structure, it clusters with those residues (and His 80) on the same side of the intermolecular β-sheet (Baquero et al., 2017) (Fig. 8).

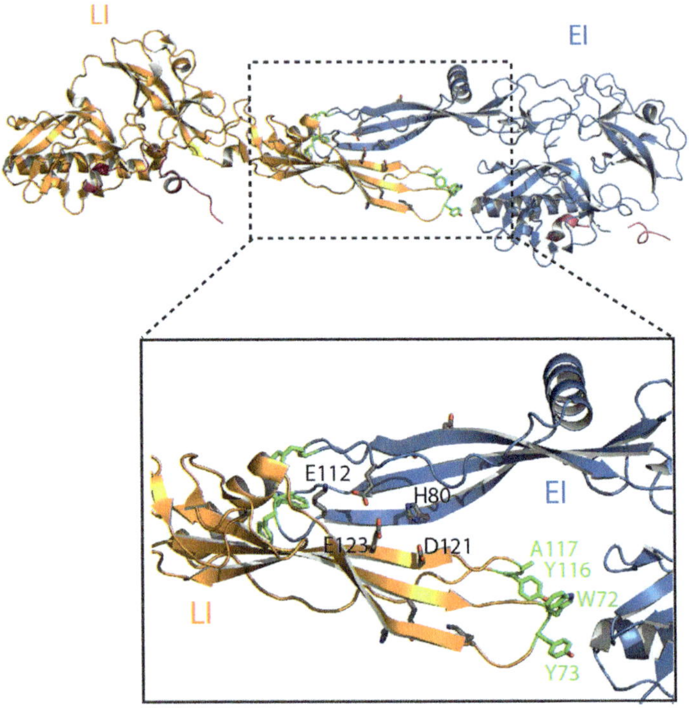

Fig. 8 Nontrimeric association of monomeric intermediate states of G. Close up view of the antiparallel β-sheet formed between EI (in blue) and LI (in yellow) FDs within the CHAV G_{th} crystal asymmetric unit. Key residues which cluster at the interface (see text) are represented in gray sticks. The residues in the fusion loops at the tip of the FD are in green sticks. Oxygen atoms are in red and nitrogen atoms in blue.

This strongly suggests that the antiparallel assembly of FDs found in CHAV G_{th} crystal is relevant for vesiculovirus fusion. Such an interaction might facilitate the insertion of the fusion loops of one protomer in the target membrane while the loops of the second attack the viral lipid bilayer.

5. Conclusions and perspectives

In recent years, our understanding of rhabdovirus entry into their host cell has considerably increased thanks to the work performed on RABV and VSV. However, besides those two model viruses, very little is known. Particularly, we know nothing about the receptors used by other rhabdoviruses for which the natural host and cellular targets are in most cases unidentified.

In terms of G structural biology, all our knowledge has been obtained on vesiculoviruses. The pathway of the structural transition, which does not correspond to what is proposed in the "unifying" classical model (Harrison, 2015), remains to be validated for other rhabdovirus genera (or for other class III viral fusion glycoproteins). Furthermore, these intermediates need to be observed during the fusion process when they interact with membranes.

Interestingly, neither a domain of G, nor a peptide has been shown to inhibit G refolding. This is in contrast with class I fusion glycoproteins, for which peptides corresponding to heptad repeats segments can inhibit the fusion process (Russell et al., 2001; Wild et al., 1994), and class II fusion glycoproteins, for which exogenously-added domain III can function as a dominant-negative fusion inhibitor (Liao and Kielian, 2005).

It is also worth noting that the TM domain of VSV G has been demonstrated to be involved at some stages of the transition from hemifusion toward complete fusion (Ci et al., 2018; Cleverley and Lenard, 1998). Therefore, structures of full length (i.e., including the TM domain) glycoproteins are necessary to complete our understanding of G-mediated membrane fusion.

Finally, both RABV G and VSV G are tools for researchers and in biotechnology. Indeed, RABV, thanks to RABV G properties, is the best tool available for transneuronal tracing and studies of monosynaptic connections (Ginger et al., 2013; Haberl et al., 2015). Nevertheless, a better understanding of RABV biology and its receptors is still needed to identify the limitations of those approaches (Albisetti et al., 2017). Similarly, VSV G is widely used for pseudotyping lentiviruses (Cronin et al., 2005; Matrai et al., 2010; Naldini et al., 1996) and VSV is a promising oncolytic virus for the treatment of cancer (Bishnoi et al., 2018; Durham et al., 2017; Fernandez et al., 2002; Hastie et al., 2013). The recent demonstration that it is possible to uncouple G receptor recognition and fusion activity paves the way to develop recombinant glycoproteins with modified tropism (Nikolic et al., 2018). As an example, a glycoprotein having (i) a point mutation that ablates the natural receptor tropism and (ii) an insertion of a protein domain targeting specifically a tumor cell antigen might allow the design of fully retargeted oncolytic VSVs. Such viruses should be able to eliminate cancerous cells while sparing normal ones. Here again, fundamental research is needed to demonstrate that VSV G can be fully functional in endocytic pathways which, depending on the targeted receptor, may not be clathrin-mediated.

References

Abou-Hamdan, A., Belot, L., Albertini, A., Gaudin, Y., 2018. Monomeric intermediates formed by vesiculovirus glycoprotein during its low-pH-induced structural transition. J. Mol. Biol. 430 (12), 1685–1695.

Albertini, A.A., Merigoux, C., Libersou, S., Madiona, K., Bressanelli, S., Roche, S., Lepault, J., Melki, R., Vachette, P., Gaudin, Y., 2012. Characterization of monomeric intermediates during VSV glycoprotein structural transition. PLoS Pathog. 8 (2) e1002556.

Albisetti, G.W., Ghanem, A., Foster, E., Conzelmann, K.K., Zeilhofer, H.U., Wildner, H., 2017. Identification of two classes of somatosensory neurons that display resistance to retrograde infection by rabies virus. J. Neurosci. 37 (43), 10358–10371.

Amirache, F., Levy, C., Costa, C., Mangeot, P.E., Torbett, B.E., Wang, C.X., Negre, D., Cosset, F.L., Verhoeyen, E., 2014. Mystery solved: VSV-G-LVs do not allow efficient gene transfer into unstimulated T cells, B cells, and HSCs because they lack the LDL receptor. Blood 123 (9), 1422–1424.

Backovic, M., Longnecker, R., Jardetzky, T.S., 2009. Structure of a trimeric variant of the Epstein-Barr virus glycoprotein B. Proc. Natl. Acad. Sci. U. S. A. 106 (8), 2880–2885.

Baquero, E., Albertini, A.A., Gaudin, Y., 2015a. Recent mechanistic and structural insights on class III viral fusion glycoproteins. Curr. Opin. Struct. Biol. 33, 52–60.

Baquero, E., Albertini, A.A., Raux, H., Buonocore, L., Rose, J.K., Bressanelli, S., Gaudin, Y., 2015b. Structure of the low pH conformation of Chandipura virus G reveals important features in the evolution of the vesiculovirus glycoprotein. PLoS Pathog. 11 (3), e1004756.

Baquero, E., Albertini, A.A., Raux, H., Abou-Hamdan, A., Boeri-Erba, E., Ouldali, M., Buonocore, L., Rose, J.K., Lepault, J., Bressanelli, S., Gaudin, Y., 2017. Structural intermediates in the fusion-associated transition of vesiculovirus glycoprotein. EMBO J. 36 (5), 679–692.

Bearzotti, M., Delmas, B., Lamoureux, A., Loustau, A.M., Chilmonczyk, S., Bremont, M., 1999. Fish rhabdovirus cell entry is mediated by fibronectin. J. Virol. 73 (9), 7703–7709.

Beglova, N., Blacklow, S.C., 2005. The LDL receptor: how acid pulls the trigger. Trends Biochem. Sci. 30 (6), 309–317.

Benmansour, A., Leblois, H., Coulon, P., Tuffereau, C., Gaudin, Y., Flamand, A., Lafay, F., 1991. Antigenicity of rabies virus glycoprotein. J. Virol. 65 (8), 4198–4203.

Bishnoi, S., Tiwari, R., Gupta, S., Byrareddy, S.N., Nayak, D., 2018. Oncotargeting by vesicular stomatitis virus (VSV): advances in cancer therapy. Viruses 10 (2), e90.

Blasdell, K.R., Voysey, R., Bulach, D., Joubert, D.A., Tesh, R.B., Boyle, D.B., Walker, P.J., 2012. Kotonkan and Obodhiang viruses: African ephemeroviruses with large and complex genomes. Virology 425 (2), 143–153.

Bloor, S., Maelfait, J., Krumbach, R., Beyaert, R., Randow, F., 2010. Endoplasmic reticulum chaperone gp96 is essential for infection with vesicular stomatitis virus. Proc. Natl. Acad. Sci. U. S. A. 107 (15), 6970–6975.

Brown, M.S., Goldstein, J.L., 1986. A receptor-mediated pathway for cholesterol homeostasis. Science 232 (4746), 34–47.

Bu, G., Marzolo, M.P., 2000. Role of rap in the biogenesis of lipoprotein receptors. Trends Cardiovasc. Med. 10 (4), 148–155.

Burke, H.G., Heldwein, E.E., 2015. Crystal structure of the human cytomegalovirus glycoprotein B. PLoS Pathog. 11 (10), e1005227.

Burrage, T.G., Tignor, G.H., Smith, A.L., 1985. Rabies virus binding at neuromuscular junctions. Virus Res. 2 (3), 273–289.

Carneiro, F.A., Stauffer, F., Lima, C.S., Juliano, M.A., Juliano, L., Da Poian, A.T., 2003. Membrane fusion induced by vesicular stomatitis virus depends on histidine protonation. J. Biol. Chem. 278 (16), 13789–13794.

Carr, C.M., Chaudhry, C., Kim, P.S., 1997. Influenza hemagglutinin is spring-loaded by a metastable native conformation. Proc. Natl. Acad. Sci. U. S. A. 94 (26), 14306–14313.
Castellanos, J.E., Castaneda, D.R., Velandia, A.E., Hurtado, H., 1997. Partial inhibition of the in vitro infection of adult mouse dorsal root ganglion neurons by rabies virus using nicotinic antagonists. Neurosci. Lett. 229 (3), 198–200.
Chandramouli, S., Ciferri, C., Nikitin, P.A., Calo, S., Gerrein, R., Balabanis, K., Monroe, J., Hebner, C., Lilja, A.E., Settembre, E.C., Carfi, A., 2015. Structure of HCMV glycoprotein B in the postfusion conformation bound to a neutralizing human antibody. Nat. Commun. 6, 8176.
Cheng, C.Y., Shih, W.L., Huang, W.R., Chi, P.I., Wu, M.H., Liu, H.J., 2012. Bovine ephemeral fever virus uses a clathrin-mediated and dynamin 2-dependent endocytosis pathway that requires Rab5 and Rab7 as well as microtubules. J. Virol. 86 (24), 13653–13661.
Chernomordik, L.V., Kozlov, M.M., 2005. Membrane hemifusion: crossing a chasm in two leaps. Cell 123 (3), 375–382.
Chernomordik, L., Kozlov, M.M., Zimmerberg, J., 1995. Lipids in biological membrane fusion. J. Membr. Biol. 146 (1), 1–14.
Chernomordik, L.V., Frolov, V.A., Leikina, E., Bronk, P., Zimmerberg, J., 1998. The pathway of membrane fusion catalyzed by influenza hemagglutinin: restriction of lipids, hemifusion, and lipidic fusion pore formation. J. Cell Biol. 140 (6), 1369–1382.
Ci, Y., Yang, Y., Xu, C., Shi, L., 2018. Vesicular stomatitis virus G protein transmembrane region is crucial for the hemi-fusion to full fusion transition. Sci. Rep. 8 (1), 10669.
Clague, M.J., Schoch, C., Zech, L., Blumenthal, R., 1990. Gating kinetics of pH-activated membrane fusion of vesicular stomatitis virus with cells: stopped-flow measurements by dequenching of octadecylrhodamine fluorescence. Biochemistry 29 (5), 1303–1308.
Cleverley, D.Z., Lenard, J., 1998. The transmembrane domain in viral fusion: essential role for a conserved glycine residue in vesicular stomatitis virus G protein. Proc. Natl. Acad. Sci. U. S. A. 95 (7), 3425–3430.
Cohen, F.S., Melikyan, G.B., 2004. The energetics of membrane fusion from binding, through hemifusion, pore formation, and pore enlargement. J. Membr. Biol. 199 (1), 1–14.
Coil, D.A., Miller, A.D., 2004. Phosphatidylserine is not the cell surface receptor for vesicular stomatitis virus. J. Virol. 78 (20), 10920–10926.
Conti, C., Superti, F., Tsiang, H., 1986. Membrane carbohydrate requirement for rabies virus binding to chicken embryo related cells. Intervirology 26 (3), 164–168.
Coulon, P., Ternaux, J.P., Flamand, A., Tuffereau, C., 1998. An avirulent mutant of rabies virus is unable to infect motoneurons in vivo and in vitro. J. Virol. 72 (1), 273–278.
Crimmins, D.L., Mehard, W.B., Schlesinger, S., 1983. Physical properties of a soluble form of the glycoprotein of vesicular stomatitis virus at neutral and acidic pH. Biochemistry 22 (25), 5790–5796.
Cronin, J., Zhang, X.Y., Reiser, J., 2005. Altering the tropism of lentiviral vectors through pseudotyping. Curr. Gene Ther. 5 (4), 387–398.
Cureton, D.K., Massol, R.H., Saffarian, S., Kirchhausen, T.L., Whelan, S.P., 2009. Vesicular stomatitis virus enters cells through vesicles incompletely coated with clathrin that depend upon actin for internalization. PLoS Pathog. 5 (4), e1000394.
Cureton, D.K., Massol, R.H., Whelan, S.P., Kirchhausen, T., 2010. The length of vesicular stomatitis virus particles dictates a need for actin assembly during clathrin-dependent endocytosis. PLoS Pathog. 6 (9), e1001127.
Dietzgen, R.G., Kondo, H., Goodin, M.M., Kurath, G., Vasilakis, N., 2017. The family Rhabdoviridae: mono- and bipartite negative-sense RNA viruses with diverse genome organization and common evolutionary origins. Virus Res. 227, 158–170.

Dietzschold, B., Gore, M., Marchadier, D., Niu, H.S., Bunschoten, H.M., Otvos Jr., L., Wunner, W.H., Ertl, H.C., Osterhaus, A.D., Koprowski, H., 1990. Structural and immunological characterization of a linear virus-neutralizing epitope of the rabies virus glycoprotein and its possible use in a synthetic vaccine. J. Virol. 64 (8), 3804–3809.

Doms, R.W., Keller, D.S., Helenius, A., Balch, W.E., 1987. Role for adenosine triphosphate in regulating the assembly and transport of vesicular stomatitis virus G protein trimers. J. Cell Biol. 105 (5), 1957–1969.

Durham, N.M., Mulgrew, K., McGlinchey, K., Monks, N.R., Ji, H., Herbst, R., Suzich, J., Hammond, S.A., Kelly, E.J., 2017. Oncolytic VSV primes differential responses to immuno-oncology therapy. Mol. Ther. 25 (8), 1917–1932.

Edson, M.A., Lin, Y.N., Matzuk, M.M., 2010. Deletion of the novel oocyte-enriched gene, Gpr149, leads to increased fertility in mice. Endocrinology 151 (1), 358–368.

Efrat, A., Chernomordik, L.V., Kozlov, M.M., 2007. Point-like protrusion as a prestalk intermediate in membrane fusion pathway. Biophys. J. 92 (8), L61–L63.

Fallahi, F., Wandeler, A.I., Nadin-Davis, S.A., 2016. Characterization of epitopes on the rabies virus glycoprotein by selection and analysis of escape mutants. Virus Res. 220, 161–171.

Ferlin, A., Raux, H., Baquero, E., Lepault, J., Gaudin, Y., 2014. Characterization of pH-sensitive molecular switches that trigger the structural transition of vesicular stomatitis virus glycoprotein from the postfusion state toward the prefusion state. J. Virol. 88 (22), 13396–13409.

Fernandez, M., Porosnicu, M., Markovic, D., Barber, G.N., 2002. Genetically engineered vesicular stomatitis virus in gene therapy: application for treatment of malignant disease. J. Virol. 76 (2), 895–904.

Finkelshtein, D., Werman, A., Novick, D., Barak, S., Rubinstein, M., 2013. LDL receptor and its family members serve as the cellular receptors for vesicular stomatitis virus. Proc. Natl. Acad. Sci. U. S. A. 110 (18), 7306–7311.

Flamand, A., Raux, H., Gaudin, Y., Ruigrok, R.W., 1993. Mechanisms of rabies virus neutralization. Virology 194 (1), 302–313.

Fredericksen, B.L., Whitt, M.A., 1996. Mutations at two conserved acidic amino acids in the glycoprotein of vesicular stomatitis virus affect pH-dependent conformational changes and reduce the pH threshold for membrane fusion. Virology 217 (1), 49–57.

Garry, C.E., Garry, R.F., 2009. Proteomics computational analyses suggest that the bornavirus glycoprotein is a class III viral fusion protein (gamma penetrene). Virol. J. 6, 145.

Gastka, M., Horvath, J., Lentz, T.L., 1996. Rabies virus binding to the nicotinic acetylcholine receptor alpha subunit demonstrated by virus overlay protein binding assay. J. Gen. Virol. 77 (Pt. 10), 2437–2440.

Gaudin, Y., 1997. Folding of rabies virus glycoprotein: epitope acquisition and interaction with endoplasmic reticulum chaperones. J. Virol. 71 (5), 3742–3750.

Gaudin, Y., 2000a. Rabies virus-induced membrane fusion pathway. J. Cell Biol. 150 (3), 601–612.

Gaudin, Y., 2000b. Reversibility in fusion protein conformational changes. The intriguing case of rhabdovirus-induced membrane fusion. Subcell. Biochem. 34, 379–408.

Gaudin, Y., Tuffereau, C., Segretain, D., Knossow, M., Flamand, A., 1991. Reversible conformational changes and fusion activity of rabies virus glycoprotein. J. Virol. 65 (9), 4853–4859.

Gaudin, Y., Ruigrok, R.W., Knossow, M., Flamand, A., 1993. Low-pH conformational changes of rabies virus glycoprotein and their role in membrane fusion. J. Virol. 67 (3), 1365–1372.

Gaudin, Y., Tuffereau, C., Durrer, P., Flamand, A., Ruigrok, R.W., 1995. Biological function of the low-pH, fusion-inactive conformation of rabies virus glycoprotein (G): G is transported in a fusion-inactive state-like conformation. J. Virol. 69 (9), 5528–5534.

Gaudin, Y., Raux, H., Flamand, A., Ruigrok, R.W., 1996. Identification of amino acids controlling the low-pH-induced conformational change of rabies virus glycoprotein. J. Virol. 70 (11), 7371–7378.

Gaudin, Y., de Kinkelin, P., Benmansour, A., 1999. Mutations in the glycoprotein of viral haemorrhagic septicaemia virus that affect virulence for fish and the pH threshold for membrane fusion. J. Gen. Virol. 80 (Pt. 5), 1221–1229.

Ge, P., Tsao, J., Schein, S., Green, T.J., Luo, M., Zhou, Z.H., 2010. Cryo-EM model of the bullet-shaped vesicular stomatitis virus. Science 327 (5966), 689–693.

Ginger, M., Haberl, M., Conzelmann, K.K., Schwarz, M.K., Frick, A., 2013. Revealing the secrets of neuronal circuits with recombinant rabies virus technology. Front. Neural Circuits 7, 2.

Gluska, S., Zahavi, E.E., Chein, M., Gradus, T., Bauer, A., Finke, S., Perlson, E., 2014. Rabies virus hijacks and accelerates the p75NTR retrograde axonal transport machinery. PLoS Pathog. 10 (8), e1004348.

Guo, Y., Duan, M., Wang, X., Gao, J., Guan, Z., Zhang, M., 2019. Early events in rabies virus infection-attachment, entry, and intracellular trafficking. Virus Res. 263, 217–225.

Haberl, M.G., Viana da Silva, S., Guest, J.M., Ginger, M., Ghanem, A., Mulle, C., Oberlaender, M., Conzelmann, K.K., Frick, A., 2015. An anterograde rabies virus vector for high-resolution large-scale reconstruction of 3D neuron morphology. Brain Struct. Funct. 220 (3), 1369–1379.

Hammond, C., Helenius, A., 1994. Folding of VSV G protein: sequential interaction with BiP and calnexin. Science 266 (5184), 456–458.

Harrison, S.C., 2015. Viral membrane fusion. Virology 479–480, 498–507.

Hastie, E., Cataldi, M., Marriott, I., Grdzelishvili, V.Z., 2013. Understanding and altering cell tropism of vesicular stomatitis virus. Virus Res. 176 (1–2), 16–32.

Heldwein, E.E., Lou, H., Bender, F.C., Cohen, G.H., Eisenberg, R.J., Harrison, S.C., 2006. Crystal structure of glycoprotein B from herpes simplex virus 1. Science 313 (5784), 217–220.

Herrmann, A., Clague, M.J., Puri, A., Morris, S.J., Blumenthal, R., Grimaldi, S., 1990. Effect of erythrocyte transbilayer phospholipid distribution on fusion with vesicular stomatitis virus. Biochemistry 29 (17), 4054–4058.

Johannsdottir, H.K., Mancini, R., Kartenbeck, J., Amato, L., Helenius, A., 2009. Host cell factors and functions involved in vesicular stomatitis virus entry. J. Virol. 83 (1), 440–453.

Kadlec, J., Loureiro, S., Abrescia, N.G., Stuart, D.I., Jones, I.M., 2008. The postfusion structure of baculovirus gp64 supports a unified view of viral fusion machines. Nat. Struct. Mol. Biol. 15 (10), 1024–1030.

Kim, I.S., Jenni, S., Stanifer, M.L., Roth, E., Whelan, S.P., van Oijen, A.M., Harrison, S.C., 2017. Mechanism of membrane fusion induced by vesicular stomatitis virus G protein. Proc. Natl. Acad. Sci. U. S. A. 114 (1), E28–E36.

Klingen, Y., Conzelmann, K.K., Finke, S., 2008. Double-labeled rabies virus: live tracking of enveloped virus transport. J. Virol. 82 (1), 237–245.

Kuzmin, P.I., Zimmerberg, J., Chizmadzhev, Y.A., Cohen, F.S., 2001. A quantitative model for membrane fusion based on low-energy intermediates. Proc. Natl. Acad. Sci. U. S. A. 98 (13), 7235–7240.

Lafay, F., Coulon, P., Astic, L., Saucier, D., Riche, D., Holley, A., Flamand, A., 1991. Spread of the CVS strain of rabies virus and of the avirulent mutant AvO1 along the olfactory pathways of the mouse after intranasal inoculation. Virology 183 (1), 320–330.

Lafay, F., Benmansour, A., Chebli, K., Flamand, A., 1996. Immunodominant epitopes defined by a yeast-expressed library of random fragments of the rabies virus glycoprotein map outside major antigenic sites. J. Gen. Virol. 77 (Pt. 2), 339–346.

Lafon, M., Wiktor, T.J., Macfarlan, R.I., 1983. Antigenic sites on the CVS rabies virus glycoprotein: analysis with monoclonal antibodies. J. Gen. Virol. 64 (Pt. 4), 843–851.
Le Blanc, I., Luyet, P.P., Pons, V., Ferguson, C., Emans, N., Petiot, A., Mayran, N., Demaurex, N., Faure, J., Sadoul, R., Parton, R.G., Gruenberg, J., 2005. Endosome-to-cytosol transport of viral nucleocapsids. Nat. Cell Biol. 7 (7), 653–664.
Lee, J., Lentz, B.R., 1998. Secretory and viral fusion may share mechanistic events with fusion between curved lipid bilayers. Proc. Natl. Acad. Sci. U. S. A. 95 (16), 9274–9279.
Lee, A.S., Burdeinick-Kerr, R., Whelan, S.P., 2014. A genome-wide small interfering RNA screen identifies host factors required for vesicular stomatitis virus infection. J. Virol. 88 (15), 8355–8360.
Lentz, T.L., Burrage, T.G., Smith, A.L., Crick, J., Tignor, G.H., 1982. Is the acetylcholine receptor a rabies virus receptor? Science 215 (4529), 182–184.
Lentz, T.L., Wilson, P.T., Hawrot, E., Speicher, D.W., 1984. Amino acid sequence similarity between rabies virus glycoprotein and snake venom curaremimetic neurotoxins. Science 226 (4676), 847–848.
Lentz, T.L., Benson, R.J., Klimowicz, D., Wilson, P.T., Hawrot, E., 1986. Binding of rabies virus to purified Torpedo acetylcholine receptor. Brain Res. 387 (3), 211–219.
Lewis, P., Lentz, T.L., 1998. Rabies virus entry into cultured rat hippocampal neurons. J. Neurocytol. 27 (8), 559–573.
Lewis, P., Fu, Y., Lentz, T.L., 1998. Rabies virus entry into endosomes in IMR-32 human neuroblastoma cells. Exp. Neurol. 153 (1), 65–73.
Liao, M., Kielian, M., 2005. Domain III from class II fusion proteins functions as a dominant-negative inhibitor of virus membrane fusion. J. Cell Biol. 171, 111–120.
Libersou, S., Albertini, A.A., Ouldali, M., Maury, V., Maheu, C., Raux, H., de Haas, F., Roche, S., Gaudin, Y., Lepault, J., 2010. Distinct structural rearrangements of the VSV glycoprotein drive membrane fusion. J. Cell Biol. 191 (1), 199–210.
Liu, H., Liu, Y., Liu, S., Pang, D.W., Xiao, G., 2011. Clathrin-mediated endocytosis in living host cells visualized through quantum dot labeling of infectious hematopoietic necrosis virus. J. Virol. 85 (13), 6252–6262.
Luo, L.H., Li, Y., Snyder, R.M., Wagner, R.R., 1988. Point mutations in glycoprotein gene of vesicular stomatitis virus (New Jersey serotype) selected by resistance to neutralization by epitope-specific monoclonal antibodies. Virology 163 (2), 341–348.
Luyet, P.P., Falguieres, T., Pons, V., Pattnaik, A.K., Gruenberg, J., 2008. The ESCRT-I subunit TSG101 controls endosome-to-cytosol release of viral RNA. Traffic 9 (12), 2279–2290.
Lyles, D.S., Varela, V.A., Parce, J.W., 1990. Dynamic nature of the quaternary structure of the vesicular stomatitis virus envelope glycoprotein. Biochemistry 29 (10), 2442–2449.
Lyles, D.S., McKenzie, M., Parce, J.W., 1992. Subunit interactions of vesicular stomatitis virus envelope glycoprotein stabilized by binding to viral matrix protein. J. Virol. 66 (1), 349–358.
Marsh, M., Helenius, A., 2006. Virus entry: open sesame. Cell 124 (4), 729–740.
Mathieu, M.E., Grigera, P.R., Helenius, A., Wagner, R.R., 1996. Folding, unfolding, and refolding of the vesicular stomatitis virus glycoprotein. Biochemistry 35 (13), 4084–4093.
Matrai, J., Chuah, M.K., VandenDriessche, T., 2010. Recent advances in lentiviral vector development and applications. Mol. Ther. 18 (3), 477–490.
Mazarakis, N.D., Azzouz, M., Rohll, J.B., Ellard, F.M., Wilkes, F.J., Olsen, A.L., Carter, E.E., Barber, R.D., Baban, D.F., Kingsman, S.M., Kingsman, A.J., O'Malley, K., Mitrophanous, K.A., 2001. Rabies virus glycoprotein pseudotyping of lentiviral vectors enables retrograde axonal transport and access to the nervous system after peripheral delivery. Hum. Mol. Genet. 10 (19), 2109–2121.
McGehee, D.S., Role, L.W., 1995. Physiological diversity of nicotinic acetylcholine receptors expressed by vertebrate neurons. Annu. Rev. Physiol. 57, 521–546.

Mire, C.E., White, J.M., Whitt, M.A., 2010. A spatio-temporal analysis of matrix protein and nucleocapsid trafficking during vesicular stomatitis virus uncoating. PLoS Pathog. 6 (7), e1000994.

Motley, A., Bright, N.A., Seaman, M.N., Robinson, M.S., 2003. Clathrin-mediated endocytosis in AP-2-depleted cells. J. Cell Biol. 162 (5), 909–918.

Munis, A.M., Tijani, M., Hassall, M., Mattiuzzo, G., Collins, M.K., Takeuchi, Y., 2018. Characterization of antibody interactions with the G protein of vesicular stomatitis virus Indiana strain and other vesiculovirus G proteins. J. Virol. 92 (23), pii: e00900-18.

Naldini, L., Blomer, U., Gallay, P., Ory, D., Mulligan, R., Gage, F.H., Verma, I.M., Trono, D., 1996. In vivo gene delivery and stable transduction of nondividing cells by a lentiviral vector. Science 272 (5259), 263–267.

Nikolic, J., Belot, L., Raux, H., Legrand, P., Gaudin, Y., Albertini, A.A., 2018. Structural basis for the recognition of LDL-receptor family members by VSV glycoprotein. Nat. Commun. 9 (1), 1029.

Ohishi, H., Shigemoto, R., Nakanishi, S., Mizuno, N., 1993. Distribution of the messenger RNA for a metabotropic glutamate receptor, mGluR2, in the central nervous system of the rat. Neuroscience 53 (4), 1009–1018.

Pak, C.C., Puri, A., Blumenthal, R., 1997. Conformational changes and fusion activity of vesicular stomatitis virus glycoprotein: [125I]iodonaphthyl azide photolabeling studies in biological membranes. Biochemistry 36 (29), 8890–8896.

Pasqual, G., Rojek, J.M., Masin, M., Chatton, J.Y., Kunz, S., 2011. Old world arenaviruses enter the host cell via the multivesicular body and depend on the endosomal sorting complex required for transport. PLoS Pathog. 7 (9), e1002232.

Peng, R., Zhang, S., Cui, Y., Shi, Y., Gao, G.F., Qi, J., 2017. Structures of human-infecting Thogotovirus fusogens support a common ancestor with insect baculovirus. Proc. Natl. Acad. Sci. U. S. A. 114 (42), E8905–E8912.

Piccinotti, S., Whelan, S.P., 2016. Rabies internalizes into primary peripheral neurons via clathrin coated pits and requires fusion at the cell body. PLoS Pathog. 12 (7), e1005753.

Piccinotti, S., Kirchhausen, T., Whelan, S.P., 2013. Uptake of rabies virus into epithelial cells by clathrin-mediated endocytosis depends upon actin. J. Virol. 87 (21), 11637–11647.

Prehaud, C., Coulon, P., LaFay, F., Thiers, C., Flamand, A., 1988. Antigenic site II of the rabies virus glycoprotein: structure and role in viral virulence. J. Virol. 62 (1), 1–7.

Prehaud, C., Wolff, N., Terrien, E., Lafage, M., Megret, F., Babault, N., Cordier, F., Tan, G.S., Maitrepierre, E., Menager, P., Chopy, D., Hoos, S., England, P., Delepierre, M., Schnell, M.J., Buc, H., Lafon, M., 2010. Attenuation of rabies virulence: takeover by the cytoplasmic domain of its envelope protein. Sci. Signal. 3 (105), ra5.

Puri, A., Winick, J., Lowy, R.J., Covell, D., Eidelman, O., Walter, A., Blumenthal, R., 1988. Activation of vesicular stomatitis virus fusion with cells by pretreatment at low pH. J. Biol. Chem. 263 (10), 4749–4753.

Puri, A., Grimaldi, S., Blumenthal, R., 1992. Role of viral envelope sialic acid in membrane fusion mediated by the vesicular stomatitis virus envelope glycoprotein. Biochemistry 31 (41), 10108–10113.

Raux, H., Coulon, P., Lafay, F., Flamand, A., 1995. Monoclonal antibodies which recognize the acidic configuration of the rabies glycoprotein at the surface of the virion can be neutralizing. Virology 210 (2), 400–408.

Reiter, R.E., Gu, Z., Watabe, T., Thomas, G., Szigeti, K., Davis, E., Wahl, M., Nisitani, S., Yamashiro, J., Le Beau, M.M., Loda, M., Witte, O.N., 1998. Prostate stem cell antigen: a cell surface marker overexpressed in prostate cancer. Proc. Natl. Acad. Sci. U. S. A. 95 (4), 1735–1740.

Roche, S., Gaudin, Y., 2002. Characterization of the equilibrium between the native and fusion-inactive conformation of rabies virus glycoprotein indicates that the fusion complex is made of several trimers. Virology 297 (1), 128–135.

Roche, S., Gaudin, Y., 2004. Evidence that rabies virus forms different kinds of fusion machines with different pH thresholds for fusion. J. Virol. 78 (16), 8746–8752.
Roche, S., Bressanelli, S., Rey, F.A., Gaudin, Y., 2006. Crystal structure of the low-pH form of the vesicular stomatitis virus glycoprotein G. Science 313 (5784), 187–191.
Roche, S., Rey, F.A., Gaudin, Y., Bressanelli, S., 2007. Structure of the prefusion form of the vesicular stomatitis virus glycoprotein G. Science 315 (5813), 843–848.
Roche, S., Albertini, A.A., Lepault, J., Bressanelli, S., Gaudin, Y., 2008. Structures of vesicular stomatitis virus glycoprotein: membrane fusion revisited. Cell. Mol. Life Sci. 65 (11), 1716–1728.
Roth, S.L., Whittaker, G.R., 2011. Promotion of vesicular stomatitis virus fusion by the endosome-specific phospholipid bis(monoacylglycero)phosphate (BMP). FEBS Lett. 585 (6), 865–869.
Rubinstein, M., 2018. Is Lgr4 essential for VSV- and VSV-G-pseudotyped lentiviral vector entry to cells? J. Biol. Chem. 293 (1), 112.
Rudenko, G., Henry, L., Henderson, K., Ichtchenko, K., Brown, M.S., Goldstein, J.L., Deisenhofer, J., 2002. Structure of the LDL receptor extracellular domain at endosomal pH. Science 298 (5602), 2353–2358.
Russell, C.J., Jardetzky, T.S., Lamb, R.A., 2001. Membrane fusion machines of paramyxoviruses: capture of intermediates of fusion. EMBO J. 20, 4024–4034.
Schlegel, R., Tralka, T.S., Willingham, M.C., Pastan, I., 1983. Inhibition of VSV binding and infectivity by phosphatidylserine: is phosphatidylserine a VSV-binding site? Cell 32 (2), 639–646.
Seif, I., Coulon, P., Rollin, P.E., Flamand, A., 1985. Rabies virulence: effect on pathogenicity and sequence characterization of rabies virus mutations affecting antigenic site III of the glycoprotein. J. Virol. 53 (3), 926–934.
Shao, L., Zhao, J., Zhang, H., 2016. Spring viraemia of carp virus enters grass carp ovary cells via clathrin-mediated endocytosis and macropinocytosis. J. Gen. Virol. 97 (11), 2824–2836.
Si, Z., Zhang, J., Shivakoti, S., Atanasov, I., Tao, C.L., Hui, W.H., Zhou, K., Yu, X., Li, W., Luo, M., Bi, G.Q., Zhou, Z.H., 2018. Different functional states of fusion protein gB revealed on human cytomegalovirus by cryo electron tomography with Volta phase plate. PLoS Pathog. 14 (12), e1007452.
Sieczkarski, S.B., Whittaker, G.R., 2003. Differential requirements of Rab5 and Rab7 for endocytosis of influenza and other enveloped viruses. Traffic 4 (5), 333–343.
Sinibaldi, L., Goldoni, P., Seganti, L., Superti, F., Tsiang, H., Orsi, N., 1985. Gangliosides in early interactions between vesicular stomatitis virus and CER cells. Microbiologica 8 (4), 355–365.
Stanifer, M.L., Cureton, D.K., Whelan, S.P., 2011. A recombinant vesicular stomatitis virus bearing a lethal mutation in the glycoprotein gene uncovers a second site suppressor that restores fusion. J. Virol. 85 (16), 8105–8115.
Sun, X., Yau, V.K., Briggs, B.J., Whittaker, G.R., 2005. Role of clathrin-mediated endocytosis during vesicular stomatitis virus entry into host cells. Virology 338 (1), 53–60.
Sun, X., Belouzard, S., Whittaker, G.R., 2008. Molecular architecture of the bipartite fusion loops of vesicular stomatitis virus glycoprotein G, a class III viral fusion protein. J. Biol. Chem. 283 (10), 6418–6427.
Superti, F., Derer, M., Tsiang, H., 1984a. Mechanism of rabies virus entry into CER cells. J. Gen. Virol. 65 (Pt. 4), 781–789.
Superti, F., Seganti, L., Tsiang, H., Orsi, N., 1984b. Role of phospholipids in rhabdovirus attachment to CER cells. Brief report. Arch. Virol. 81 (3–4), 321–328.
Superti, F., Hauttecoeur, B., Morelec, M.J., Goldoni, P., Bizzini, B., Tsiang, H., 1986. Involvement of gangliosides in rabies virus infection. J. Gen. Virol. 67 (Pt. 1), 47–56.

Superti, F., Seganti, L., Ruggeri, F.M., Tinari, A., Donelli, G., Orsi, N., 1987. Entry pathway of vesicular stomatitis virus into different host cells. J. Gen. Virol. 68 (Pt. 2), 387–399.

Thoulouze, M.I., Lafage, M., Schachner, M., Hartmann, U., Cremer, H., Lafon, M., 1998. The neural cell adhesion molecule is a receptor for rabies virus. J. Virol. 72 (9), 7181–7190.

Tuffereau, C., Leblois, H., Benejean, J., Coulon, P., Lafay, F., Flamand, A., 1989. Arginine or lysine in position 333 of ERA and CVS glycoprotein is necessary for rabies virulence in adult mice. Virology 172 (1), 206–212.

Tuffereau, C., Benejean, J., Blondel, D., Kieffer, B., Flamand, A., 1998. Low-affinity nerve-growth factor receptor (P75NTR) can serve as a receptor for rabies virus. EMBO J. 17 (24), 7250–7259.

Tuffereau, C., Desmezieres, E., Benejean, J., Jallet, C., Flamand, A., Tordo, N., Perrin, P., 2001. Interaction of lyssaviruses with the low-affinity nerve-growth factor receptor p75NTR. J. Gen. Virol. 82 (Pt. 12), 2861–2867.

Tuffereau, C., Schmidt, K., Langevin, C., Lafay, F., Dechant, G., Koltzenburg, M., 2007. The rabies virus glycoprotein receptor p75NTR is not essential for rabies virus infection. J. Virol. 81 (24), 13622–13630.

Vandepol, S.B., Lefrancois, L., Holland, J.J., 1986. Sequences of the major antibody binding epitopes of the Indiana serotype of vesicular stomatitis virus. Virology 148 (2), 312–325.

Walker, P.J., Firth, C., Widen, S.G., Blasdell, K.R., Guzman, H., Wood, T.G., Paradkar, P.N., Holmes, E.C., Tesh, R.B., Vasilakis, N., 2015. Evolution of genome size and complexity in the Rhabdoviridae. PLoS Pathog. 11 (2), e1004664.

Walker, P.J., Blasdell, K.R., Calisher, C.H., Dietzgen, R.G., Kondo, H., Kurath, G., Longdon, B., Stone, D.M., Tesh, R.B., Tordo, N., Vasilakis, N., Whitfield, A.E., Ictv Report, C., 2018. ICTV virus taxonomy profile: Rhabdoviridae. J. Gen. Virol. 99 (4), 447–448.

Wang, J., Wang, Z., Liu, R., Shuai, L., Wang, X., Luo, J., Wang, C., Chen, W., Wang, X., Ge, J., He, X., Wen, Z., Bu, Z., 2018. Metabotropic glutamate receptor subtype 2 is a cellular receptor for rabies virus. PLoS Pathog. 14 (7) e1007189.

Weekes, M.P., Antrobus, R., Talbot, S., Hor, S., Simecek, N., Smith, D.L., Bloor, S., Randow, F., Lehner, P.J., 2012. Proteomic plasma membrane profiling reveals an essential role for gp96 in the cell surface expression of LDLR family members, including the LDL receptor and LRP6. J. Proteome Res. 11 (3), 1475–1484.

Wild, C.T., Shugars, D.C., Greenwell, T.K., McDanal, C.B., Matthews, T.J., 1994. Peptides corresponding to a predictive alpha-helical domain of human immunodeficiency virus type 1 gp41 are potent inhibitors of virus infection. Proc. Natl. Acad. Sci. U. S. A. 91, 9770–9774.

Willnow, T.E., 1998. Receptor-associated protein (RAP): a specialized chaperone for endocytic receptors. Biol. Chem. 379 (8-9), 1025–1031.

Wimley, W.C., White, S.H., 1992. Partitioning of tryptophan side-chain analogs between water and cyclohexane. Biochemistry 31 (51), 12813–12818.

Yamada, S., Ohnishi, S., 1986. Vesicular stomatitis virus binds and fuses with phospholipid domain in target cell membranes. Biochemistry 25 (12), 3703–3708.

Zagouras, P., Ruusala, A., Rose, J.K., 1991. Dissociation and reassociation of oligomeric viral glycoprotein subunits in the endoplasmic reticulum. J. Virol. 65 (4), 1976–1984.

Zaitseva, E., Mittal, A., Griffin, D.E., Chernomordik, L.V., 2005. Class II fusion protein of alphaviruses drives membrane fusion through the same pathway as class I proteins. J. Cell Biol. 169 (1), 167–177.

Zhang, N., Huang, H., Tan, B., Wei, Y., Xiong, Q., Yan, Y., Hou, L., Wu, N., Siwko, S., Cimarelli, A., Xu, J., Han, H., Qian, M., Liu, M., Du, B., 2017. Leucine-rich repeat-containing G protein-coupled receptor 4 facilitates vesicular stomatitis virus infection by binding vesicular stomatitis virus glycoprotein. J. Biol. Chem. 292 (40), 16527–16538.

CHAPTER SIX

Hantavirus entry: Perspectives and recent advances

Eva Mittler†, Maria Eugenia Dieterle†, Lara M. Kleinfelter†, Megan M. Slough†, Kartik Chandran*, Rohit K. Jangra*

Department of Microbiology and Immunology, Albert Einstein College of Medicine, Bronx, NY, United States

*Corresponding authors: e-mail address: kartik.chandran@einstein.yu.edu; rohit.jangra@einstein.yu.edu

Contents

Abstract

Hantaviruses are important zoonotic pathogens of public health importance that are found on all continents except Antarctica and are associated with hemorrhagic fever with renal syndrome (HFRS) in the Old World and hantavirus pulmonary syndrome (HPS) in the New World. Despite the significant disease burden they cause, no FDA-approved specific therapeutics or vaccines exist against these lethal viruses. The lack of available interventions is largely due to an incomplete understanding of hantavirus pathogenesis and molecular mechanisms of virus replication, including cellular entry. Hantavirus Gn/Gc glycoproteins are the only viral proteins exposed on the surface of virions and are necessary and sufficient to orchestrate virus attachment and entry. *In vitro* studies have implicated integrins (β_{1-3}), DAF/CD55, and gC1qR as candidate receptors that mediate viral attachment for both Old World and New World hantaviruses.

† These authors contributed equally to this manuscript.

Advances in Virus Research, Volume 104
ISSN 0065-3527
https://doi.org/10.1016/bs.aivir.2019.07.002

Recently, protocadherin-1 (PCDH1) was demonstrated as a requirement for cellular attachment and entry of New World hantaviruses *in vitro* and lethal HPS *in vivo*, making it the first clade-specific host factor to be identified. Attachment of hantavirus particles to cellular receptors induces their internalization by clathrin-mediated, dynamin-independent, or macropinocytosis-like mechanisms, followed by particle trafficking to an endosomal compartment where the fusion of viral and endosomal membranes can occur. Following membrane fusion, which requires cholesterol and acid pH, viral nucleocapsids escape into the cytoplasm and launch genome replication. In this review, we discuss the current mechanistic understanding of hantavirus entry, highlight gaps in our existing knowledge, and suggest areas for future inquiry.

Abbreviations

Ad5V	adenovirus serotype-5
ADCC	antibody-dependent cellular cytotoxicity
AdV	adenovirus
ANDV	Andes virus
BCCV	Black Creek Canal virus
cDNA	complementary DNA
CHO	Chinese hamster ovary
CRISPR	clustered regularly interspaced short palindromic repeats
CryoEM	cryo-electron microscopy
DAF/CD55	decay accelerating factor/cluster of differentiation-55
DOBV	Dobrava-Belgrade virus
EC1	first extracellular cadherin repeat
EGF	epidermal growth factor
ER	endoplasmic reticulum
Fc	fragment crystallizable region
FDA	United States Food and Drug Administration
Gc	C-terminal glycoprotein subunit
gC1qR	globular head of complement component 1 Q-subcomponent receptor
Gn	N-terminal glycoprotein subunit
GPC	glycoprotein precursor
GPI	glycosylphosphatidylinositol
HFRS	hemorrhagic fever with renal syndrome
HPS	hantavirus pulmonary syndrome
HTNV	Hantaan virus
HUVEC	human umbilical vein endothelial cells
KO	knockout
mAb	monoclonal antibody
MPRLV	Maporal virus
N	nucleocapsid protein
nAb	neutralizing antibody
NIAID	United States National Institute of Allergy and Infectious Diseases
NSs	hantavirus non-structural protein
NY-1V	New York-1 virus

PCDH1	protocadherin-1
PHV	Prospect Hill virus
PSI	plexin-semaphorin-integrin
PUUV	Puumala virus
RdRp	RNA-dependent RNA polymerase
RGD	arginine-glycine-aspartate
SANGV	Sangassou virus
SEOV	Seoul virus
siRNA	small interfering RNA
SNV	Sin Nombre virus
TMIR	target membrane-interacting region
TULV	Tula virus
VEGF	vascular endothelial growth factor
VEGFR2	vascular endothelial growth factor receptor-2
VLP	virus-like particle
vRNA	viral RNA
VSV	vesicular stomatitis virus
WT	wild-type

1. Introduction

Hantaviruses are emerging rodent-borne pathogens whose accidental human transmission causes Hemorrhagic Fever with Renal Syndrome (HFRS) in the Old World and Hantavirus Pulmonary Syndrome (HPS) in the New World (Vaheri et al., 2013). HPS-causing hantaviruses are listed as Category A pathogens by the US National Institute of Allergy and Infectious Diseases (NIAID). Over 50,000 cases occur globally each year with fatality rates of up to 12% (HFRS) and 40% (HPS), depending on the infecting virus (Watson et al., 2014). The number of cases is predicted to increase as climate change and deforestation cause increases in rodent populations, bringing them into close proximity with human settlements (Guterres and de Lemos, 2018; Klempa, 2009). Despite the likelihood that the burden of hantavirus disease will increase, no FDA-approved vaccines or therapeutics exist against HFRS or HPS, in part due to a limited understanding of the viral life cycle. Although the discovery of HPS-causing hantaviruses in the 1990s significantly accelerated hantavirus research, the specific roles of host factors and the details of entry mechanisms remain poorly characterized. Here, we summarize the current understanding of hantavirus entry, including the structures of viral surface glycoproteins, the roles of candidate receptors in entry, pathogenesis, and host range. We also discuss surrogate systems to study hantavirus entry, and the targeting of hantavirus glycoproteins for the development of vaccines and antibody-based therapeutics. Finally, we highlight key questions that remain unanswered in the field.

2. Hantavirus ecology and diseases

Hantaviruses are members of the family *Hantaviridae* within the order *Bunyavirales*, which includes many viruses of medical, agricultural, and veterinary importance. Similar to some other bunyaviruses, hantaviruses are maintained and transmitted through many mammalian host reservoirs including bats (order *Chiroptera*), shrews and moles (order *Soricomorpha*), and rodents (order *Rodentia*) (Yanagihara et al., 2014; Zhang, 2014). Virus-host relationships appear to be specific—in general, each hantavirus associates with only one reservoir host (Table 1). Rodent-borne hantaviruses are classified into three clades based on their reservoir hosts: (1) New World viruses [e.g., Sin Nombre virus (SNV), New York-1 virus (NY-1V), and Andes virus (ANDV)] causing HPS that are carried by members of the *Sigmodontinae* sub-family, (2) Old World viruses [e.g., Hantaan virus (HTNV), Seoul virus (SEOV), and Dobrava-Belgrade virus (DOBV)] causing HFRS that are harbored in *Murinae* rodents, and (3) viruses found in the New or Old World that are associated with mild disease [Puumala virus (PUUV)] or are non-virulent [e.g., Prospect Hill virus (PHV) and Tula virus (TULV)] and are carried by *Arvicolinae* rodents (Meyer and Schmaljohn, 2000). Phylogenetic analyses suggest that hantaviruses originated in bats, shrews, or moles and later established persistent infections in rodents. Hantaviruses further diverged as rodent populations underwent geographic separation and divergent evolution (Witkowski et al., 2016; Yanagihara et al., 2014; Zhang, 2014). It is assumed that this co-evolution of virus and host ultimately led to the emergence of virulent hantaviruses, as only rodent-borne hantaviruses, found in the *Orthohantavirus* genus, are recognized to cause human disease (Forbes et al., 2018). However, there is evidence supporting the occurrence of shrew-borne hantavirus infection in humans (Heinemann et al., 2016; Okumura et al., 2007).

Hantaviruses establish a persistent infection in their rodent hosts that is generally recognized as asymptomatic and can persist for months to the lifetime of the animal (Ermonval et al., 2016; Lee et al., 1978; Yanagihara et al., 1985). Once infected, the host will shed virus through its saliva, urine, and

Table 1 Selected hantaviruses and their reservoir hosts.

Virus	Host sub-family	Host species	Distribution	Associated disease (fatality rate)
Hantaan	*Murinae*	Striped field mouse (*Apodemus agrarius*)	Asia and Russia	HFRS (10–15%)
Dobrava-Belgrade	*Murinae*	Yellow-necked mouse (*Apodemus flavicollis*)	Europe	HFRS (7–12%)
Seoul	*Murinae*	Norway rat (*Rattus norvegicus*)	Worldwide	HFRS (1–5%)
Puumala	*Arvicolinae*	Bank vole (*Clethrionomys glareolus*)	Europe	HFRS (0.1–0.4%)
Khabarovsk	*Arvicolinae*	Reed vole (*Microtus fortis*)	Russia	N/A
Prospect Hill	*Arvicolinae*	Meadow vole (*Microtus pennsylvanicus*)	North America	N/A
Tula	*Arvicolinae*	Common vole (*Microtus arvalis*)	Europe	Unknown
Isla Vista	*Arvicolinae*	California vole (*Microtus californicus*)	North America	N/A
Sin Nombre	*Sigmodontinae*	Deer mouse (*Peromyscus maniculatus*)	North America	HPS (40–60%)
Andes	*Sigmodontinae*	Long-tailed pygmy rice rat (*Oligoryzomys longicaudatus*)	South America	HPS (43–56%)
Black Creek Canal	*Sigmodontinae*	Hispid cotton rat (*Sigmodon hispidus*)	North America	HPS (>40%)
Rio Segundo	*Sigmodontinae*	Mexican harvest mouse (*Reithrodontomys mexicanus*)	Central America	Unknown
Laguna Negra	*Sigmodontinae*	Large vesper mouse (*Calomys callosus*)	South America	HPS (15%)

Red lines separate the rodent-borne hantavirus clades. Viruses in each clade are carried by rodents of the same sub-family. Hantavirus species are found on most continents and display varied virulence in humans (Hardcastle et al., 2016; Jiang et al., 2017; Meyer and Schmaljohn, 2000).

feces, allowing it to infect other rodents through bites or through its excrement, which can form aerosols inhaled by other rodents (Nuzum et al., 1988; Voutilainen et al., 2015). Human infection is considered a spillover event and the main transmission route is through inhalation of virus particles from contaminated rodent excrement (Johnson, 2001). There is some evidence that infection can also occur by the gastrointestinal route (Witkowski et al., 2017). Human-to-human transmission has only been documented for ANDV and occurs among those with fairly intimate contact, such as those who sleep in close proximity over a prolonged period (Martinez et al., 2005; Martinez-Valdebenito et al., 2014). It is suspected that ANDV can be transmitted through bodily fluids from infected individuals. Human infection is characterized by fever, muscle aches, gastrointestinal symptoms, and, most notably, vascular leakage (Sargianou et al., 2012). Due to similar underlying pathology and clinical features such as vascular leakage and thrombocytopenia, some have argued that HFRS and HPS are one disease, and not two (Clement et al., 2014; Rasmuson et al., 2011). However, renal involvement is more prominent in HFRS cases, resulting in oliguria and renal failure, and pulmonary symptoms are more pronounced in HPS cases and many patients experience pulmonary edema (Sargianou et al., 2012), raising the possibility that viral differences in host factor usage account, in part, for these differences in hantavirus pathogenesis.

3. Hantavirus particle and genes

The infectious virions of hantaviruses are enveloped spherical particles with a diameter of 80–120 nm (McCormick et al., 1982; Muyangwa et al., 2015) (Fig. 1A). The negative-strand RNA genomes of hantaviruses comprise three segments: a 1.8–2.1 kb small segment (S), a 3.7–3.8 kb medium segment (M), and a 6.5–6.6 kb large segment (L) (Fig. 1B) (Hepojoki et al., 2012; Schmaljohn et al., 1983). Each segment has highly conserved 3′ and 5′ untranslated regions that contain complementary nucleotides (Chizhikov et al., 1995; Schmaljohn and Dalrymple, 1983). These regions recognize each other to form panhandle structures that give the RNA segments a circular appearance in some microscopic images. These panhandle structures are a hallmark of bunyaviruses (Elliott et al., 1991) and are thought to regulate viral transcription and replication, as demonstrated for other negative-strand RNA viruses, such as vesicular stomatitis virus (Wertz et al., 1994) and influenza A virus (Lee and Seong, 1996; Tiley et al., 1994). Viral RNA (vRNA) segments are each associated with an RNA-dependent

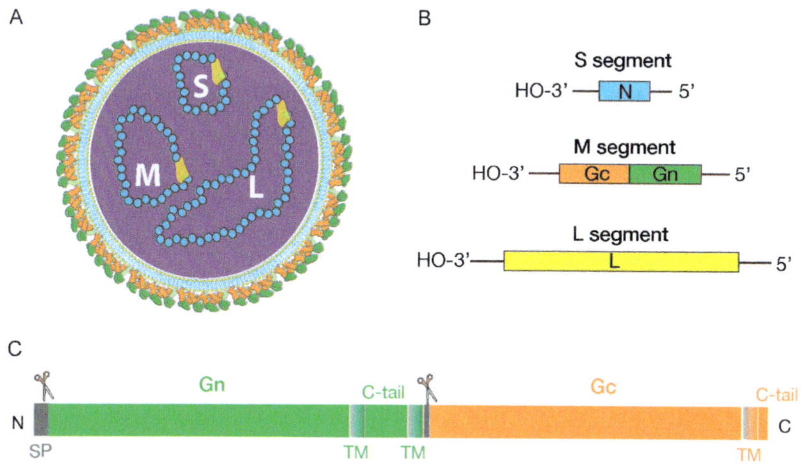

Fig. 1 Schematic representation of hantavirus particle and genes. (A) Hantaviruses carry a tri-segmented negative-sense RNA genome encased in a lipid envelope that is studded with surface spikes comprising the Gn and Gc glycoproteins. (B) Small (S), medium (M) and large (L) genome segments encode nucleoprotein (N), Gn/Gc glycoproteins and RNA-dependent RNA polymerase, respectively. (C) Linear representation of hantavirus Gn/Gc. Gn is proteolytically cleaved from Gc at the conserved WAASA site by the cellular signal protease complex (represented by scissors). SP, signal peptide; TM, transmembrane region (shown as shaded block); C-tail, cytoplasmic tail.

RNA polymerase (RdRp) and are encased in nucleocapsid protein (Hepojoki et al., 2012; Mir et al., 2006). These are contained in a host-derived lipid envelope that is decorated with two glycoproteins, Gn (N-terminal subunit) and Gc (C-terminal subunit) (Fig. 1B).

The L and S segments encode the 250-kDa RdRp and the 50-kDa nucleocapsid protein (N), respectively (Schmaljohn et al., 1983, 1986). The RdRp mediates transcription and replication of the hantavirus genome with its transcriptase, replicase, and endonuclease activities (Kukkonen et al., 2005; Muyangwa et al., 2015). The N protein coats the vRNA to protect it from cellular nucleases and works together with RdRp to ensure efficient replication of the viral genome (Mir and Panganiban, 2006; Mir et al., 2006). In addition, ANDV N protein enhances microvascular endothelial cell permeability (Gorbunova et al., 2016). It was also reported that N proteins help mediate viral immune evasion by protecting infected endothelial cells from T cell attack and by down-regulating the type I interferon response (Gupta et al., 2013; Pan et al., 2015). Akin to other bunyaviruses, hantaviruses harbored in *Arvicolinae* and *Sigmodontinae* rodents have an overlapping open reading frame (+1) in the S segment that encodes for a putative

7–10 kDa non-structural protein NS_S. However, only PUUV, TULV and ANDV are known to express it during infection (Jääskeläinen et al., 2007; Vera-Otarola et al., 2012). The reduced frequency of nucleotide substitutions in the predicted NS_S coding region likely indicates a functional role for this protein—NS_S has been proposed to inhibit the host's interferon response (Jääskeläinen et al., 2007). However, the contribution of NS_S during hantavirus pathogenesis and the reasons for its presence only in viruses infecting arvicoline and sigmodontine rodents are still unknown.

The M segment encodes a glycoprotein precursor (GPC) of 1133–1158 amino acids (Schmaljohn et al., 1987). A signal peptide at the N-terminus of GPC directs the translating ribosomes to the endoplasmic reticulum (ER), where GPC is co-translationally cleaved by the cellular signal peptidase complex at a conserved WAASA sequence to yield Gn and Gc (Fig. 1C) (Kamrud and Schmaljohn, 1994; Löber et al., 2001). Gn and Gc undergo N- and O-glycosylations and traffic together to the Golgi apparatus prior to their incorporation into viral particles. Gn and Gc are the only viral proteins on the surface of hantavirus virions and orchestrate viral entry into susceptible cells.

4. Hantavirus entry pathway

Although hantaviruses cause human diseases worldwide, our knowledge of their entry mechanisms remains limited. This lack of understanding presents one of the major obstacles to the development of new therapies. Viral attachment to the cell surface is the first essential step to establish infection; several cellular proteins have been implicated in mediating Old World and New World hantavirus attachment to cells (see Section 6 for more details, Fig. 2).

Following attachment, hantaviruses are internalized, possibly through a variety of endocytic strategies including clathrin-mediated endocytosis and dynamin-independent pathways, depending on cell type and experimental conditions. Some studies suggest that Old World hantaviruses (HTNV, SEOV) are internalized via clathrin-mediated endocytosis (Jin et al., 2002), whereas studies with the New World hantavirus ANDV suggest a more complex picture (Chiang et al., 2016; Ramanathan and Jonsson, 2008). Specifically, Ramanathan and Jonsson described dynamin-independent uptake of ANDV, whereas a small interfering RNA (siRNA) screen in human lung microvascular cells identified genes encoding for dynamin, the clathrin heavy chain, and the adaptor protein AP2, three major

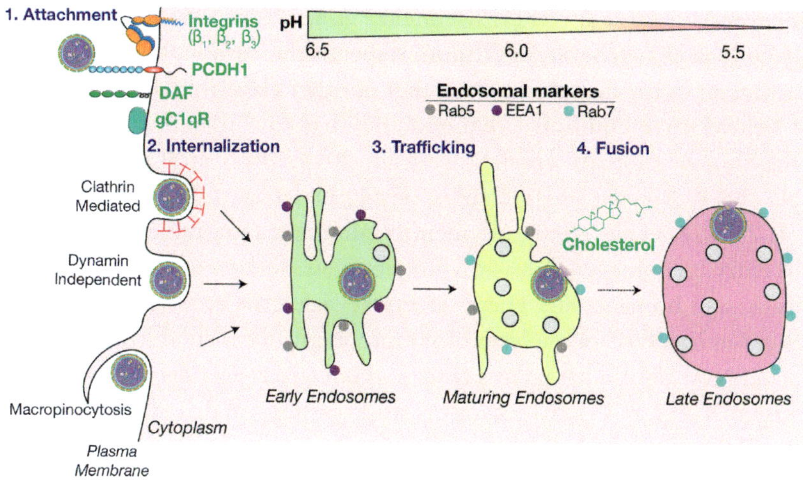

Fig. 2 An overview of the hantavirus entry pathway. Host factors demonstrated or proposed to play a role in four stages of hantavirus entry, i.e., attachment, internalization, trafficking and membrane fusion are indicated. See the text for details.

components of the clathrin-dependent endocytosis pathway, as major hits for ANDV internalization (Chiang et al., 2016; Ramanathan and Jonsson, 2008). Further, ANDV was shown to be dependent on host factors implicated in dynamin-independent macropinocytosis, including cholesterol and the Rho GTPase Rac11 (Chiang et al., 2016; Torriani et al., 2019).

After internalization, viral particles are delivered to early endosomes and eventually trafficked to late endosomal and lysosomal compartments. One of the hallmarks of endosomal maturation is the decrease of their intra-luminal pH, from mildly acidic in early endosomes to strongly acidic in endo-lysosomes. Because hantaviruses require acid pH (5.8–6.3) to undergo membrane fusion (Arikawa et al., 1985; Cifuentes-Muñoz et al., 2011; Kleinfelter et al., 2015), they are generally assumed to traffic to early or late endosomes (Albornoz et al., 2016). However, details on the molecular mechanisms of hantavirus endocytic trafficking remain limited.

Viral entry culminates in fusion of the viral membrane with the endosomal membrane, followed by escape of the viral nucleocapsid payload into the cytoplasm. Hantavirus membrane fusion is driven by large conformational changes in the Gc protein. The site of hantavirus membrane fusion and the endosomal conditions required for fusion, other than luminal acid pH, are not well defined. Membrane fusion mediated by Gn/Gc of New World hantavirus ANDV has a pH threshold of 5.8, whereas that by Old World hantavirus HTNV Gn/Gc is 6.3 (Arikawa et al., 1985;

Cifuentes-Muñoz et al., 2011; Kleinfelter et al., 2015), suggesting late and early endosomes as the sites of fusion, respectively. Although pH is a major determinant of the fusion site, a number of other class II viral glycoproteins (see below) have additional requirements that can influence the identity of the fusion compartment (Bitto et al., 2016; Bron et al., 1993; Dubé et al., 2014; Kielian and Helenius, 1984; Lu et al., 1999; Nieva et al., 1994; Zaitseva et al., 2010). To identify such host requirements for hantavirus entry, genome-wide loss-of-function genetic screens were performed in insect and human cells. These screens revealed the requirement of multiple genes involved in cholesterol sensing, regulation, and biosynthesis for Old and New World hantavirus entry and infection (Kleinfelter et al., 2015; Petersen et al., 2014). Mechanistic studies showed that the cholesterol content of target membranes strongly influences the efficiency of hantavirus Gn/Gc-mediated membrane fusion (Kleinfelter et al., 2015). While alphaviruses, such as Semliki Forest virus and Sindbis virus, require cholesterol in the target membrane for virus fusion (Bron et al., 1993; Kielian and Helenius, 1984; Lu et al., 1999), hantavirus membrane fusion and entry is exquisitely sensitive to small reductions in cellular cholesterol levels, underlining this key requirement (Kleinfelter et al., 2015). Following membrane fusion, the viral nucleocapsid is presumed to escape into the cytoplasm. The molecular details of viral "uncoating" and the delivery of nucleocapsid cores to intracellular sites of transcription and replication remain obscure, as is the potential involvement of host factors in these processes.

5. Structure and function of hantavirus glycoproteins

Hantavirus Gn/Gc are necessary and sufficient for cellular entry. On the viral membrane, Gn and Gc form a grid-like pattern that is a characteristic of hantaviruses (Huiskonen et al., 2010; Martin et al., 1985). Each square-shaped Gn/Gc spike has a fourfold symmetry and is composed of four copies each of Gn and Gc (Fig. 3A). The viral Gn is a ~70 kDa protein composed of anti-parallel β sheets and α helices (Rissanen et al., 2017) (Fig. 3A). Fitting the crystal structure of PUUV Gn onto a cryo-electron tomography reconstruction of the TULV surface showed that Gn is distal to the viral membrane and sits atop Gc, probably protecting the target membrane-interacting region (TMIR) (Li et al., 2016). Gn contains seven disulfide bonds that stabilize the protein and that are conserved among hantaviruses (Li et al., 2016). In addition, Gn contains multiple N-linked glycans that project toward neighboring Gn proteins and may be important for

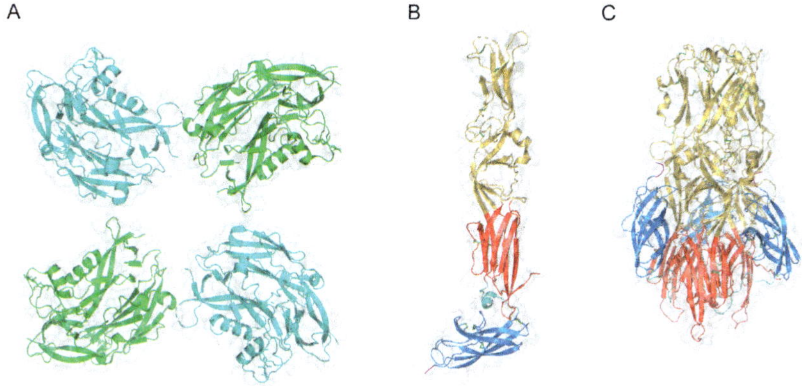

Fig. 3 Structure of hantavirus Gn and Gc. (A) Partial structural model for Gn tetramer derived by fitting the X-ray crystal structure of the N-terminal region of PUUV Gn (amino acid residues 29–383; PDB ID: 5FXU) into the cryo-electron tomography (cryo-ET)-derived density map of the TULV glycoprotein spike (PDB ID: 5FYN) (Li et al., 2016). Adjacent subunits are colored in cyan and green. Gray outlines in (A)–(C) represent surface-shaded views. *En face* view relative to the surface of the viral particle is shown. (B) X-ray crystal structure of the monomeric ectodomain of HTNV Gc in its putative pre-fusion conformation (PDB ID: 5LJY) (Guardado-Calvo et al., 2016). (C) X-ray crystal structure of the ectodomain of HTNV Gc in its trimeric post-fusion conformation (PDB ID: 5LKO) (Guardado-Calvo et al., 2016). (B) and (C) Gc is colored according to the convention for Class II fusion proteins: domains I, II, and III are shown in red, yellow, and blue, respectively. The DI–DIII linker is shown in cyan, and the beginning of the stem region at the C-terminus of the crystallized ectodomain constructs is shown in magenta. Disulfide bonds are shown as sticks, with sulfur atoms colored green. (C) The trimer is oriented axially to the viral and target membranes, with the fusion loops in domain II at the top of the structure.

Gn-Gn interactions. Although soluble Gn is monomeric and has limited Gn-Gn interactions, acidification is proposed to lead to the formation of compact Gn tetramers by locking a β-hairpin into a pocket on the adjacent Gn subunit (Rissanen et al., 2017). Gn is less conserved in amino acid sequence among hantaviruses than is Gc (e.g., 50% and 60% amino acid sequence identity between HTNV and PUUV Gn and Gc, respectively). Nevertheless, divergent Gn (and Gc) proteins share a common fold, suggesting that the overall architecture of Gn/Gc complexes is conserved among hantaviruses (Rissanen et al., 2017). Gn is proposed to bind to cell-surface receptors during entry based on its membrane-distal, solvent-exposed position on the spike and information derived from other bunyaviruses, although this has not been demonstrated experimentally. Given their variability in Gn sequence, different clades of hantaviruses are likely to use different attachment factors and/or receptors (Gavrilovskaya et al., 1998, 1999; Jangra et al., 2018).

The Gn cytoplasmic tail is 110 amino acids long and carries a single structured domain, a zinc finger, which likely plays a role in interactions with the N protein during virus assembly (Estrada et al., 2009, 2011; Wang et al., 2010), possibly in a manner functionally analogous to the matrix proteins of other viruses (Hepojoki et al., 2010a,b). The Gn cytoplasmic tails of virulent hantaviruses can also block the type I interferon response to limit the host immune response to infection (Alff et al., 2008; Shim et al., 2011).

Crystal structures of the pre-fusion ectodomain of Gc proteins from multiple hantaviruses reveal that Gc is a class II fusion protein similar in structure to the glycoproteins of flaviviruses, alphaviruses, Rubella virus, and phleboviruses (Guardado-Calvo and Rey, 2017; Guardado-Calvo et al., 2016; Willensky et al., 2016). Like these latter class II fusion proteins, hantavirus Gc comprises three domains of β strands and sheets: Domain I (DI) is the central domain, Domain II (DII) is located in the membrane-distal region and contains the TMIR, and Domain III (DIII) resides proximal to the viral membrane and connects to the stem and transmembrane domain (Fig. 3B). The hantavirus Gc structure also possesses certain unique features. The hantavirus *ij* loop on DII is unusually long, allowing the TMIR to consist of three fusion loops involving residues on the *cd* loop, *ij* loop, and *bc* loop (termed the α-αA loop on the PUUV Gc structure) (Guardado-Calvo et al., 2016; Willensky et al., 2016) (Fig. 3B). Gc also contains a conserved N-linked glycosylation site on DII, which is thought to help stabilize Gc interactions with DIII of neighboring Gc protomers during fusion (Willensky et al., 2016). Gc protomers mediate interactions at twofold Gc:Gc contacts between neighboring Gn/Gc spikes on the virion surface (Bignon et al., 2019; Guardado-Calvo et al., 2016; Willensky et al., 2016). Even at neutral pH, these spikes alternate between "open" (in which the TMIR is exposed) and "closed" (in which the TMIR is protected by Gn) conformations in a temperature-dependent manner and only the "closed" form is capable of inducing acid-pH-mediated membrane fusion (Bignon et al., 2019). Hantavirus Gc contains 26 conserved cysteine residues that form disulfide bonds, providing rigidity and stability to the TMIR (Willensky et al., 2016) (Fig. 3B). The Gc cytoplasmic tail is short (9–26 amino acid residues), and has no known function.

The post-fusion Gc structure also resembles those of other class II fusion proteins. DIII packs against the DI and DII junction, and DII hinges to project the fusion loop toward the target membrane (Fig. 3C). At acid pH, Gc forms trimers that are stabilized by salt bridges and a unique N-terminal tail (Acuña et al., 2015; Guardado-Calvo et al., 2016; Willensky et al., 2016).

The Gc stem is vital for the formation of post-fusion conformation (Guardado-Calvo and Rey, 2017) and peptides corresponding to it block ANDV fusion (Barriga et al., 2016). The Gc sequence is fairly well-conserved among hantaviruses, consistent with its direct role in viral membrane fusion. ANDV and PUUV share 76% sequence identity and 89% sequence similarity, suggesting that the features seen in Old World hantavirus Gc crystal structures are also present in Gc proteins of New World hantaviruses. This is also supported by three-dimensional (3D) molecular modeling of ANDV Gc (Tischler et al., 2005).

As mentioned above, hantavirus Gn/Gc proteins undergo acid-induced conformational changes during fusion. Associations between Gn and Gc are pH-sensitive and dissociation begins at pH 6.4 for PUUV Gn/Gc (Hepojoki et al., 2010a,b). However, Gn/Gc dissociation is reversible and does not represent virus fusion triggering. Rather, Gn/Gc dissociation likely acts as a priming step that allows Gc to undergo subsequent fusion triggering. In the post-fusion state, spike proteins on the viral surface are composed of Gn tetramers and Gc trimers. Based on a comparison of TULV cryoEM structures at neutral and acid pH, and HTNV Gn crystal structure at acid pH, Rissanen and co-workers proposed that the formation of acid-induced, narrower, compact Gn tetramers detaches it from Gc and exposes otherwise-shielded hydrophobic fusion loops allowing the formation of an extended intermediate (Rissanen et al., 2017). However, structures of the Gn/Gc complexes at neutral and acid pH need to be solved to further test this model. Whether host factors, including the candidate receptors, play a role in the Gn/Gc conformational changes that eventually lead to the fusion of viral and endosomal membranes needs to be determined.

6. Candidate hantavirus receptors

The first proteins characterized as hantavirus entry factors were $\alpha_5\beta_1$ and $\alpha_V\beta_3$ integrins. Integrins are heterodimeric transmembrane glycoproteins composed of alpha and beta subunits. To fulfill their cellular function, integrins bind to components of the extracellular matrix and link them to the actin cytoskeleton at focal adhesion sites, thus affording bidirectional signaling (Campbell and Humphries, 2011). Both non-enveloped (Guerrero et al., 2000; Jackson et al., 2000; Maginnis et al., 2006; Wickham et al., 1993) and enveloped viruses (Chu and Ng, 2004; Cseke et al., 2009; Feire et al., 2004; Gianni et al., 2010) use integrins for particle attachment or internalization into cells. Integrin-specific ligands and antibodies were evaluated for their effect on hantavirus infection (Gavrilovskaya et al., 1998, 1999). Pre-incubating human umbilical vein endothelial cells (HUVEC) or VeroE6 grivet monkey kidney epithelial cells with the $\alpha_V\beta_3$ ligand vitronectin reduced cellular susceptibility to multiple virulent New World (ANDV, SNV and NY-1V) and Old World hantaviruses (HTNV, SEOV and PUUV). Conversely, infection by the avirulent PHV was only reduced by the $\alpha_5\beta_1$ ligand fibronectin (Gavrilovskaya et al., 1998, 1999). However, binding of vitronectin and fibronectin is not limited to $\alpha_V\beta_3$ and $\alpha_5\beta_1$ integrins, respectively. For instance, $\alpha_V\beta_3$ integrin can recognize both fibronectin and vitronectin depending on its conformation (Boettiger et al., 2001; Van Agthoven et al., 2014). To provide additional specificity, antibodies against $\alpha_V\beta_3$ and $\alpha_5\beta_1$ integrins were employed in infection-blocking experiments in HUVEC and VeroE6 cells, and were shown to selectively inhibit infections by virulent hantaviruses (ANDV, SNV, HTNV, DOBV and NY-1V), and avirulent PHV, respectively (Gavrilovskaya et al., 1998, 1999; Popugaeva et al., 2012). Many integrins recognize a tripeptide arginine-glycine-aspartate (RGD) motif that is essential for integrin-dependent cell adhesion. The RGD motif is found in fibronectin, vitronectin, fibrinogen, von Willebrand factor, and many other large glycoproteins (Ruoslahti, 1996). However, the lack of inhibition of NY-1V and SNV infection by soluble integrin-blocking RGD peptides and the absence of RGD motifs in hantavirus Gn/Gc proteins suggested a non-RGD-mediated interaction. Moreover, Chinese hamster ovary (CHO) cells expressing a mutated $\alpha_V\beta_3$ integrin were rendered incapable of binding its physiological ligands and were susceptible to hantavirus infection, further suggesting an RGD-independent mechanism (Gavrilovskaya et al., 1998, 1999).

Ectodomains of the β_3 integrin subunit consist of a β-I domain, a hybrid domain, a plexin-semaphorin-integrin (PSI) domain, and four epidermal growth factor (EGF) modules (Campbell and Humphries, 2011; Xiong et al., 2001, 2002). The PSI domain is located at the apex of the β_3 integrin protein in its bent/inactive conformation (Xiong et al., 2001). *In vitro* genetic complementation and neutralization studies mapped the $\alpha_V\beta_3$ integrin requirement for ANDV and NY-1V infection to the PSI domain (Matthys et al., 2010; Raymond et al., 2005). However, overexpression of human $\alpha_V\beta_3$ integrin in poorly permissive SupT1 cells failed to restore HTNV and PUUV infectivity (Higa et al., 2012), and a clear correlation between levels of $\alpha_V\beta_3$ expression and susceptibility to HTNV and PUUV infections was lacking (Higa et al., 2012; Müller et al., 2019) suggesting the existence of integrin-independent entry pathways. Although many studies have outlined the importance of β_1/β_3 integrins for cellular susceptibility to hantavirus infection, a direct interaction between hantavirus Gn/Gc and β_1/β_3 integrins has not yet been reported. Additionally, efforts to test the roles of $\alpha_V\beta_3$ and $\alpha_5\beta_1$ integrins *in vivo* have been inconclusive. Treatment of newborn murine pups with antibodies targeting $\alpha_V\beta_3$ integrin after HTNV challenge afforded only a modest increase in survival time (Song et al., 2005). A similar, modest, increase in survival time was also reported when mice were injected with β_1 integrin-specific antibodies (Song et al., 2005).

Raftery et al. suggested that hantaviruses use the $\alpha_M\beta_2$ and $\alpha_X\beta_2$ integrins to activate neutrophils resulting in the formation of neutrophil extracellular traps (Raftery et al., 2014). Overexpression of $\alpha_M\beta_2$ or $\alpha_X\beta_2$ integrin in CHO cells modestly increased HTNV infection suggesting their potential role in virus entry and infection (Raftery et al., 2014). Given that β_2 integrins are exclusively expressed on leukocytes, their relevance as hantavirus receptors is unclear since hantavirus infection of neutrophils is nonproductive (Raftery et al., 2014).

Given their pulmonary route of infection in humans, initial infections by hantaviruses are likely to occur at the apical surface of the respiratory epithelium. *In vitro* studies showed that HTNV, PUUV and Black Creek Canal virus (BCCV) infected polarized epithelial cells predominantly via the apical surface (Krautkrämer and Zeier, 2008; Ravkov et al., 1997), whereas ANDV could infect via both, apical and basolateral surfaces (Rowe and Pekosz, 2006). It is unclear if $\alpha_V\beta_3$ integrins are expressed at the apical surface (Aplin et al., 1996; Conforti et al., 1992; Schoenenberger et al., 1994) suggesting other cell-surface receptors may mediate virus entry. Experiments to identify such molecules were carried out in polarized cells. Removal of glycosylphosphatidylinositol (GPI)-anchored proteins from the cell surface via phosphatidylinositol-specific phospholipase C treatment led to the identification of decay-accelerating factor (DAF/CD55) as a candidate receptor (Krautkrämer and Zeier, 2008). DAF/CD55 is a 70-kDa GPI-anchored protein localized exclusively to the apical surface of polarized epithelial cells and has been described as an important co-receptor for a number of enteroviruses (Bergelson et al., 1994; Karnauchow et al., 1996; Shafren et al., 1995). Treatment of polarized cells with DAF/CD55-specific antibodies or human recombinant DAF/CD55 ectodomain reduced infection of HTNV and PUUV in a concentration-dependent manner. A separate study confirmed the importance of DAF/CD55 for DOBV (Popugaeva et al., 2012). Other studies have described high-affinity binding of inactivated SNV particles to DAF/CD55 expressed on Tanoue B cells (Buranda et al., 2010, 2014). However, the relevance of DAF/CD55 as a hantavirus receptor *in vivo* and its role in viral pathogenesis remain unknown.

Proteomic approaches to identify HTNV-binding host factors revealed gC1qR as a candidate cell attachment factor (Choi et al., 2008). gC1qR, a 32-kDa glycoprotein, has no transmembrane domain, but associates with the cell surface through its interaction with other proteins, including integrins, and it can bind to the globular head domain of C1q, a complement protein (Feng et al., 2002). RNA interference-mediated suppression of gC1qR in human lung epithelial A549 cells reduced HTNV binding and infection. Similarly, ectopic expression of gC1qR in CHO cells conferred susceptibility to HTNV infection. However, the precise role of gC1qR in hantavirus entry and its *in vivo* relevance to viral infection and pathogenesis are unknown.

Using a genome-wide loss-of-function genetic screen in human haploid cells, we recently identified *PCDH1* as a New World hantavirus receptor (Jangra et al., 2018). This type I transmembrane protein is a member of the cadherin superfamily and is composed of seven extracellular cadherin (EC) repeats, a transmembrane domain, and a cytoplasmic tail (Sotomayor et al., 2014). PCDH1 is highly expressed in neuronal cells during development, but is also expressed in airway endothelial cells (Faura Tellez et al., 2016; Koppelman et al., 2009; Kozu et al., 2015), which are the major target cells of hantavirus infection *in vivo* (Nolte et al., 1995; Zaki et al., 1995). Genetic knockout (KO) of *PCDH1* in human cell lines and primary endothelial cells significantly reduced their susceptibility to multiple New World viruses, including ANDV, SNV, PHV and Maporal virus (MPRLV), but not to the Old World viruses HTNV and SEOV, providing evidence that PCDH1 is a hantavirus clade-specific entry factor (Jangra et al., 2018). The first extracellular cadherin repeat (EC1) of PCDH1 was found to be required for New World hantavirus entry. Further, treatment of cells with an antibody against the EC1 domain prior to virus exposure or preincubation of virus with a soluble version of PCDH1 reduced New World hantavirus infection in a dose-dependent manner. Importantly, multiple assays showed that New World hantavirus Gn/Gc proteins directly recognize EC1, and a biolayer interferometry-based assay with purified components indicated that these proteins interact with nanomolar affinity (Jangra et al., 2018). To test the *in vivo* relevance of PCDH1, *PCDH1*-KO Syrian golden hamsters were generated by CRISPR-Cas9 genome engineering, and WT and *PCDH1*-KO hamsters were challenged intranasally with a lethal dose of ANDV. Unlike their WT counterparts, the *PCDH1*-KO hamsters largely survived viral challenge and exhibited limited pathology. The *PCDH1*-KO hamsters also had lower viral loads in the lung than WT hamsters. Taken together, these findings validated the requirement of PCDH1 for ANDV entry, infection, and pathogenesis *in vivo*. However, genetic disruption of PCDH1-hantavirus Gn/Gc interaction did not completely inhibit New World hantavirus attachment and entry *in vitro*, and *PCDH1* KO failed to completely eliminate viral replication and disease in hamsters, suggesting that PCDH1-independent entry pathways exist (Jangra et al., 2018). This work further underlines the complexity of the hantavirus entry process and points to the need to evaluate the interplay of different candidate receptors and other host factors in hantavirus entry.

7. Role of candidate receptors in pathogenesis

Although capillary endothelial cells are the major targets of hantavirus infection, viral replication does not appear to directly damage these cells or the vascular endothelium (Nolte et al., 1995; Zaki et al., 1995). Rather, the impairment of endothelial barrier integrity due to an excessive innate immune response is proposed to be central to pathogenesis (Hepojoki et al., 2014; Schönrich et al., 2015) and is a hallmark of hantavirus disease (Duchin et al., 1994; Spiropoulou and Srikiatkhachorn, 2013).

Vascular endothelial growth factor (VEGF), also known as vascular permeability factor, is a key regulator of endothelial permeability (Rahimi, 2017; Senger et al., 1983, 1986). β_3 Integrin has been proposed as a host determinant of hantavirus pathogenesis (Mackow and Gavrilovskaya, 2009) given its link to the disruption of endothelial barrier function. VEGF binding to its receptor, VEGFR2, triggers a cascade of signaling events leading to the internalization of VE-cadherin, an adherens junctions protein, which then increases endothelial cell permeability and vascular leakage (Gavard and Gutkind, 2006). On endothelial cells, $\alpha_V\beta_3$ integrins regulate the permeabilizing effects of VEGF by forming a complex with VEGFR2 (Borges et al., 2000; Soldi et al., 1999). Angiogenic responses to hypoxia and vascular endothelial growth factor (VEGF) are enhanced when β_3 integrins are knocked out *in vitro* and *in vivo* (Hodivala-Dilke et al., 1999; Reynolds et al., 2002; Su et al., 2012). Further, endothelial cells infected with pathogenic hantaviruses are able to recruit platelets via $\alpha_{II}\beta_3$ integrin *in vitro* (Gavrilovskaya et al., 2010), which might contribute to the development of thrombocytopenia, another hallmark of hantavirus disease. However, β_3 integrin-dependent hantaviruses differ greatly in their virulence in humans as well as experimental animals and cause distinct pathologies, suggesting a more complex picture (Hepojoki et al., 2014). Use of $\alpha_5\beta_1$ integrin instead of $\alpha_V\beta_3$ integrin by some hantaviruses has been proposed as an explanation for their lack of virulence (Gavrilovskaya et al., 1998, 1999). However, Sangassou virus (SANGV), a β_1 integrin-dependent hantavirus (Klempa et al., 2012), is capable of causing human infections, if not disease, in Africa (Klempa et al., 2010). Thus, β_1 integrin usage by avirulent PHV and TULV may not be generalizable to all non-pathogenic hantaviruses, and evaluation of integrin usage of a larger group of hantaviruses is warranted. Engagement of another candidate receptor, β_2 integrin, by hantaviruses on neutrophils has also been proposed and was implicated as the mechanism of recruitment and activation of neutrophils to the site of hantavirus replication and pathogenesis (Raftery et al., 2014; Schönrich et al., 2015).

In vivo studies with *PCDH1*-KO Syrian hamsters suggest that PCDH1 is an essential determinant of the severity of HPS (Jangra et al., 2018). As compared to WT animals, lungs from KO animals bore much lower levels of viral RNA and antigen following a lethal ANDV challenge. Moreover, only mild inflammation and tissue damage was seen in the *PCDH1*-KO animals (Jangra et al., 2018). The lower inflammation and tissue damage in *PCDH1*-KO animals may simply be a result of lower viral loads, but it is also conceivable that PCDH1 plays a more direct role in the pathogenesis of HPS. PCDH1 localizes to cell-cell contacts of epithelial cells and its knockdown compromises epithelial barrier function (Faura Tellez et al., 2016; Kozu et al., 2015). Thus, perturbation of PCDH1 expression or localization by viral infection may alter vascular permeability. Further, PCDH1 carries a long cytoplasmic tail that is involved in signaling (Kim et al., 2011). Those signaling events, poorly defined at present, may contribute to hantavirus infection and/or HPS pathogenesis. Regardless of the role of PCDH1 and other viral receptors and entry factors in the pathogenesis of HPS, current evidence indicates that the M genome segment, encoding the Gn/Gc proteins, is not the sole determinant of the difference in viral virulence between ANDV and SNV in Syrian hamsters (McElroy et al., 2004). Indeed, the capacity of virulent hantaviruses to delay the early induction of the interferon response (Geimonen et al., 2002; Hepojoki et al., 2014; Matthys and Mackow, 2012; Spiropoulou et al., 2007) is likely to play a role.

8. Candidate receptors as determinants of host range

Host range at a cellular level is determined by a combination of susceptibility, the ability of cells to allow entry of virions into the cytoplasm, and permissiveness, the capacity of cells to support cytoplasmic viral replication. Together with antiviral host factors that restrict viral replication, host factors required for viral entry and replication, including viral receptors, are key molecular determinants of host range for viruses. However, the roles of candidate hantavirus receptors as potential determinants of host range are poorly understood. An aspartate to asparagine (D39N) mutation at position 39 in human β_3 integrin impedes NY-1V infection (Raymond et al., 2005). Moreover, a leucine residue at position 33 (L33) in the PSI domain of human β_3 integrin has been proposed as a key determinant of ANDV infection and a proline mutation (L33P) in its bovine ortholog impairs ANDV infection (Matthys et al., 2010). Single-nucleotide polymorphisms (SNPs) with changes to valine and proline residues at L33 have been reported in human β_3 integrin (Joutsi-Korhonen et al., 2004; Watkins et al., 2002). Based on a limited data set, ANDV-infected individuals were recently shown to have a higher prevalence of L33 (89%) than exposed but uninfected close-household contacts (60%) (Martínez-Valdebenito et al., 2019). By contrast, another study observed no effect of SNPs in α_V or β_3 integrins on HFRS susceptibility (Chen et al., 2017). However, the *in vivo* relevance of β_3 integrin as a candidate receptor, its direct interaction with hantavirus glycoproteins and any effect of β_3 integrin mutations on host range remain to be conclusively determined. Robust PUUV and HTNV infection of kidney cells derived from bank voles, the PUUV reservoir host, despite undetectable levels of β_3 integrin (Müller et al., 2019) is suggestive of distinct receptor usage in these reservoir hosts and the identification of such factors will enhance our understanding of hantavirus biology.

9. Model systems to study hantavirus entry

The need for high-biocontainment facilities has challenged the study of entry by highly lethal hantaviruses. This roadblock has been eased by the development of surrogate viral systems bearing the hantavirus glycoproteins, Gn/Gc, which are necessary and sufficient for hantavirus entry. When co-expressed independent of other hantavirus proteins, Gn/Gc assemble into virus-like particles (VLPs) bearing heteromeric spikes similar to those on authentic virions, and resemble the latter in antigenicity (Acuña et al., 2014; Betenbaugh et al., 1995; Li et al., 2010). Although hantavirus VLPs have proven highly useful for studies of Gn/Gc organization and function (Acuña et al., 2014; Betenbaugh et al., 1995; Li et al., 2010), current approaches do not allow for the packaging of a reporter genome, and thus VLPs cannot be employed to assay productive viral entry at present.

To overcome this limitation, researchers have developed various surrogate systems that rely on replacing the native viral entry proteins of model viruses with hantavirus Gn/Gc. The latter can be provided either *in trans* (by ectopic Gn/Gc expression in infected cells) or *in cis* (by engineering the hantavirus M segment encoding Gn/Gc into the viral genome) to generate "sheep in wolf's clothing" viral pseudotypes capable of single or multiple rounds of entry and infection, respectively. Several viruses have been employed to generate these pseudotypes, most notably the rhabdovirus vesicular stomatitis virus (VSV) (Brown et al., 2011; Jangra et al., 2018; Kleinfelter et al., 2015; Petersen et al., 2014; Slough et al., 2019) and several retroviruses (Cifuentes-Muñoz et al., 2010, 2011; Ma et al., 1999; Yu et al., 2013). Importantly, data obtained with such surrogate viral systems must be corroborated with the respective hantaviruses, and indeed, VLPs and pseudotyped viruses have both been shown to behave similarly to their authentic viral counterparts (Higa et al., 2012; Jangra et al., 2018; Kleinfelter et al., 2015; Paneth Iheozor-Ejiofor et al., 2016; Petersen et al., 2014; Ray et al., 2010).

The use of surrogate viruses has afforded the discovery of new mechanistic information regarding hantavirus entry. Because VSVs are cytolytic, they have been used to uncover host factors required for hantavirus entry in comprehensive genetic screens in human cells using cell survival as a selection and to dissect the entry-related roles of these host factors (Jangra et al., 2018; Kleinfelter et al., 2015; Petersen et al., 2014). This approach has uncovered the critical roles of host factors, including the New World hantavirus receptor, PCDH1 (Jangra et al., 2018), and cholesterol during virus entry and infection (Kleinfelter et al., 2015; Petersen et al., 2014). Pseudotyped VSVs have also enabled the production of large viral preparations necessary to efficiently evaluate neutralizing antibodies against both Old World and New World hantaviruses (Higa et al., 2012; Kallio-Kokko et al., 2001; Lee et al., 2006; Levanov et al., 2019; Ogino et al., 2003; Paneth Iheozor-Ejiofor et al., 2016) and immunization with these viruses induces production of neutralizing antibodies and protection in animal models (see Section 10 for details) (Brown et al., 2011; Lee et al., 2006; McClain et al., 2000; Prescott et al., 2014). Pseudotyped lentiviruses bearing hantavirus Gn/Gc proteins have been used to characterize the fusion loops within Gc (Cifuentes-Muñoz et al., 2011) and to selectively transduce vascular endothelial cells (Qian et al., 2006).

Not all hantavirus glycoproteins are amenable to efficient pseudotyping in VSV vectors, however, likely due to a mismatch in the sites of VSV and hantavirus budding within cells. Specifically, the VSV glycoprotein is targeted to the plasma membrane, which is the site of VSV budding (Brown and Lyles, 2003; Green et al., 1981; Johnson et al., 1998), whereas hantavirus Gn/Gc proteins are largely targeted to the Golgi complex (Deyde et al., 2005; Pensiero and Hay, 1992; Ruusala et al., 1992; Spiropoulou, 2001; Spiropoulou et al., 2003) and Old World hantaviruses bud into this organelle (Hung, 1988; Hung et al., 1983). Despite this mismatch, it has been possible to generate VSVs bearing Gn/Gc from ANDV and other New World hantaviruses (Brown et al., 2011; Jangra et al., 2018; Kleinfelter et al., 2015), likely because some Gn/Gc localizes to the plasma membrane (Deyde et al., 2005; Spiropoulou et al., 2003) and can be incorporated into budding virions (Goldsmith et al., 1995; Ravkov et al., 1997). By contrast, generating such VSVs bearing Old World hantavirus Gn/Gc has been more challenging, because little or none of these proteins localize to the plasma membrane (Ogino et al., 2003, 2004; Shi and Elliott, 2002). Recently, replication-competent VSVs were used in a forward genetic approach to identify key mutations in Old World HTNV and DOBV Gn/Gc that enhanced viral multiplication. These studies found that mutations in the cytoplasmic domain of Gn and in the membrane-proximal region of the Gc ectodomain acted synergistically to enhance Gn/Gc incorporation into VSV particles, primarily by increasing Gn/Gc cell surface expression (Slough et al., 2019). The use of similar strategies to generate VSVs expressing other hantavirus Gn/Gc that have been hard to produce is expected to facilitate studies on hantavirus entry. Moreover, similar forward genetic approaches with replication-competent VSVs could be employed to investigate the mechanisms of action of antiviral drugs and neutralizing antibodies through the selection of drug- and antibody-neutralization escape mutants.

10. Vaccines and neutralizing antibodies

No FDA-approved vaccines or specific therapeutics are available to prevent or treat hantavirus infection and/or disease. Studies in HPS (ANDV- or SNV-infected) as well as HFRS (PUUV-infected) convalescent patients suggest a direct correlation between the early mounting of an efficient neutralizing antibody (nAb) response and milder disease outcomes (Bharadwaj et al., 2000; Pettersson et al., 2014; Valdivieso et al., 2006). The protective efficacy of hyperimmune sera in animal models and humans further suggests that Gn/Gc-specific antibodies elicited by active or passive immunization may have utility in the treatment of hantavirus disease (Brocato et al., 2012, 2014; Hooper et al., 2008, 2014a,b; Klingström et al., 2008; Vial et al., 2015). Although numerous studies highlight the importance of nAbs for protection against hantaviruses, cDNA-based vaccination could protect Syrian hamsters from lethal ANDV challenge in the absence of any nAb activity suggesting that nAbs are sufficient but not essential (Brocato et al., 2013).

Vaccination is the most successful approach to induce a potent antiviral state and afford protection against viral infections and diseases. Over the last 20 years, numerous vaccine candidates and modalities have been evaluated for their efficiency against hantaviruses. Since 1990, a formalin-inactivated HTNV vaccine grown in suckling mouse brains (Hantavax™) has been marketed in the Republic of Korea (Cho et al., 2002; Yamanishi et al., 1988). However, due to the lack of well-designed clinical trials, its clinical efficacy and long-term immunogenicity have remained unclear. In a recent phase III clinical trial, high seroconversion but poor nAb generation was reported (Song et al., 2016). Another case-control study observed a moderate protective efficacy of Hantavax™, especially in high-risk populations (Jung et al., 2018). Despite the availability of Hantavax™, therefore, there is a clear and urgent need for new and more efficacious hantavirus vaccines.

As an alternative to inactivated vaccines, live-attenuated vaccines using recombinant VSV or adenovirus (AdV) vectors expressing hantavirus Gn/Gc have been developed. A single injection of a replication-competent VSV vector expressing ANDV Gn/Gc protected Syrian hamsters against lethal ANDV challenge as early as 3 days post immunization (Brown et al., 2011; Prescott et al., 2014). Protection lasted for up to 6 months and loss of protection at later time points correlated with a loss of nAb responses (Prescott et al., 2014). Vaccination of hamsters at 24 h post-ANDV challenge was largely protective, suggesting that induction of a potent antiviral state by the recombinant vector itself may have contributed to protection (Brown et al., 2011; Prescott et al., 2014). Single-cycle VSVs carrying HTNV Gn/Gc also afforded protection against HTNV infection in a non-lethal Balb/c mouse model (Lee et al., 2006). Recombinant human AdV serotype 5 (Ad5V)-based vectors expressing ANDV Gn or Gc and recombinant canine AdV serotype 2-based vectors expressing SEOV Gn afforded protection of Syrian hamsters and Balb/c mice against lethal virus challenge, respectively (Safronetz et al., 2009; Yuan et al., 2009). Syrian hamsters vaccinated with a vaccinia virus vector expressing HTNV Gn/Gc and N proteins showed no viremia when challenged with HTNV or SEOV, but were partially susceptible to PUUV infection suggesting partial cross-protection between closely related hantaviruses, and the possibility that broadly protective hantavirus vaccines can be designed (Chu et al., 1995; Schmaljohn et al., 1992).

Although virus-based vaccine platforms have shown promise in animal models, their translation to human use has been complicated by several factors (Tatsis and Ertl, 2004). Pre-existing immunity to Ad5V and other common AdV serotypes might reduce vaccine efficacy. This problem can be avoided by establishing vectors based on rare human (Barouch et al., 2004; Lemckert et al., 2005) or animal adenovirus serotypes (Kobinger et al., 2006). Poor vaccine efficacy due to pre-existing immunity was also observed with vaccinia virus-based vectors: vaccinia viruses expressing HTNV Gn/Gc elicited HTNV-specific nAb titers in 72% of vaccinia virus-naïve patients, but only in 26% of vaccinia virus-immune patients (McClain et al., 2000). Although there is little pre-existing immunity against VSV in most human populations, the increasing popularity of VSV-based vaccines might alter this scenario in the future (Lévy et al., 2018; Regules et al., 2017; Suder et al., 2018).

DNA-based subunit vaccines comprising cDNA vectors encoding hantavirus Gn/Gc have also been evaluated. Immunization of Syrian hamsters with DNA vaccines encoding SEOV or HTNV Gn/Gc rendered them resistant to challenges with HTNV, SEOV and DOBV, but not with the more divergent PUUV (Hooper et al., 1999, 2001). PUUV and SNV Gn/Gc-based cDNA vaccines were also protective in Syrian hamsters (Brocato et al., 2013; Hooper et al., 2013). Unexpectedly, an equivalent ANDV cDNA vaccine was not immunogenic or protective in Syrian hamsters despite the generation of high nAb levels against ANDV as well as heterologous SNV in rhesus macaques (Custer et al., 2003). Passive transfer of sera conferred complete protection to Syrian hamsters even when carried out 4–5 days post-ANDV challenge and delayed HPS development when injected 1 day pre-challenge (Custer et al., 2003). In two phase I clinical studies, three vaccinations with cDNA vaccines expressing HTNV or PUUV Gn/Gc by particle-mediated epidermal delivery showed limited immunogenicity: only 30–56% of individuals developed a nAb response (Boudreau et al., 2012; Hooper et al., 2014a,b). cDNA electroporation improved nAb generation and resulted in nAb responses in up to 78% of individuals (Hooper et al., 2014a,b).

Passive transfer of polyclonal antibodies from animals immunized with hantavirus Gn/Gc-based vaccines conferred protection against lethal ANDV challenge in Syrian hamsters (Brocato et al., 2012, 2014; Haese et al., 2015; Hooper et al., 2008, 2014a,b). Importantly, passive transfusion of hyperimmune sera from ANDV convalescent donors to acute HPS patients improved disease outcomes and nAb titers were found to correlate inversely with HPS severity (Vial et al., 2015) suggesting that mAb-based anti-hantavirus therapeutics could be developed as a complement to vaccines.

For the development of mAb-based therapies, Garrido et al. isolated ANDV Gn/Gc-specific memory B cells from a HPS convalescent patient to generate recombinant human IgG antibodies and evaluated two potently neutralizing mAbs (JL16 and MIB22) for their post-exposure efficacy in Syrian hamsters (Garrido et al., 2018). All hamsters treated with the mAbs, alone or in combination, at 3 and 8 days post lethal ANDV challenge were protected. One of the two mAbs, JL16, cleared virus completely from the hamster lung despite its lower *in vitro* neutralization capacity, suggesting that antibody Fc-dependent effector functions, including antibody-dependent cellular cytotoxicity (ADCC), may play roles in hantavirus clearance *in vivo*.

11. Conclusions and future directions

Despite significant progress in understanding the biology of hantaviruses, the mechanisms by which these viruses enter cells and establish infection remain poorly understood. Many of the experiments describing hantavirus candidate receptors such as integrins (β_1, β_2 and β_3), DAF/CD55, and gC1qR were performed before the advent of powerful experimental approaches for the genetic manipulation of cells and hosts, including RNA interference-mediated gene silencing and CRISPR-Cas9 genome engineering. Thus, genetic evidence to support a direct role of these and other candidate entry host factors is lacking. Further, little biochemical or structural evidence is currently available to support hypotheses that these candidate receptors directly engage the hantavirus Gn/Gc glycoproteins. Recent *in vitro* and *in vivo* studies provided genetic evidence that the cadherin-superfamily protein, PCDH1, plays a critical role in cell entry, infection, and disease by New World hantaviruses, and used a variety of approaches to demonstrate that PCDH1 directly interacts with hantavirus Gn and/or Gc.

Elucidation of the structures of hantavirus particles via cryo-electron microscopy and the X-ray crystal structures of isolated Gn and Gc subunits has helped characterize the conformational rearrangements required for hantavirus membrane fusion. However, structures of the entire Gn/Gc complex alone and bound to candidate receptors are needed to deepen our understanding of hantavirus entry. The lack of suitable animal models for most hantaviruses and genetic engineering tools to generate gene knockouts in Syrian hamsters, the only available small animal model for human hantavirus disease, until recently have further hampered the *in vivo* evaluation of host factors in hantavirus infection and disease. However, the wider availability of CRISPR-Cas9 tools, newly generated pseudotyped surrogate viruses for a number of hantaviruses, and recent discoveries of novel hantaviruses from a variety of non-rodent hosts have set the stage for more extensive investigations of hantavirus entry, infection and disease and may pave the way for the development of improved animal models for HPS and HFRS. Despite significant progress in the development of hantavirus vaccines, key features of the immunological mechanisms by which these vaccines afford protection, including the definition of protective epitopes in hantavirus Gn/Gc, remain poorly characterized. We anticipate that the new tools for hantavirus research mentioned above will accelerate the design of improved vaccine immunogens and the development of highly effective antibody therapies to prevent and treat hantavirus disease.

Box 1 Open questions

Despite significant advances, many gaps exist in the understanding of how hantaviruses hijack host machinery to enter their target cells. Some key questions are:

- What is the *in vivo* relevance of the candidate receptors β_1, β_2, and β_3 integrins, DAF/CD55, and gC1qR, in hantavirus infection and disease?
- What, if any, is the interplay among these candidate receptors and PCDH1 in hantavirus entry *in vitro* and *in vivo*?
- What is the structure of the hantavirus Gn/Gc complex in the oligomeric state found in virions?
- What are the structural bases of the interactions between Gn/Gc and its receptor(s)?
- Are there conformational changes in hantavirus Gn and/or Gc upon receptor binding, and if so, are these changes required for subsequent membrane fusion?
- How are the various pathways for hantavirus internalization into cells influenced by the type of virus, host, or cell type?
- Do hantavirus receptors play a direct role in mediating the internalization of virus particles?
- What are the endosomal sites of hantavirus membrane fusion and are they dependent on the type of virus or cell?
- Does hantavirus membrane fusion require host factors other than acid pH and cholesterol?
- Does host factor usage differ between human and natural reservoir hosts?
- What are the differences in cellular entry between rodent-borne and non-rodent-borne hantaviruses?

Continued

Box 1 Open questions—cont'd

- Which, if any, entry host factors influence hantavirus host range and tissue tropism? How can we employ this knowledge to develop better animal models for hantaviruses?
- What are the major antigenic sites in hantavirus Gn/Gc proteins, how do they differ among different hantaviruses, and can we exploit this knowledge to design broadly efficacious vaccines and therapeutics?
- What is the relationship between nAbs and cell-mediated immunity in conferring protection against hantavirus infection and what role(s) do antibody Fc-effector functions play?

Acknowledgments

This work is supported by U.S. National Institute of Health grants R01 AI132633 and U19 AI142777 (to K.C.). M. Eugenia Dieterle, Ph.D., is a Latin American Fellow in the Biomedical Sciences, supported by the Pew Charitable Trusts.

References

Acuña, R., Cifuentes-Muñoz, N., Márquez, C.L., Bulling, M., Klingström, J., Mancini, R., Lozach, P.-Y., Tischler, N.D., 2014. Hantavirus Gn and Gc glycoproteins self-assemble into virus-like particles. J. Virol. 88, 2344–2348. https://doi.org/10.1128/JVI.03118-13.

Acuña, R., Bignon, E.A., Mancini, R., Lozach, P.-Y., Tischler, N.D., 2015. Acidification triggers Andes hantavirus membrane fusion and rearrangement of Gc into a stable post-fusion homotrimer. J. Gen. Virol. 96, 3192–3197. https://doi.org/10.1099/jgv.0.000269.

Albornoz, A., Hoffmann, A.B., Lozach, P.-Y., Tischler, N.D., 2016. Early bunyavirus-host cell interactions. Viruses 8, 143. https://doi.org/10.3390/v8050143.

Alff, P.J., Sen, N., Gorbunova, E., Gavrilovskaya, I.N., Mackow, E.R., 2008. The NY-1 hantavirus Gn cytoplasmic tail coprecipitates TRAF3 and inhibits cellular interferon responses by disrupting TBK1-TRAF3 complex formation. J. Virol. 82, 9115–9122. https://doi.org/10.1128/JVI.00290-08.

Aplin, J.D., Spanswick, C., Behzad, F., Kimber, S.J., Vićovac, L., 1996. Integrins beta 5, beta 3 and alpha v are apically distributed in endometrial epithelium. Mol. Hum. Reprod. 2, 527–534.

Arikawa, J., Takashima, I., Hashimoto, N., 1985. Cell fusion by haemorrhagic fever with renal syndrome (HFRS) viruses and its application for titration of virus infectivity and neutralizing antibody. Arch. Virol. 86, 303–313.

Barouch, D.H., Pau, M.G., Custers, J.H.H.V., Koudstaal, W., Kostense, S., Havenga, M.J.E., Truitt, D.M., Sumida, S.M., Kishko, M.G., Arthur, J.C., Korioth-Schmitz, B., Newberg, M.H., Gorgone, D.A., Lifton, M.A., Panicali, D.L., Nabel, G.J., Letvin, N.L., Goudsmit, J., 2004. Immunogenicity of recombinant adenovirus serotype 35 vaccine in the presence of pre-existing anti-Ad5 immunity. J. Immunol. 172, 6290–6297.

Barriga, G.P., Villalón-Letelier, F., Márquez, C.L., Bignon, E.A., Acuña, R., Ross, B.H., Monasterio, O., Mardones, G.A., Vidal, S.E., Tischler, N.D., 2016. Inhibition of the hantavirus fusion process by predicted domain III and stem peptides from glycoprotein Gc. PLoS Negl. Trop. Dis. 10 e0004799https://doi.org/10.1371/journal.pntd.0004799.

Bergelson, J.M., Chan, M., Solomon, K.R., St John, N.F., Lin, H., Finberg, R.W., 1994. Decay-accelerating factor (CD55), a glycosylphosphatidylinositol-anchored complement regulatory protein, is a receptor for several echoviruses. Proc. Natl. Acad. Sci. U. S. A. 91, 6245–6248.

Betenbaugh, M., Yu, M., Kuehl, K., White, J., Pennock, D., Spik, K., Schmaljohn, C., 1995. Nucleocapsid- and virus-like particles assemble in cells infected with recombinant baculoviruses or vaccinia viruses expressing the M and the S segments of Hantaan virus. Virus Res. 38, 111–124. https://doi.org/10.1016/0168-1702(95)00053-S.

Bharadwaj, M., Nofchissey, R., Goade, D., Koster, F., Hjelle, B., 2000. Humoral immune responses in the hantavirus cardiopulmonary syndrome. J. Infect. Dis. 182, 43–48. https://doi.org/10.1086/315657.

Bignon, E.A., Albornoz, A., Guardado-Calvo, P., Rey, F.A., Tischler, N.D., 2019. Molecular organization and dynamics of the fusion protein Gc at the hantavirus surface. Elife 8, e46028. https://doi.org/10.7554/eLife.46028.

Bitto, D., Halldorsson, S., Caputo, A., Huiskonen, J.T., 2016. Low pH and anionic lipid-dependent fusion of uukuniemi phlebovirus to liposomes. J. Biol. Chem. 291, 6412–6422. https://doi.org/10.1074/jbc.M115.691113.

Boettiger, D., Lynch, L., Blystone, S., Huber, F., 2001. Distinct ligand-binding modes for integrin alpha(v)beta(3)-mediated adhesion to fibronectin versus vitronectin. J. Biol. Chem. 276, 31684–31690. https://doi.org/10.1074/jbc.M103997200.

Borges, E., Jan, Y., Ruoslahti, E., 2000. Platelet-derived growth factor receptor beta and vascular endothelial growth factor receptor 2 bind to the beta 3 integrin through its extracellular domain. J. Biol. Chem. 275, 39867–39873. https://doi.org/10.1074/jbc.M007040200.

Boudreau, E.F., Josleyn, M., Ullman, D., Fisher, D., Dalrymple, L., Sellers-Myers, K., Loudon, P., Rusnak, J., Rivard, R., Schmaljohn, C., Hooper, J.W., 2012. A phase 1 clinical trial of Hantaan virus and Puumala virus M-segment DNA vaccines for hemorrhagic fever with renal syndrome. Vaccine 30, 1951–1958. https://doi.org/10.1016/j.vaccine.2012.01.024.

Brocato, R., Josleyn, M., Ballantyne, J., Vial, P., Hooper, J.W., 2012. DNA vaccine-generated duck polyclonal antibodies as a postexposure prophylactic to prevent hantavirus pulmonary syndrome (HPS). PLoS One 7, e35996. https://doi.org/10.1371/journal.pone.0035996.

Brocato, R.L., Josleyn, M.J., Wahl-Jensen, V., Schmaljohn, C.S., Hooper, J.W., 2013. Construction and nonclinical testing of a Puumala virus synthetic M gene-based DNA vaccine. Clin. Vaccine Immunol. 20, 218–226. https://doi.org/10.1128/CVI.00546-12.

Brocato, R.L., Hammerbeck, C.D., Bell, T.M., Wells, J.B., Queen, L.A., Hooper, J.W., 2014. A lethal disease model for hantavirus pulmonary syndrome in immunosuppressed Syrian hamsters infected with Sin Nombre virus. J. Virol. 88, 811–819. https://doi.org/10.1128/JVI.02906-13.

Bron, R., Wahlberg, J.M., Garoff, H., Wilschut, J., 1993. Membrane fusion of Semliki forest virus in a model system: correlation between fusion kinetics and structural changes in the envelope glycoprotein. EMBO J. 12, 693–701.

Brown, E.L., Lyles, D.S., 2003. Organization of the vesicular stomatitis virus glycoprotein into membrane microdomains occurs independently of intracellular viral components. J. Virol. 77, 3985–3992.

Brown, K.S., Safronetz, D., Marzi, A., Ebihara, H., Feldmann, H., 2011. Vesicular stomatitis virus-based vaccine protects hamsters against lethal challenge with Andes virus. J. Virol. 85, 12781–12791. https://doi.org/10.1128/JVI.00794-11.

Buranda, T., Wu, Y., Perez, D., Jett, S.D., BonduHawkins, V., Ye, C., Edwards, B., Hall, P., Larson, R.S., Lopez, G.P., Sklar, L.A., Hjelle, B., 2010. Recognition of decay accelerating factor and alpha(v)beta(3) by inactivated hantaviruses: toward the development of high-throughput screening flow cytometry assays. Anal. Biochem. 402, 151–160. https://doi.org/10.1016/j.ab.2010.03.016.

Buranda, T., Swanson, S., Bondu, V., Schaefer, L., Maclean, J., Mo, Z., Wycoff, K., Belle, A., Hjelle, B., 2014. Equilibrium and kinetics of Sin Nombre hantavirus binding at DAF/CD55 functionalized bead surfaces. Viruses 6, 1091–1111. https://doi.org/10.3390/v6031091.

Campbell, I.D., Humphries, M.J., 2011. Integrin structure, activation, and interactions. Cold Spring Harb. Perspect. Biol. 3. https://doi.org/10.1101/cshperspect.a004994.

Chen, X.-P., Xiong, H.-R., Zhu, N., Chen, Q.-Z., Wang, H., Zhong, C.-J., Wang, M.-R., Lu, S., Luo, F., Hou, W., 2017. Lack of association between integrin αvβ3 gene polymorphisms and hemorrhagic fever with renal syndrome in Han Chinese from Hubei, China. Virol. Sin. 32, 73–79. https://doi.org/10.1007/s12250-016-3888-0.

Chiang, C.-F., Flint, M., Lin, J.-M.S., Spiropoulou, C.F., 2016. Endocytic pathways used by Andes virus to enter primary human lung endothelial cells. PLoS One 11, e0164768. https://doi.org/10.1371/journal.pone.0164768.

Chizhikov, V.E., Spiropoulou, C.F., Morzunov, S.P., Monroe, M.C., Peters, C.J., Nichol, S.T., 1995. Complete genetic characterization and analysis of isolation of Sin Nombre virus. J. Virol. 69, 8132–8136.

Cho, H.-W., Howard, C.R., Lee, H.-W., 2002. Review of an inactivated vaccine against hantaviruses. Intervirology 45, 328–333. https://doi.org/10.1159/000067925.

Choi, Y., Kwon, Y.-C., Kim, S.-I., Park, J.-M., Lee, K.-H., Ahn, B.-Y., 2008. A hantavirus causing hemorrhagic fever with renal syndrome requires gC1qR/p32 for efficient cell binding and infection. Virology 381, 178–183. https://doi.org/10.1016/j.virol.2008.08.035.

Chu, J.J.-H., Ng, M.-L., 2004. Interaction of West Nile virus with alpha v beta 3 integrin mediates virus entry into cells. J. Biol. Chem. 279, 54533–54541. https://doi.org/10.1074/jbc.M410208200.

Chu, Y.K., Jennings, G.B., Schmaljohn, C.S., 1995. A vaccinia virus-vectored Hantaan virus vaccine protects hamsters from challenge with Hantaan and Seoul viruses but not Puumala virus. J. Virol. 69, 6417–6423.

Cifuentes-Muñoz, N., Darlix, J.-L., Tischler, N.D., 2010. Development of a lentiviral vector system to study the role of the Andes virus glycoproteins. Virus Res. 153, 29–35. https://doi.org/10.1016/j.virusres.2010.07.001.

Cifuentes-Muñoz, N., Barriga, G.P., Valenzuela, P.D.T., Tischler, N.D., 2011. Aromatic and polar residues spanning the candidate fusion peptide of the Andes virus Gc protein are essential for membrane fusion and infection. J. Gen. Virol. 92, 552–563. https://doi.org/10.1099/vir.0.027235-0.

Clement, J., Maes, P., Van Ranst, M., 2014. Hemorrhagic fever with renal syndrome in the new, and hantavirus pulmonary syndrome in the old world: paradi(se)gm lost or regained? Virus Res. 187, 55–58. https://doi.org/10.1016/j.virusres.2013.12.036.

Conforti, G., Dominguez-Jimenez, C., Zanetti, A., Gimbrone, M.A., Cremona, O., Marchisio, P.C., Dejana, E., 1992. Human endothelial cells express integrin receptors on the luminal aspect of their membrane. Blood 80, 437–446.

Cseke, G., Maginnis, M.S., Cox, R.G., Tollefson, S.J., Podsiad, A.B., Wright, D.W., Dermody, T.S., Williams, J.V., 2009. Integrin alphavbeta1 promotes infection by human metapneumovirus. Proc. Natl. Acad. Sci. U. S. A. 106, 1566–1571. https://doi.org/10.1073/pnas.0801433106.

Custer, D.M., Thompson, E., Schmaljohn, C.S., Ksiazek, T.G., Hooper, J.W., 2003. Active and passive vaccination against hantavirus pulmonary syndrome with Andes virus M genome segment-based DNA vaccine. J. Virol. 77, 9894–9905. https://doi.org/10.1128/JVI.77.18.9894-9905.2003.

Deyde, V.M., Rizvanov, A.A., Chase, J., Otteson, E.W., St Jeor, S.C., 2005. Interactions and trafficking of Andes and Sin Nombre Hantavirus glycoproteins G1 and G2. Virology 331, 307–315. https://doi.org/10.1016/j.virol.2004.11.003.

Dubé, M., Rey, F.A., Kielian, M., 2014. Rubella virus: first calcium-requiring viral fusion protein. PLoS Pathog. 10, e1004530. https://doi.org/10.1371/journal.ppat.1004530.

Duchin, J.S., Koster, F.T., Peters, C.J., Simpson, G.L., Tempest, B., Zaki, S.R., Ksiazek, T.G., Rollin, P.E., Nichol, S., Umland, E.T., 1994. Hantavirus pulmonary syndrome: a clinical description of 17 patients with a newly recognized disease. The Hantavirus Study Group. N. Engl. J. Med. 330, 949–955. https://doi.org/10.1056/NEJM199404073301401.

Elliott, R.M., Schmaljohn, C.S., Collett, M.S., 1991. Bunyaviridae genome structure and gene expression. Curr. Top. Microbiol. Immunol. 169, 91–141.

Ermonval, M., Baychelier, F., Tordo, N., 2016. What do we know about how hantaviruses interact with their different hosts? Viruses 8, 223. https://doi.org/10.3390/v8080223.

Estrada, D.F., Boudreaux, D.M., Zhong, D., St Jeor, S.C., De Guzman, R.N., 2009. The Hantavirus glycoprotein G1 tail contains dual CCHC-type classical zinc fingers. J. Biol. Chem. 284, 8654–8660. https://doi.org/10.1074/jbc.M808081200.

Estrada, D.F., Conner, M., Jeor, S.C., Guzman, R.N.D., 2011. The structure of the hantavirus zinc finger domain is conserved and represents the only natively folded region of the Gn cytoplasmic tail. Front. Microbiol. 2, 251. https://doi.org/10.3389/fmicb.2011.00251.

Faura Tellez, G., Willemse, B.W.M., Brouwer, U., Nijboer-Brinksma, S., Vandepoele, K., Noordhoek, J.A., Heijink, I., de Vries, M., Smithers, N.P., Postma, D.S., Timens, W., Wiffen, L., van Roy, F., Holloway, J.W., Lackie, P.M., Nawijn, M.C., Koppelman, G.H., 2016. Protocadherin-1 localization and cell-adhesion function in airway epithelial cells in asthma. PLoS One 11, e0163967. https://doi.org/10.1371/journal.pone.0163967.

Feire, A.L., Koss, H., Compton, T., 2004. Cellular integrins function as entry receptors for human cytomegalovirus via a highly conserved disintegrin-like domain. Proc. Natl. Acad. Sci. U. S. A. 101, 15470–15475. https://doi.org/10.1073/pnas.0406821101.

Feng, X., Tonnesen, M.G., Peerschke, E.I.B., Ghebrehiwet, B., 2002. Cooperation of C1q receptors and integrins in C1q-mediated endothelial cell adhesion and spreading. J. Immunol. 168, 2441–2448.

Forbes, K.M., Sironen, T., Plyusnin, A., 2018. Hantavirus maintenance and transmission in reservoir host populations. Curr. Opin. Virol. 28, 1–6. https://doi.org/10.1016/j.coviro.2017.09.003.

Garrido, J.L., Prescott, J., Calvo, M., Bravo, F., Alvarez, R., Salas, A., Riquelme, R., Rioseco, M.L., Williamson, B.N., Haddock, E., Feldmann, H., Barria, M.I., 2018. Two recombinant human monoclonal antibodies that protect against lethal Andes hantavirus infection in vivo. Sci. Transl. Med. 10, eaat6420. https://doi.org/10.1126/scitranslmed.aat6420.

Gavard, J., Gutkind, J.S., 2006. VEGF controls endothelial-cell permeability by promoting the beta-arrestin-dependent endocytosis of VE-cadherin. Nat. Cell Biol. 8, 1223–1234. https://doi.org/10.1038/ncb1486.

Gavrilovskaya, I.N., Shepley, M., Shaw, R., Ginsberg, M.H., Mackow, E.R., 1998. Beta3 integrins mediate the cellular entry of hantaviruses that cause respiratory failure. Proc. Natl. Acad. Sci. U. S. A. 95, 7074–7079.

Gavrilovskaya, I.N., Brown, E.J., Ginsberg, M.H., Mackow, E.R., 1999. Cellular entry of hantaviruses which cause hemorrhagic fever with renal syndrome is mediated by beta3 integrins. J. Virol. 73, 3951–3959.

Gavrilovskaya, I.N., Gorbunova, E.E., Mackow, E.R., 2010. Pathogenic hantaviruses direct the adherence of quiescent platelets to infected endothelial cells. J. Virol. 84, 4832–4839. https://doi.org/10.1128/JVI.02405-09.

Geimonen, E., Neff, S., Raymond, T., Kocer, S.S., Gavrilovskaya, I.N., Mackow, E.R., 2002. Pathogenic and nonpathogenic hantaviruses differentially regulate endothelial cell responses. Proc. Natl. Acad. Sci. U. S. A. 99, 13837–13842. https://doi.org/10.1073/pnas.192298899.

Gianni, T., Cerretani, A., Dubois, R., Salvioli, S., Blystone, S.S., Rey, F., Campadelli-Fiume, G., 2010. Herpes simplex virus glycoproteins H/L bind to cells independently of {alpha}V{beta}3 integrin and inhibit virus entry, and their constitutive expression restricts infection. J. Virol. 84, 4013–4025. https://doi.org/10.1128/JVI.02502-09.

Goldsmith, C.S., Elliott, L.H., Peters, C.J., Zaki, S.R., 1995. Ultrastructural characteristics of Sin Nombre virus, causative agent of hantavirus pulmonary syndrome. Arch. Virol. 140, 2107–2122. https://doi.org/10.1007/BF01323234.

Gorbunova, E.E., Simons, M.J., Gavrilovskaya, I.N., Mackow, E.R., 2016. The Andes virus nucleocapsid protein directs basal endothelial cell permeability by activating RhoA. MBio 7, e01747–e01816. https://doi.org/10.1128/mBio.01747-16.

Green, R.F., Meiss, H.K., Rodriguez-Boulan, E., 1981. Glycosylation does not determine segregation of viral envelope proteins in the plasma membrane of epithelial cells. J. Cell Biol. 89, 230–239. https://doi.org/10.1083/jcb.89.2.230.

Guardado-Calvo, P., Rey, F.A., 2017. The envelope proteins of the bunyavirales. Adv. Virus Res. 98, 83–118. https://doi.org/10.1016/bs.aivir.2017.02.002.

Guardado-Calvo, P., Bignon, E.A., Stettner, E., Jeffers, S.A., Pérez-Vargas, J., Pehau-Arnaudet, G., Tortorici, M.A., Jestin, J.-L., England, P., Tischler, N.D., Rey, F.A., 2016. Mechanistic insight into bunyavirus-induced membrane fusion from structure-function analyses of the hantavirus envelope glycoprotein Gc. PLoS Pathog. 12, e1005813. https://doi.org/10.1371/journal.ppat.1005813.

Guerrero, C.A., Méndez, E., Zárate, S., Isa, P., López, S., Arias, C.F., 2000. Integrin alpha(v)beta(3) mediates rotavirus cell entry. Proc. Natl. Acad. Sci. U. S. A. 97, 14644–14649. https://doi.org/10.1073/pnas.250299897.

Gupta, S., Braun, M., Tischler, N.D., Stoltz, M., Sundström, K.B., Björkström, N.K., Ljunggren, H.-G., Klingström, J., 2013. Hantavirus-infection confers resistance to cytotoxic lymphocyte-mediated apoptosis. PLoS Pathog. 9, e1003272. https://doi.org/10.1371/journal.ppat.1003272.

Guterres, A., de Lemos, E.R.S., 2018. Hantaviruses and a neglected environmental determinant. One Health 5, 27–33. https://doi.org/10.1016/j.onehlt.2017.12.002.

Haese, N., Brocato, R.L., Henderson, T., Nilles, M.L., Kwilas, S.A., Josleyn, M.D., Hammerbeck, C.D., Schiltz, J., Royals, M., Ballantyne, J., Hooper, J.W., Bradley, D.S., 2015. Antiviral biologic produced in DNA vaccine/goose platform protects hamsters against hantavirus pulmonary syndrome when administered post-exposure. PLoS Negl. Trop. Dis. 9, e0003803. https://doi.org/10.1371/journal.pntd.0003803.

Hardcastle, K., Scott, D., Safronetz, D., Brining, D.L., Ebihara, H., Feldmann, H., LaCasse, R.A., 2016. Laguna Negra virus infection causes hantavirus pulmonary syndrome in Turkish hamsters (*Mesocricetus brandti*). Vet. Pathol. 53, 182–189. https://doi.org/10.1177/0300985815570071.

Heinemann, P., Tia, M., Alabi, A., Anon, J.-C., Auste, B., Essbauer, S., Gnionsahe, A., Kigninlman, H., Klempa, B., Kraef, C., Kruger, N., Leendertz, F.H., Ndhatz-Sanogo, M., Schaumburg, F., Witkowski, P.T., Akoua-Koffi, C.G., Kruger, D.H., 2016. Human infections by non-rodent-associated hantaviruses in Africa. J. Infect. Dis. 214, 1507–1511. https://doi.org/10.1093/infdis/jiw401.

Hepojoki, J., Strandin, T., Vaheri, A., Lankinen, H., 2010a. Interactions and oligomerization of hantavirus glycoproteins. J. Virol. 84, 227–242. https://doi.org/10.1128/JVI.00481-09.

Hepojoki, J., Strandin, T., Wang, H., Vapalahti, O., Vaheri, A., Lankinen, H., 2010b. Cytoplasmic tails of hantavirus glycoproteins interact with the nucleocapsid protein. J. Gen. Virol. 91, 2341–2350. https://doi.org/10.1099/vir.0.021006-0.

Hepojoki, J., Strandin, T., Lankinen, H., Vaheri, A., 2012. Hantavirus structure–molecular interactions behind the scene. J. Gen. Virol. 93, 1631–1644. https://doi.org/10.1099/vir.0.042218-0.

Hepojoki, J., Vaheri, A., Strandin, T., 2014. The fundamental role of endothelial cells in hantavirus pathogenesis. Front. Microbiol. 5, 727. https://doi.org/10.3389/fmicb.2014.00727.

Higa, M.M., Petersen, J., Hooper, J., Doms, R.W., 2012. Efficient production of Hantaan and Puumala pseudovirions for viral tropism and neutralization studies. Virology 423, 134–142. https://doi.org/10.1016/j.virol.2011.08.012.

Hodivala-Dilke, K.M., McHugh, K.P., Tsakiris, D.A., Rayburn, H., Crowley, D., Ullman-Cullere, M., Ross, F.P., Coller, B.S., Teitelbaum, S., Hynes, R.O., 1999. Beta3-integrin-deficient mice are a model for Glanzmann thrombasthenia showing placental defects and reduced survival. J. Clin. Invest. 103, 229–238. https://doi.org/10.1172/JCI5487.

Hooper, J.W., Kamrud, K.I., Elgh, F., Custer, D., Schmaljohn, C.S., 1999. DNA vaccination with hantavirus M segment elicits neutralizing antibodies and protects against Seoul virus infection. Virology 255, 269–278. https://doi.org/10.1006/viro.1998.9586.

Hooper, J.W., Custer, D.M., Thompson, E., Schmaljohn, C.S., 2001. DNA vaccination with the Hantaan virus M gene protects Hamsters against three of four HFRS hantaviruses and elicits a high-titer neutralizing antibody response in Rhesus monkeys. J. Virol. 75, 8469–8477. https://doi.org/10.1128/JVI.75.18.8469-8477.2001.

Hooper, J.W., Ferro, A.M., Wahl-Jensen, V., 2008. Immune serum produced by DNA vaccination protects hamsters against lethal respiratory challenge with Andes virus. J. Virol. 82, 1332–1338. https://doi.org/10.1128/JVI.01822-07.

Hooper, J.W., Josleyn, M., Ballantyne, J., Brocato, R., 2013. A novel Sin Nombre virus DNA vaccine and its inclusion in a candidate pan-hantavirus vaccine against hantavirus pulmonary syndrome (HPS) and hemorrhagic fever with renal syndrome (HFRS). Vaccine 31, 4314–4321. https://doi.org/10.1016/j.vaccine.2013.07.025.

Hooper, J.W., Brocato, R.L., Kwilas, S.A., Hammerbeck, C.D., Josleyn, M.D., Royals, M., Ballantyne, J., Wu, H., Jiao, J., Matsushita, H., Sullivan, E.J., 2014a. DNA vaccine-derived human IgG produced in transchromosomal bovines protect in lethal models of hantavirus pulmonary syndrome. Sci. Transl. Med. 6, 264ra162. https://doi.org/10.1126/scitranslmed.3010082.

Hooper, J.W., Moon, J.E., Paolino, K.M., Newcomer, R., McLain, D.E., Josleyn, M., Hannaman, D., Schmaljohn, C., 2014b. A phase 1 clinical trial of Hantaan virus and Puumala virus M-segment DNA vaccines for haemorrhagic fever with renal syndrome delivered by intramuscular electroporation. Clin. Microbiol. Infect. 20 (Suppl. 5), 110–117. https://doi.org/10.1111/1469-0691.12553.

Huiskonen, J.T., Hepojoki, J., Laurinmäki, P., Vaheri, A., Lankinen, H., Butcher, S.J., Grünewald, K., 2010. Electron cryotomography of Tula hantavirus suggests a unique assembly paradigm for enveloped viruses. J. Virol. 84, 4889–4897. https://doi.org/10.1128/JVI.00057-10.

Hung, T., 1988. Atlas of Hemorrhagic Fever With Renal Syndrome. Science Press, Beijing.

Hung, T., Xia, S.M., Zhao, T.X., Zhou, J.Y., Song, G., Liao, G.X.H., Ye, W.W., Chu, Y.L., Hang, C.S., 1983. Morphological evidence for identifying the viruses of hemorrhagic fever with renal syndrome as candidate members of the bunyaviridae family. Arch. Virol. 78, 137–144. https://doi.org/10.1007/BF01310869.

Jääskeläinen, K.M., Kaukinen, P., Minskaya, E.S., Plyusnina, A., Vapalahti, O., Elliott, R.M., Weber, F., Vaheri, A., Plyusnin, A., 2007. Tula and Puumala hantavirus NSs ORFs are functional and the products inhibit activation of the interferon-beta promoter. J. Med. Virol. 79, 1527–1536. https://doi.org/10.1002/jmv.20948.

Jackson, T., Sheppard, D., Denyer, M., Blakemore, W., King, A.M., 2000. The epithelial integrin alphavbeta6 is a receptor for foot-and-mouth disease virus. J. Virol. 74, 4949–4956.

Jangra, R.K., Herbert, A.S., Li, R., Jae, L.T., Kleinfelter, L.M., Slough, M.M., Barker, S.L., Guardado-Calvo, P., Román-Sosa, G., Dieterle, M.E., Kuehne, A.I., Muena, N.A.,

Wirchnianski, A.S., Nyakatura, E.K., Fels, J.M., Ng, M., Mittler, E., Pan, J., Bharrhan, S., Wec, A.Z., Chandran, K., 2018. Protocadherin-1 is essential for cell entry by New World hantaviruses. Nature 563, 559–563. https://doi.org/10.1038/s41586-018-0702-1.

Jiang, H., Zheng, X., Wang, L., Du, H., Wang, P., Bai, X., 2017. Hantavirus infection: a global zoonotic challenge. Virol. Sin. 32, 32–43. https://doi.org/10.1007/s12250-016-3899-x.

Jin, M., Park, J., Lee, S., Park, B., Shin, J., Song, K.-J., Ahn, T.-I., Hwang, S.-Y., Ahn, B.-Y., Ahn, K., 2002. Hantaan virus enters cells by clathrin-dependent receptor-mediated endocytosis. Virology 294, 60–69. https://doi.org/10.1006/viro.2001.1303.

Johnson, K.M., 2001. Hantaviruses: history and overview. Curr. Top. Microbiol. Immunol. 256, 1–14.

Johnson, J.E., Rodgers, W., Rose, J.K., 1998. A plasma membrane localization signal in the HIV-1 envelope cytoplasmic domain prevents localization at sites of vesicular stomatitis virus budding and incorporation into VSV virions. Virology 251, 244–252. https://doi.org/10.1006/viro.1998.9429.

Joutsi-Korhonen, L., Preston, S., Smethurst, P.A., Ijsseldijk, M., Schaffner-Reckinger, E., Armour, K.L., Watkins, N.A., Clark, M.R., de Groot, P.G., Farndale, R.W., Ouwehand, W.H., Williamson, L.M., 2004. The effect of recombinant IgG antibodies against the leucine-33 form of the platelet beta3 integrin (HPA-1a) on platelet function. Thromb. Haemost. 91, 743–754. https://doi.org/10.1160/TH03-07-0484.

Jung, J., Ko, S.-J., Oh, H.S., Moon, S.M., Song, J.-W., Huh, K., 2018. Protective effectiveness of inactivated hantavirus vaccine against hemorrhagic fever with renal syndrome. J. Infect. Dis. 217, 1417–1420. https://doi.org/10.1093/infdis/jiy037.

Kallio-Kokko, H., Leveelahti, R., Brummer-Korvenkontio, M., Lundkvist, A., Vaheri, A., Vapalahti, O., 2001. Human immune response to Puumala virus glycoproteins and nucleocapsid protein expressed in mammalian cells. J. Med. Virol. 65, 605–613. https://doi.org/10.1002/jmv.2079.

Kamrud, K.I., Schmaljohn, C.S., 1994. Expression strategy of the M genome segment of Hantaan virus. Virus Res. 31, 109–121.

Karnauchow, T.M., Tolson, D.L., Harrison, B.A., Altman, E., Lublin, D.M., Dimock, K., 1996. The HeLa cell receptor for enterovirus 70 is decay-accelerating factor (CD55). J. Virol. 70, 5143–5152.

Kielian, M.C., Helenius, A., 1984. Role of cholesterol in fusion of Semliki Forest virus with membranes. J. Virol. 52, 281–283.

Kim, S.-Y., Yasuda, S., Tanaka, H., Yamagata, K., Kim, H., 2011. Non-clustered protocadherin. Cell Adh. Migr. 5, 97–105. https://doi.org/10.4161/cam.5.2.14374.

Kleinfelter, L.M., Jangra, R.K., Jae, L.T., Herbert, A.S., Mittler, E., Stiles, K.M., Wirchnianski, A.S., Kielian, M., Brummelkamp, T.R., Dye, J.M., Chandran, K., 2015. Haploid genetic screen reveals a profound and direct dependence on cholesterol for hantavirus membrane fusion. MBio 6, e00801. https://doi.org/10.1128/mBio.00801-15.

Klempa, B., 2009. Hantaviruses and climate change. Clin. Microbiol. Infect. 15, 518–523. https://doi.org/10.1111/j.1469-0691.2009.02848.x.

Klempa, B., Koivogui, L., Sylla, O., Koulemou, K., Auste, B., Krüger, D.H., ter Meulen, J., 2010. Serological evidence of human hantavirus infections in Guinea, West Africa. J. Infect. Dis. 201, 1031–1034. https://doi.org/10.1086/651169.

Klempa, B., Witkowski, P.T., Popugaeva, E., Auste, B., Koivogui, L., Fichet-Calvet, E., Strecker, T., Ter Meulen, J., Krüger, D.H., 2012. Sangassou virus, the first hantavirus isolate from Africa, displays genetic and functional properties distinct from those of other murinae-associated hantaviruses. J. Virol. 86, 3819–3827. https://doi.org/10.1128/JVI.05879-11.

Klingström, J., Stoltz, M., Hardestam, J., Ahlm, C., Lundkvist, A., 2008. Passive immunization protects cynomolgus macaques against Puumala hantavirus challenge. Antivir. Ther. 13, 125–133.

Kobinger, G.P., Feldmann, H., Zhi, Y., Schumer, G., Gao, G., Feldmann, F., Jones, S., Wilson, J.M., 2006. Chimpanzee adenovirus vaccine protects against Zaire Ebola virus. Virology 346, 394–401. https://doi.org/10.1016/j.virol.2005.10.042.
Koppelman, G.H., Meyers, D.A., Howard, T.D., Zheng, S.L., Hawkins, G.A., Ampleford, E.J., Xu, J., Koning, H., Bruinenberg, M., Nolte, I.M., van Diemen, C.C., Boezen, H.M., Timens, W., Whittaker, P.A., Stine, O.C., Barton, S.J., Holloway, J.W., Holgate, S.T., Graves, P.E., Martinez, F.D., Postma, D.S., 2009. Identification of PCDH1 as a novel susceptibility gene for bronchial hyperresponsiveness. Am. J. Respir. Crit. Care Med. 180, 929–935. https://doi.org/10.1164/rccm.200810-1621OC.
Kozu, Y., Gon, Y., Maruoka, S., Kazumichi, K., Sekiyama, A., Kishi, H., Nomura, Y., Ikeda, M., Hashimoto, S., 2015. Protocadherin-1 is a glucocorticoid-responsive critical regulator of airway epithelial barrier function. BMC Pulm. Med. 15, 80. https://doi.org/10.1186/s12890-015-0078-z.
Krautkrämer, E., Zeier, M., 2008. Hantavirus causing hemorrhagic fever with renal syndrome enters from the apical surface and requires decay-accelerating factor (DAF/CD55). J. Virol. 82, 4257–4264. https://doi.org/10.1128/JVI.02210-07.
Kukkonen, S.K.J., Vaheri, A., Plyusnin, A., 2005. L protein, the RNA-dependent RNA polymerase of hantaviruses. Arch. Virol. 150, 533–556. https://doi.org/10.1007/s00705-004-0414-8.
Lee, Y.S., Seong, B.L., 1996. Mutational analysis of influenza B virus RNA transcription in vitro. J. Virol. 70, 1232–1236.
Lee, H.W., Lee, P.W., Johnson, K.M., 1978. Isolation of the etiologic agent of Korean hemorrhagic fever. J. Infect. Dis. 137, 298–308. https://doi.org/10.1093/infdis/137.3.298.
Lee, B.-H., Yoshimatsu, K., Araki, K., Okumura, M., Nakamura, I., Arikawa, J., 2006. A pseudotype vesicular stomatitis virus containing Hantaan virus envelope glycoproteins G1 and G2 as an alternative to hantavirus vaccine in mice. Vaccine 24, 2928–2934. https://doi.org/10.1016/j.vaccine.2005.12.040.
Lemckert, A.A.C., Sumida, S.M., Holterman, L., Vogels, R., Truitt, D.M., Lynch, D.M., Nanda, A., Ewald, B.A., Gorgone, D.A., Lifton, M.A., Goudsmit, J., Havenga, M.J.E., Barouch, D.H., 2005. Immunogenicity of heterologous prime-boost regimens involving recombinant adenovirus serotype 11 (Ad11) and Ad35 vaccine vectors in the presence of anti-ad5 immunity. J. Virol. 79, 9694–9701. https://doi.org/10.1128/JVI.79.15.9694-9701.2005.
Levanov, L., Iheozor-Ejiofor, R.P., Lundkvist, Å., Vapalahti, O., Plyusnin, A., 2019. Defining of MAbs-neutralizing sites on the surface glycoproteins Gn and Gc of a hantavirus using vesicular stomatitis virus pseudotypes and site-directed mutagenesis. J. Gen. Virol. 100, 145–155. https://doi.org/10.1099/jgv.0.001202.
Lévy, Y., Lane, C., Piot, P., Beavogui, A.H., Kieh, M., Leigh, B., Doumbia, S., D'Ortenzio, E., Lévy-Marchal, C., Pierson, J., Watson-Jones, D., Nguyen, V.-K., Larson, H., Lysander, J., Lacabaratz, C., Thiebaut, R., Augier, A., Ishola, D., Kennedy, S., Chêne, G., Yazdanpanah, Y., 2018. Prevention of Ebola virus disease through vaccination: where we are in 2018. Lancet 392, 787–790. https://doi.org/10.1016/S0140-6736(18)31710-0.
Li, C., Liu, F., Liang, M., Zhang, Q., Wang, X., Wang, T., Li, J., Li, D., 2010. Hantavirus-like particles generated in CHO cells induce specific immune responses in C57BL/6 mice. Vaccine 28, 4294–4300. https://doi.org/10.1016/j.vaccine.2010.04.025.
Li, S., Rissanen, I., Zeltina, A., Hepojoki, J., Raghwani, J., Harlos, K., Pybus, O.G., Huiskonen, J.T., Bowden, T.A., 2016. A molecular-level account of the antigenic hantaviral surface. Cell Rep. 15, 959–967. https://doi.org/10.1016/j.celrep.2016.03.082.
Löber, C., Anheier, B., Lindow, S., Klenk, H.D., Feldmann, H., 2001. The Hantaan virus glycoprotein precursor is cleaved at the conserved pentapeptide WAASA. Virology 289, 224–229. https://doi.org/10.1006/viro.2001.1171.

Lu, Y.E., Cassese, T., Kielian, M., 1999. The cholesterol requirement for sindbis virus entry and exit and characterization of a spike protein region involved in cholesterol dependence. J. Virol. 73, 4272–4278.

Ma, M., Kersten, D.B., Kamrud, K.I., Wool-Lewis, R.J., Schmaljohn, C., González-Scarano, F., 1999. Murine leukemia virus pseudotypes of La Crosse and Hantaan Bunyaviruses: a system for analysis of cell tropism. Virus Res. 64, 23–32. https://doi.org/10.1016/S0168-1702(99)00070-2.

Mackow, E.R., Gavrilovskaya, I.N., 2009. Hantavirus regulation of endothelial cell functions. Thromb. Haemost. 102, 1030–1041. https://doi.org/10.1160/TH09-09-0640.

Maginnis, M.S., Forrest, J.C., Kopecky-Bromberg, S.A., Dickeson, S.K., Santoro, S.A., Zutter, M.M., Nemerow, G.R., Bergelson, J.M., Dermody, T.S., 2006. Beta1 integrin mediates internalization of mammalian reovirus. J. Virol. 80, 2760–2770. https://doi.org/10.1128/JVI.80.6.2760-2770.2006.

Martin, M.L., Lindsey-Regnery, H., Sasso, D.R., McCormick, J.B., Palmer, E., 1985. Distinction between Bunyaviridae genera by surface structure and comparison with Hantaan virus using negative stain electron microscopy. Arch. Virol. 86, 17–28. https://doi.org/10.1007/BF01314110.

Martinez, V.P., Bellomo, C., San Juan, J., Pinna, D., Forlenza, R., Elder, M., Padula, P.J., 2005. Person-to-person transmission of Andes virus. Emerg. Infect. Dis. 11, 1848–1853. https://doi.org/10.3201/eid1112.050501.

Martinez-Valdebenito, C., Calvo, M., Vial, C., Mansilla, R., Marco, C., Palma, R.E., Vial, P.A., Valdivieso, F., Mertz, G., Ferrés, M., 2014. Person-to-person household and nosocomial transmission of Andes hantavirus, Southern Chile, 2011. Emerg. Infect. Dis. 20, 1629–1636. https://doi.org/10.3201/eid2010.140353.

Martínez-Valdebenito, C., Angulo, J., Le Corre, N., Marco, C., Vial, C., Miquel, J.F., Cerda, J., Mertz, G., Vial, P., Lopez-Lastra, M., Ferrés, M., 2019. A single-nucleotide polymorphism of αVβ□ integrin is associated with the Andes virus infection susceptibility. Viruses 11, 169. https://doi.org/10.3390/v11020169.

Matthys, V.S., Gorbunova, E.E., Gavrilovskaya, I.N., Mackow, E.R., 2010. Andes virus recognition of human and Syrian hamster beta3 integrins is determined by an L33P substitution in the PSI domain. J. Virol. 84, 352–360. https://doi.org/10.1128/JVI.01013-09.

Matthys, V., Mackow, E.R., 2012. Hantavirus regulation of type I interferon responses. Adv. Virol. 2012, 524024. https://doi.org/10.1155/2012/524024.

McClain, D.J., Summers, P.L., Harrison, S.A., Schmaljohn, A.L., Schmaljohn, C.S., 2000. Clinical evaluation of a vaccinia-vectored Hantaan virus vaccine. J. Med. Virol. 60, 77–85. https://doi.org/10.1002/(SICI)1096-9071(200001)60:1<77::AID-JMV13->3.0.CO;2-S.

McCormick, J.B., Sasso, D.R., Palmer, E.L., Kiley, M.P., 1982. Morphological identification of the agent of Korean haemorrhagic fever (Hantaan virus) as a member of the Bunyaviridae. Lancet 1, 765–768.

McElroy, A.K., Smith, J.M., Hooper, J.W., Schmaljohn, C.S., 2004. Andes virus M genome segment is not sufficient to confer the virulence associated with Andes virus in Syrian hamsters. Virology 326, 130–139. https://doi.org/10.1016/j.virol.2004.05.018.

Meyer, B.J., Schmaljohn, C.S., 2000. Persistent hantavirus infections: characteristics and mechanisms. Trends Microbiol. 8, 61–67. https://doi.org/10.1016/S0966-842X(99)01658-3.

Mir, M.A., Panganiban, A.T., 2006. Characterization of the RNA chaperone activity of hantavirus nucleocapsid protein. J. Virol. 80, 6276–6285. https://doi.org/10.1128/JVI.00147-06.

Mir, M.A., Brown, B., Hjelle, B., Duran, W.A., Panganiban, A.T., 2006. Hantavirus N protein exhibits genus-specific recognition of the viral RNA panhandle. J. Virol. 80, 11283–11292. https://doi.org/10.1128/JVI.00820-06.

Müller, A., Baumann, A., Essbauer, S., Radosa, L., Krüger, D.H., Witkowski, P.T., Zeier, M., Krautkrämer, E., 2019. Analysis of the integrin β3 receptor for pathogenic orthohantaviruses in rodent host species. Virus Res. 267, 36–40. https://doi.org/10.1016/j.virusres.2019.04.009.

Muyangwa, M., Martynova, E.V., Khaiboullina, S.F., Morzunov, S.P., Rizvanov, A.A., 2015. Hantaviral proteins: structure, functions, and role in hantavirus infection. Front. Microbiol. 6, 1326. https://doi.org/10.3389/fmicb.2015.01326.

Nieva, J.L., Bron, R., Corver, J., Wilschut, J., 1994. Membrane fusion of Semliki Forest virus requires sphingolipids in the target membrane. EMBO J. 13, 2797–2804.

Nolte, K.B., Feddersen, R.M., Foucar, K., Zaki, S.R., Koster, F.T., Madar, D., Merlin, T.L., McFeeley, P.J., Umland, E.T., Zumwalt, R.E., 1995. Hantavirus pulmonary syndrome in the United States: a pathological description of a disease caused by a new agent. Hum. Pathol. 26, 110–120. https://doi.org/10.1016/0046-8177(95)90123-X.

Nuzum, E.O., Rossi, C.A., Stephenson, E.H., LeDuc, J.W., 1988. Aerosol transmission of Hantaan and related viruses to laboratory rats. Am. J. Trop. Med. Hyg. 38, 636–640. https://doi.org/10.4269/ajtmh.1988.38.636.

Ogino, M., Ebihara, H., Lee, B.-H., Araki, K., Lundkvist, A., Kawaoka, Y., Yoshimatsu, K., Arikawa, J., 2003. Use of vesicular stomatitis virus pseudotypes bearing hantaan or Seoul virus envelope proteins in a rapid and safe neutralization test. Clin. Diagn. Lab. Immunol. 10, 154–160.

Ogino, M., Yoshimatsu, K., Ebihara, H., Araki, K., Lee, B.-H., Okumura, M., Arikawa, J., 2004. Cell fusion activities of Hantaan virus envelope glycoproteins. J. Virol. 78, 10776–10782. https://doi.org/10.1128/JVI.78.19.10776-10782.2004.

Okumura, M., Yoshimatsu, K., Kumperasart, S., Nakamura, I., Ogino, M., Taruishi, M., Sungdee, A., Pattamadilok, S., Ibrahim, I.N., Erlina, S., Agui, T., Yanagihara, R., Arikawa, J., 2007. Development of serological assays for Thottapalayam virus, an insectivore-borne Hantavirus. Clin. Vaccine Immunol. 14, 173–181. https://doi.org/10.1128/CVI.00347-06.

Pan, W., Bian, G., Wang, K., Feng, T., Dai, J., 2015. Effects of different doses of nucleocapsid protein from Hantaan virus A9 strain on regulation of interferon signaling. Viral Immunol. 28, 448–454. https://doi.org/10.1089/vim.2015.0004.

Paneth Iheozor-Ejiofor, R., Levanov, L., Hepojoki, J., Strandin, T., Lundkvist, Å., Plyusnin, A., Vapalahti, O., 2016. Vaccinia virus-free rescue of fluorescent replication-defective vesicular stomatitis virus and pseudotyping with Puumala virus glycoproteins for use in neutralization tests. J. Gen. Virol. 97, 1052–1059. https://doi.org/10.1099/jgv.0.000437.

Pensiero, M.N., Hay, J., 1992. The Hantaan virus M-segment glycoproteins G1 and G2 can be expressed independently. J. Virol. 66, 1907–1914.

Petersen, J., Drake, M.J., Bruce, E.A., Riblett, A.M., Didigu, C.A., Wilen, C.B., Malani, N., Male, F., Lee, F.-H., Bushman, F.D., Cherry, S., Doms, R.W., Bates, P., Briley, K., 2014. The major cellular sterol regulatory pathway is required for Andes virus infection. PLoS Pathog. 10, e1003911. https://doi.org/10.1371/journal.ppat.1003911.

Pettersson, L., Thunberg, T., Rocklöv, J., Klingström, J., Evander, M., Ahlm, C., 2014. Viral load and humoral immune response in association with disease severity in Puumala hantavirus-infected patients—implications for treatment. Clin. Microbiol. Infect. 20, 235–241. https://doi.org/10.1111/1469-0691.12259.

Popugaeva, E., Witkowski, P.T., Schlegel, M., Ulrich, R.G., Auste, B., Rang, A., Krüger, D.H., Klempa, B., 2012. Dobrava-Belgrade hantavirus from Germany shows receptor usage and innate immunity induction consistent with the pathogenicity of the virus in humans. PLoS One 7, e35587. https://doi.org/10.1371/journal.pone.0035587.

Prescott, J., DeBuysscher, B.L., Brown, K.S., Feldmann, H., 2014. Long-term single-dose efficacy of a vesicular stomatitis virus-based Andes virus vaccine in Syrian hamsters. Viruses 6, 516–523. https://doi.org/10.3390/v6020516.
Qian, Z., Haessler, M., Lemos, J.A., Arsenault, J.R., Aguirre, J.E., Gilbert, J.R., Bowler, R.P., Park, F., 2006. Targeting vascular injury using Hantavirus-pseudotyped lentiviral vectors. Mol. Ther. 13, 694–704. https://doi.org/10.1016/j.ymthe.2005.11.016.
Raftery, M.J., Lalwani, P., Krautkrämer, E., Peters, T., Scharffetter-Kochanek, K., Krüger, R., Hofmann, J., Seeger, K., Krüger, D.H., Schönrich, G., 2014. β2 integrin mediates hantavirus-induced release of neutrophil extracellular traps. J. Exp. Med. 211, 1485–1497. https://doi.org/10.1084/jem.20131092.
Rahimi, N., 2017. Defenders and challengers of endothelial barrier function. Front. Immunol. 8, 1847. https://doi.org/10.3389/fimmu.2017.01847.
Ramanathan, H.N., Jonsson, C.B., 2008. New and old world hantaviruses differentially utilize host cytoskeletal components during their life cycles. Virology 374, 138–150. https://doi.org/10.1016/j.virol.2007.12.030.
Rasmuson, J., Andersson, C., Norrman, E., Haney, M., Evander, M., Ahlm, C., 2011. Time to revise the paradigm of hantavirus syndromes? Hantavirus pulmonary syndrome caused by European hantavirus. Eur. J. Clin. Microbiol. Infect. Dis. 30, 685–690. https://doi.org/10.1007/s10096-010-1141-6.
Ravkov, E.V., Nichol, S.T., Compans, R.W., 1997. Polarized entry and release in epithelial cells of Black Creek Canal virus, a new world hantavirus. J. Virol. 71, 1147–1154.
Ray, N., Whidby, J., Stewart, S., Hooper, J.W., Bertolotti-Ciarlet, A., 2010. Study of Andes virus entry and neutralization using a pseudovirion system. J. Virol. Methods 163, 416–423. https://doi.org/10.1016/j.jviromet.2009.11.004.
Raymond, T., Gorbunova, E., Gavrilovskaya, I.N., Mackow, E.R., 2005. Pathogenic hantaviruses bind plexin-semaphorin-integrin domains present at the apex of inactive, bent alphavbeta3 integrin conformers. Proc. Natl. Acad. Sci. U. S. A. 102, 1163–1168. https://doi.org/10.1073/pnas.0406743102.
Regules, J.A., Beigel, J.H., Paolino, K.M., Voell, J., Castellano, A.R., Hu, Z., Muñoz, P., Moon, J.E., Ruck, R.C., Bennett, J.W., Twomey, P.S., Gutiérrez, R.L., Remich, S.A., Hack, H.R., Wisniewski, M.L., Josleyn, M.D., Kwilas, S.A., Van Deusen, N., Mbaya, O.T., Zhou, Y., rVSVΔG-ZEBOV-GP Study Group, 2017. A recombinant vesicular stomatitis virus ebola vaccine. N. Engl. J. Med. 376, 330–341. https://doi.org/10.1056/NEJMoa1414216.
Reynolds, L.E., Wyder, L., Lively, J.C., Taverna, D., Robinson, S.D., Huang, X., Sheppard, D., Hynes, R.O., Hodivala-Dilke, K.M., 2002. Enhanced pathological angiogenesis in mice lacking beta3 integrin or beta3 and beta5 integrins. Nat. Med. 8, 27–34. https://doi.org/10.1038/nm0102-27.
Rissanen, I., Stass, R., Zeltina, A., Li, S., Hepojoki, J., Harlos, K., Gilbert, R.J.C., Huiskonen, J.T., Bowden, T.A., 2017. Structural transitions of the conserved and metastable hantaviral glycoprotein envelope. J. Virol. 91, e00378-17. https://doi.org/10.1128/JVI.00378-17.
Rowe, R.K., Pekosz, A., 2006. Bidirectional virus secretion and nonciliated cell tropism following Andes virus infection of primary airway epithelial cell cultures. J. Virol. 80, 1087–1097. https://doi.org/10.1128/JVI.80.3.1087-1097.2006.
Ruoslahti, E., 1996. RGD and other recognition sequences for integrins. Annu. Rev. Cell Dev. Biol. 12, 697–715. https://doi.org/10.1146/annurev.cellbio.12.1.697.
Ruusala, A., Persson, R., Schmaljohn, C.S., Pettersson, R.F., 1992. Coexpression of the membrane glycoproteins G1 and G2 of Hantaan virus is required for targeting to the Golgi complex. Virology 186, 53–64.
Safronetz, D., Hegde, N.R., Ebihara, H., Denton, M., Kobinger, G.P., St Jeor, S., Feldmann, H., Johnson, D.C., 2009. Adenovirus vectors expressing hantavirus proteins

protect hamsters against lethal challenge with andes virus. J. Virol. 83, 7285–7295. https://doi.org/10.1128/JVI.00373-09.
Sargianou, M., Watson, D.C., Chra, P., Papa, A., Starakis, I., Gogos, C., Panos, G., 2012. Hantavirus infections for the clinician: from case presentation to diagnosis and treatment. Crit. Rev. Microbiol. 38, 317–329. https://doi.org/10.3109/1040841X.2012.673553.
Schmaljohn, C.S., Dalrymple, J.M., 1983. Analysis of Hantaan virus RNA: evidence for a new genus of bunyaviridae. Virology 131, 482–491. https://doi.org/10.1016/0042-6822(83)90514-7.
Schmaljohn, C.S., Hasty, S.E., Harrison, S.A., Dalrymple, J.M., 1983. Characterization of Hantaan virions, the prototype virus of hemorrhagic fever with renal syndrome. J. Infect. Dis. 148, 1005–1012. https://doi.org/10.1093/infdis/148.6.1005.
Schmaljohn, C.S., Jennings, G.B., Hay, J., Dalrymple, J.M., 1986. Coding strategy of the S genome segment of Hantaan virus. Virology 155, 633–643.
Schmaljohn, C.S., Schmaljohn, A.L., Dalrymple, J.M., 1987. Hantaan virus M RNA: coding strategy, nucleotide sequence, and gene order. Virology 157, 31–39.
Schmaljohn, C.S., Hasty, S.E., Dalrymple, J.M., 1992. Preparation of candidate vaccinia-vectored vaccines for haemorrhagic fever with renal syndrome. Vaccine 10, 10–13.
Schoenenberger, C.A., Zuk, A., Zinkl, G.M., Kendall, D., Matlin, K.S., 1994. Integrin expression and localization in normal MDCK cells and transformed MDCK cells lacking apical polarity. J. Cell Sci. 107 (Pt. 2), 527–541.
Schönrich, G., Krüger, D.H., Raftery, M.J., 2015. Hantavirus-induced disruption of the endothelial barrier: neutrophils are on the payroll. Front. Microbiol. 6, 222. https://doi.org/10.3389/fmicb.2015.00222.
Senger, D.R., Galli, S.J., Dvorak, A.M., Perruzzi, C.A., Harvey, V.S., Dvorak, H.F., 1983. Tumor cells secrete a vascular permeability factor that promotes accumulation of ascites fluid. Science 219, 983–985. https://doi.org/10.1126/science.6823562.
Senger, D.R., Perruzzi, C.A., Feder, J., Dvorak, H.F., 1986. A highly conserved vascular permeability factor secreted by a variety of human and rodent tumor cell lines. Cancer Res. 46, 5629–5632.
Shafren, D.R., Bates, R.C., Agrez, M.V., Herd, R.L., Burns, G.F., Barry, R.D., 1995. Coxsackieviruses B1, B3, and B5 use decay accelerating factor as a receptor for cell attachment. J. Virol. 69, 3873–3877.
Shi, X., Elliott, R.M., 2002. Golgi localization of Hantaan virus glycoproteins requires coexpression of G1 and G2. Virology 300, 31–38. https://doi.org/10.1006/viro.2002.1414.
Shim, S.H., Park, M.-S., Moon, S., Park, K.S., Song, J.-W., Song, K.-J., Baek, L.J., 2011. Comparison of innate immune responses to pathogenic and putative non-pathogenic hantaviruses in vitro. Virus Res. 160, 367–373. https://doi.org/10.1016/j.virusres.2011.07.013.
Slough, M.M., Chandran, K., Jangra, R.K., 2019. Two point mutations in old world hantavirus glycoproteins afford the generation of highly infectious recombinant vesicular stomatitis virus vectors. MBio 10, e02372–e02418. https://doi.org/10.1128/mBio.02372-18.
Soldi, R., Mitola, S., Strasly, M., Defilippi, P., Tarone, G., Bussolino, F., 1999. Role of alphavbeta3 integrin in the activation of vascular endothelial growth factor receptor-2. EMBO J. 18, 882–892. https://doi.org/10.1093/emboj/18.4.882.
Song, J.-W., Song, K.-J., Baek, L.-J., Frost, B., Poncz, M., Park, K., 2005. In vivo characterization of the integrin beta3 as a receptor for Hantaan virus cellular entry. Exp. Mol. Med. 37, 121–127. https://doi.org/10.1038/emm.2005.16.
Song, J.Y., Woo, H.J., Cheong, H.J., Noh, J.Y., Baek, L.J., Kim, W.J., 2016. Long-term immunogenicity and safety of inactivated Hantaan virus vaccine (Hantavax^{TM}) in healthy adults. Vaccine 34, 1289–1295. https://doi.org/10.1016/j.vaccine.2016.01.031.

Sotomayor, M., Gaudet, R., Corey, D.P., 2014. Sorting out a promiscuous superfamily: towards cadherin connectomics. Trends Cell Biol. 24, 524–536. https://doi.org/10.1016/j.tcb.2014.03.007.

Spiropoulou, C.F., 2001. Hantavirus maturation. Curr. Top. Microbiol. Immunol. 256, 33–46.

Spiropoulou, C.F., Albariño, C.G., Ksiazek, T.G., Rollin, P.E., 2007. Andes and Prospect Hill hantaviruses differ in early induction of interferon although both can downregulate interferon signaling. J. Virol. 81, 2769–2776. https://doi.org/10.1128/JVI.02402-06.

Spiropoulou, C.F., Srikiatkhachorn, A., 2013. The role of endothelial activation in dengue hemorrhagic fever and hantavirus pulmonary syndrome. Virulence 4, 525–536. https://doi.org/10.4161/viru.25569.

Spiropoulou, C.F., Goldsmith, C.S., Shoemaker, T.R., Peters, C.J., Compans, R.W., 2003. Sin Nombre virus glycoprotein trafficking. Virology 308, 48–63. https://doi.org/10.1016/S0042-6822(02)00092-2.

Su, G., Atakilit, A., Li, J.T., Wu, N., Bhattacharya, M., Zhu, J., Shieh, J.E., Li, E., Chen, R., Sun, S., Su, C.P., Sheppard, D., 2012. Absence of integrin αvβ3 enhances vascular leak in mice by inhibiting endothelial cortical actin formation. Am. J. Respir. Crit. Care Med. 185, 58–66. https://doi.org/10.1164/rccm.201108-1381OC.

Suder, E., Furuyama, W., Feldmann, H., Marzi, A., de Wit, E., 2018. The vesicular stomatitis virus-based Ebola virus vaccine: from concept to clinical trials. Hum. Vaccin. Immunother. 14, 2107–2113. https://doi.org/10.1080/21645515.2018.1473698.

Tatsis, N., Ertl, H.C.J., 2004. Adenoviruses as vaccine vectors. Mol. Ther. 10, 616–629. https://doi.org/10.1016/j.ymthe.2004.07.013.

Tiley, L.S., Hagen, M., Matthews, J.T., Krystal, M., 1994. Sequence-specific binding of the influenza virus RNA polymerase to sequences located at the 5' ends of the viral RNAs. J. Virol. 68, 5108–5116.

Tischler, N.D., Gonzalez, A., Perez-Acle, T., Rosemblatt, M., Valenzuela, P.D.T., 2005. Hantavirus Gc glycoprotein: evidence for a class II fusion protein. J. Gen. Virol. 86, 2937–2947. https://doi.org/10.1099/vir.0.81083-0.

Torriani, G., Mayor, J., Zimmer, G., Kunz, S., Rothenberger, S., Engler, O., 2019. Macropinocytosis contributes to hantavirus entry into human airway epithelial cells. Virology 531, 57–68. https://doi.org/10.1016/j.virol.2019.02.013.

Vaheri, A., Strandin, T., Hepojoki, J., Sironen, T., Henttonen, H., Mäkelä, S., Mustonen, J., 2013. Uncovering the mysteries of hantavirus infections. Nat. Rev. Microbiol. 11, 539–550. https://doi.org/10.1038/nrmicro3066.

Valdivieso, F., Vial, P., Ferres, M., Ye, C., Goade, D., Cuiza, A., Hjelle, B., 2006. Neutralizing antibodies in survivors of Sin Nombre and Andes hantavirus infection. Emerg. Infect. Dis. 12, 166–168. https://doi.org/10.3201/eid1201.050930.

Van Agthoven, J.F., Xiong, J.-P., Alonso, J.L., Rui, X., Adair, B.D., Goodman, S.L., Arnaout, M.A., 2014. Structural basis for pure antagonism of integrin αVβ3 by a high-affinity form of fibronectin. Nat. Struct. Mol. Biol. 21, 383–388. https://doi.org/10.1038/nsmb.2797.

Vera-Otarola, J., Solis, L., Soto-Rifo, R., Ricci, E.P., Pino, K., Tischler, N.D., Ohlmann, T., Darlix, J.-L., López-Lastra, M., 2012. The Andes hantavirus NSs protein is expressed from the viral small mRNA by a leaky scanning mechanism. J. Virol. 86, 2176–2187. https://doi.org/10.1128/JVI.06223-11.

Vial, P.A., Valdivieso, F., Calvo, M., Rioseco, M.L., Riquelme, R., Araneda, A., Tomicic, V., Graf, J., Paredes, L., Florenzano, M., Bidart, T., Cuiza, A., Marco, C., Hjelle, B., Ye, C., Hanfelt-Goade, D., Vial, C., Rivera, J.C., Delgado, I., Mertz, G.J., Hantavirus Study Group in Chile, 2015. A non-randomized multicentre trial of human immune plasma for treatment of hantavirus cardiopulmonary syndrome caused by Andes virus. Antivir. Ther. 20, 377–386. https://doi.org/10.3851/IMP2875.

Voutilainen, L., Sironen, T., Tonteri, E., Bäck, A.T., Razzauti, M., Karlsson, M., Wahlström, M., Niemimaa, J., Henttonen, H., Lundkvist, Å., 2015. Life-long shedding of Puumala hantavirus in wild bank voles (*Myodes glareolus*). J. Gen. Virol. 96, 1238–1247. https://doi.org/10.1099/vir.0.000076.

Wang, H., Alminaite, A., Vaheri, A., Plyusnin, A., 2010. Interaction between hantaviral nucleocapsid protein and the cytoplasmic tail of surface glycoprotein Gn. Virus Res. 151, 205–212. https://doi.org/10.1016/j.virusres.2010.05.008.

Watkins, N.A., Smethurst, P.A., Allen, D., Smith, G.A., Ouwehand, W.H., 2002. Platelet alphaIIbbeta3 recombinant autoantibodies from the B-cell repertoire of a post-transfusion purpura patient. Br. J. Haematol. 116, 677–685.

Watson, D.C., Sargianou, M., Papa, A., Chra, P., Starakis, I., Panos, G., 2014. Epidemiology of Hantavirus infections in humans: a comprehensive, global overview. Crit. Rev. Microbiol. 40, 261–272. https://doi.org/10.3109/1040841X.2013.783555.

Wertz, G.W., Whelan, S., LeGrone, A., Ball, L.A., 1994. Extent of terminal complementarity modulates the balance between transcription and replication of vesicular stomatitis virus RNA. Proc. Natl. Acad. Sci. U. S. A. 91, 8587–8591. https://doi.org/10.1073/pnas.91.18.8587.

Wickham, T.J., Mathias, P., Cheresh, D.A., Nemerow, G.R., 1993. Integrins $\alpha v\beta 3$ and $\alpha v\beta 5$ promote adenovirus internalization but not virus attachment. Cell 73, 309–319. https://doi.org/10.1016/0092-8674(93)90231-E.

Willensky, S., Bar-Rogovsky, H., Bignon, E.A., Tischler, N.D., Modis, Y., Dessau, M., 2016. Crystal structure of glycoprotein C from a Hantavirus in the post-fusion conformation. PLoS Pathog. 12, e1005948. https://doi.org/10.1371/journal.ppat.1005948.

Witkowski, P.T., Drexler, J.F., Kallies, R., Ličková, M., Bokorová, S., Mananga, G.D., Szemes, T., Leroy, E.M., Krüger, D.H., Drosten, C., Klempa, B., 2016. Phylogenetic analysis of a newfound bat-borne hantavirus supports a laurasiatherian host association for ancestral mammalian hantaviruses. Infect. Genet. Evol. 41, 113–119. https://doi.org/10.1016/j.meegid.2016.03.036.

Witkowski, P.T., Perley, C.C., Brocato, R.L., Hooper, J.W., Jürgensen, C., Schulzke, J.-D., Krüger, D.H., Bücker, R., 2017. Gastrointestinal tract as entry route for hantavirus infection. Front. Microbiol. 8, 1721. https://doi.org/10.3389/fmicb.2017.01721.

Xiong, J.P., Stehle, T., Diefenbach, B., Zhang, R., Dunker, R., Scott, D.L., Joachimiak, A., Goodman, S.L., Arnaout, M.A., 2001. Crystal structure of the extracellular segment of integrin alpha Vbeta3. Science 294, 339–345. https://doi.org/10.1126/science.1064535.

Xiong, J.-P., Stehle, T., Zhang, R., Joachimiak, A., Frech, M., Goodman, S.L., Arnaout, M.A., 2002. Crystal structure of the extracellular segment of integrin alpha Vbeta3 in complex with an Arg-Gly-Asp ligand. Science 296, 151–155. https://doi.org/10.1126/science.1069040.

Yamanishi, K., Tanishita, O., Tamura, M., Asada, H., Kondo, K., Takagi, M., Yoshida, I., Konobe, T., Fukai, K., 1988. Development of inactivated vaccine against virus causing haemorrhagic fever with renal syndrome. Vaccine 6, 278–282.

Yanagihara, R., Amyx, H.L., Gajdusek, D.C., 1985. Experimental infection with Puumala virus, the etiologic agent of nephropathia epidemica, in bank voles (*Clethrionomys glareolus*). J. Virol. 55, 34–38.

Yanagihara, R., Gu, S.H., Arai, S., Kang, H.J., Song, J.-W., 2014. Hantaviruses: rediscovery and new beginnings. Virus Res. 187, 6–14. https://doi.org/10.1016/j.virusres.2013.12.038.

Yu, L., Bai, W., Wu, X., Zhang, L., Zhang, L., Li, P., Wang, F., Liu, Z., Zhang, F., Xu, Z., 2013. A recombinant pseudotyped lentivirus expressing the envelope glycoprotein of hantaan virus induced protective immunity in mice. Virol. J. 10, 301. https://doi.org/10.1186/1743-422X-10-301.

Yuan, Z.-G., Li, X.-M., Mahmmod, Y.S., Wang, X.-H., Xu, H.-J., Zhang, X.-X., 2009. A single immunization with a recombinant canine adenovirus type 2 expressing the seoul virus Gn glycoprotein confers protective immunity against seoul virus in mice. Vaccine 27, 5247–5251. https://doi.org/10.1016/j.vaccine.2009.06.062.

Zaitseva, E., Yang, S.-T., Melikov, K., Pourmal, S., Chernomordik, L.V., 2010. Dengue virus ensures its fusion in late endosomes using compartment-specific lipids. PLoS Pathog. 6, e1001131. https://doi.org/10.1371/journal.ppat.1001131.

Zaki, S.R., Greer, P.W., Coffield, L.M., Goldsmith, C.S., Nolte, K.B., Foucar, K., Feddersen, R.M., Zumwalt, R.E., Miller, G.L., Khan, A.S., 1995. Hantavirus pulmonary syndrome. Pathogenesis of an emerging infectious disease. Am. J. Pathol. 146, 552–579.

Zhang, Y.-Z., 2014. Discovery of hantaviruses in bats and insectivores and the evolution of the genus Hantavirus. Virus Res. 187, 15–21. https://doi.org/10.1016/j.virusres.2013.12.035.

CHAPTER SEVEN

Common characteristics and unique features: A comparison of the fusion machinery of the alphaherpesviruses Pseudorabies virus and Herpes simplex virus

Melina Vallbracht[a,*], Marija Backovic[b], Barbara G. Klupp[a], Felix A. Rey[b], Thomas C. Mettenleiter[a]

[a]Institute of Molecular Virology and Cell Biology, Friedrich-Loeffler-Institut, Greifswald – Insel Riems, Germany

[b]Institut Pasteur, Unité de Virologie Structurale, UMR3569 (CNRS), Paris, France

*Corresponding author: e-mail address: melina.vallbracht@fli.de

Contents

Advances in Virus Research, Volume 104
ISSN 0065-3527
https://doi.org/10.1016/bs.aivir.2019.05.007

Abstract

Membrane fusion is a fundamental biological process that allows different cellular compartments delimited by a lipid membrane to release or exchange their respective contents. Similarly, enveloped viruses such as alphaherpesviruses exploit membrane fusion to enter and infect their host cells. For infectious entry the prototypic human Herpes simplex viruses 1 and 2 (HSV-1 and -2, collectively termed HSVs) and the porcine Pseudorabies virus (PrV) utilize four different essential envelope glycoproteins (g): the *bona fide* fusion protein gB and the regulatory heterodimeric gH/gL complex that constitute the "core fusion machinery" conserved in all members of the *Herpesviridae*; and the subfamily specific receptor binding protein gD. These four components mediate attachment and fusion of the virion envelope with the host cell plasma membrane through a tightly regulated sequential activation process. Although PrV and the HSVs are closely related and employ the same set of glycoproteins for entry, they show remarkable differences in the requirements for fusion. Whereas the HSVs strictly require all four components for membrane fusion, PrV can mediate cell-cell fusion without gD. Moreover, in contrast to the HSVs, PrV provides a unique opportunity for reversion analyses of gL-negative mutants by serial cell culture passaging, due to a limited cell-cell spread capacity of gL-negative PrV not observed in the HSVs. This allows a more direct analysis of the function of gH/gL during membrane fusion. Unraveling the molecular mechanism of herpesvirus fusion has been a goal of fundamental research for years, and yet important mechanistic details remain to be uncovered. Nevertheless, the elucidation of the crystal structures of all key players involved in PrV and HSV membrane fusion, coupled with a wealth of functional data, has shed some light on this complex puzzle. In this review, we summarize and discuss the contemporary knowledge on the molecular mechanism of entry and membrane fusion utilized by the alphaherpesvirus PrV, and highlight similarities but also remarkable differences in the requirements for fusion between PrV and the HSVs.

1. Introduction

The *Herpesviridae* within the order *Herpesvirales* form a large family of enveloped, double-stranded DNA viruses infecting mammals, birds and reptiles (Davison, 2010). This family is classified into the three subfamilies *Alpha-*, *Beta-*, and *Gammaherpesvirinae,* which differ in their host range and properties of the viral life cycle. Alphaherpesviruses display the broadest host range and include the important human pathogens Herpes simplex virus 1 and 2 (HSV-1, -2, *Human alphaherpesvirus-1, -2*) and Varicella-zoster virus (VZV, *Human alphaherpesvirus* 3), the causative agents of cold sores, genital ulcer disease, and chickenpox or shingles, respectively (Davison, 2010). Moreover, this family contains important veterinary viruses with high economic impact, such as *Bovine alphaherpesviruses 1* and 5

(BoHV-1, -5), the *Equid alphaherpesviruses* (EHV-1, -3, -4, -8, -9), the *Gallid alphaherpesviruses* (GaHV-1, -2, -3) and Pseudorabies virus (PrV, *Suid alphaherpesvirus 1*) (Davison, 2010; McGeoch et al., 1995). Key features of this group of viruses are their pronounced neurotropism and the ability to establish lifelong latency in sensory ganglia of their hosts (Davison, 2010). This review will primarily focus on PrV and the HSVs, which are well-studied viruses that have served as useful models for understanding the molecular details of alphaherpesvirus entry, spread and pathogenesis both *in vitro* and *in vivo* (Kramer and Enquist, 2013; Mettenleiter, 1996, 2008; Mettenleiter et al., 2009; Pomeranz et al., 2005; Subak-Sharpe and Dargan, 1998; Whitley and Roizman, 2001). In contrast to the human pathogens HSV-1 and -2, PrV is an alphaherpesvirus of swine and the causative agent of Aujeszky's disease. Whereas its natural host is the pig, PrV has a remarkably broad host range and can lethally infect nearly all mammals including cattle, goats, sheep, cats and dogs (Mettenleiter, 2000). In contrast, equids as well as higher primates including humans show a high degree of resistance (Mettenleiter, 2008; Pomeranz et al., 2005).

Despite their differences in tropism and pathogenesis, alphaherpesviruses have a common and unique virion morphology (Davison et al., 2009; Mettenleiter, 2002). The linear double-stranded DNA genome is enclosed in an icosahedral capsid composed of 150 hexons and 12 pentons ($T=16$), forming the nucleocapsid. The nucleocapsid is embedded in a proteinaceous layer, termed tegument, which in turn is surrounded by a host cell derived lipid-bilayer, the envelope. The viral envelope contains multiple copies of different virally encoded, and mostly glycosylated proteins (Davison et al., 2009; Granzow et al., 1997; Mettenleiter et al., 2009). PrV and the HSVs encode at least 11 different envelope glycoproteins, including a specific set that is essential for virus entry (Campadelli-Fiume and Menotti, 2007; Mettenleiter, 1994, 2008).

Entry of herpesviruses into the host cells and the subsequent nucleocapsid delivery into the cytosol involves the fusion of the virion envelope with a target cell membrane. This central membrane fusion event requires merger of the inner and outer layer of both membranes. Although fusion of two separate lipid membranes into a single continuous bilayer is thermodynamically favored, several energy barriers such as lipid head dehydration and repulsion between membrane surfaces with repulsive electrostatic potentials must be overcome (Chernomordik et al., 1987; Chernomordik and Kozlov, 2003, 2008; Francois-Martin et al., 2017; Kielian, 2014; Kozlovsky et al., 2002; Leikin et al., 1993). For the majority of enveloped viruses, the expression of a single specialized viral surface glycoprotein, termed fusion protein or protein fusogen, is sufficient to overcome these barriers and mediate the fusion process at the appropriate time and place (Chernomordik and Kozlov, 2008; Harrison, 2015). These protein fusogens are thought to provide the energy for fusion by refolding from a metastable, high-energy "prefusion form," to a low energy, stable "postfusion form" (Harrison, 2015; Kielian and Rey, 2006; White et al., 2008).

In contrast, herpesvirus entry and membrane fusion is more complex and requires the coordinated action of at least three conserved virion glycoproteins. The herpesvirus core fusion machinery is composed of the heterodimeric gH/gL complex, which is thought to regulate fusion mediated by the *bona fide* fusion glycoprotein gB. These three components are conserved throughout the *Herpesviridae* (Akhtar and Shukla, 2009; Atanasiu et al., 2010a; Campadelli-Fiume et al., 2012; Connolly et al., 2011; Eisenberg et al., 2012). In addition, entry of PrV, the HSVs and most other alphaherpesviruses requires a fourth protein, gD, which has receptor binding function. Several lines of evidence suggest that these four components mediate entry through a tightly regulated sequential activation process (Atanasiu et al., 2010a; Eisenberg et al., 2012). Current models propose that binding of gD to a specific cellular receptor, such as nectin-1, induces a conformational change in the C-terminal part of the gD ectodomain, enabling it to activate the regulatory gH/gL complex which, in turn, triggers gB. Activated gB undergoes a fusogenic conformational change from the pre- to the postfusion form, thereby promoting fusion (Cooper and Heldwein, 2015; Eisenberg et al., 2012) (Fig. 1). Recent structural and functional studies on the entry glycoproteins of PrV, the HSVs and other alphaherpesviruses have shed light on the complex fusion process of these viruses. In this review, we will describe the key players involved in PrV receptor binding and fusion with a particular focus on their structural features imparting viral binding and fusion function, and also highlight some remarkable differences in the requirements for fusion between PrV and the HSVs.

2. Attachment to cells

PrV and the HSVs employ glycoprotein C (gC) to attach to heparan sulfate moieties of cell surface proteoglycans (HSPG) of the target cell. This process of viral tethering is reversible and involves relatively nonspecific, charge-based interactions that serve to concentrate virus particles at the cell

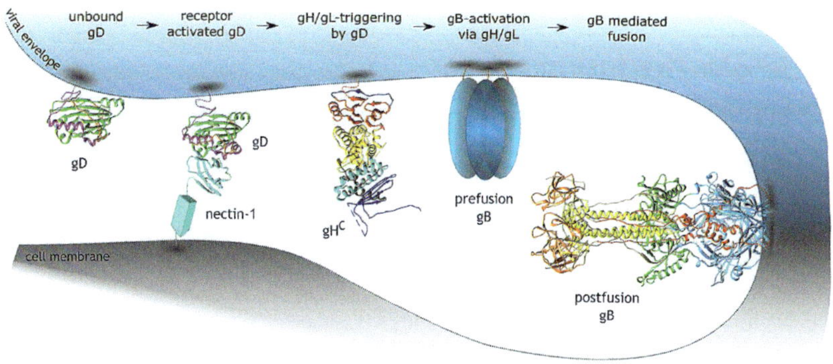

Fig. 1 The fusion machinery of PrV. The minimal set of PrV glycoproteins required for entry is depicted to scale, indicating essential steps in membrane fusion. The receptor binding protein gD determines tropism by binding to cellular receptors such as nectin-1. Receptor-bound gD may transmit a fusion triggering signal to gH/gL. Here, the structure of the core-fragment of PrV gH (gH^C) is shown in which the N-terminal domain I and complex partner gL are absent. Triggered gH/gL presumably activates gB, which then refolds from a presumed prefusion to postfusion form resulting in merger of virus and cell membrane. Coloring of the glycoproteins corresponds to their structural regions as defined previously for PrV gD, in unbound state (PDB ID: 5X5V) or in complex with nectin-1 (PDB ID: 5X5W) (Li et al., 2017a), PrV gH^C (PDB ID: 2XQY) (Backovic et al., 2010) and PrV gB (PDB ID: 6ESC) (Vallbracht et al., 2017a). Figure was made with UCSF Chimera (Pettersen et al., 2004).

surface without triggering membrane fusion. Although gC mediated attachment is beneficial for productive infection of the HSVs and PrV, it is not essential (Flynn and Ryan, 1996; Herold et al., 1991; Karger and Mettenleiter, 1993; Laquerre et al., 1998; Mettenleiter, 1989; Mettenleiter et al., 1990; Robbins et al., 1986; Shieh et al., 1992; Trybala et al., 1996). In the absence of gC, HSV-1 can bind to HSPG *via* gB (Herold et al., 1994), an interaction which is also not required for membrane fusion (Laquerre et al., 1998). Binding of PrV gB to heparin has been described in *in vitro* assays (Sawitzky et al., 1990) but a PrV mutant lacking gC infected cells by a heparan-sulfate independent pathway (Klupp et al., 1997b; Mettenleiter et al., 1990). Another attachment factor for HSV-1 gB and gC expressed on dendritic cells is the "dendritic cell-specific C-type lectin" (DC-SIGN) (de Jong et al., 2008). None of these interactions can trigger the fusion process, which instead requires binding of gD to a specific cell surface receptor (Karger and Mettenleiter, 1993; McClain and Fuller, 1994; Spear et al., 2006).

3. gD: The fusion trigger

3.1 Receptor binding and fusion initiation

Many alphaherpesviruses bind to a variety of cell surface molecules *via* gD (Spear et al., 2000), in part explaining their ability to infect a wide range of different species and cell types. PrV engages several members of the nectin family from humans and animals as entry receptors, but may also interact with other so far unidentified cell surface molecules (Geraghty et al., 1998; Milne et al., 2001; Nixdorf et al., 1999; Spear et al., 2000; Spear and Longnecker, 2003; Warner et al., 1998). For the HSVs, three classes of entry receptors have been identified (Spear et al., 2000): the tumor necrosis factor (TNF) receptor-related protein designated as herpesvirus entry mediator (HVEM; HveA) (Carfi et al., 2001; Montgomery et al., 1996; Warner et al., 1998); 3-O-sulfonated-heparan sulfate (Shukla et al., 1999); and three members of the immunoglobulin superfamily, namely, the poliovirus receptor related protein (HveD; CD155) (Geraghty et al., 1998; Mendelsohn et al., 1989), nectin-2 (HveB) (Warner et al., 1998) and nectin-1 (HveC) (Di Giovine et al., 2011; Geraghty et al., 1998; Li et al., 2017a). Although alphaherpesvirus gD homologs, including the HSVs, PrV and BoHV-1, have low amino acid sequence identity (22–33%), they all can utilize nectin-1 as entry receptor (Connolly et al., 2001). PrV gD binds to both human and swine nectin-1 with similar affinities but is unable to bind to HVEM (Geraghty et al., 1998; Li et al., 2017a; Roller and Rauch, 1998).

Binding of gD to one of its cellular receptors initiates the membrane fusion cascade (Krummenacher et al., 2013), as demonstrated by the ability of the soluble gD ectodomain of HSV-1 to trigger fusion of cells expressing the receptor and the core fusion machinery components gB, gH and gL (Atanasiu et al., 2010a). Soluble HSV gD also rescues the infectivity of gD-deficient virus mutants, showing again that receptor engagement by gD surpasses mere tethering of the virus particles onto the cell surface, and is essential for fusion triggering (Cocchi et al., 2004; Tsvitov et al., 2007). Conversely, the soluble nectin-1 ectodomain can support entry of HSV-1 into resistant CHO cells (Kwon et al., 2006). While great progress has been made in understanding the series of events leading to alphaherpesvirus entry (Eisenberg et al., 2012), the exact molecular mechanism by which receptor-activated gD transmits the fusion triggering signal to the fusion protein complex remains elusive, most likely due to the short-lived nature of these transient interactions.

3.2 gD: A molecule with a compact core and mobile ends

PrV gD, previously designated gp50, is a ~50 kDa type I transmembrane protein composed of an N-terminal ectodomain, a C-terminal transmembrane anchor, and a cytoplasmic domain (Wathen and Wathen, 1984). In contrast to HSV-1 gD, it does not carry N-linked carbohydrates and is solely modified by *O*-glycosylation (Petrovskis et al., 1986). The crystal structures of the unbound PrV gD ectodomain and in complex with nectin-1 have been determined at atomic resolution recently (Li et al., 2017a), allowing comparative analyses with the HSV homologs (Carfi et al., 2001; Di Giovine et al., 2011; Krummenacher et al., 2005; Lazear et al., 2014; Lu et al., 2014; Zhang et al., 2011) (Fig. 2).

PrV gD possesses a core-domain displaying an immunoglobulin superfamily fold, which is flanked by long, surface exposed N- and C-terminal extensions, both of which interact with nectin-1. The gD core-domain is composed of two α-helices and a nine stranded β-barrel (Fig. 2, gray). The same core-domain fold has been observed for HSV-1 and -2 gD, indicating that it may represent a common feature of all alphaherpesvirus gD (Carfi et al., 2001; Di Giovine et al., 2011; Li et al., 2017a; Lu et al., 2014). The N- and C-terminal extensions (Fig. 2 green and red, respectively) cover the core-domain. While the N-terminal extension forms a large loop structure and additionally contains a short β-strand, the C-terminal extension is longer and contains several α-helices and a short β-strand, stretching over about half of the central core outer surface.

The overall structure of PrV gD is similar to HSV-1 and -2 gD as demonstrated by superposition of the respective structures, revealing a well-aligned core-domain (Li et al., 2017a). PrV gD contains three disulfide bonds, which are conserved in HSV gD (Li et al., 2017a; Petrovskis et al., 1986). While two of them connect the C-terminal extension to the core, one disulfide bridge links two β-strands of the core domain (Fig. 2). The PrV gD/nectin-1 binding mode resembles that of its HSV homologs, with essentially the same surface footprints (Di Giovine et al., 2011; Li et al., 2017a; Lu et al., 2014). Both PrV and HSV gD utilize the terminal extensions to interact with the nectin-1 dimerization interface. As cell adhesion molecules, the nectins participate in tissue organization by mediating cell adhesion *via* essential *cis*- and *trans*-dimerization (Narita et al., 2011; Takai et al., 2008). Thus, gD directly perturbs nectin-1 mediated cell-adhesion by binding to the nectin-1 dimerization interface. The affinity of PrV gD for human nectin-1 is around 100-fold higher than the dimer affinity of

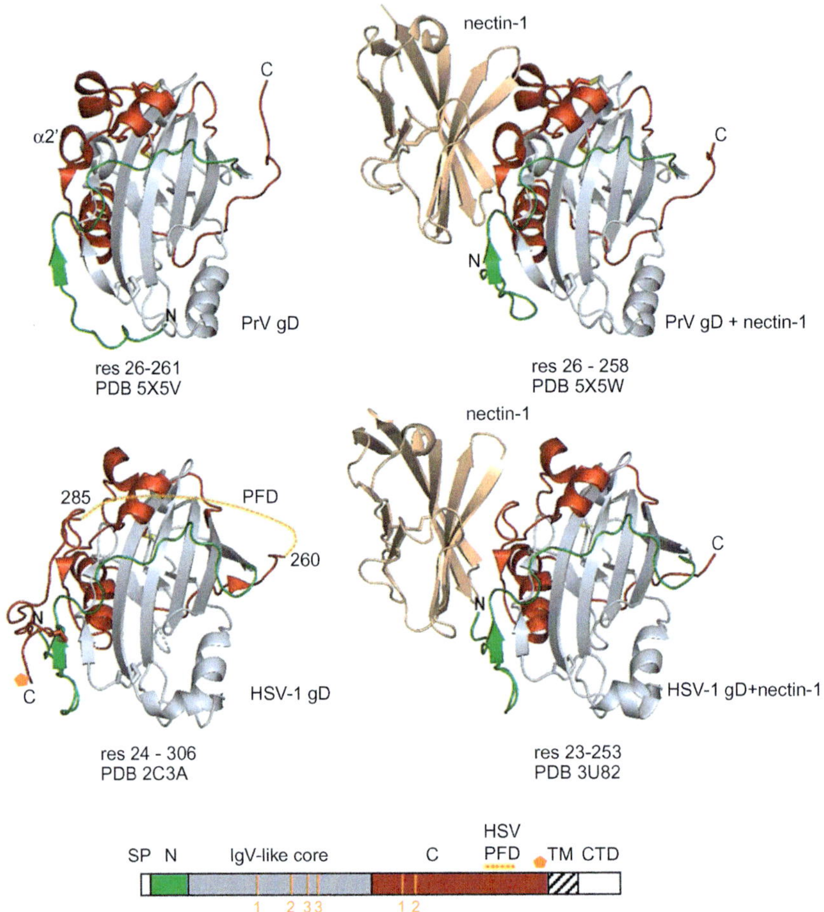

Fig. 2 Structural comparison of PrV and HSV-1 gD in unbound and receptor-bound states. Ribbon diagrams of unbound PrV and HSV-1 gD and their complexes with nectin-1. The IgV-like core is colored in gray, with the N- and C-terminal extensions highlighted in green and red, respectively. The PDB IDs and boundaries of the crystallized constructs are displayed below each structure. The PrV gD C-terminal extension contains the extra helix labeled as α2′, representing the main structural difference between PrV and HSV-1 gD. The unresolved pro-fusion domain (PFD) in HSV-1 gD is drawn as a yellow dashed line to indicate its putative location. The construct contained an engineered cysteine residue at the C-terminus (aa 306, marked with an orange pentagon symbol). A schematic diagram of gD is shown at the bottom, colored according to structural elements as in the ribbon diagram, and including the signal peptide (SP), transmembrane (TM) and cytoplasmic domain (CTD). Conserved disulfide bridges are indicated in yellow and numbered 1–3. The structure images were generated using PyMOL (DeLano, 2002).

nectin-1, making gD a potent competitor (Harrison et al., 2012; Li et al., 2017a). Indeed, HSV-1 gD can disrupt these complexes and was found to induce a rapid redistribution of nectin-1 away from cell junctions. This receptor relocation is hypothesized to serve as prelude to virus internalization by endocytosis (Bhargava et al., 2016; Stiles and Krummenacher, 2010). Despite their structural similarity, PrV gD cannot substitute for its HSV counterpart in cell fusion assays (Zago et al., 2004, Böhm et al., 2016).

The main structural difference between PrV and HSV gD is the presence of an extra α-helix (α2′) in the C-terminal extension in PrV, which projects outward and makes multiple contacts with the receptor (Fig. 2). The locations of the N- and C-terminal extensions relative to the core are also different, which is not surprising considering that these segments are highly flexible. Compared to PrV, the N-terminal extension of HSV gD is 19 amino acids longer and contains all residues critical for interaction with HVEM (Carfi et al., 2001; Lazear et al., 2014; Li et al., 2017a). Upon binding to HVEM, the HSV gD N-terminal extension folds into a hairpin loop (Lazear et al., 2014). In contrast, PrV gD does not utilize HVEM as an entry receptor (Geraghty et al., 1998; Li et al., 2017a; Roller and Rauch, 1998), and it was hypothesized that the shorter N-terminal extension precludes PrV gD binding to HVEM (Li et al., 2017a).

The proline-rich region that is a part of the C-terminal extension of the gD ectodomain is required for its ability to trigger the fusion machinery and has been termed "pro-fusion domain" (PFD) (residues 260–285 of mature gD, HSV-1 strain F) (Cocchi et al., 2004; Fusco et al., 2005). The PFD is thought to be released upon receptor binding, and is then able to interact with the core fusion components gB and/or gH/gL (Carfi et al., 2001; Fusco et al., 2005; Krummenacher et al., 2005; Lazear et al., 2008), although HSV-1 gD mutants carrying deletions or amino acid substitutions within the PFD are still functional in fusion (Zago et al., 2004). This observation suggests that the PFD does not constitute a specific interaction domain with gB or gH/gL, but that it may be necessary for gD to adopt the correct conformation (Zago et al., 2004). The significance of the corresponding proline-rich region in PrV gD, which is remarkable in carrying several proline/arginine repeats (5–7, depending on the virus strain), has not been investigated thus far.

Structural analyses of the HSV gD ectodomain in the unbound state revealed that the N- and C-terminal extensions interact with each other, constraining the PFD (Krummenacher et al., 2005) (Fig. 2). In this state, the C-terminal segment of the gD ectodomain also masks the receptor-binding site and must be displaced for receptor engagement (Cocchi et al., 2004; Fusco et al., 2005; Krummenacher et al., 2005). A model aimed at explaining how gD engages the receptor when its binding site is covered by the C-terminal segment suggested that gD exhibits an equilibrium between an "open conformation," in which the gD-core is partially exposed and would allow for initial receptor interaction, and a "closed conformation," in which the receptor binding site is masked (Lazear et al., 2014). Thus, "breathing" (subtle in-and-out motion) of the gD ectodomain C-terminal segment would facilitate the initial interaction with the receptor but prevent full activation of gD in the absence of receptor. Full activation of gD and displacement of the C-terminal segment requires receptor binding, which induces a new conformation of the gD N-terminal extension, thereby preventing premature activation of gD, and ultimately of the fusion machinery.

Overall, gD with its two flexible arms appears to have two roles, one in receptor binding and the other in triggering the fusion machinery for entry. Interestingly, engineered disulfide bonds in HSV-1 gD locking the C-terminal arm at different positions to the gD-core separates these two functions, as demonstrated by a gD mutant that binds to the receptors, but is unable to trigger cell-cell fusion or to complement the infectivity of a gD-negative virus (Atanasiu et al., 2018; Lazear et al., 2008). Based on these findings two portions of the gD C-terminus were proposed to move to allow for the different gD functions, residues close to the transmembrane domain (286–306 of mature gD, HSV-1 F) for receptor binding and residues of the PFD (at or before residue 279) for triggering fusion.

Together, the current structural and functional data on HSVs and PrV gD suggest a model in which receptor binding of gD to nectin-1 (PrV and HSVs) or HVEM (HSVs) induces a new conformation of the gD N-terminal extension and displaces the C-terminal segment of the gD ectodomain. This conformational change in gD leads to the exposure of a fusion-activating interface in gD, enabling it to cooperate with gH/gL and/or gB to promote fusion (Connolly et al., 2001; Fan et al., 2014; Fusco et al., 2005; Gallagher et al., 2013; Krummenacher et al., 2005; Lazear et al., 2008, 2014).

3.3 Interactions of gD with the core fusion machinery

To serve as the initial fusion trigger gD must cross-talk with at least one of the downstream gB and gH/gL. The identification of functional interaction site(s) is thus crucial for understanding the nature, function and regulation of the complex cascade of events leading to virus entry and membrane fusion.

Different functional studies on HSV-1 provide evidence for physical contacts between the entry proteins (Atanasiu et al., 2007, 2010a; Avitabile et al., 2007; Cairns et al., 2011; Chowdary et al., 2010; Fan et al., 2014, 2015; Gallagher et al., 2013). Interaction of gD with gH/gL and gB has been demonstrated using bimolecular fluorescence complementation (BiFC or BiMC), in which HSV-1 gD, gH/gL and gB were fused to inactive, complementary fragments of fluorescent proteins (EGFP or EYFP). When the two fragments are brought into physical proximity by direct interaction between the transiently expressed entry glycoproteins they restore a fluorescent signal. These studies revealed that gD interacts with gH and gB independently of each other. Moreover, gH/gL and gB interacted and induced syncytia formation only in the presence of receptor activated gD (Atanasiu et al., 2007, 2010b; Avitabile et al., 2007), again indicating that binding of gD to its receptor triggers the core fusion machinery to form a fusion active complex. However, a stable complex involving gD and gH/gL and/or gB has never been captured in alphaherpesviruses, and whether the contacts detected by BiFC reflect functional interactions remains to be determined.

The association of HSV-1 gD with gH/gL and gB in infected and transfected cells was demonstrated by co-immunoprecipitation and pull-down experiments. The C-terminal segment of the gD ectodomain, including residues located in the PFD, harbors regions critical for complex formation with gB and gH/gL, as demonstrated by mutants carrying substitutions or deletions in the PDF weakening the interactions between the glycoproteins. gD-gB and gD-gH/gL interactive sites in gD were allocated to the C-terminal part of the gD ectodomain (Gianni et al., 2009). None of the gD variants with mutations at these sites, however, was completely blocked in their ability to bind to gB or gH/gL, suggesting that the binding sites in gD for gB and for gH/gL comprise multiple contacts.

Further evidence that gD interacts with gH/gL for fusion comes from studies in which the entry glycoproteins of HSV-1 and Saimiriine herpesvirus 1 (SaHV-1) were exchanged, and a functional homotypic interaction between gD and gH/gL was observed (Fan et al., 2014). In a follow up study, further defining the gH/gL requirements for the homotypic interaction, HSV-1 gH domains I and II were found to functionally interact with gD (Fan et al., 2015).

Two monoclonal antibodies (MAbs) specific for gD, designated MC2 and MC5, can neutralize HSVs infection and block cell-cell fusion without interfering with gD receptor binding, indicating that they must block a distinct step in the downstream fusion pathway, which was presumed to be activation of gH/gL (Lazear et al., 2012). Detection of gD mutants resistant to neutralization by these MAbs pointed to a putative gH/gL binding site on HSV-1 gD, involving residues located in the gD N-terminal extension close to the gD-core (R67 and P54 in mature HSV-1 strain F gD) (Atanasiu et al., 2018).

Together these studies suggest a sequence of events leading to membrane fusion: starting with gD binding to a receptor, and followed by a conformational change and activation of gH/gL *via* residues located in the C- and/or N-terminus of the gD ectodomain and gH domains I and II. The current data provide ample, but only indirect evidence that gD associates with gH/gL for fusion activation. Obviously, co-crystallization of the corresponding entry glycoproteins would be the best way to demonstrate and detail the interaction. Thus, a direct demonstration of a gD-gH/gL interaction for fusion and identification of the sites of interaction between the entry glycoproteins remains a challenge, which is difficult to address in light of the apparently transient nature of the interactions.

3.4 Requirement for gD varies among different alphaherpesviruses and fusion events

During infection within the host, PrV and the HSVs are able to enter cells by at least two entry routes, either through entry of extracellular virions (cell-free spread) or through direct spread from an infected cell to a non-infected neighboring cell *via* direct cell-cell contacts. The latter mode, which is referred to as "direct cell-to-cell spread," is mechanistically poorly understood (Cocchi et al., 2000; Miranda-Saksena et al., 2018; Mothes et al., 2010; Sattentau, 2008). By spreading directly from cell-to-cell, herpesviruses can evade immunological as well as physical barriers. In fact, an early evidence for direct cell-to-cell spread of enveloped viruses was the finding that spread occurred in the presence of neutralizing antibodies that completely block spread by extracellular virus (Dingwell et al., 1994, 1995; Gupta et al., 1989; Sattentau, 2008).

One of the most intriguing differences between PrV and the HSVs is the requirement for gD during direct viral cell-to-cell spread. Whereas gD is absolutely required for HSVs spread *via* extracellular infectious virus and directly from cell to cell (Pertel et al., 2001), direct spread of PrV, including trans-synaptic spread in animals, can occur independently of gD (Ch'ng et al., 2007; Heffner et al., 1993; Mulder et al., 1996; Peeters et al., 1992;

Rauh and Mettenleiter, 1991). This remarkable difference between the closely related alphaherpesviruses raises the question as to why the HSVs, in contrast to PrV, are reliant on gD and more importantly, how PrV is able to bypass the requirement for gD. One possibility is that additional, so far unidentified receptors for PrV gH/gL or gB exist, and that binding of either of the core fusion machinery components to these receptors could also trigger fusion. The molecular mechanism behind this process, however, remains to be understood.

A very useful surrogate assay to study the herpesvirus induced membrane fusion in the absence of infection are cell-cell fusion assays, in which the herpesvirus entry glycoproteins are transiently expressed in susceptible cells *in vitro* (Fan et al., 2002, Klupp et al., 2000, McShane and Longnecker, 2005, Muggeridge, 2000, Turner et al., 1998, Vallbracht et al., 2017c). Co-expression of the PrV core fusion machinery components gB and gH/gL in cultured RK13 cells *in vitro* is sufficient to induce cell-cell fusion (Klupp et al., 2000; Schröter et al., 2016; Vallbracht et al., 2017c). Since gD function is required for PrV entry but not for direct cell-to-cell spread, the transient transfection-based cell fusion system more closely resembles cell-to-cell spread than fusion during viral entry (Klupp et al., 2000). In contrast to PrV, the induction of syncytia formation cannot be achieved by co-expression of wild-type HSV gB and gH/gL, but co-expression of HSV gD and its receptor is required in addition (Fan et al., 2002; Klupp et al., 2000; McShane and Longnecker, 2005; Muggeridge, 2000; Turner et al., 1998). By generation of vesicular stomatitis virus (VSV) virions, pseudotyped with HSV-1 gD, gH/gL, and gB but lacking the VSV autologous fusion glycoprotein G, it was demonstrated that the four HSV-1 glycoproteins are not only required, but are also sufficient for membrane fusion during entry (Rogalin and Heldwein, 2016). Because of the different requirements for gD in entry and cell-to-cell spread of PrV, it would be interesting to investigate whether entry of VSV virions, pseudotyped with the PrV glycoproteins, requires the presence of gD as well.

The efficient PrV cell-to-cell-spread in the absence of gD (Ch'ng et al., 2007; Mulder et al., 1996; Peeters et al., 1992; Rauh and Mettenleiter, 1991) has been exploited for reversion analyses of a gD-deficient PrV mutant in cell culture, by looking for selection of additional mutations that could rescue infectivity in the normal context (Schmidt et al., 1997, 2001). Repeated passaging of cells infected with gD-negative mutants led to the isolation of PrV-gD$^-$ Pass and PrV-376Pass, which were infectious in the absence of gD (Schmidt et al., 1997, 2001). PrV-gD$^-$ Pass was shown to contain compensatory mutations in gH and gB (Schmidt et al., 2001). Interestingly, the PrV-gD$^-$ Pass mutant utilizes receptors different from those recognized by the wild-type virus, as demonstrated by infection studies with a bovine kidney cell (MDBK) derived cell clone that was susceptible to infection by wild-type PrV but not by PrV-gD$^-$ Pass (Karger et al., 1998). Moreover, receptor binding and infectivity of PrV-gD$^-$ Pass were also independent of the presence of gC (Karger et al., 1998), supporting the idea of tertiary receptor interactions presumably mediated by gH and/or gB. In conclusion, these studies demonstrated that gD itself is not an integral component of the PrV fusion machinery, since essential functions encoded by gD can be compensated. Mutations in gH and/or gB of BoHV-1 can also compensate for absence of gD during BoHV-1 entry and cell-cell fusion (Schröder et al., 1997, Schröder and Keil, 1999). Similarly, mutant forms of HSV-2 gH/gL, with deletion of the 28 N-terminal gH residues, bypassed the requirement for gD and/or receptor and induced low levels of gB mediated cell-cell fusion in *in vitro* assays (Atanasiu et al., 2013, 2016). However, whether the same N-terminal gH deletion would allow for gD-independent entry during infection or direct cell-to-cell spread, has not been determined. In addition, specific mutations in HSV-1 gH and gB have been reported to circumvent the requirement for known gD receptors (Uchida et al., 2013). Although these studies demonstrated that mutations in gH and gB can compensate for the absence of known gD receptors in the HSVs, gD was still required. Thus, in contrast to PrV (Schmidt et al., 1997, 2001), these mutant forms of HSV-1 gH and gB were not able to carry out entry independent of gD (Uchida et al., 2013). Interestingly, insertion of a single chain antibody variable domain (scFv) directed against the cancer receptor human epidermal growth factor receptor 2 (HER2) into HSV gD (Menotti et al., 2008), gH (Gatta et al., 2015), or gB (Petrovic et al., 2017) allowed HSV entry and retargeting to HER2-positive cancer cells, which was independent of receptors for gD and gH/gL but still dependent on the presence of the glycoproteins. Together, these studies imply that gD and/or gH/gL, in addition to their receptor binding functions, may play a structural role important for membrane fusion.

In contrast to PrV and the HSVs, other important alphaherpesviruses such as VZV completely lack a gD homolog (Davison and Scott, 1986). In the case of VZV, the receptor binding function is presumably executed by gE (Cole and Grose, 2003; Li et al., 2007, 2010). VZV is predominantly cell associated and disseminates mainly *via* cell-to-cell spread *in vivo,* as manifested by very low levels of free virus detected for all VZV isolates, which form syncytia in culture (Cole and Grose, 2003; Sattentau, 2008). Similar to VZV, Marek's disease virus (MDV) is also highly cell-associated, and although gD is encoded in the MDV genome, it is not transcribed and expressed in cultured chicken embryo cells and is not involved in MDV membrane fusion (Chi et al., 2013; Tan et al., 2001; Tischer et al., 2005; Tulman et al., 2000). It is thus tempting to speculate that the lack of a gD homolog in VZV and MDV might relate to the different mechanisms of viral dissemination. Nevertheless, MDV gD is expressed in feather follicles of infected chickens shedding large numbers of free infectious virions (Niikura et al., 1999).

Comparable to PrV, *in vitro* fusion could be achieved by transient expression of the VZV glycoproteins gB and gH/gL, with the caveat that the last eight residues of the VZV gH cytoplasmic domain were deleted (Suenaga et al., 2010; Yang et al., 2014). Overall, with respect to the requirement for gD, PrV appears to be intermediate between the HSVs, in which gD is essential for both fusion during entry and cell-to-cell spread, MDV, which expresses gD only for transmission between animals, and VZV, which completely lacks a gD homolog (Chi et al., 2013; Karger et al., 1998; Tischer et al., 2005).

4. The gH/gL complex and its role in fusion

4.1 Functional studies and structural features of gH/gL

PrV gH (strain Kaplan) is a 686 aa type-I single pass transmembrane protein comprising an N-terminal signal peptide, a large ectodomain, a transmembrane domain (TM), and a short cytoplasmic domain (CTD) (Klupp et al., 1992). The five functional N-linked glycosylation sites on PrV gH appear to play a modulatory, but nonessential role for gH function (Fuchs et al., 2012; Klupp et al., 1992; Vallbracht et al., 2018b). HSV-1 gH is 838 aa long containing seven predicted *N*-glycosylation sites with unknown functional significance (Buckmaster et al., 1984; Galdiero et al., 1997).

The small gL lacks a membrane anchor and is noncovalently associated with the N-terminal part of the gH ectodomain. gL relies on this intimate interaction with gH for incorporation into virions in herpesviruses in general (Backovic et al., 2010; Chowdary et al., 2010; Eisenberg et al., 2012; Hutchinson et al., 1992; Kaye et al., 1992; Klupp et al., 1992, 1994; Liu et al., 1993; Matsuura et al., 2010; Xing et al., 2015). Whereas gH of HSV-1 and VZV require gL for correct folding, trafficking, and incorporation into the virion envelope (Chowdary et al., 2010; Duus and Grose,

1996; Roop et al., 1993; Yaswen et al., 1993), gL is dispensable for trafficking and virion localization of gH in PrV, BoHV-4 and certain HSV-2 gH variants (see Section 4.3) (Cairns et al., 2007; Klupp et al., 1997a; Lete et al., 2012).

Much effort has been directed toward understanding the function of the gH/gL complex during membrane fusion, and the ideas related to the role(s) the complex plays in entry have changed in the last decade. Initially, gH was thought to be a fusion protein, since helical regions in gH resemble heptad repeats and fusion peptides typical for class I fusion proteins (Galdiero et al., 2007, 2008; Gianni et al., 2005a,b). Fusion peptides are hydrophobic sequences that allow protein fusogens to insert into the target membrane (Harrison, 2015). Although synthetic peptides corresponding to the predicted fusion peptides of HSV gH induced fusion of liposomes (Galdiero et al., 2007, 2008; Gianni et al., 2006), the full-length protein and forms of gH or gH/gL without the TM were not able to associate with lipid bilayers (Bender et al., 2003; Cairns et al., 2011). Studies on VZV gH/gL indicated that the heterodimer may have inherent fusogenic properties, since transiently expressed gH/gL resulted in formation of syncytia, albeit at a low level (Cole and Grose, 2003; Kinzler and Compton, 2005; Maresova et al., 2001).

With the elucidation of the X-ray structures of the ectodomains of gH homologs of PrV, HSV-2, VZV, and EBV (a gammaherpesvirus), it became clear that gH/gL does not resemble any known viral fusion protein (Backovic et al., 2010; Chowdary et al., 2010; Matsuura et al., 2010; Xing et al., 2015). The structural data demonstrated that the previously identified putative fusogenic and membrane-interacting gH peptides are buried within the molecule and are unlikely to become available for interaction with membranes upon a conformational change, arguing against a direct role of these segments in fusion (Backovic et al., 2010; Chowdary et al., 2010).

Recent studies instead suggest that gH/gL functions as a regulator of fusion (Atanasiu et al., 2013, 2016; Backovic et al., 2009, 2010; Chowdary et al., 2010; Matsuura et al., 2010; Roche et al., 2006; Xing et al., 2015). The complex was proposed to activate or upregulate gB, the fusion effector protein, through direct interactions (Atanasiu et al., 2010a). Co-floatation of soluble HSV gH/gL with gB after low pH treatment was detected in liposome binding assays, pointing to a gH/gL-gB interaction, but isolation of this tripartite complex was not successful (Cairns et al., 2011). In BiFC and co-immunoprecipitation studies, the interaction between gH/gL and gB could be detected only in the presence of receptor activated gD in agreement with the notion that gD is required to trigger the core fusion machinery (Atanasiu et al., 2007; Avitabile et al., 2007, 2009). In line with these results, LP11, the most potent neutralizing MAb targeting HSV-1 gH, not only affected entry but also abrogated *in vitro* cell fusion, presumably by blocking interaction of gB with gH/gL (Chowdary et al., 2010). In addition, anti-HSV gB antibodies inhibited fusion without disrupting gH/gL interaction with gB, suggesting that cross-talk between gH/gL and gB is required, and that it precedes membrane fusion (Atanasiu et al., 2010b). Unfortunately, in contrast to the HSVs for which a plethora of neutralizing Mabs is available, no neutralizing PrV anti-gH or gH/gL Mabs have been described.

The core fragment of the PrV gH ectodomain (gH^C; aa 107–639, counting the signal sequence) folds into three domains, designated II-IV, while the N-terminal domain I, predicted to interact with gL, is rich in proline and glycine residues rendering it highly mobile. It was omitted to allow crystallization (Fig. 3) (Backovic et al., 2010). The gH^C indeed represents the minimal, "core" functional unit that can support gB mediated membrane fusion without the need for gL, as shown by reversion analysis and mutagenesis studies (see Section 4.3) (Klupp and Mettenleiter, 1999; Vallbracht et al., 2017b).

Although gH homologs share only low sequence similarity even within the alphaherpesvirus subfamily, the structures of PrV gH^C (Backovic et al., 2010), HSV-2 gH/gL (Chowdary et al., 2010), and VZV gH/gL (Xing et al., 2015) reveal a similar four domain organization (Fig. 3) (Backovic et al., 2010; Chowdary et al., 2010; Matsuura et al., 2010; Xing et al., 2015).

The N-terminal gH domain I is the least conserved, varying significantly in length and amino acid sequence even between herpesviruses of the same subfamily. In the HSVs and VZV, this domain was shown to form tight contacts with gL, which are unique for each gH/gL pair, and are most extensive in the case of HSV-2 gH and gL (Chowdary et al., 2010; Matsuura et al., 2010; Xing et al., 2015). The gH domains I and II in HSV-1 have been proposed to interact with gD (see Section 3.3) (Atanasiu et al., 2018; Fan et al., 2015). Moreover, the integrin binding site maps to domain I of HSV-2 gH (Backovic et al., 2010; Galdiero et al., 1997; Gianni et al., 2010).

Functional evidence for the involvement of gH domain I in species-specific interactions with gL came from studies of chimeric gH variants, containing domain I of HSV-1 and domains II-IV of PrV (Böhm et al., 2016). In combination with the matching HSV-1 gL and PrV gB and gD, this chimeric gH variant was active in fusion (Böhm et al., 2016). PrV mutants lacking sequences in the gH N-terminus were unable to bind

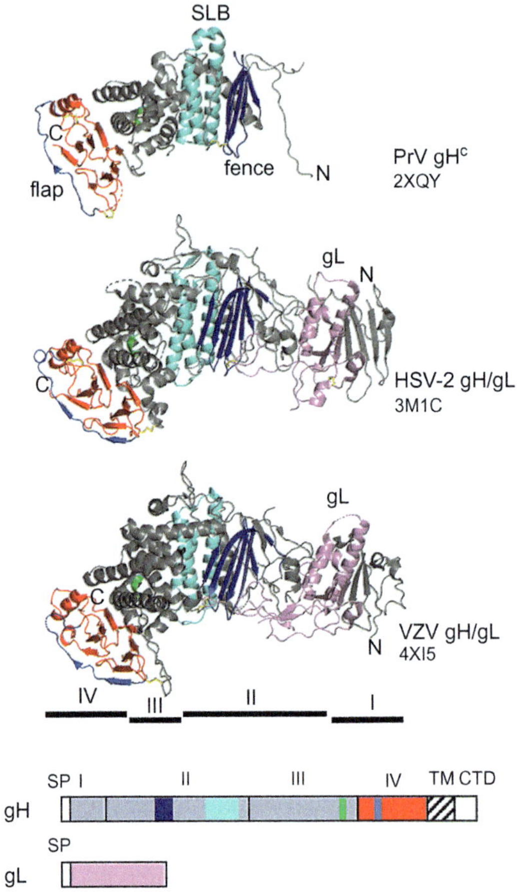

Fig. 3 Linear or boot-shaped gH ectodomains. Side-by-side comparison of the PrV gH^C fragment, HSV-2 gH/gL and VZV gH/gL. The PDB IDs of the crystallized constructs are given below each structure. The structures were superposed on the syntaxin-like bundle (SLB; cyan) and are displayed in the same orientation. The N-terminal gH domain I, absent from PrV gH^C, makes extensive contacts with gL (pink). Domain II contains conserved structural elements, the fence (blue) and the SLB (cyan). Domain III contains the strictly conserved SPC motif depicted in green. Domain IV is colored in red with the flap highlighted in blue. The disulfide bonds are displayed as yellow sticks. Schematic diagrams of gH and gL are shown below, colored according to domains and elements as in the ribbon diagrams. The structure images were generated using PyMOL (DeLano, 2002).

gL or to target it into virions (Vallbracht et al., 2017b). Several studies on the role of gL during membrane fusion indicated that neither PrV gL and nor gH domain I were actively involved in fusion *per se* (see Section 4.3) (Klupp and Mettenleiter, 1999; Schröter et al., 2016; Vallbracht et al., 2017b).

Domain II of gH contains two prominent structural elements, the "fence" (Fig. 3, blue) and the "syntaxin-like bundle" (SLB) (Fig. 3, cyan), which are structurally conserved (Backovic et al., 2010). The fence is a planar sheet of five antiparallel β-strands separating domain I from the downstream gH domains. It is followed by a helix that connects it to a bundle of three α-helices, which constitute the SLB. The term "SLB" was chosen because of the structural similarities with the N-terminal domain of the SNARE proteins syntaxin 1A and 6 (Backovic et al., 2010). Syntaxins are involved in vesicle fusion during neuronal exocytosis (Han et al., 2017). These trans-membrane proteins form a hetero-tetrameric parallel alpha-helical coiled-coil, the SNARE motif, with the individual subunits anchored to both membranes to be fused, pulling the membranes toward each other and mediating fusion (Sutton et al., 1998; Weimbs et al., 1997). Although this similarity with eukaryotic fusion proteins is intriguing, there is no evidence for a conformational change in gH that would lead to exposure of the SLB for oligomerization or interaction with membranes (Backovic et al., 2010). The structural similarity inspired studies on the functional relevance of potential conformational changes of the SLB during membrane fusion. The helices of the SLB in PrV gH were either disrupted by introduction of proline residues or were covalently joined by introduction of intramolecular disulfide bonds between themselves, to the adjacent fence region, or to gH domain III. These studies revealed that both the integrity and a potential mobility of the SLB are important for PrV gH function during membrane fusion (Böhm et al., 2015).

The segments forming domains II and III in HSV-2 and VZV gH are connected by a long α-helix that is absent from PrV gH (Fig. 3). This helix alters the domain II/III interface and thereby contributes to a different packing angle when compared to PrV gH^C. Thus, although gH domains II and III are individually similar, the packing of domains II-IV in PrV gH results in a straighter, more rod-like shape, whereas HSV-2 and VZV gH is boot-shaped, with the domain II/III interface forming the heel of the boot (Fig. 3). Studies using chimeric gH molecules exchanging PrV and HSV-1 gH domains showed that separate substitutions of gH domains II and III between HSV-1 and PrV always led to non-functional proteins (Böhm et al., 2016). These results suggested that the structural variations in the domain II/III interfaces of the different gH homologs may represent specific adaptations allowing optimal interactions with the homologous fusion protein gB and/or other accessory fusion-promoting proteins like gD (Böhm et al., 2016). Substitution and insertion mutations at specific residues located in domain II-III prevent binding of the neutralizing LP11 antibody, which blocks the interaction between gH/gL with gB and inhibits fusion (Chowdary et al., 2010), suggesting that the gB interaction site on HSV gH overlaps with the LP11 epitope; *i.e.*, that it is located at the domain II-III interface (Chowdary et al., 2010; Galdiero et al., 1997; Gompels et al., 1991).

The PrV gH domain III consists of eight α-helices, with the central one containing an serine-proline-cysteine (SPC) motif that is highly conserved across herpesviruses (in PrV $^{437}SPC^{439}$; Fig. 3, green) (Backovic et al., 2010; Klupp and Mettenleiter, 1991). The P^{438} in PrV introduces a kink in an α-helix that allows C^{439} to form a disulfide bond with C^{404} (Backovic et al., 2010). The formation of this conserved disulfide bond, together with contacting amino acids in domain III, is important for gH function in PrV, HSV-1, VZV and EBV (Backovic et al., 2010; Cairns et al., 2005; Möhl et al., 2015; Schröter et al., 2014; Vleck et al., 2011).

The membrane proximal domain IV is the most conserved region of gH, both in sequence and structure, consisting of a β-sandwich made of two opposed four-stranded β-sheets. The most striking feature of this domain is the flap (Fig. 3, blue), a negatively charged surface loop that connects the two β-sheets and is flanked by intramolecular disulfide bonds (Backovic et al., 2010). The flap is present in all gH structures, and together with a highly conserved N-linked glycan (N^{627} in PrV), covers a conserved patch of hydrophobic amino acids (Backovic et al., 2010; Galdiero et al., 1997; Klupp and Mettenleiter, 1991). Based on the structural data, it was suggested that this underlying hydrophobic patch could become available for interactions with the viral membrane for fusion upon movement of the flap, in turn induced by a receptor triggered conformational change of gH (Backovic et al., 2010; Fuchs et al., 2012). In support of this hypothesis, it was shown that: (i) multiple alanine substitutions within the flap or the hydrophobic patch; (ii) prevention of flap movement by introduction of artificial disulfide bonds; and (iii) disruption of conserved disulfide bonds important for positioning of the flap severely affected gH function (Fuchs et al., 2012). Binding of recombinant PrV gH/gL ectodomains to liposomes, as membrane bilayer mimics, has not been observed (unpublished data), and it remains to be determined if removal of the flap would allow the interactions to take place.

The N-glycan at position 627 in PrV gH appears to support the flap in masking the hydrophobic patch. Mutation of this site in PrV significantly impaired gH trafficking in transfected cells and resulted in ER retention and in reduced *in vitro* fusion activity. A corresponding virus mutant exhibited delayed penetration kinetics and smaller virus plaques (Fuchs et al., 2012; Vallbracht et al., 2018b). While these studies on PrV indicated that N^{627} glycosylation is important for gH cell surface expression and *in vitro* fusion, and is beneficial for virus entry and cell-cell spread, mutation of the corresponding site in HSV-1 had no effect on viral replication properties (Galdiero et al., 1997).

Recent studies demonstrated that the HSV-1 gH domain IV could functionally substitute for the corresponding domain in PrV gH, demonstrating its functional conservation (Böhm et al., 2016). When the chimeric gH molecule, composed of PrV gH domains I-III and HSV-1 domain IV (P^{I-III}/H^{IV}), was coexpressed with PrV gD, gB and gL, efficient *in vitro* fusion was achieved, which was substantially enhanced when PrV gB was replaced by HSV-1 gB. These results suggest homologous interaction of the C-terminal domain of the gH ectodomain with gB to drive fusion (Böhm et al., 2016). Remarkably, fusion activity of the gH chimera PrV^{I-III}/HSV^{IV} was abolished by omission or replacement of PrV gD by the homologous HSV-1 gD. This suggests that homotypic gD-gH interaction involves domains I-III. In light of the fact that gD normally is not required for *in vitro* fusion in PrV (Klupp et al., 2000; Schröter et al., 2016), this further indicates that PrV gH domain IV and/or the gH cytoplasmic domain imparts gD independence (Böhm et al., 2016).

4.2 Functional relevance of the gH transmembrane domain and cytoplasmic tail

In addition to the gH ectodomain, important roles have been ascribed to the structurally uncharacterized TM and the short CTD of gH (predicted 19 aa in PrV; 14 aa in HSV-1; 18 aa in VZV). The TM of PrV or HSV-1 gH cannot be substituted by analogous domains from other viral glycoproteins, cellular proteins or by a glycosyl-phosphatidylinositol (GPI) anchor without compromising function (Harman et al., 2002; Jones and Geraghty, 2004; Vallbracht et al., 2018a). These studies demonstrated that the authentic gH TM is essential for gH function. Only four residues are conserved between the HSV-1 and PrV TM, yet, the HSV-1 gH TM-CTD can functionally substitute for the corresponding domain of PrV gH during cell fusion and virus replication in cell culture (Vallbracht et al., 2018a). The conserved residues were predicted to map to the same face of an α-helix and three (A808, S809 and G812 in PrV; A651, S652, G665 in HSV-1) were crucial for gH function in HSV-1. This suggests that the gH TM has an intrinsic property to specifically interact with membrane components such as lipids or (viral glyco-) proteins as a prerequisite for triggering membrane fusion (Harman et al., 2002; Vallbracht et al., 2018a).

The CTDs of HSV-1, VZV and EBV gH have been suggested to regulate gB fusogenicity by direct interactions with the gB CTD (Browne et al., 1996; Chen et al., 2016; Harman et al., 2002; Pasieka et al., 2003; Rogalin and Heldwein, 2015; Suenaga et al., 2010; Yang et al., 2014). A transient, steric mechanism was proposed for the short HSV-1 gH CTD. Specifically, it was hypothesized that the HSV-1 gH CTD acts as a wedge, releasing the fusion-restricting gB CTD clamp (see Section 5.3), thereby enabling gB to undergo fusogenic refolding (Rogalin and Heldwein, 2015). Similar to HSV-1, truncation of approximately half of the PrV gH CTD (10 of 19 aa in PrV, 8 of 14 aa in HSV) did not affect *in vitro* fusion or complementation of gH-negative PrV. Further deletion of the PrV and HSV-1 CTD, however, significantly reduced fusion activity, suggesting a regulatory role of the gH CTD (Rogalin and Heldwein, 2015; Vallbracht et al., 2018a). It is therefore likely that the PrV gH CTD regulates fusion in a similar manner as proposed for its HSV-1 counterpart. In contrast to HSV-1 gH, in which the presence of the CTD is required for correct trafficking and virion incorporation of gH (Harman et al., 2002; Rogalin and Heldwein, 2015), PrV gH with a complete deletion of the CTD was still functional and complemented gH negative virus, albeit to around 10-fold reduced final titers (Vallbracht et al., 2018a). Together, these studies suggested that the gH CTD may have a regulatory function in fusion. However, in contrast to HSV-1, the PrV gH CTD is nonessential for gH function.

Remarkably, and in contrast to PrV and HSV-1, truncation of the VZV gH CTD resulted in an excessive fusion in infected cells concomitant with a significant reduction in virus titers, indicating that the VZV gH CTD restrains fusion instead of promoting it (Suenaga et al., 2010; Yang et al., 2014). This loss of regulation, resulting in the formation of larger syncytia, was hypothesized to limit the ability of VZV to effectively spread, impeding its penetration of the basal layer that allows for expansion of the lesion. These studies on VZV suggested that the gH CTD acts as a gate keeper, controlling access to functional domains of neighboring proteins, thereby, *e.g.*, allowing or preventing phosphorylation of the gB CTD (Yang et al., 2014).

4.3 Lessons from reversion analyses of gL-deficient virus mutants

Several interesting and valuable findings on the role of the gH/gL complex have emerged from reversion analyses of gL-deleted, entry-deficient PrV mutants. Although gL is required for entry of both PrV and the HSVs, in PrV there is limited cell-to-cell spread in the absence of gL (Klupp et al., 1997a; Klupp and Mettenleiter, 1999). Thus, PrV provides a unique opportunity for reversion analyses of gL-negative mutants by serial cell culture passaging. This feature of PrV has been exploited in several independent experiments with the aim to understand the function of gL, and the actual role of gH during membrane fusion (Klupp and Mettenleiter, 1999; Schröter et al., 2016).

4.3.1 Chimeric gD/gH compensates the absence of gL

In a first passaging experiment, an infectious gL-negative revertant virus could be isolated in which gL function was compensated by formation of a gDgH hybrid protein (Klupp and Mettenleiter, 1999). In this gDgH chimera, the N-terminal 271 residues of gD including the receptor binding domain were fused to a truncated gH, lacking the first N-terminal 96 aa, which encompass the predicted gL binding domain. This chimeric polypeptide not only compensated for gL function (Klupp and Mettenleiter, 1999), but was also sufficient to complement virus mutants lacking gD, gH, and gL, individually or in combination (Klupp et al., 2008), and to induce syncytia formation in cell-cell fusion assays when coexpressed with gB only (Klupp et al., 2000). These results demonstrated that gH and gB possess the inherent functional requirements to allow PrV entry and cell-cell fusion. More importantly, the results indicated that gL itself is not directly involved in the activation stages of the fusion process.

The PrV gDgH chimera inspired studies on HSV-1 to investigate whether an HSV-1 gDgH hybrid protein is able to exert similar functions (Cairns et al., 2003). However, the engineered HSV-1 gDgH chimeras did not functionally mimic the PrV chimera, as they failed to substitute for gL during membrane fusion. Nevertheless, in contrast to wild-type gH, some of the constructs (lacking gH residues 1–388 and 1–432) exhibited gL-independent transport to the cell surface, and two of the HSV-1 chimeras could substitute for gD in at least one functional assay (Cairns et al., 2003). The N-termini of PrV and HSV-1 gH ectodomains lack any sequence similarity and at the time the HSV-1 gH chimeric molecules were engineered, structural data on gH of PrV and the HSVs were not available. Thus, it cannot be excluded that the lack of functionality of the HSV-1 gDgH chimera could have been due to a structurally mismatched gD-gH junction in the construct used. Although an HSV-1 gDgH mutant lacking gH residues 1–432 did not form a complex with gL, it was able to function in fusion when gL was co-transfected, indicating a functional role for the soluble, unbound HSV-1 gL. Recently, binding of HSV-1 gH/gL to integrins has been shown to promote gL dissociation from gH in the presence of receptor-bound gD and gB (Gianni et al., 2015). In addition, dissociated HSV-1 gL was suggested to accelerate the recovery of membrane asymmetry upon membrane fusion through binding to phospholipid scramblase-1, thereby preventing prolonged exposure of phosphatidylserines on the outer leaflet of the plasma membrane, which is an apoptotic signal (Cheshenko et al., 2018).

Similarities in the organization of the PrV gDgH hybrid protein and the γ-retrovirus envelope fusion protein (Env) were pointed out (Rey, 2006). In the PrV gDgH chimera, the gD portion (residues 1–271) including the receptor binding domain is connected through a proline rich region (PRR) to the gH core fragment. This is similar to murine leukemia virus Env, where the receptor-binding domain at the N-terminus of the protein is separated from the downstream segment by a PRR region that carries out the fusion function. However, as discussed above, despite these features and the presence of the SLB, the gH structure did not reveal any other element characteristic of a *bona fide* viral fusion protein (see Section 4.1) (Backovic et al., 2010).

4.3.2 Different ways to compensate for gL function: Mutations in the gB, gH, gD triad

A second, independent passaging experiment resulted in isolation of revertant PrV-ΔgLPassB4.1 that also replicated efficiently without gL and had acquired mutations in the other entry glycoproteins (designated $gH^{B4.1}$, $gB^{B4.1}$, $gD^{B4.1}$) (Schröter et al., 2016). While two point mutations in the N-terminal gL-binding domain of $gH^{B4.1}$ (L70P and W103R) were sufficient to compensate for gL function in *in vitro* fusion assays and during entry in complementation assays, two mutations in $gB^{B4.1}$ (G672R and ΔK883, mapping to domain IV and the CTD, respectively) strongly enhanced fusogenicity in the fusion assays. The $gH^{B4.1}$ and $gB^{B4.1}$ variants were hyperfusogenic when co-expressed in the same cell, while the additional co-expression of $gD^{B4.1}$ completely blocked cell-cell fusion. This observation was surprising since PrV gD is normally dispensable for *in vitro* fusion or for direct viral cell-to-cell spread (Klupp et al., 2000; Peeters et al., 1992; Rauh and Mettenleiter, 1991). The fusion inhibition by $gD^{B4.1}$ could be attributed to a single amino acid substitution (A106V), which resides in the gD-core domain. The molecular mechanism behind fusion inhibition by $gD^{B4.1}$ remains to be uncovered but the appearance of compensatory mutations in all of the components of the fusion machinery observed in this study indicates that the interplay between the various herpesviral entry glycoproteins to mediate fusion is tightly regulated. Although gL is an essential component of the herpesvirus fusion machinery, its function can be compensated by mutations in gH (as in gDgH) or in gH plus gB (as in PrV-ΔgLPassB4.1) implying a regulatory role for gL during membrane fusion.

Both gL-independently replicating mutants had acquired compensatory mutations affecting the predicted gL-binding domain I in gH. In a follow-up study, the functional role of this domain was investigated by construction of an N-terminally truncated gH variant ($gH^{32/98}$). This mutant lacked domain I (aa32–98) and was therefore similar to the crystallized gH-core fragment (gH^{C}) and to the gH moiety of the gDgH hybrid protein. These studies demonstrated that the N-terminal domain of PrV gH is essential for gH function during fusion, since $gH^{32/98}$ was unable to promote fusion of wild-type gB in fusion assays and exhibited a total loss of function in the viral context (Vallbracht et al., 2017b). With respect to the gDgH fusion protein, these results suggested that the gD moiety in gDgH is critical for proper function of the gH-core fragment, and may adopt a stabilizing or modulating influence on the gH structure. In this regard, it was hypothesized that these stabilizing functions are normally executed by gL and are important for interaction of gH with gB. Notably, the functionality of $gH^{32/98}$ in both viral and cellular contexts, was rescued by substitution of wild-type gB for the more fusogenic gB of PrV-ΔgLPassB4.1. Additional expression of gL in *in vitro* fusion assays or in the viral background had no impact on fusion. These findings suggested that $gB^{B4.1}$ had been explicitly selected for interaction with a gH lacking its interaction partner gL, suggesting a direct interaction between PrV $gH^{32/98}$ and $gB^{B4.1}$ (Vallbracht et al., 2018b).

Overall, these results demonstrate that not only gL but also the gL-binding domain of PrV gH are not strictly required for membrane fusion during virus entry and spread. Nevertheless, compensatory mutations must be present in gB or gH to restore a fully functional fusion machinery, supporting the notion of a functional gH-gB interaction as a prerequisite for membrane fusion.

4.4 gH/gL and its functions in receptor binding

HSV-1 has been shown to use αvβ3-, αvβ6- and αvβ8-integrins as receptors to regulate the route and timing of infection (Cooper and Heldwein, 2015; Gianni et al., 2013). An RGD motif, the recognition site for several integrins, is present in HSV-1 and -2 gH, and was suggested to mediate binding to αvβ3-integrins (Galdiero et al., 1997; Gianni et al., 2010). The integrin binding activity of gH/gL was, however, found to be non-essential for membrane fusion, and mutation of the RGD motif in HSV-1 gH had no influence on entry, indicating that this interaction does not play a major role in triggering fusion (Galdiero et al., 1997). In contrast to HSV gH/gL, cellular receptors for PrV gH/gL have not been described so far. Reversion analysis of a gD-deficient PrV mutant in cell culture led to the isolation of a revertant with mutations in gH and gB that was capable of infecting cells in the absence of gD, strongly suggesting the existence of other cellular receptors recognized by either gH/gL or gB (Schmidt et al., 1997, 2001).

5. gB: The fusion executor

5.1 Structural features of the gB ectodomain

5.1.1 gB: A class III fusion protein

Glycoprotein B (gB) is the most conserved entry glycoprotein among herpesviruses with an amino acid sequence identity of around ~50% within each subfamily. gB is a type I transmembrane protein consisting of a signal peptide of ~60 residues in PrV, which is longer than average, a large ectodomain, a single-spanning TM, a membrane proximal region (MPR), and a long CTD comprising more than 90 residues in PrV and the HSVs (Robbins et al., 1987).

gB is the *bona fide* herpesvirus fusion protein. The earliest evidence pointing to its involvement in membrane fusion came from the observation that syncytial HSV-1 strains harbored mutations in the gB gene (Bzik et al., 1984). PrV gB, previously designated as gII, was discovered as an HSV-1 gB homolog based on its 52% amino acid sequence identity (Robbins et al., 1987; Whealy et al., 1990). Early functional studies revealed that gB-negative PrV and HSV virions could still bind to target cells, but were unable to enter or to form plaques (Cai et al., 1988; Peeters et al., 1992; Rauh and Mettenleiter, 1991). At the same time BoHV-1 gB, originally termed gI, was found to functionally complement a gB-negative PrV mutant, indicating functional equivalence (Rauh et al., 1991), although PrV gB could not substitute for BoHV-1 gB due to differences in the C-terminal half of the molecule (Miethke et al., 1995). Similarly, PrV gB complemented the lethal defect of a gB-negative HSV-1 mutant, but not *vice versa* (Mettenleiter and Spear, 1994). In line with these results, PrV gB could substitute for HSV-1 gB in *in vitro* fusion assays, and only low fusion activity was observed when HSV-1 gB was coexpressed with PrV gH, gL and gD (Böhm et al., 2016).

Although the functional studies indicated that PrV and HSV gB act as fusion proteins, the protein itself was unable to mediate the merger of lipid membranes. The first crystal structure of gB from HSV-1 provided the most direct evidence that it is indeed the *bona fide* protein fusogen of herpesviruses (Heldwein et al., 2006). Despite a lack of sequence conservation and having significantly different sizes (~700 *vs* ~400 aa), the HSV-1 gB ectodomain was shown to have a conserved secondary, tertiary and quaternary structure with the fusion protein G of the otherwise unrelated vesicular stomatitis virus (VSV) (Roche et al., 2006). Soon after, the baculovirus fusion glycoprotein 64 (gp64) was shown to exhibit a similar structure (Kadlec et al., 2008). Since these proteins displayed a novel three-dimensional arrangement distinct from class I and class II fusion proteins, they were grouped into to the newly defined class III of viral fusogens (Backovic and Jardetzky, 2009, 2011; Heldwein et al., 2006; Li et al., 2017b; Steven and Spear, 2006; Vallbracht et al., 2017a). Specific for herpesvirus gB is the fact that it is necessary but not sufficient for membrane fusion, whereas the other class III fusion proteins studied are both. Indeed, although PrV and HSV-1 gB appear to contain all features of the other class III fusion proteins required to drive fusion (Cooper and Heldwein, 2015), they strictly depend on the presence of gH/gL for their fusogenic activity. This raises questions regarding the structural features of gB that make it reliant on these accessory proteins. So far, structures of gB ectodomains from alphaherpesviruses HSV-1 and PrV, the betaherpesvirus human cytomegalovirus (HCMV) and the gammaherpesvirus EBV have been determined, revealing a highly conserved fold (Backovic et al., 2009; Burke and Heldwein, 2015; Heldwein et al., 2006; Li et al., 2017b; Vallbracht et al., 2017a).

VSV G is the only class III fusion protein for which the structure in the prefusion state has been determined, and the only protein fusogen known whose conformational change is reversible (Roche et al., 2006, 2007). The structural similarity between herpesvirus gB and VSV G indicated that gB may also undergo a structural rearrangement for fusion. The fusion trigger for VSV G is low pH in the endocytic compartment (Roche et al., 2006, 2007). In contrast, entry of PrV and the HSVs can occur at neutral pH at the plasma membrane as well as after endocytosis at neutral or low pH (Butcher et al., 1990; Dollery et al., 2010; Falke et al., 1985; Mettenleiter et al., 2009; Miller et al., 2019; Milne et al., 2005; Nicola and Straus, 2004; Stampfer et al., 2010). Local structural changes induced by low pH were observed in HSV-1 gB, but were found to be insufficient to trigger the fusogenic global rearrangement, indicating that gB mediated fusion is not triggered by acidification of the milieu (Dollery et al., 2010; Stampfer et al., 2010).

5.1.2 gB fusion loops: The business end of the molecule

In general, viral fusion proteins reside in a metastable, high-energy prefusion conformation on the viral envelope (Harrison, 2015). Binding to a receptor, or to protons in the acidic endosomal environment, induces a conformational change that results in exposure of an initially shielded hydrophobic polypeptide segment, termed fusion peptide (FP) when it is located N-terminally, or fusion loop (FL) when it is internal as in gB (Harrison, 2015; Kielian and Rey, 2006; White et al., 2008). These short segments are typically apolar and project outward, anchoring the virus to the target cell membrane (White et al., 2008), and forming an unstable extended intermediate, termed prehairpin (Kielian, 2014; Kim et al., 2011; Li and Modis, 2014). In this state, the fusion protein is bound to both the viral membrane (*via* the TM) and the cellular membrane (*via* the FP or FL). This intermediate rapidly collapses into an energetically stable postfusion conformation, in which the FPs/FLs are located at the same end of the molecule as the TM (Kielian and Rey, 2006). The FP/FL insertion and the subsequent folding-back are proposed to drive distortion of the two membranes and to pull them toward each other, causing first the merger of the outer leaflets of the membrane (hemifusion), followed by merger of the inner leaflets and formation of a fusion pore (Harrison, 2015; Kielian and Rey, 2006; Plonsky et al., 2008).

The crystal structure of the gB ectodomains reveals an elongated spike-like trimer in which the protomers adopt a hairpin conformation corresponding to that observed in the postfusion form of VSV G. Each gB protomer consists of five distinct domains (domain I–V) that twist around each other around the molecular threefold axis, and stabilize the trimeric structure by multiple inter-subunit contacts (Fig. 4A) (Heldwein et al., 2006; Li et al., 2017b; Vallbracht et al., 2017a).

Domain I, also termed the fusion-domain, contains two FLs at the tip of the molecule adjacent to the C-terminus of the ectodomain (Fig. 4A). In general, FLs are rich in apolar and aromatic residues conducive to insertion into the lipid bilayer. However, due to the lack of amino acid sequence conservation, the identification of herpesvirus gB FLs was difficult until structural data became available. The position of the two FLs (FL1 and FL2) in

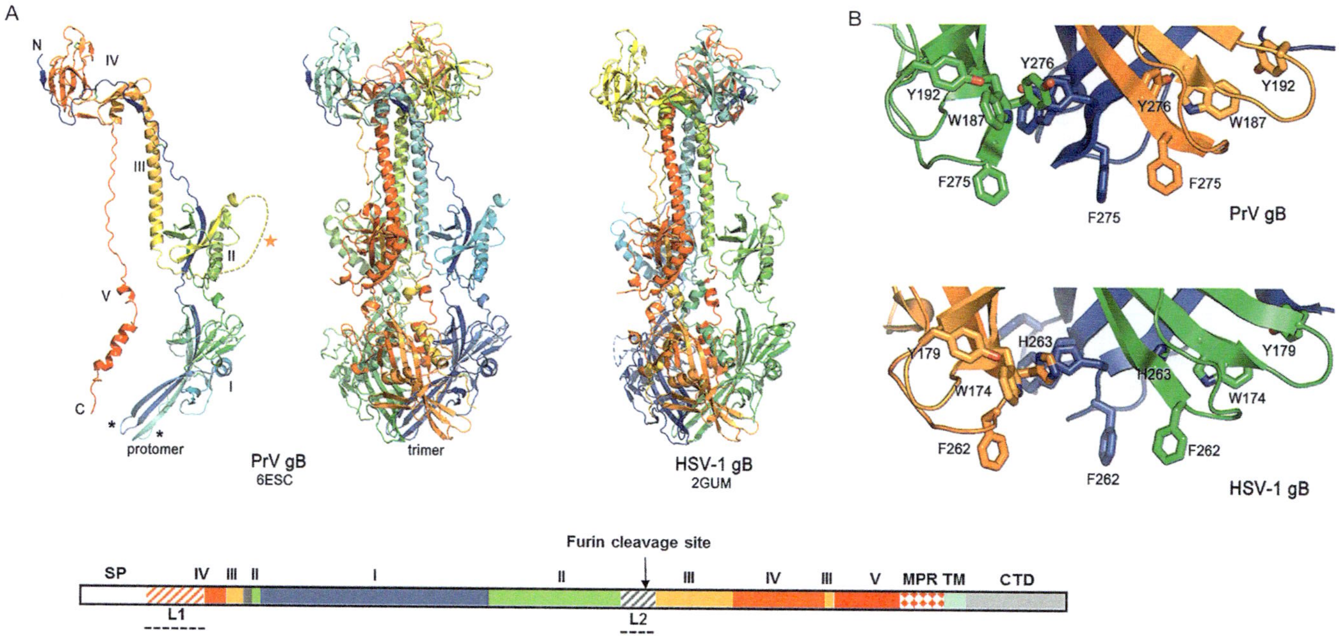

Fig. 4 The fusion protein gB. (A) Ribbon diagrams of a single PrV gB protomer, and PrV and HSV-1 gB trimers are shown. The PDB IDs of the crystallized constructs are shown below each structure. Domain I (blue) contains the bipartite fusion loops (FLs) magnified in panel (B). Domains II and III are connected by a flexible linker (L2), which in PrV gB contains the furin cleavage site (*orange star*). This segment was not resolved in the structure and is depicted as a dashed line. The first 50 residues at the N-terminus were not resolved either (L1). Domain III (yellow) forms the coiled-coil core, which extends to the top domain IV (orange). The C-terminal domain V (red) packs against the coiled-coil core and terminates in close proximity of the fusion loops. (B) Close-up view of the PrV and HSV-1 gB fusion loops. Key residues are labeled and their side chains are shown as sticks. Schematic representation of gB is shown below, showing the signal peptide (SP), which encompasses around 60 residues and is unusually long in PrV, the five domains forming the ectodomain, followed by the membrane proximal regions (MPR), transmembrane domain (TM) and C-terminal domain (CTD). The flexible regions not resolved in the structure are indicated with the dashed line, below. The structure images were generated using PyMOL (DeLano, 2002).

HSV-1 gB was proposed by analogy with the well-defined FLs of VSV G (Baquero et al., 2015; Heldwein et al., 2006; Sun et al., 2008). In support of this hypothesis, it was shown that several hydrophobic residues within these two loops, and, surprisingly, charged residues adjacent to FL2 are important for membrane binding and fusion (Hannah et al., 2007, 2009; Lin and Spear, 2007; Maurer et al., 2013). Based on the results from structure-based functional analyses of the PrV gB FLs combined with previous results reported for gB from other herpesviruses, a model was proposed on how gB may interact with the target membrane. This model suggests that hydrophobic residues exposed in FL2 penetrate deeper into the hydrocarbon core of the lipid bilayer, while residues from both FLs form a rim above, which would insert into the interfacial region of the membrane to catalyze the fusion process (Vallbracht et al., 2017a). Specifically, four residues in the PrV gB FLs, W187, Y192, F275, and Y276 (corresponding to W174, Y179, F262, and H263 in HSV-1 gB) (Fig. 4B) were crucial for gB function by forming a continuous hydrophobic and electrostatically neutral patch at the surface of the trimeric gB postfusion spike.

Although PrV and HSV-1 gB seem to share principles for insertion of aromatic and hydrophobic residues, the HSV-1 gB FLs contain a greater number of charged residues, some of which were shown to play an important role in membrane binding (Hannah et al., 2009). Mutation of the corresponding residues in PrV gB did not affect membrane binding (Vallbracht et al., 2017a). A shallow membrane insertion of the FLs is proposed for both PrV and HSV-1 gB (Maurer et al., 2013; Vallbracht et al., 2017a), but the presence of additional charged residues in the HSV-1 gB FLs could point to intrinsic differences in the overall interaction with the membrane. Comparative structural analyses with gB from the beta- and gammaherpesviruses suggests that a common mode of gB insertion into membranes may have evolved within each subfamily (Vallbracht et al., 2017a). Although the FL insertion mode may be similar between PrV and HSV-1, residues comprising FL1 and FL2 cannot be exchanged without loss of function (data not published). Surprisingly, FLs of the gammaherpesvirus EBV can functionally substitute for the HSV-1 FLs, as shown in *in vitro* fusion assays (Zago et al., 2013). However, when the less hydrophobic HSV-1 fusion loops were grafted into EBV gB, the resulting mutants failed to function in fusion (Backovic et al., 2007). These studies indicate that although the insertion mode is similar, FLs *per se* are not rigid entities that insert into membranes independent from the protein background presenting them.

5.1.3 Role of gB domains II–V

The fold of gB domain II (Fig. 4B, green) is reminiscent of pleckstrin homology (pH) domains (Eisenberg et al., 2012; Heldwein et al., 2006), which, in general, provide binding sites for phosphoinositide and peptide moieties of diverse cellular proteins (Blomberg et al., 1999; Lemmon, 2007). The epitopes of several neutralizing anti-HSV-1 gB MAbs map to this domain, some of which block association of gB with gH/gL (Atanasiu et al., 2010b; Bender et al., 2007; Cairns et al., 2014; Eisenberg et al., 2012; Heldwein et al., 2006).

Processing of PrV gB (Whealy et al., 1990) and gB of other alphaherpesviruses such as VZV, BoHV-1 (van Drunen Littel-Van Den Hurk and Babiuk, 1986) or EHV-1 (Meredith et al., 1989), but not HSV-1 gB, involves proteolytic cleavage by cellular furin. The cleavage site of PrV gB (residues ^{501}RRARR^{505}) is located in a flexible linker connecting domains II and III (Fig. 4A, orange star) (Hosaka et al., 1991; Li et al., 2017b; Vallbracht et al., 2017a). Cleavage of gB, however, is dispensable for viral replication in cell culture (Kopp et al., 1994; Okazaki, 2007; Strive et al., 2002; Vallbracht et al., 2017a), in contrast to class I fusion proteins, such as influenza HA, where furin cleavage is essential as it liberates the FP to interact with membranes (Stieneke-Grober et al., 1992; Webster and Rott, 1987).

Domain III forms a trimeric α-helical coiled-coil, also known as the central stalk, that is a prominent structural feature of class III and class I fusion proteins in the postfusion state (Backovic and Jardetzky, 2009). By analogy with the prefusion form of VSV G, the N-terminal part of this long helix was hypothesized to be unfolded in prefusion gB. However, mutations designed to impede its refolding showed that it is remarkably stable since introduction of glycine, proline, phenylalanine or tryptophan residues resulted in the same coiled-coil conformation as in the native protein (Silverman et al., 2010; Vitu et al., 2013). Simultaneous introduction of double phenylalanine and double tryptophan mutations, thereby adding six bulky side chains at the coiled-coil interface, destabilized the postfusion trimer to some extent, but did not prevent spontaneous folding of the gB ectodomain into the postfusion conformation (Vitu et al., 2013).

The central stalk extends to the top or crown domain of the molecule (domain IV, Fig. 4A, orange) (Backovic et al., 2009; Burke and Heldwein, 2015; Chandramouli et al., 2015; Heldwein et al., 2006; Li et al., 2017b; Vallbracht et al., 2017a). The most noticeable difference between the gB ectodomain structures is the location of domain IV relative to the rest of the ectodomain. When the structures are superimposed on domains I and II (Vallbracht et al., 2017a), domain IV shows a different rotation with respect to the rest of the molecule suggesting that it could undergo structural changes during the pre- to postfusion transition that may require intrinsic structural plasticity (Heldwein et al., 2006; Zeev-Ben-Mordehai et al., 2016). A hyperfusogenic PrV gB variant carries a mutation in domain IV (G672R), suggesting that the mutation could lower the energy required for movement of this domain during pre- to postfusion refolding (Schröter et al., 2016). Epitopes for several virus-neutralizing anti-HSV-1 MAbs reside within this domain, indicating that they are likely to be exposed in the prefusion form (Heldwein et al., 2006).

Domain V (Fig. 4A, red) forms a long extension that stretches from the top to the bottom end of the spike, terminating in close proximity to the fusion domain and FLs. The long arm formed by one protomer packs tightly into the groove between the core domains of the other two protomers, resulting in an extensive trimerization interface (Backovic and Jardetzky, 2009; Heldwein et al., 2006; Vallbracht et al., 2017a). This arm, running antiparallel to the central, trimeric stalk is reminiscent of the six-helix bundle fold that is typical for the postfusion conformation of class I fusion proteins (White et al., 2008). This coil-arm complex was therefore proposed to perform a similar function in gB (Connolly and Longnecker, 2012; Fan et al., 2017; Heldwein et al., 2006). Mutations in the arm reduce gB fusion activity without affecting protein processing and were therefore suggested to either stabilize the prefusion conformation or to destabilize the postfusion form of gB, making the conformational change less likely (Fan et al., 2017).

Different conformations of gB before and after membrane fusion have been reported using FRET based analyses of gB labeled with fluorescent proteins (Gallagher et al., 2014). Consistent with the concept of gB undergoing a major conformational change to drive the membrane fusion reaction, mutations that abrogated fusion have been found in all five domains of HSV-1 gB (Bender et al., 2007; Hannah et al., 2007; Lin and Spear, 2007).

It has been hypothesized that the energy required for fusion pore formation and propagation requires the concerted action of several fusion protein trimers (Harrison, 2015; Kielian and Rey, 2006; Plonsky et al., 2008). PrV and HSV-1 gB have an intrinsic propensity to cluster on membranes. Individual PrV gB postfusion ectodomain trimers establish an extensive network of lateral interactions when bound to liposomes, giving rise to an array with the appearance of a glycoprotein coat (Vallbracht et al., 2017a). Likewise, soluble HSV-1 gB trimers were shown to form lateral contacts. Lateral assembly of full-length gB trimers, presumably in a postfusion conformation, was also observed on HSV-1 virions (Maurer et al., 2013). It will be interesting to investigate the biological significance of these trimer-trimer contacts for the membrane fusion process.

5.2 Molecular gymnastics: Understanding the conformational change of gB during pre- to postfusion transition

While the postfusion gB ectodomain is well characterized, no high resolution structure of gB in its prefusion state is available. No neutralizing antibodies specific for the prefusion form of gB, which could be used to arrest gB in the this state, have been described (Bender et al., 2007; Cairns et al., 2014). This strategy was successfully used for crystallization of the fusion (F) protein of respiratory syncytial virus or Ebola GP in prefusion state (Lee et al., 2008; Lee and Saphire, 2009; McLellan, 2015; Mclellan et al., 2011, 2013). Other approaches to stabilize the prefusion trimers of class I proteins, such as introduction of proline residues into the flexible regions that refold into the coiled-coil in the postfusion form (Krarup et al., 2015; Sanders et al., 2002; Torrents de la Pena and Sanders, 2018), or appending exogenous soluble trimerization domains (Lee and Saphire, 2009; Stevens et al., 2004; Yin et al., 2006), were not successful for gB. The modified HSV-1 gB ectodomains all folded into the postfusion conformation (Baquero et al., 2015; Cooper and Heldwein, 2015; Vitu et al., 2013). Therefore, the MPR, the TM, and/or the CTD were proposed to be necessary to establish and maintain the metastable prefusion conformation of gB (Vitu et al., 2013).

Indeed, a shorter and more condensed form of full-length gB was observed in exosomes using cryo electron tomography (cryoET) (Zeev-Ben-Mordehai et al., 2016). While the postfusion gB ectodomain has a length of 16 nm, the short-form gB measured 10 nm. In this conformation, interpreted as a prefusion form, domain I was assigned to the top of the spike, furthest away from the MPR and TM, with exposed FLs pointing away and being apart from each other (Zeev-Ben-Mordehai et al., 2016). Based on these results a conformational change in domain III was proposed that would result in the formation of an extended intermediate. Concomitantly, domain I would undergo a rotation so that the FLs come together to form a hydrophobic patch compatible with insertion into the target membrane. As a last step, the long arms of domain V would fold back providing the force for membrane fusion (Zeev-Ben-Mordehai et al., 2016).

Another condensed and distinct form of full-length HSV-1 gB has been reported in which the FLs point toward the anchoring membrane (Fontana et al., 2017), resembling the prefusion form reported for VSV G (Fontana et al., 2017; Roche et al., 2007). For human cytomegalovirus (HCMV), a betaherpesvirus, a low-resolution structure of gB was obtained *in situ* by cryoET of intact particles and using sub-tomogram averaging. This structure revealed an organization of prefusion gB on virions at a resolution of 21 Å that could also be interpreted in terms of the atomic models for the individual domains of gB extracted from the postfusion form, but rearranged to make a Christmas tree trimer, according to the electron density. The putative prefusion trimer is about 2/3 of the length of the postfusion form, which was also present at the surface of the HCMV particles studied, and could be directly compared. In the derived model for prefusion HCMV gB, the FLs are away from each other but point downward toward the viral membrane (SI et al., 2018) as in prefusion VSV G, confirming the model deduced by Fontana and colleagues (Fontana et al., 2017). This work also showed that a subset of the gB spikes in prefusion conformation interact with densities attributed to gH/gL at the particle surface. The suspected gB-gH/gL complex was visualized at low resolution only (around 30 Å resolution). The inferred contact sites between gB and gH/gL were proposed to reside in domain I of gB and domain III of gH, which is in line with proposed interaction sites in HSV-1 (Atanasiu et al., 2010b; Chowdary et al., 2010). Further structural data are required to understand the gH-gB interactions on the viral particle in more detail. However, given the structural similarity of postfusion gB across herpesviruses, the model proposed for the fusogenic conformational change of HCMV gB (Si et al., 2018) is expected to be valid also for alphaherpesvirus gB (Fig. 5).

Together, these studies provided snap-shots of possible conformational gB rearrangements, providing a starting point for a molecular understanding of the mechanism of gB mediated membrane fusion.

5.3 Functional relevance of the gB membrane proximal region, the transmembrane and the cytoplasmic domain

The conformation and associated fusogenic activity of the gB ectodomain is proposed to be influenced by the adjoining MPR, the TM and the CTD. As discussed above, the MPR, TM and CTD may be required to maintain gB in its metastable prefusion form. The structure of the full-length HSV-1 gB trimer was recently determined, demonstrating that these three domains

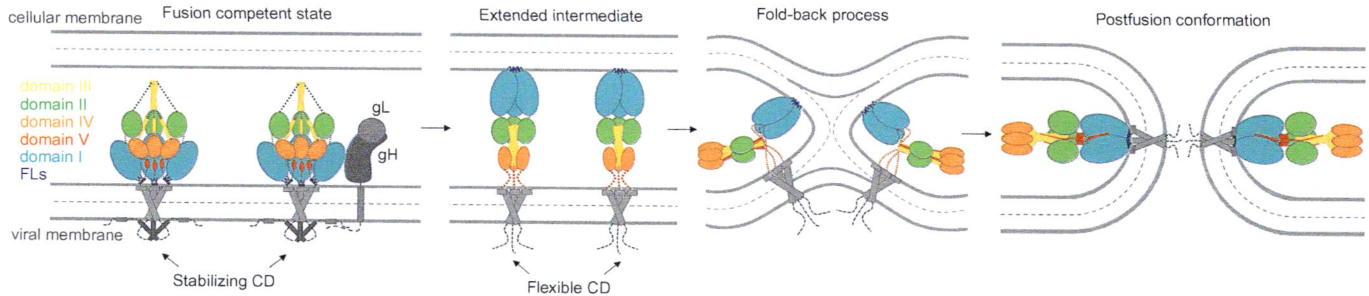

Fig. 5 Model of presumed conformational changes in gB during fusion. Proposed transitions of gB from the pre- to the postfusion conformation are schematically shown. The fusion loops (FLs) in prefusion gB most likely point toward the viral membrane and are highlighted in blue. Interactions of the cytoplasmic domain (CD) of gB with the membrane and the gH/gL complex stabilize the prefusion conformation. Destabilization of prefusion gB by activated gH/gL leads to a gB conformational change into an extended intermediate. This step might be induced or preceded by rearrangements in the gB CD. The initially shielded FLs are subsequently exposed and interact with the target membrane. Zippering of the long arms of domain V along the trimer pulls the membranes together and leads to formation of the energetically stable postfusion conformation in which the FLs and the C-terminal part of the molecule come together leading to membrane fusion. *Modified after Fontana, J., Atanasiu, D., Saw, W.T., Gallagher, J.R., Cox, R.G., Whitbeck, J.C., Brown, L.M., Eisenberg, R.J., Cohen, G. H. 2017. The fusion loops of the initial prefusion conformation of herpes simplex virus 1 fusion protein point toward the membrane. MBio, 8.Zeev-Ben-Mordehai, T., Vasishtan, D., Hernandez Duran, A., Vollmer, B., White, P., Prasad Pandurangan, A., Siebert, C.A., Topf, M., Grunewald, K. 2016. Two distinct trimeric conformations of natively membrane-anchored full-length herpes simplex virus 1 glycoprotein B. Proc. Natl. Acad. Sci. U.S.A., 113, 4176–81.SI, Z., Zhang, J., Shivakoti, S., Atanasov, I., Tao, C.L., Hui, W.H., Zhou, K., Yu, X., LI, W., Luo, M., Bi, G.Q., Zhou, Z.H. 2018. Different functional states of fusion protein gB revealed on human cytomegalovirus by cryo electron tomography with Volta phase plate. PLoS Pathog., 14, e1007452.*

form a trimeric pedestal underneath the ectodomain (Cooper et al., 2018). In the structure of full-length HCMV gB the CTD was found to be ordered/stable in the prefusion form but not in the gB postfusion form, suggesting that the CTD becomes disordered/flexible during the pre- to postfusion transition (Si et al., 2018). The crystal structure of HSV-1 full-length gB appears to show a hybrid form, with the ectodomain in postfusion and the CTD in prefusion state. Full-length HSV-1 gB was crystallized in a mixture of detergents for solubilization and short amphipathic polymers in order to stabilize the membrane regions that were otherwise largely unresolved. As discussed by the authors of that report, detergent solubilization of the membrane may be responsible for the postfusion conformation of its ectodomain (Cooper et al., 2018).

The ectodomain and the TM are connected *via* the MPR that forms the top of the pedestal underneath the ectodomain and is composed of three symmetry related helices as demonstrated by crystallography studies (Cooper et al., 2018). The structure of the MPR was determined in the presence of detergent micelles as lipid membrane mimics and its position in the membrane was inferred from the surface distribution of hydrophobic, polar, and charged residues. The hydrophobic MPR helix was proposed to be embedded in the hydrocarbon core and to run parallel to the membrane bilayer. The side chains of three serine residues, two of which are conserved between PrV and HSV-1, presumably protrude into the lipid head groups of the outer leaflet. Such deep placement of the MPR helices within the membrane suggests that they may contribute directly to the fusion process by facilitating lipid mixing. Insertion and deletion mutations within this region in HSV-1 gB resulted mostly in a non-functional protein or in a protein with very low fusion activity, demonstrating its importance for membrane fusion (Lin and Spear, 2007; Shelly et al., 2012). Several residues within the 43 aa MPR are conserved among herpesviruses, and HSV-1 gB is even functional with the gammaherpesvirus EBV gB MPR pointing to functional conservation (Zago et al., 2013). The structures of the N- and C-terminal boundaries of the MPR remain unresolved possibly due to their flexible nature (Cooper et al., 2018). The unresolved N-terminus of the MPR was proposed to shield the fusion loops in the prefusion form to prevent their unproductive insertion into the viral membrane (Fig. 5) (Maurer et al., 2013; Shelly et al., 2012). In line with this notion, the HSV-1 gB ectodomain with a C-terminal extension to include the first residues of the MPR did not associate with liposomes. The hinge region connecting the MPR with the TM was proposed to be a highly dynamic region essential for proper positioning of the MPR and folding of gB as was shown by structural and mutagenesis studies (Cooper et al., 2018; Wanas et al., 1999). The observed flexibility of the MPR also could have been a consequence of the prefusion-postfusion hybrid structure reported for the full-length gB. In the postfusion form, the MPRs are likely to be apart from the fusion loops, and in the prefusion form they should connect to the CTD *via* the TM region, suggesting that the MPR is required to be flexible to adapt to both configurations.

Like the MPR, the gB TM plays an essential role for gB function. The native TM is important for correct folding, transport and virion incorporation of HSV-1 and PrV gB (Cai et al., 1988; Gilbert et al., 1994; Lin and Spear, 2007; Nixdorf et al., 2000; Rasile et al., 1993; Silverman et al., 2012). Replacement of the authentic HSV-1 TM by a GPI-anchor was proposed to result in fusion arrest at the hemifusion stage, pointing to a critical importance of the TM during later stages of fusion (Jones and Geraghty, 2004). These results suggested that the function of the TM is more complex than serving as a mere membrane anchor and implied an active role of the TM during membrane fusion. Biophysical studies revealed that the HSV-1 gB TM forms a trimer with a unique inverted teepee shape (Cooper et al., 2018). Although the structure of the TM of PrV gB has not been determined, a similar arrangement for all alphaherpesviruses has been proposed since the residues important for the inverted teepee arrangement of the TM helices are strictly conserved in this subfamily (Cooper et al., 2018). The conditions used for crystallization of the TM region do not represent a native-like environment for TM and further studies on the mechanistic contributions of the TM are necessary to decipher its precise role in fusion.

The CTDs of HSV-1 gB form an intertwined trimer as shown by X-ray crystallography of full-length HSV-1 gB (Chowdary and Heldwein, 2010; Cooper et al., 2018). While the CTD-core is composed of two α-helices and one 3_{10} helix, the structure of the C-terminus of the CTD (residues 866–904) was unresolved in the crystal but inferred by continuous-wave electron spin resonance measurements revealing that it contains a putative amphipathic helix that is formed only in the presence of membranes. Soluble CTDs were found to bind to synthetic membranes *via* this helix (Cooper et al., 2018; Silverman et al., 2012). In fact, the intrinsic ability of the gB CTD to stably bind to lipid membranes was postulated to be a key feature necessary for regulation of gB fusogenicity (Chowdary and Heldwein, 2010; Cooper et al., 2018; Silverman et al., 2012). The CTD of PrV and the HSVs gB negatively regulate fusion, since engineered or naturally occurring mutations, insertions or truncations in the CTD increase gB fusogenicity (Diakidi-Kosta et al., 2003; Fan et al., 2002; Foster et al., 2001; Gage et al., 1993; Klupp et al., 2000; Nixdorf et al., 2000; Ruel et al., 2006). Based on the available structural and functional data, a model was suggested in which the gB CTD functions as an inhibitory membrane-dependent clamp that braces against the viral membrane, stabilizing the prefusion state, which could explain why the ectodomain adopts the postfusion conformation when expressed on its own (Chowdary and Heldwein, 2010; Cooper et al., 2018). The side chains of several residues in the TM and the CTD form a positively charged belt compatible with interactions with the negatively charged phospholipid head groups of the membrane and this interaction could brace the CTD against the membrane. The high conservation of residues forming this belt (six out of seven between PrV and HSV gB) suggest a similar function for this region in PrV membrane fusion. Fusion enhancing mutations mapping to the belt have been proposed to act by weakening or disrupting the interaction of the CTD with the membrane by reducing the positive charge in the vicinity of the membrane (Cooper et al., 2018), and thus decreasing the kinetic energy barrier for the conformational change. For example, deletion of a single conserved residue K^{883} (Schröter et al., 2016) as well as truncation of the last 28 residues of the PrV gB CTD (Nixdorf et al., 2000) significantly increased fusogenicity (Klupp et al., 2000), supporting the notion that the PrV gB CTD plays a similar role as its HSV-1 counterpart. Other mutations outside of the basic belt were proposed to act either by disrupting inter-protomer bonds or by destabilizing the trimeric CTD interfaces preventing the CTD from folding into a functionally active conformation. Several residues comprising the trimeric CTD interface are conserved between PrV and the HSVs (Chowdary and Heldwein, 2010; Cooper et al., 2018; Rogalin and Heldwein, 2015; Vitu et al., 2013).

The gB CTDs of PrV and the HSVs contain functional endocytosis motifs important for regulation of gB surface expression. C-terminal truncations that remove these motifs lead to enhanced surface expression and fusion activity (Fan et al., 2002; Nixdorf et al., 2000). However, increased surface expression and enhanced fusion activity were found to be unrelated, since mutation of only the endocytosis motif did not increase fusion despite enhanced surface expression (Fan et al., 2002). Interestingly, truncation of the last 60 C-terminal residues of PrV gB resulted in a variant (gB-007) that was unable to support *in vitro* fusion and PrV entry but was able to mediate cell-to-cell spread of a gB-negative PrV mutant (Nixdorf et al., 2000, 2001). This feature was exploited for reversion analysis of a PrV mutant expressing gB-007 (Nixdorf et al., 2001). This led to the isolation of PrV-007Pass, which had acquired two point mutations in the gB ectodomain (D305A and D744E), sufficient to restore the defects in entry and cell-to-cell spread of the PrV mutant lacking gB (Nixdorf et al., 2001). The fact that defects in membrane fusion resulting from a truncated gB CTD can be corrected by mutations in the gB ectodomain supports the idea that there is a cross-talk between the CTD and the gB ectodomain, probably mediated by the MPR and TMs. One could imagine that gB acts like a push-up puppet, whose ectodomain conformation is influenced by the status of the MPR, TM, and CTD.

Together, the functional data obtained for the gB CTD of PrV and HSV-1 are in agreement, and indicate that the gB CTD is of critical importance for the conformation and, concomitantly, for a tight regulation of the gB fusogenicity during the membrane fusion process. Nevertheless, even extremely hyperfusogenic PrV gB ($gB^{B4.1}$) (Schröter et al., 2016) and HSV-1 gB variants (Rogalin and Heldwein, 2015) require activation *via* gH/gL. As discussed in Section 4.2 the gH CTD was proposed to activate gB by acting as a wedge that releases the fusion restricting gB CTD clamp, allowing gB to undergo the fusogenic conformational change (Rogalin and Heldwein, 2015). However, although *in vitro* fusion mediated by PrV gB was similarly progressively reduced when the PrV gH CTD was truncated by more than half of the CTD residues as observed for HSV-1, PrV gH lacking the CTD was still able to support entry of a gH-deficient PrV mutant (Vallbracht et al., 2018a), demonstrating that the CTD of PrV gH is not necessary to activate gB. In addition, soluble HSV-1 gH/gL was shown to induce low levels of membrane fusion (Atanasiu et al., 2010a). These results indicate that gH/gL lacking the CTD maintains, at least partially, the ability to trigger gB fusogenicity. Thus, although gB-mediated fusion might be in part regulated by interactions between the gH and gB CTDs, liberation of the fusion-restricting CTD clamp of gB by the gH CTD wedge appears not to be the sole gB fusion activator. Overall, elucidation of the mechanism of gB fusion activation by the heterodimeric gH/gL complex still remains a challenge.

5.4 gB as a tropism determinant?

HSV-1 and PrV gB can bind to paired immunoglobulin-like type 2 receptor α (PILRα), and this interaction is involved in viral entry into certain cell lines *in vitro* (Arii et al., 2009; Satoh et al., 2008; Satoh and Arase, 2008). PILRα-mediated HSV-1 infection was dependent on the presence of gD, indicating that it functions as coreceptor. Binding of gB to PILRα required the presence of sialylated *O*-glycans on gB (Wang et al., 2009), as reported for other cellular ligands (Sun et al., 2012). In addition to PILRα, PrV gB can also bind porcine PILRβ, an activating natural killer cell receptor (De Pelsmaeker et al., 2019). Whether PILRβ may serve as a gB receptor for PrV fusion and entry into host cells remains to be investigated. For HSV-1 gB, non-muscle myosin heavy chain IIA (NMHC-IIA) and myelin-associated glycoprotein expressed on neural tissues have been identified as additional receptors (Arii et al., 2010; Suenaga et al., 2010). Although binding of gB to PILRα was found to mediate infection of cell lines *in vitro*, the significance of receptor binding by gB *in vivo* remains to be determined. It would be interesting to investigate whether gB binding to PILRα or alternative receptors in non-immune cells induces conformational changes in gB contributing to its activation for fusion. In the current entry models, gD acts as the predominant receptor binding protein triggering membrane fusion. Thus, the idea that gB can engage other receptors to mediate entry appears interesting, particularly, when considering that PrV does not need gD for *in vitro* fusion or cell-cell spread *in vitro* and *in vivo*, and that mutants lacking gD can regain infectivity correlating with mutations in gB and gH.

6. Concluding remarks and open questions

The crystal structures of the four key players involved in PrV and HSVs membrane fusion are now known and supported by a great body of functional data, providing important insights into the overall mechanism of alphaherpesvirus fusion. Yet, several key questions and essential mechanistic details remain unresolved.

It is clear how gD binds to its receptor nectin-1 and that this interaction leads to structural rearrangements, resulting in exposure of the previously hidden receptor binding residues and of a fusion activating interface of gD. Activated gD triggers the conserved core fusion machinery consisting of gH/gL and gB to promote fusion. Although substantial evidence exists that gD transmits the fusion triggering signal by direct interaction with gH/gL which, in turn, activates the fusion executor gB, for the most part the corresponding complexes have not been captured and the exact binding sites remain speculative. The identification and characterization of the interaction sites could provide a basis for the design of potential therapeutic targets blocking the fusion process.

PrV and the HSVs share many features, and membrane fusion mediated by the HSVs strongly resembles PrV membrane fusion. For viral entry, they all require the same set of proteins, gD, gH/gL and gB whose respective molecular structures and functions are similar. Nevertheless, there are some intriguing differences: whereas PrV and the HSVs both require gD for entry, PrV has an edge over the HSVs, enabling it to mediate cell-to-cell spread and *in vitro* fusion without the need for gD. This raises the question as to why the HSVs are reliant on gD, or which features enable PrV to spread from cell-to-cell independent of gD. Answers to these key questions could shed light on the actual role and precise mechanism of fusion activation by gD and the differences between fusion during entry and cell-to-cell spread. PrV provides the unique opportunity for reversion analyses of gD negative virus mutants that already resulted in PrV mutants capable of infecting cells independent of gD, harboring mutations in gH and gB (Karger et al., 1998; Schmidt et al., 1997, 2001). In this case, the viral particles must bind and enter the target cell through alternative entry receptors. It is thus conceivable that entry receptors for the other glycoproteins may exist as well and binding to them could bypass the requirement for gD. Future studies aiming to identify such receptors could elucidate the mechanism of gD independent infectivity.

In contrast to the HSVs in which gL is strictly required for gH processing and function, PrV is able to perform limited cell-to-cell spread in the absence of gL. Reversion analyses of gL-deficient mutants resulted in revertants capable of infecting cells independent of gL, suggesting that gL plays a modulatory nonessential role for PrV gH function (Klupp and Mettenleiter, 1999; Schröter et al., 2016; Vallbracht et al., 2017b). Whether gL plays only a chaperone-like structural role for gH of the HSVs or is actively involved in the fusion process needs to be addressed in future studies.

One of the ultimate key questions, *i.e.*, why alphaherpesvirus gB, a class III fusion protein that at first glance appears to contain all features necessary to mediate membrane fusion alone, is reliant on gH/gL, remains unanswered. The gH/gL complex must interact with gB to promote fusion but the exact interaction sites remain to be identified and further studies are necessary to define the precise role of the fusion regulator gH/gL. Does gH/gL promote gB fusogenicity, or is it rather required to stabilize gB in its metastable prefusion form? By cryoET of intact virus particles, HCMV gB was partially captured in complex with gH/gL in a form that was interpreted as gB prefusion conformation, suggesting that in addition to the fusion restricting gB CTD clamp, interactions with the gH ectodomain may stabilize the metastable prefusion gB ectodomain (Si et al., 2018). Identification and characterization of such potential gB-gH/gL triggering complexes in a native environment would be the next challenge in structural biology of alphaherpesvirus entry. Moreover, a complete and accurate structure of alphaherpesvirus prefusion gB could unravel how gB changes its conformation to promote membrane fusion.

Acknowledgments

This work was supported by the Deutsche Forschungsgemeinschaft (Grant Me 854/11-2) and from the French Government's *Investissement d'Avenir* program *Laboratoire d'Excellence* (LABEX) entitled "Integrative Biology of Emerging Infectious Diseases" (grant n°ANR-10-LABX-62-IBEID). We thank former and current members of the Klupp/Mettenleiter and Rey laboratories for their contributions to elucidate herpesvirus entry.

References

Akhtar, J., shukla, D., 2009. Viral entry mechanisms: cellular and viral mediators of herpes simplex virus entry. FEBS J. 276, 7228–7236.

Arii, J., Uema, M., Morimoto, T., Sagara, H., Akashi, H., Ono, E., Arase, H., Kawaguchi, Y., 2009. Entry of herpes simplex virus 1 and other alphaherpesviruses via the paired immunoglobulin-like type 2 receptor alpha. J. Virol. 83, 4520–4527.

Arii, J., Goto, H., Suenaga, T., Oyama, M., Kozuka-Hata, H., Imai, T., Minowa, A., Akashi, H., Arase, H., Kawaoka, Y., Kawaguchi, Y., 2010. Non-muscle myosin IIA is a functional entry receptor for herpes simplex virus-1. Nature 467, 859–862.

Atanasiu, D., Whitbeck, J.C., Cairns, T.M., Reilly, B., Cohen, G.H., Eisenberg, R.J., 2007. Bimolecular complementation reveals that glycoproteins gB and gH/gL of herpes simplex virus interact with each other during cell fusion. Proc. Natl. Acad. Sci. U.S.A. 104, 18718–18723.

Atanasiu, D., Saw, W.T., Cohen, G.H., Eisenberg, R.J., 2010a. Cascade of events governing cell-cell fusion induced by herpes simplex virus glycoproteins gD, gH/gL, and gB. J. Virol. 84, 12292–12299.

Atanasiu, D., Whitbeck, J.C., De Leon, M.P., Lou, H., Hannah, B.P., Cohen, G.H., Eisenberg, R.J., 2010b. Bimolecular complementation defines functional regions of herpes simplex virus gB that are involved with gH/gL as a necessary step leading to cell fusion. J. Virol. 84, 3825–3834.

Atanasiu, D., Cairns, T.M., Whitbeck, J.C., Saw, W.T., Rao, S., Eisenberg, R.J., Cohen, G.H., 2013. Regulation of herpes simplex virus gB-induced cell-cell fusion by mutant forms of gH/gL in the absence of gD and cellular receptors. MBio 4, e00046-13.

Atanasiu, D., Saw, W.T., Eisenberg, R.J., Cohen, G.H., 2016. Regulation of HSV glycoprotein induced cascade of events governing cell-cell fusion. J. Virol. 90, 10535–10544.

Atanasiu, D., Saw, W.T., Lazear, E., Whitbeck, J.C., Cairns, T.M., Lou, H., Eisenberg, R.J., Cohen, G.H., 2018. Using antibodies and mutants to localize the presumptive gH/gL binding site on herpes simplex virus gD. J. Virol. 92, e01694-18.
Avitabile, E., Forghieri, C., Campadelli-Fiume, G., 2007. Complexes between herpes simplex virus glycoproteins gD, gB, and gH detected in cells by complementation of split enhanced green fluorescent protein. J. Virol. 81, 11532–11537.
Avitabile, E., Forghieri, C., Campadelli-Fiume, G., 2009. Cross talk among the glycoproteins involved in herpes simplex virus entry and fusion: the interaction between gB and gH/gL does not necessarily require gD. J. Virol. 83, 10752–10760.
Backovic, M., Jardetzky, T.S., 2009. Class III viral membrane fusion proteins. Curr. Opin. Struct. Biol. 19, 189–196.
Backovic, M., Jardetzky, T.S., 2011. Class III viral membrane fusion proteins. Adv. Exp. Med. Biol. 714, 91–101.
Backovic, M., Jardetzky, T.S., Longnecker, R., 2007. Hydrophobic residues that form putative fusion loops of Epstein-Barr virus glycoprotein B are critical for fusion activity. J. Virol. 81, 9596–9600.
Backovic, M., Longnecker, R., Jardetzky, T.S., 2009. Structure of a trimeric variant of the Epstein-Barr virus glycoprotein B. Proc. Natl. Acad. Sci. U.S.A. 106, 2880–2885.
Backovic, M., Dubois, R.M., Cockburn, J.J., Sharff, A.J., Vaney, M.C., Granzow, H., Klupp, B.G., Bricogne, G., Mettenleiter, T.C., Rey, F.A., 2010. Structure of a core fragment of glycoprotein H from pseudorabies virus in complex with antibody. Proc. Natl. Acad. Sci. U.S.A. 107, 22635–22640.
Baquero, E., Albertini, A.A., Gaudin, Y., 2015. Recent mechanistic and structural insights on class III viral fusion glycoproteins. Curr. Opin. Struct. Biol. 33, 52–60.
Bender, F.C., Whitbeck, J.C., Ponce De Leon, M., Lou, H., Eisenberg, R.J., Cohen, G.H., 2003. Specific association of glycoprotein B with lipid rafts during herpes simplex virus entry. J. Virol. 77, 9542–9552.
Bender, F.C., Samanta, M., Heldwein, E.E., De Leon, M.P., Bilman, E., Lou, H., Whitbeck, J.C., Eisenberg, R.J., Cohen, G.H., 2007. Antigenic and mutational analyses of herpes simplex virus glycoprotein B reveal four functional regions. J. Virol. 81, 3827–3841.
Bhargava, A.K., Rothlauf, P.W., Krummenacher, C., 2016. Herpes simplex virus glycoprotein D relocates nectin-1 from intercellular contacts. Virology 499, 267–277.
Blomberg, N., Baraldi, E., Nilges, M., Saraste, M., 1999. The PH superfold: a structural scaffold for multiple functions. Trends Biochem. Sci. 24, 441–445.
Böhm, S.W., Eckroth, E., Backovic, M., Klupp, B.G., Rey, F.A., Mettenleiter, T.C., Fuchs, W., 2015. Structure-based functional analyses of domains II and III of pseudorabies virus glycoprotein H. J. Virol. 89, 1364–1376.
Böhm, S.W., Backovic, M., Klupp, B.G., Rey, F.A., Mettenleiter, T.C., Fuchs, W., 2016. Functional characterization of glycoprotein H chimeras composed of conserved domains of the pseudorabies virus and herpes simplex virus 1 homologs. J. Virol. 90, 421–432.
Browne, H.M., Bruun, B.C., Minson, A.C., 1996. Characterization of herpes simplex virus type 1 recombinants with mutations in the cytoplasmic tail of glycoprotein H. J. Gen. Virol. 77 (Pt 10), 2569–2573.
Buckmaster, E.A., Gompels, U., Minson, A., 1984. Characterisation and physical mapping of an HSV-1 glycoprotein of approximately 115 X 10(3) molecular weight. Virology 139, 408–413.
Burke, H.G., Heldwein, E.E., 2015. Crystal structure of the human cytomegalovirus glycoprotein B. PLoS Pathog. 11, e1005227.
Butcher, M., Raviprakash, K., Ghosh, H.P., 1990. Acid pH-induced fusion of cells by herpes simplex virus glycoproteins gB an gD. J. Biol. Chem. 265, 5862–5868.

Bzik, D.J., Fox, B.A., Deluca, N.A., Person, S., 1984. Nucleotide sequence of a region of the herpes simplex virus type 1 gB glycoprotein gene: mutations affecting rate of virus entry and cell fusion. Virology 137, 185–190.

Cai, W.H., Gu, B., Person, S., 1988. Role of glycoprotein B of herpes simplex virus type 1 in viral entry and cell fusion. J. Virol. 62, 2596–2604.

Cairns, T.M., Milne, R.S., Ponce-De-Leon, M., Tobin, D.K., Cohen, G.H., Eisenberg, R.J., 2003. Structure-function analysis of herpes simplex virus type 1 gD and gH-gL: clues from gDgH chimeras. J. Virol. 77, 6731–6742.

Cairns, T.M., Landsburg, D.J., Whitbeck, J.C., Eisenberg, R.J., Cohen, G.H., 2005. Contribution of cysteine residues to the structure and function of herpes simplex virus gH/gL. Virology 332, 550–562.

Cairns, T.M., Friedman, L.S., Lou, H., Whitbeck, J.C., Shaner, M.S., Cohen, G.H., Eisenberg, R.J., 2007. N-terminal mutants of herpes simplex virus type 2 gH are transported without gL but require gL for function. J. Virol. 81, 5102–5111.

Cairns, T.M., Whitbeck, J.C., Lou, H., Heldwein, E.E., Chowdary, T.K., Eisenberg, R.J., Cohen, G.H., 2011. Capturing the herpes simplex virus core fusion complex (gB-gH/gL) in an acidic environment. J. Virol. 85, 6175–6184.

Cairns, T.M., Fontana, J., Huang, Z.Y., Whitbeck, J.C., Atanasiu, D., Rao, S., Shelly, S.S., Lou, H., Ponce De Leon, M., Steven, A.C., Eisenberg, R.J., Cohen, G.H., 2014. Mechanism of neutralization of herpes simplex virus by antibodies directed at the fusion domain of glycoprotein B. J. Virol. 88, 2677–2689.

Campadelli-Fiume, G., Menotti, L., 2007. Entry of alphaherpesviruses into the cell. In: Arvin, A., Campadelli-Fiume, G., Mocarski, E., Moore, P.S., Roizman, B., Whitley, R., Yamanishi, K. (Eds.), Human Herpesviruses: Biology, Therapy, and Immunoprophylaxis. Cambridge University Press, Cambridge.

Campadelli-Fiume, G., Menotti, L., Avitabile, E., Gianni, T., 2012. Viral and cellular contributions to herpes simplex virus entry into the cell. Curr. Opin. Virol. 2, 28–36.

Carfi, A., Willis, S.H., Whitbeck, J.C., Krummenacher, C., Cohen, G.H., Eisenberg, R.J., Wiley, D.C., 2001. Herpes simplex virus glycoprotein D bound to the human receptor HveA. Mol. Cell 8, 169–179.

Chandramouli, S., Ciferri, C., Nikitin, P.A., Calo, S., Gerrein, R., Balabanis, K., Monroe, J., Hebner, C., Lilja, A.E., Settembre, E.C., Carfi, A., 2015. Structure of HCMV glycoprotein B in the postfusion conformation bound to a neutralizing human antibody. Nat. Commun. 6, 8176.

Chen, J., Jardetzky, T.S., Longnecker, R., 2016. The cytoplasmic tail domain of Epstein-Barr virus gH regulates membrane fusion activity through altering gH binding to gp42 and epithelial cell attachment. MBio 7, e01871-16.

Chernomordik, L.V., Kozlov, M.M., 2003. Protein-lipid interplay in fusion and fission of biological membranes. Annu. Rev. Biochem. 72, 175–207.

Chernomordik, L.V., Kozlov, M.M., 2008. Mechanics of membrane fusion. Nat. Struct. Mol. Biol. 15, 675–683.

Chernomordik, L.V., Melikyan, G.B., Chizmadzhev, Y.A., 1987. Biomembrane fusion: a new concept derived from model studies using two interacting planar lipid bilayers. Biochim. Biophys. Acta 906, 309–352.

Cheshenko, N., Pierce, C., Herold, B.C., 2018. Herpes simplex viruses activate phospholipid scramblase to redistribute phosphatidylserines and Akt to the outer leaflet of the plasma membrane and promote viral entry. PLoS Pathog. 14, e1006766.

Chi, X.J., Lu, Y.X., Zhao, P., Li, C.G., Wang, X.J., Wang, M., 2013. Interaction domain of glycoproteins gB and gH of Marek's disease virus and identification of an antiviral peptide with dual functions. PLoS One 8, e54761.

Ch'ng, T.H., Spear, P.G., Struyf, F., Enquist, L.W., 2007. Glycoprotein D-independent spread of pseudorabies virus infection in cultured peripheral nervous system neurons in a compartmented system. J. Virol. 81, 10742–10757.

Chowdary, T.K., Heldwein, E.E., 2010. Syncytial phenotype of C-terminally truncated herpes simplex virus type 1 gB is associated with diminished membrane interactions. J. Virol. 84, 4923–4935.
Chowdary, T.K., Cairns, T.M., Atanasiu, D., Cohen, G.H., Eisenberg, R.J., Heldwein, E.E., 2010. Crystal structure of the conserved herpesvirus fusion regulator complex gH-gL. Nat. Struct. Mol. Biol. 17, 882–888.
Cocchi, F., Menotti, L., Dubreuil, P., Lopez, M., Campadelli-Fiume, G., 2000. Cell-to-cell spread of wild-type herpes simplex virus type 1, but not of syncytial strains, is mediated by the immunoglobulin-like receptors that mediate virion entry, nectin1 (PRR1/HveC/HigR) and nectin2 (PRR2/HveB). J. Virol. 74, 3909–3917.
Cocchi, F., Fusco, D., Menotti, L., Gianni, T., Eisenberg, R.J., Cohen, G.H., Campadelli-Fiume, G., 2004. The soluble ectodomain of herpes simplex virus gD contains a membrane-proximal pro-fusion domain and suffices to mediate virus entry. Proc. Natl. Acad. Sci. U.S.A. 101, 7445–7450.
Cole, N.L., Grose, C., 2003. Membrane fusion mediated by herpesvirus glycoproteins: the paradigm of varicella-zoster virus. Rev. Med. Virol. 13, 207–222.
Connolly, S.A., Longnecker, R., 2012. Residues within the C-terminal arm of the herpes simplex virus 1 glycoprotein B ectodomain contribute to its refolding during the fusion step of virus entry. J. Virol. 86, 6386–6393.
Connolly, S.A., Whitbeck, J.J., Rux, A.H., Krummenacher, C., van Drunen Littel-Van Den Hurk, S., Cohen, G.H., Eisenberg, R.J., 2001. Glycoprotein D homologs in herpes simplex virus type 1, pseudorabies virus, and bovine herpes virus type 1 bind directly to human HveC(nectin-1) with different affinities. Virology 280, 7–18.
Connolly, S.A., Jackson, J.O., Jardetzky, T.S., Longnecker, R., 2011. Fusing structure and function: a structural view of the herpesvirus entry machinery. Nat. Rev. Microbiol. 9, 369–381.
Cooper, R.S., Heldwein, E.E., 2015. Herpesvirus gB: a finely tuned fusion machine. Viruses 7, 6552–6569.
Cooper, R.S., Georgieva, E.R., Borbat, P.P., Freed, J.H., Heldwein, E.E., 2018. Structural basis for membrane anchoring and fusion regulation of the herpes simplex virus fusogen gB. Nat. Struct. Mol. Biol. 25, 416–424.
Davison, A.J., 2010. Herpesvirus systematics. Vet. Microbiol. 143, 52–69.
Davison, A.J., Scott, J.E., 1986. The complete DNA sequence of varicella-zoster virus. J. Gen. Virol. 67 (Pt 9), 1759–1816.
Davison, A.J., Eberle, R., Ehlers, B., Hayward, G.S., McGeoch, D.J., Minson, A.C., Pellett, P.E., Roizman, B., Studdert, M.J., Thiry, E., 2009. The order Herpesvirales. Arch. Virol. 154, 171–177.
De Jong, M.A., De Witte, L., Bolmstedt, A., Van Kooyk, Y., Geijtenbeek, T.B., 2008. Dendritic cells mediate herpes simplex virus infection and transmission through the C-type lectin DC-SIGN. J. Gen. Virol. 89, 2398–2409.
De Pelsmaeker, S., Dierick, E., Klupp, B., Mettenleiter, T.C., Cantoni, C., Vitale, M., Favoreel, H.W., 2019. Expression of the pseudorabies virus gB glycoprotein triggers NK cell cytotoxicity and increases binding of the activating NK cell receptor PILRbeta. J. Virol. 93, e02107–e02118.
Delano, W., 2002. The PyMOL Molecular Graphics System. DeLano Scientific, San Carlos, CA, USA.
Di Giovine, P., Settembre, E.C., Bhargava, A.K., Luftig, M.A., Lou, H., Cohen, G.H., Eisenberg, R.J., Krummenacher, C., Carfi, A., 2011. Structure of herpes simplex virus glycoprotein D bound to the human receptor nectin-1. PLoS Pathog. 7, e1002277.
Diakidi-Kosta, A., Michailidou, G., Kontogounis, G., Sivropoulou, A., Arsenakis, M., 2003. A single amino acid substitution in the cytoplasmic tail of the glycoprotein B of herpes simplex virus 1 affects both syncytium formation and binding to intracellular heparan sulfate. Virus Res. 93, 99–108.

Dingwell, K.S., Brunetti, C.R., Hendricks, R.L., Tang, Q., Tang, M., Rainbow, A.J., Johnson, D.C., 1994. Herpes simplex virus glycoproteins E and I facilitate cell-to-cell spread in vivo and across junctions of cultured cells. J. Virol. 68, 834–845.

Dingwell, K.S., Doering, L.C., Johnson, D.C., 1995. Glycoproteins E and I facilitate neuron-to-neuron spread of herpes simplex virus. J. Virol. 69, 7087–7098.

Dollery, S.J., Delboy, M.G., Nicola, A.V., 2010. Low pH-induced conformational change in herpes simplex virus glycoprotein B. J. Virol. 84, 3759–3766.

Duus, K.M., Grose, C., 1996. Multiple regulatory effects of varicella-zoster virus (VZV) gL on trafficking patterns and fusogenic properties of VZV gH. J. Virol. 70, 8961–8971.

Eisenberg, R.J., Atanasiu, D., Cairns, T.M., Gallagher, J.R., Krummenacher, C., Cohen, G.H., 2012. Herpes virus fusion and entry: a story with many characters. Viruses 4, 800–832.

Falke, D., Knoblich, A., Muller, S., 1985. Fusion from without induced by herpes simplex virus type 1. Intervirology 24, 211–219.

Fan, Z., Grantham, M.L., Smith, M.S., Anderson, E.S., Cardelli, J.A., Muggeridge, M.I., 2002. Truncation of herpes simplex virus type 2 glycoprotein B increases its cell surface expression and activity in cell-cell fusion, but these properties are unrelated. J. Virol. 76, 9271–9283.

Fan, Q., Longnecker, R., Connolly, S.A., 2014. Substitution of herpes simplex virus 1 entry glycoproteins with those of saimiriine herpesvirus 1 reveals a gD-gH/gL functional interaction and a region within the gD profusion domain that is critical for fusion. J. Virol. 88, 6470–6482.

Fan, Q., Longnecker, R., Connolly, S.A., 2015. A functional interaction between herpes simplex virus 1 glycoprotein gH/gL domains I and II and gD is defined by using Alphaherpesvirus gH and gL chimeras. J. Virol. 89, 7159–7169.

Fan, Q., Kopp, S.J., Connolly, S.A., Longnecker, R., 2017. Structure-based mutations in the herpes simplex virus 1 glycoprotein B ectodomain arm impart a slow-entry phenotype. MBio 8, e00614–e00617.

Flynn, S.J., Ryan, P., 1996. The receptor-binding domain of pseudorabies virus glycoprotein gC is composed of multiple discrete units that are functionally redundant. J. Virol. 70, 1355–1364.

Fontana, J., Atanasiu, D., Saw, W.T., Gallagher, J.R., Cox, R.G., Whitbeck, J.C., Brown, L.M., Eisenberg, R.J., Cohen, G.H., 2017. The fusion loops of the initial prefusion conformation of herpes simplex virus 1 fusion protein point toward the membrane. MBio. 8, e01268-17.

Foster, T.P., Melancon, J.M., Kousoulas, K.G., 2001. An alpha-helical domain within the carboxyl terminus of herpes simplex virus type 1 (HSV-1) glycoprotein B (gB) is associated with cell fusion and resistance to heparin inhibition of cell fusion. Virology 287, 18–29.

Francois-Martin, C., Rothman, J.E., Pincet, F., 2017. Low energy cost for optimal speed and control of membrane fusion. Proc. Natl. Acad. Sci. U.S.A. 114, 1238–1241.

Fuchs, W., Backovic, M., Klupp, B.G., Rey, F.A., Mettenleiter, T.C., 2012. Structure-based mutational analysis of the highly conserved domain IV of glycoprotein H of pseudorabies virus. J. Virol. 86, 8002–8013.

Fusco, D., Forghieri, C., Campadelli-Fiume, G., 2005. The pro-fusion domain of herpes simplex virus glycoprotein D (gD) interacts with the gD N terminus and is displaced by soluble forms of viral receptors. Proc. Natl. Acad. Sci. U.S.A. 102, 9323–9328.

Gage, P.J., Levine, M., Glorioso, J.C., 1993. Syncytium-inducing mutations localize to two discrete regions within the cytoplasmic domain of herpes simplex virus type 1 glycoprotein B. J. Virol. 67, 2191–2201.

Galdiero, M., Whiteley, A., Bruun, B., Bell, S., Minson, T., Browne, H., 1997. Site-directed and linker insertion mutagenesis of herpes simplex virus type 1 glycoprotein H. J. Virol. 71, 2163–2170.

Galdiero, S., Falanga, A., Vitiello, M., D'isanto, M., Collins, C., Orrei, V., Browne, H., Pedone, C., Galdiero, M., 2007. Evidence for a role of the membrane-proximal region of herpes simplex virus type 1 glycoprotein H in membrane fusion and virus inhibition. Chembiochem 8, 885–895.

Galdiero, S., Falanga, A., Vitiello, M., D'isanto, M., Cantisani, M., Kampanaraki, A., Benedetti, E., Browne, H., Galdiero, M., 2008. Peptides containing membrane-interacting motifs inhibit herpes simplex virus type 1 infectivity. Peptides 29, 1461–1471.

Gallagher, J.R., Saw, W.T., Atanasiu, D., Lou, H., Eisenberg, R.J., Cohen, G.H., 2013. Displacement of the C terminus of herpes simplex virus gD is sufficient to expose the fusion-activating interfaces on gD. J. Virol. 87, 12656–12666.

Gallagher, J.R., Atanasiu, D., Saw, W.T., Paradisgarten, M.J., Whitbeck, J.C., Eisenberg, R.J., Cohen, G.H., 2014. Functional fluorescent protein insertions in herpes simplex virus gB report on gB conformation before and after execution of membrane fusion. PLoS Pathog. 10, e1004373.

Gatta, V., Petrovic, B., Campadelli-Fiume, G., 2015. The engineering of a novel ligand in gH confers to HSV an expanded tropism independent of gD activation by its receptors. PLoS Pathog. 11, e1004907.

Geraghty, R.J., Krummenacher, C., Cohen, G.H., Eisenberg, R.J., Spear, P.G., 1998. Entry of alphaherpesviruses mediated by poliovirus receptor-related protein 1 and poliovirus receptor. Science 280, 1618–1620.

Gianni, T., Martelli, P.L., Casadio, R., Campadelli-Fiume, G., 2005a. The ectodomain of herpes simplex virus glycoprotein H contains a membrane alpha-helix with attributes of an internal fusion peptide, positionally conserved in the herpesviridae family. J. Virol. 79, 2931–2940.

Gianni, T., Menotti, L., Campadelli-Fiume, G., 2005b. A heptad repeat in herpes simplex virus 1 gH, located downstream of the alpha-helix with attributes of a fusion peptide, is critical for virus entry and fusion. J. Virol. 79, 7042–7049.

Gianni, T., Fato, R., Bergamini, C., Lenaz, G., Campadelli-Fiume, G., 2006. Hydrophobic alpha-helices 1 and 2 of herpes simplex virus gH interact with lipids, and their mimetic peptides enhance virus infection and fusion. J. Virol. 80, 8190–8198.

Gianni, T., Amasio, M., Campadelli-Fiume, G., 2009. Herpes simplex virus gD forms distinct complexes with fusion executors gB and gH/gL in part through the C-terminal profusion domain. J. Biol. Chem. 284, 17370–17382.

Gianni, T., Gatta, V., Campadelli-Fiume, G., 2010. {alpha}V{beta}3-integrin routes herpes simplex virus to an entry pathway dependent on cholesterol-rich lipid rafts and dynamin2. Proc. Natl. Acad. Sci. U.S.A. 107, 22260–22265.

Gianni, T., Salvioli, S., Chesnokova, L.S., Hutt-Fletcher, L.M., Campadelli-Fiume, G., 2013. alphavbeta6- and alphavbeta8-integrins serve as interchangeable receptors for HSV gH/gL to promote endocytosis and activation of membrane fusion. PLoS Pathog. 9, e1003806.

Gianni, T., Massaro, R., Campadelli-Fiume, G., 2015. Dissociation of HSV gL from gH by alphavbeta6- or alphavbeta8-integrin promotes gH activation and virus entry. Proc. Natl. Acad. Sci. U.S.A. 112, E3901–E3910.

Gilbert, R., Ghosh, K., Rasile, L., Ghosh, H.P., 1994. Membrane anchoring domain of herpes simplex virus glycoprotein gB is sufficient for nuclear envelope localization. J. Virol. 68, 2272–2285.

Gompels, U.A., Carss, A.L., Saxby, C., Hancock, D.C., Forrester, A., Minson, A.C., 1991. Characterization and sequence analyses of antibody-selected antigenic variants of herpes simplex virus show a conformationally complex epitope on glycoprotein H. J. Virol. 65, 2393–2401.

Granzow, H., Weiland, F., Jons, A., Klupp, B.G., Karger, A., Mettenleiter, T.C., 1997. Ultrastructural analysis of the replication cycle of pseudorabies virus in cell culture: a reassessment. J. Virol. 71, 2072–2082.

Gupta, P., Balachandran, R., Ho, M., Enrico, A., Rinaldo, C., 1989. Cell-to-cell transmission of human immunodeficiency virus type 1 in the presence of azidothymidine and neutralizing antibody. J. Virol. 63, 2361–2365.

Han, J., Pluhackova, K., Bockmann, R.A., 2017. The multifaceted role of SNARE proteins in membrane fusion. Front. Physiol. 8, 5.

Hannah, B.P., Heldwein, E.E., Bender, F.C., Cohen, G.H., Eisenberg, R.J., 2007. Mutational evidence of internal fusion loops in herpes simplex virus glycoprotein B. J. Virol. 81, 4858–4865.

Hannah, B.P., Cairns, T.M., Bender, F.C., Whitbeck, J.C., Lou, H., Eisenberg, R.J., Cohen, G.H., 2009. Herpes simplex virus glycoprotein B associates with target membranes via its fusion loops. J. Virol. 83, 6825–6836.

Harman, A., Browne, H., Minson, T., 2002. The transmembrane domain and cytoplasmic tail of herpes simplex virus type 1 glycoprotein H play a role in membrane fusion. J. Virol. 76, 10708–10716.

Harrison, S.C., 2015. Viral membrane fusion. Virology 479-480, 498–507.

Harrison, O.J., Vendome, J., Brasch, J., Jin, X., Hong, S., Katsamba, P.S., Ahlsen, G., Troyanovsky, R.B., Troyanovsky, S.M., Honig, B., Shapiro, L., 2012. Nectin ectodomain structures reveal a canonical adhesive interface. Nat. Struct. Mol. Biol. 19, 906–915.

Heffner, S., Kovacs, F., Klupp, B.G., Mettenleiter, T.C., 1993. Glycoprotein gp50-negative pseudorabies virus: a novel approach toward a nonspreading live herpesvirus vaccine. J. Virol. 67, 1529–1537.

Heldwein, E.E., Lou, H., Bender, F.C., Cohen, G.H., Eisenberg, R.J., Harrison, S.C., 2006. Crystal structure of glycoprotein B from herpes simplex virus 1. Science 313, 217–220.

Herold, B.C., Wudunn, D., Soltys, N., Spear, P.G., 1991. Glycoprotein C of herpes simplex virus type 1 plays a principal role in the adsorption of virus to cells and in infectivity. J. Virol. 65, 1090–1098.

Herold, B.C., Visalli, R.J., Susmarski, N., Brandt, C.R., Spear, P.G., 1994. Glycoprotein C-independent binding of herpes simplex virus to cells requires cell surface heparan sulphate and glycoprotein B. J. Gen. Virol. 75 (Pt 6), 1211–1222.

Hosaka, M., Nagahama, M., Kim, W.S., Watanabe, T., Hatsuzawa, K., Ikemizu, J., Murakami, K., Nakayama, K., 1991. Arg-X-Lys/Arg-Arg motif as a signal for precursor cleavage catalyzed by furin within the constitutive secretory pathway. J. Biol. Chem. 266, 12127–12130.

Hutchinson, L., Browne, H., Wargent, V., Davis-Poynter, N., Primorac, S., Goldsmith, K., Minson, A.C., Johnson, D.C., 1992. A novel herpes simplex virus glycoprotein, gL, forms a complex with glycoprotein H (gH) and affects normal folding and surface expression of gH. J. Virol. 66, 2240–2250.

Jones, N.A., Geraghty, R.J., 2004. Fusion activity of lipid-anchored envelope glycoproteins of herpes simplex virus type 1. Virology 324, 213–228.

Kadlec, J., Loureiro, S., Abrescia, N.G., Stuart, D.I., Jones, I.M., 2008. The postfusion structure of baculovirus gp64 supports a unified view of viral fusion machines. Nat. Struct. Mol. Biol. 15, 1024–1030.

Karger, A., Mettenleiter, T.C., 1993. Glycoproteins gIII and gp50 play dominant roles in the biphasic attachment of pseudorabies virus. Virology 194, 654–664.

Karger, A., Schmidt, J., Mettenleiter, T.C., 1998. Infectivity of a pseudorabies virus mutant lacking attachment glycoproteins C and D. J. Virol. 72, 7341–7348.

Kaye, J.F., Gompels, U.A., Minson, A.C., 1992. Glycoprotein H of human cytomegalovirus (HCMV) forms a stable complex with the HCMV UL115 gene product. J. Gen. Virol. 73 (Pt 10), 2693–2698.

Kielian, M., 2014. Mechanisms of virus membrane fusion proteins. Annu. Rev. Virol. 1, 171–189.

Kielian, M., Rey, F.A., 2006. Virus membrane-fusion proteins: more than one way to make a hairpin. Nat. Rev. Microbiol. 4, 67–76.

Kim, Y.H., Donald, J.E., Grigoryan, G., Leser, G.P., Fadeev, A.Y., Lamb, R.A., Degrado, W.F., 2011. Capture and imaging of a prehairpin fusion intermediate of the paramyxovirus PIV5. Proc. Natl. Acad. Sci. U.S.A. 108, 20992–20997.

Kinzler, E.R., Compton, T., 2005. Characterization of human cytomegalovirus glycoprotein-induced cell-cell fusion. J. Virol. 79, 7827–7837.

Klupp, B.G., Mettenleiter, T.C., 1991. Sequence and expression of the glycoprotein gH gene of pseudorabies virus. Virology 182, 732–741.

Klupp, B.G., Mettenleiter, T.C., 1999. Glycoprotein gL-independent infectivity of pseudorabies virus is mediated by a gD-gH fusion protein. J. Virol. 73, 3014–3022.

Klupp, B.G., Visser, N., Mettenleiter, T.C., 1992. Identification and characterization of pseudorabies virus glycoprotein H. J. Virol. 66, 3048–3055.

Klupp, B.G., Baumeister, J., Karger, A., Visser, N., Mettenleiter, T.C., 1994. Identification and characterization of a novel structural glycoprotein in pseudorabies virus, gL. J. Virol. 68, 3868–3878.

Klupp, B.G., Fuchs, W., Weiland, E., Mettenleiter, T.C., 1997a. Pseudorabies virus glycoprotein L is necessary for virus infectivity but dispensable for virion localization of glycoprotein H. J. Virol. 71, 7687–7695.

Klupp, B.G., Karger, A., Mettenleiter, T.C., 1997b. Bovine herpesvirus 1 glycoprotein B does not productively interact with cell surface heparan sulfate in a pseudorabies virion background. J. Virol. 71, 4838–4841.

Klupp, B.G., Nixdorf, R., Mettenleiter, T.C., 2000. Pseudorabies virus glycoprotein M inhibits membrane fusion. J. Virol. 74, 6760–6768.

Klupp, B., Altenschmidt, J., Granzow, H., Fuchs, W., Mettenleiter, T.C., 2008. Glycoproteins required for entry are not necessary for egress of pseudorabies virus. J. Virol. 82, 6299–6309.

Kopp, A., Blewett, E., Misra, V., Mettenleiter, T.C., 1994. Proteolytic cleavage of bovine herpesvirus 1 (BHV-1) glycoprotein gB is not necessary for its function in BHV-1 or pseudorabies virus. J. Virol. 68, 1667–1674.

Kozlovsky, Y., Chernomordik, L.V., Kozlov, M.M., 2002. Lipid intermediates in membrane fusion: formation, structure, and decay of hemifusion diaphragm. Biophys. J. 83, 2634–2651.

Kramer, T., Enquist, L.W., 2013. Directional spread of alphaherpesviruses in the nervous system. Viruses 5, 678–707.

Krarup, A., Truan, D., Furmanova-Hollenstein, P., Bogaert, L., Bouchier, P., Bisschop, I.J., Widjojoatmodjo, M.N., Zahn, R., Schuitemaker, H., McLellan, J.S., Langedijk, J.P., 2015. A highly stable prefusion RSV F vaccine derived from structural analysis of the fusion mechanism. Nat. Commun. 6, 8143.

Krummenacher, C., Supekar, V.M., Whitbeck, J.C., Lazear, E., Connolly, S.A., Eisenberg, R.J., Cohen, G.H., Wiley, D.C., Carfi, A., 2005. Structure of unliganded HSV gD reveals a mechanism for receptor-mediated activation of virus entry. EMBO J. 24, 4144–4153.

Krummenacher, C., Carfi, A., Eisenberg, R.J., Cohen, G.H., 2013. Entry of herpesviruses into cells: the enigma variations. Adv. Exp. Med. Biol. 790, 178–195.

Kwon, H., Bai, Q., Baek, H.J., Felmet, K., Burton, E.A., Goins, W.F., Cohen, J.B., Glorioso, J.C., 2006. Soluble V domain of Nectin-1/HveC enables entry of herpes simplex virus type 1 (HSV-1) into HSV-resistant cells by binding to viral glycoprotein D. J. Virol. 80, 138–148.

Laquerre, S., Argnani, R., Anderson, D.B., Zucchini, S., Manservigi, R., Glorioso, J.C., 1998. Heparan sulfate proteoglycan binding by herpes simplex virus type 1 glycoproteins B and C, which differ in their contributions to virus attachment, penetration, and cell-to-cell spread. J. Virol. 72, 6119–6130.

Lazear, E., Carfi, A., Whitbeck, J.C., Cairns, T.M., Krummenacher, C., Cohen, G.H., Eisenberg, R.J., 2008. Engineered disulfide bonds in herpes simplex virus type 1 gD separate receptor binding from fusion initiation and viral entry. J. Virol. 82, 700–709.
Lazear, E., Whitbeck, J.C., Ponce-De-Leon, M., Cairns, T.M., Willis, S.H., Zuo, Y., Krummenacher, C., Cohen, G.H., Eisenberg, R.J., 2012. Antibody-induced conformational changes in herpes simplex virus glycoprotein gD reveal new targets for virus neutralization. J. Virol. 86, 1563–1576.
Lazear, E., Whitbeck, J.C., Zuo, Y., Carfi, A., Cohen, G.H., Eisenberg, R.J., Krummenacher, C., 2014. Induction of conformational changes at the N-terminus of herpes simplex virus glycoprotein D upon binding to HVEM and nectin-1. Virology 448, 185–195.
Lee, J.E., Saphire, E.O., 2009. Ebolavirus glycoprotein structure and mechanism of entry. Futur. Virol. 4, 621–635.
Lee, J.E., Fusco, M.L., Hessell, A.J., Oswald, W.B., Burton, D.R., Saphire, E.O., 2008. Structure of the Ebola virus glycoprotein bound to an antibody from a human survivor. Nature 454, 177–182.
Leikin, S., Parsegian, V.A., Rau, D.C., Rand, R.P., 1993. Hydration forces. Annu. Rev. Phys. Chem. 44, 369–395.
Lemmon, M.A., 2007. Pleckstrin homology (PH) domains and phosphoinositides. Biochem. Soc. Symp. 74, 81–93.
Lete, C., Machiels, B., Stevenson, P.G., Vanderplasschen, A., Gillet, L., 2012. Bovine herpesvirus type 4 glycoprotein L is nonessential for infectivity but triggers virion endocytosis during entry. J. Virol. 86, 2653–2664.
Li, Y., Modis, Y., 2014. A novel membrane fusion protein family in Flaviviridae? Trends Microbiol. 22, 176–182.
Li, Q., Krogmann, T., Ali, M.A., Tang, W.J., Cohen, J.I., 2007. The amino terminus of varicella-zoster virus (VZV) glycoprotein E is required for binding to insulin-degrading enzyme, a VZV receptor. J. Virol. 81, 8525–8532.
Li, Q., Ali, M.A., Wang, K., Sayre, D., Hamel, F.G., Fischer, E.R., Bennett, R.G., Cohen, J.I., 2010. Insulin degrading enzyme induces a conformational change in varicella-zoster virus gE, and enhances virus infectivity and stability. PLoS One 5, e11327.
Li, A., Lu, G., Qi, J., Wu, L., Tian, K., Luo, T., Shi, Y., Yan, J., Gao, G.F., 2017a. Structural basis of nectin-1 recognition by pseudorabies virus glycoprotein D. PLoS Pathog. 13, e1006314.
Li, X., Yang, F., HU, X., Tan, F., Qi, J., Peng, R., Wang, M., Chai, Y., Hao, L., Deng, J., Bai, C., Wang, J., Song, H., Tan, S., Lu, G., Gao, G.F., Shi, Y., Tian, K., 2017b. Two classes of protective antibodies against pseudorabies virus variant glycoprotein B: implications for vaccine design. PLoS Pathog. 13, e1006777.
Lin, E., Spear, P.G., 2007. Random linker-insertion mutagenesis to identify functional domains of herpes simplex virus type 1 glycoprotein B. Proc. Natl. Acad. Sci. U.S.A. 104, 13140–13145.
Liu, D.X., Gompels, U.A., Nicholas, J., Lelliott, C., 1993. Identification and expression of the human herpesvirus 6 glycoprotein H and interaction with an accessory 40K glycoprotein. J. Gen. Virol. 74 (Pt 9), 1847–1857.
Lu, G., Zhang, N., Qi, J., Li, Y., Chen, Z., Zheng, C., Gao, G.F., Yan, J., 2014. Crystal structure of herpes simplex virus 2 gD bound to nectin-1 reveals a conserved mode of receptor recognition. J. Virol. 88, 13678–13688.
Maresova, L., Pasieka, T.J., Grose, C., 2001. Varicella-zoster virus gB and gE coexpression, but not gB or gE alone, leads to abundant fusion and syncytium formation equivalent to those from gH and gL coexpression. J. Virol. 75, 9483–9492.
Matsuura, H., Kirschner, A.N., Longnecker, R., Jardetzky, T.S., 2010. Crystal structure of the Epstein-Barr virus (EBV) glycoprotein H/glycoprotein L (gH/gL) complex. Proc. Natl. Acad. Sci. U.S.A. 107, 22641–22646.

Maurer, U.E., Zeev-Ben-Mordehai, T., Pandurangan, A.P., Cairns, T.M., Hannah, B.P., Whitbeck, J.C., Eisenberg, R.J., Cohen, G.H., Topf, M., Huiskonen, J.T., Grunewald, K., 2013. The structure of herpesvirus fusion glycoprotein B-bilayer complex reveals the protein-membrane and lateral protein-protein interaction. Structure 21, 1396–1405.
McClain, D.S., Fuller, A.O., 1994. Cell-specific kinetics and efficiency of herpes simplex virus type 1 entry are determined by two distinct phases of attachment. Virology 198, 690–702.
McGeoch, D.J., Cook, S., Dolan, A., Jamieson, F.E., Telford, E.A., 1995. Molecular phylogeny and evolutionary timescale for the family of mammalian herpesviruses. J. Mol. Biol. 247, 443–458.
McLellan, J.S., 2015. Neutralizing epitopes on the respiratory syncytial virus fusion glycoprotein. Curr. Opin. Virol. 11, 70–75.
Mclellan, J.S., Yang, Y., Graham, B.S., Kwong, P.D., 2011. Structure of respiratory syncytial virus fusion glycoprotein in the postfusion conformation reveals preservation of neutralizing epitopes. J. Virol. 85, 7788–7796.
McLellan, J.S., Chen, M., Leung, S., Graepel, K.W., Du, X., Yang, Y., Zhou, T., Baxa, U., Yasuda, E., Beaumont, T., Kumar, A., Modjarrad, K., Zheng, Z., Zhao, M., Xia, N., Kwong, P.D., Graham, B.S., 2013. Structure of RSV fusion glycoprotein trimer bound to a prefusion-specific neutralizing antibody. Science 340, 1113–1117.
McShane, M.P., Longnecker, R., 2005. Analysis of fusion using a virus-free cell fusion assay. Methods Mol. Biol. 292, 187–196.
Mendelsohn, C.L., Wimmer, E., Racaniello, V.R., 1989. Cellular receptor for poliovirus: molecular cloning, nucleotide sequence, and expression of a new member of the immunoglobulin superfamily. Cell 56, 855–865.
Menotti, L., Cerretani, A., Hengel, H., Campadelli-Fiume, G., 2008. Construction of a fully retargeted herpes simplex virus 1 recombinant capable of entering cells solely via human epidermal growth factor receptor 2. J. Virol. 82, 10153–10161.
Meredith, D.M., Stocks, J.M., Whittaker, G.R., Halliburton, I.W., Snowden, B.W., Killington, R.A., 1989. Identification of the gB homologues of equine herpesvirus types 1 and 4 as disulphide-linked heterodimers and their characterization using monoclonal antibodies. J. Gen. Virol. 70 (Pt 5), 1161–1172.
Mettenleiter, T.C., 1989. Glycoprotein gIII deletion mutants of pseudorabies virus are impaired in virus entry. Virology 171, 623–625.
Mettenleiter, T.C., 1994. Pseudorabies (Aujeszky's disease) virus: state of the art. August 1993. Acta Vet. Hung. 42, 153–177.
Mettenleiter, T.C., 1996. Immunobiology of pseudorabies (Aujeszky's disease). Vet. Immunol. Immunopathol. 54, 221–229.
Mettenleiter, T.C., 2000. Aujeszky's disease (pseudorabies) virus: the virus and molecular pathogenesis–state of the art, June 1999. Vet. Res. 31, 99–115.
Mettenleiter, T.C., 2002. Herpesvirus assembly and egress. J. Virol. 76, 1537–1547.
Mettenleiter, T.C., 2008. Pseudorabies virus. In: Mahy, B.W.J., Van Regenmortel, M.H.V. (Eds.), Encyclopedia of Virology. Elsevier, Oxford.
Mettenleiter, T.C., Spear, P.G., 1994. Glycoprotein gB (gII) of pseudorabies virus can functionally substitute for glycoprotein gB in herpes simplex virus type 1. J. Virol. 68, 500–504.
Mettenleiter, T.C., Zsak, L., Zuckermann, F., Sugg, N., Kern, H., Ben-Porat, T., 1990. Interaction of glycoprotein gIII with a cellular heparinlike substance mediates adsorption of pseudorabies virus. J. Virol. 64, 278–286.
Mettenleiter, T.C., Klupp, B.G., Granzow, H., 2009. Herpesvirus assembly: an update. Virus Res. 143, 222–234.
Miethke, A., Keil, G.M., Weiland, F., Mettenleiter, T.C., 1995. Unidirectional complementation between glycoprotein B homologues of pseudorabies virus and bovine

herpesvirus 1 is determined by the carboxy-terminal part of the molecule. J. Gen. Virol. 76 (Pt 7), 1623–1635.

Miller, J.L., Weed, D.J., Lee, B.H., Pritchard, S.M., Nicola, A.V., 2019. Low-pH endocytic entry of the porcine alphaherpesvirus pseudorabies virus. J. Virol. 93, e01849-18.

Milne, R.S., Connolly, S.A., Krummenacher, C., Eisenberg, R.J., Cohen, G.H., 2001. Porcine HveC, a member of the highly conserved HveC/nectin 1 family, is a functional alphaherpesvirus receptor. Virology 281, 315–328.

Milne, R.S., Nicola, A.V., Whitbeck, J.C., Eisenberg, R.J., Cohen, G.H., 2005. Glycoprotein D receptor-dependent, low-pH-independent endocytic entry of herpes simplex virus type 1. J. Virol. 79, 6655–6663.

Miranda-Saksena, M., Denes, C.E., Diefenbach, R.J., Cunningham, A.L., 2018. Infection and transport of herpes simplex virus type 1 in neurons: role of the cytoskeleton. Viruses 10, E92.

Möhl, B.S., Schröter, C., Klupp, B.G., Fuchs, W., Mettenleiter, T.C., Jardetzky, T.S., Longnecker, R., 2015. Comparative mutagenesis of pseudorabies virus and Epstein-Barr virus gH identifies a structural determinant within domain III of gH required for surface expression and entry function. J. Virol. 90, 2285–2293.

Montgomery, R.I., Warner, M.S., Lum, B.J., Spear, P.G., 1996. Herpes simplex virus-1 entry into cells mediated by a novel member of the TNF/NGF receptor family. Cell 87, 427–436.

Mothes, W., Sherer, N.M., Jin, J., Zhong, P., 2010. Virus cell-to-cell transmission. J. Virol. 84, 8360–8368.

Muggeridge, M.I., 2000. Characterization of cell-cell fusion mediated by herpes simplex virus 2 glycoproteins gB, gD, gH and gL in transfected cells. J. Gen. Virol. 81, 2017–2027.

Mulder, W., Pol, J., Kimman, T., Kok, G., Priem, J., Peeters, B., 1996. Glycoprotein D-negative pseudorabies virus can spread transneuronally via direct neuron-to-neuron transmission in its natural host, the pig, but not after additional inactivation of gE or gI. J. Virol. 70, 2191–2200.

Narita, H., Yamamoto, Y., Suzuki, M., Miyazaki, N., Yoshida, A., Kawai, K., Iwasaki, K., Nakagawa, A., Takai, Y., Sakisaka, T., 2011. Crystal structure of the cis-Dimer of Nectin-1: implications for the architecture of cell-cell junctions. J. Biol. Chem. 286, 12659–12669.

Nicola, A.V., Straus, S.E., 2004. Cellular and viral requirements for rapid endocytic entry of herpes simplex virus. J. Virol. 78, 7508–7517.

Niikura, M., Witter, R.L., Jang, H.K., Ono, M., Mikami, T., Silva, R.F., 1999. MDV glycoprotein D is expressed in the feather follicle epithelium of infected chickens. Acta Virol. 43, 159–163.

Nixdorf, R., Schmidt, J., Karger, A., Mettenleiter, T.C., 1999. Infection of Chinese hamster ovary cells by pseudorabies virus. J. Virol. 73, 8019–8026.

Nixdorf, R., Klupp, B.G., Karger, A., Mettenleiter, T.C., 2000. Effects of truncation of the carboxy terminus of pseudorabies virus glycoprotein B on infectivity. J. Virol. 74, 7137–7145.

Nixdorf, R., Klupp, B.G., Mettenleiter, T.C., 2001. Restoration of function of carboxy-terminally truncated pseudorabies virus glycoprotein B by point mutations in the ectodomain. J. Virol. 75, 11526–11533.

Okazaki, K., 2007. Proteolytic cleavage of glycoprotein B is dispensable for in vitro replication, but required for syncytium formation of pseudorabies virus. J. Gen. Virol. 88, 1859–1865.

Pasieka, T.J., Maresova, L., Grose, C., 2003. A functional YNKI motif in the short cytoplasmic tail of varicella-zoster virus glycoprotein gH mediates clathrin-dependent and antibody-independent endocytosis. J. Virol. 77, 4191–4204.

Peeters, B., De Wind, N., Hooisma, M., Wagenaar, F., Gielkens, A., Moormann, R., 1992. Pseudorabies virus envelope glycoproteins gp50 and gII are essential for virus penetration, but only gII is involved in membrane fusion. J. Virol. 66, 894–905.
Pertel, P.E., Fridberg, A., Parish, M.L., Spear, P.G., 2001. Cell fusion induced by herpes simplex virus glycoproteins gB, gD, and gH-gL requires a gD receptor but not necessarily heparan sulfate. Virology 279, 313–324.
Petrovic, B., Gianni, T., Gatta, V., Campadelli-Fiume, G., 2017. Insertion of a ligand to HER2 in gB retargets HSV tropism and obviates the need for activation of the other entry glycoproteins. PLoS Pathog. 13, e1006352.
Petrovskis, E.A., Timmins, J.G., Armentrout, M.A., Marchioli, C.C., Yancey Jr., R.J., Post, L.E., 1986. DNA sequence of the gene for pseudorabies virus gp50, a glycoprotein without N-linked glycosylation. J. Virol. 59, 216–223.
Pettersen, E.F., Goddard, T.D., Huang, C.C., Couch, G.S., Greenblatt, D.M., Meng, E.C., Ferrin, T.E., 2004. UCSF chimera–a visualization system for exploratory research and analysis. J. Comput. Chem. 25, 1605–1612.
Plonsky, I., Kingsley, D.H., Rashtian, A., Blank, P.S., Zimmerberg, J., 2008. Initial size and dynamics of viral fusion pores are a function of the fusion protein mediating membrane fusion. Biol. Cell 100, 377–386.
Pomeranz, L.E., Reynolds, A.E., Hengartner, C.J., 2005. Molecular biology of pseudorabies virus: impact on neurovirology and veterinary medicine. Microbiol. Mol. Biol. Rev. 69, 462–500.
Rasile, L., Ghosh, K., Raviprakash, K., Ghosh, H.P., 1993. Effects of deletions in the carboxy-terminal hydrophobic region of herpes simplex virus glycoprotein gB on intracellular transport and membrane anchoring. J. Virol. 67, 4856–4866.
Rauh, I., Mettenleiter, T.C., 1991. Pseudorabies virus glycoproteins gII and gp50 are essential for virus penetration. J. Virol. 65, 5348–5356.
Rauh, I., Weiland, F., Fehler, F., Keil, G.M., Mettenleiter, T.C., 1991. Pseudorabies virus mutants lacking the essential glycoprotein gII can be complemented by glycoprotein gI of bovine herpesvirus 1. J. Virol. 65, 621–631.
Rey, F.A., 2006. Molecular gymnastics at the herpesvirus surface. EMBO Rep. 7, 1000–1005.
Robbins, A.K., Whealy, M.E., Watson, R.J., Enquist, L.W., 1986. Pseudorabies virus gene encoding glycoprotein gIII is not essential for growth in tissue culture. J. Virol. 59, 635–645.
Robbins, A.K., Dorney, D.J., Wathen, M.W., Whealy, M.E., Gold, C., Watson, R.J., Holland, L.E., Weed, S.D., Levine, M., Glorioso, J.C., et al., 1987. The pseudorabies virus gII gene is closely related to the gB glycoprotein gene of herpes simplex virus. J. Virol. 61, 2691–2701.
Roche, S., Bressanelli, S., Rey, F.A., Gaudin, Y., 2006. Crystal structure of the low-pH form of the vesicular stomatitis virus glycoprotein G. Science 313, 187–191.
Roche, S., Rey, F.A., Gaudin, Y., Bressanelli, S., 2007. Structure of the prefusion form of the vesicular stomatitis virus glycoprotein G. Science 315, 843–848.
Rogalin, H.B., Heldwein, E.E., 2015. Interplay between the herpes simplex virus 1 gB cytodomain and the gH cytotail during cell-cell fusion. J. Virol. 89, 12262–12272.
Rogalin, H.B., Heldwein, E.E., 2016. Characterization of vesicular stomatitis virus pseudotypes bearing essential entry glycoproteins gB, gD, gH, and gL of herpes simplex virus 1. J. Virol. 90, 10321–10328.
Roller, R.J., Rauch, D., 1998. Herpesvirus entry mediator HVEM mediates cell-cell spread in BHK(TK-) cell clones. J. Virol. 72, 1411–1417.
Roop, C., Hutchinson, L., Johnson, D.C., 1993. A mutant herpes simplex virus type 1 unable to express glycoprotein L cannot enter cells, and its particles lack glycoprotein H. J. Virol. 67, 2285–2297.

Ruel, N., Zago, A., Spear, P.G., 2006. Alanine substitution of conserved residues in the cytoplasmic tail of herpes simplex virus gB can enhance or abolish cell fusion activity and viral entry. Virology 346, 229–237.

Sanders, R.W., Vesanen, M., Schuelke, N., Master, A., Schiffner, L., Kalyanaraman, R., Paluch, M., Berkhout, B., Maddon, P.J., Olson, W.C., Lu, M., Moore, J.P., 2002. Stabilization of the soluble, cleaved, trimeric form of the envelope glycoprotein complex of human immunodeficiency virus type 1. J. Virol. 76, 8875–8889.

Satoh, T., Arase, H., 2008. HSV-1 infection through inhibitory receptor, *PILRalpha*. Uirusu 58, 27–36.

Satoh, T., Arii, J., Suenaga, T., Wang, J., Kogure, A., Uehori, J., Arase, N., Shiratori, I., Tanaka, S., Kawaguchi, Y., Spear, P.G., Lanier, L.L., Arase, H., 2008. PILRalpha is a herpes simplex virus-1 entry coreceptor that associates with glycoprotein B. Cell 132, 935–944.

Sattentau, Q., 2008. Avoiding the void: cell-to-cell spread of human viruses. Nat. Rev. Microbiol. 6, 815–826.

Sawitzky, D., Hampl, H., Habermehl, K.O., 1990. Comparison of heparin-sensitive attachment of pseudorabies virus (PRV) and herpes simplex virus type 1 and identification of heparin-binding PRV glycoproteins. J. Gen. Virol. 71 (Pt 5), 1221–1225.

Schmidt, J., Klupp, B.G., Karger, A., Mettenleiter, T.C., 1997. Adaptability in herpesviruses: glycoprotein D-independent infectivity of pseudorabies virus. J. Virol. 71, 17–24.

Schmidt, J., Gerdts, V., Beyer, J., Klupp, B.G., Mettenleiter, T.C., 2001. Glycoprotein D-independent infectivity of pseudorabies virus results in an alteration of in vivo host range and correlates with mutations in glycoproteins B and H. J. Virol. 75, 10054–10064.

Schröder, C., Keil, G.M., 1999. Bovine herpesvirus 1 requires glycoprotein H for infectivity and direct spreading and glycoproteins gH(W450) and gB for glycoprotein D-independent cell-to-cell spread. J. Gen. Virol. 80 (Pt 1), 57–61.

Schröder, C., Linde, G., Fehler, F., Keil, G.M., 1997. From essential to beneficial: glycoprotein D loses importance for replication of bovine herpesvirus 1 in cell culture. J. Virol. 71, 25–33.

Schröter, C., Klupp, B.G., Fuchs, W., Gerhard, M., Backovic, M., Rey, F.A., Mettenleiter, T.C., 2014. The highly conserved proline at position 438 in pseudorabies virus gH is important for regulation of membrane fusion. J. Virol. 88, 13064–13072.

Schröter, C., Vallbracht, M., Altenschmidt, J., Kargoll, S., Fuchs, W., Klupp, B.G., Mettenleiter, T.C., 2016. Mutations in pseudorabies virus glycoproteins gB, gD, and gH functionally compensate for the absence of gL. J. Virol. 90, 2264–2272.

Shelly, S.S., Cairns, T.M., Whitbeck, J.C., Lou, H., Krummenacher, C., Cohen, G.H., Eisenberg, R.J., 2012. The membrane-proximal region (MPR) of herpes simplex virus gB regulates association of the fusion loops with lipid membranes. MBio 3, e00429-12.

Shieh, M.T., Wudunn, D., Montgomery, R.I., Esko, J.D., Spear, P.G., 1992. Cell surface receptors for herpes simplex virus are heparan sulfate proteoglycans. J. Cell Biol. 116, 1273–1281.

Shukla, D., Liu, J., Blaiklock, P., Shworak, N.W., Bai, X., Esko, J.D., Cohen, G.H., Eisenberg, R.J., Rosenberg, R.D., Spear, P.G., 1999. A novel role for 3-O-sulfated heparan sulfate in herpes simplex virus 1 entry. Cell 99, 13–22.

Si, Z., Zhang, J., Shivakoti, S., Atanasov, I., Tao, C.L., Hui, W.H., Zhou, K., Yu, X., LI, W., Luo, M., Bi, G.Q., Zhou, Z.H., 2018. Different functional states of fusion protein gB revealed on human cytomegalovirus by cryo electron tomography with Volta phase plate. PLoS Pathog. 14, e1007452.

Silverman, J.L., Sharma, S., Cairns, T.M., Heldwein, E.E., 2010. Fusion-deficient insertion mutants of herpes simplex virus type 1 glycoprotein B adopt the trimeric postfusion conformation. J. Virol. 84, 2001–2012.

Silverman, J.L., Greene, N.G., King, D.S., Heldwein, E.E., 2012. Membrane requirement for folding of the herpes simplex virus 1 gB cytodomain suggests a unique mechanism of fusion regulation. J. Virol. 86, 8171–8184.

Spear, P.G., Longnecker, R., 2003. Herpesvirus entry: an update. J. Virol. 77, 10179–10185.

Spear, P.G., Eisenberg, R.J., Cohen, G.H., 2000. Three classes of cell surface receptors for alphaherpesvirus entry. Virology 275, 1–8.

Spear, P.G., Manoj, S., Yoon, M., Jogger, C.R., Zago, A., Myscofski, D., 2006. Different receptors binding to distinct interfaces on herpes simplex virus gD can trigger events leading to cell fusion and viral entry. Virology 344, 17–24.

Stampfer, S.D., Lou, H., Cohen, G.H., Eisenberg, R.J., Heldwein, E.E., 2010. Structural basis of local, pH-dependent conformational changes in glycoprotein B from herpes simplex virus type 1. J. Virol. 84, 12924–12933.

Steven, A.C., Spear, P.G., 2006. Biochemistry. Viral glycoproteins and an evolutionary conundrum. Science 313, 177–178.

Stevens, J., Corper, A.L., Basler, C.F., Taubenberger, J.K., Palese, P., Wilson, I.A., 2004. Structure of the uncleaved human H1 hemagglutinin from the extinct 1918 influenza virus. Science 303, 1866–1870.

Stieneke-Grober, A., Vey, M., Angliker, H., Shaw, E., Thomas, G., Roberts, C., Klenk, H.D., Garten, W., 1992. Influenza virus hemagglutinin with multibasic cleavage site is activated by furin, a subtilisin-like endoprotease. EMBO J. 11, 2407–2414.

Stiles, K.M., Krummenacher, C., 2010. Glycoprotein D actively induces rapid internalization of two nectin-1 isoforms during herpes simplex virus entry. Virology 399, 109–119.

Strive, T., Borst, E., Messerle, M., Radsak, K., 2002. Proteolytic processing of human cytomegalovirus glycoprotein B is dispensable for viral growth in culture. J. Virol. 76, 1252–1264.

Subak-Sharpe, J.H., Dargan, D.J., 1998. HSV molecular biology: general aspects of herpes simplex virus molecular biology. Virus Genes 16, 239–251.

Suenaga, T., Satoh, T., Somboonthum, P., Kawaguchi, Y., Mori, Y., Arase, H., 2010. Myelin-associated glycoprotein mediates membrane fusion and entry of neurotropic herpesviruses. Proc. Natl. Acad. Sci. U.S.A. 107, 866–871.

Sun, X., Belouzard, S., Whittaker, G.R., 2008. Molecular architecture of the bipartite fusion loops of vesicular stomatitis virus glycoprotein G, a class III viral fusion protein. J. Biol. Chem. 283, 6418–6427.

Sun, Y., Senger, K., Baginski, T.K., Mazloom, A., Chinn, Y., Pantua, H., Hamidzadeh, K., Ramani, S.R., Luis, E., Tom, I., Sebrell, A., Quinones, G., MA, Y., Mukhyala, K., Sai, T., Ding, J., Haley, B., Shadnia, H., Kapadia, S.B., Gonzalez, L.C., Hass, P.E., Zarrin, A.A., 2012. Evolutionarily conserved paired immunoglobulin-like receptor alpha (PILRalpha) domain mediates its interaction with diverse sialylated ligands. J. Biol. Chem. 287, 15837–15850.

Sutton, R.B., Fasshauer, D., Jahn, R., Brunger, A.T., 1998. Crystal structure of a SNARE complex involved in synaptic exocytosis at 2.4 A resolution. Nature 395, 347–353.

Takai, Y., Ikeda, W., Ogita, H., Rikitake, Y., 2008. The immunoglobulin-like cell adhesion molecule nectin and its associated protein afadin. Annu. Rev. Cell Dev. Biol. 24, 309–342.

Tan, X., Brunovskis, P., Velicer, L.F., 2001. Transcriptional analysis of Marek's disease virus glycoprotein D, I, and E genes: gD expression is undetectable in cell culture. J. Virol. 75, 2067–2075.

Tischer, B.K., Schumacher, D., Chabanne-Vautherot, D., Zelnik, V., Vautherot, J.F., Osterrieder, N., 2005. High-level expression of Marek's disease virus glycoprotein C is detrimental to virus growth in vitro. J. Virol. 79, 5889–5899.

Torrents De LA Pena, A., Sanders, R.W., 2018. Stabilizing HIV-1 envelope glycoprotein trimers to induce neutralizing antibodies. Retrovirology 15, 63.

Trybala, E., Bergstrom, T., Spillmann, D., Svennerholm, B., Olofsson, S., Flynn, S.J., Ryan, P., 1996. Mode of interaction between pseudorabies virus and heparan sulfate/heparin. Virology 218, 35–42.

Tsvitov, M., Frampton Jr., A.R., Shah, W.A., Wendell, S.K., Ozuer, A., Kapacee, Z., Goins, W.F., Cohen, J.B., Glorioso, J.C., 2007. Characterization of soluble glycoprotein D-mediated herpes simplex virus type 1 infection. Virology 360, 477–491.

Tulman, E.R., Afonso, C.L., Lu, Z., Zsak, L., Rock, D.L., Kutish, G.F., 2000. The genome of a very virulent Marek's disease virus. J. Virol. 74, 7980–7988.

Turner, A., Bruun, B., Minson, T., Browne, H., 1998. Glycoproteins gB, gD, and gHgL of herpes simplex virus type 1 are necessary and sufficient to mediate membrane fusion in a Cos cell transfection system. J. Virol. 72, 873–875.

Uchida, H., Chan, J., Shrivastava, I., Reinhart, B., Grandi, P., Glorioso, J.C., Cohen, J.B., 2013. Novel mutations in gB and gH circumvent the requirement for known gD receptors in herpes simplex virus 1 entry and cell-to-cell spread. J. Virol. 87, 1430–1442.

Vallbracht, M., Brun, D., Tassinari, M., Vaney, M.C., Pehau-Arnaudet, G., Guardado-Calvo, P., Haouz, A., Klupp, B.G., Mettenleiter, T.C., Rey, F.A., Backovic, M., 2017a. Structure-function dissection of the pseudorabies virus glycoprotein B fusion loops. J. Virol. 92, e01203–e01217.

Vallbracht, M., Rehwaldt, S., Klupp, B.G., Mettenleiter, T.C., Fuchs, W., 2017b. Functional relevance of the N-terminal domain of pseudorabies virus envelope glycoprotein H and its interaction with glycoprotein L. J. Virol. 91, e00061-17.

Vallbracht, M., Schröter, C., Klupp, B.G., Mettenleiter, T.C., 2017c. Transient transfection-based fusion assay for viral proteins. Bio-Protocol 7, e2162.

Vallbracht, M., Fuchs, W., Klupp, B.G., Mettenleiter, T.C., 2018a. Functional relevance of the transmembrane domain and cytoplasmic tail of the pseudorabies virus glycoprotein H for membrane fusion. J. Virol. 92, e00376-18.

Vallbracht, M., Rehwaldt, S., Klupp, B.G., Mettenleiter, T.C., Fuchs, W., 2018b. Functional role of N-linked glycosylation in pseudorabies virus glycoprotein gH. J. Virol. 92, e00084-18.

van Drunen Littel-Van Den Hurk, S., Babiuk, L.A., 1986. Synthesis and processing of bovine herpesvirus 1 glycoproteins. J. Virol. 59, 401–410.

Vitu, E., Sharma, S., Stampfer, S.D., Heldwein, E.E., 2013. Extensive mutagenesis of the HSV-1 gB ectodomain reveals remarkable stability of its postfusion form. J. Mol. Biol. 425, 2056–2071.

Vleck, S.E., Oliver, S.L., Brady, J.J., Blau, H.M., Rajamani, J., Sommer, M.H., Arvin, A.M., 2011. Structure-function analysis of varicella-zoster virus glycoprotein H identifies domain-specific roles for fusion and skin tropism. Proc. Natl. Acad. Sci. U.S.A. 108, 18412–18417.

Wanas, E., Efler, S., Ghosh, K., Ghosh, H.P., 1999. Mutations in the conserved carboxy-terminal hydrophobic region of glycoprotein gB affect infectivity of herpes simplex virus. J. Gen. Virol. 80 (Pt 12), 3189–3198.

Wang, J., Fan, Q., Satoh, T., Arii, J., Lanier, L.L., Spear, P.G., Kawaguchi, Y., Arase, H., 2009. Binding of herpes simplex virus glycoprotein B (gB) to paired immunoglobulin-like type 2 receptor alpha depends on specific sialylated O-linked glycans on gB. J. Virol. 83, 13042–13045.

Warner, M.S., Geraghty, R.J., Martinez, W.M., Montgomery, R.I., Whitbeck, J.C., Xu, R., Eisenberg, R.J., Cohen, G.H., Spear, P.G., 1998. A cell surface protein with herpesvirus entry activity (HveB) confers susceptibility to infection by mutants of herpes simplex virus type 1, herpes simplex virus type 2, and pseudorabies virus. Virology 246, 179–189.

Wathen, M.W., Wathen, L.M., 1984. Isolation, characterization, and physical mapping of a pseudorabies virus mutant containing antigenically altered gp50. J. Virol. 51, 57–62.

Webster, R.G., Rott, R., 1987. Influenza virus A pathogenicity: the pivotal role of hemagglutinin. Cell 50, 665–666.

Weimbs, T., Low, S.H., Chapin, S.J., Mostov, K.E., Bucher, P., Hofmann, K., 1997. A conserved domain is present in different families of vesicular fusion proteins: a new superfamily. Proc. Natl. Acad. Sci. U.S.A. 94, 3046–3051.

Whealy, M.E., Robbins, A.K., Enquist, L.W., 1990. The export pathway of the pseudorabies virus gB homolog gII involves oligomer formation in the endoplasmic reticulum and protease processing in the Golgi apparatus. J. Virol. 64, 1946–1955.

White, J.M., Delos, S.E., Brecher, M., Schornberg, K., 2008. Structures and mechanisms of viral membrane fusion proteins: multiple variations on a common theme. Crit. Rev. Biochem. Mol. Biol. 43, 189–219.

Whitley, R.J., Roizman, B., 2001. Herpes simplex virus infections. Lancet 357, 1513–1518.

Xing, Y., Oliver, S.L., Nguyen, T., Ciferri, C., Nandi, A., Hickman, J., Giovani, C., Yang, E., Palladino, G., Grose, C., Uematsu, Y., Lilja, A.E., Arvin, A.M., Carfi, A., 2015. A site of varicella-zoster virus vulnerability identified by structural studies of neutralizing antibodies bound to the glycoprotein complex gHgL. Proc. Natl. Acad. Sci. U.S.A. 112, 6056–6061.

Yang, E., Arvin, A.M., Oliver, S.L., 2014. The cytoplasmic domain of varicella-zoster virus glycoprotein H regulates syncytia formation and skin pathogenesis. PLoS Pathog. 10, e1004173.

Yaswen, L.R., Stephens, E.B., Davenport, L.C., Hutt-Fletcher, L.M., 1993. Epstein-Barr virus glycoprotein gp85 associates with the BKRF2 gene product and is incompletely processed as a recombinant protein. Virology 195, 387–396.

Yin, H.S., Wen, X., Paterson, R.G., Lamb, R.A., Jardetzky, T.S., 2006. Structure of the parainfluenza virus 5 F protein in its metastable, prefusion conformation. Nature 439, 38–44.

Zago, A., Jogger, C.R., Spear, P.G., 2004. Use of herpes simplex virus and pseudorabies virus chimeric glycoprotein D molecules to identify regions critical for membrane fusion. Proc. Natl. Acad. Sci. U.S.A. 101, 17498–17503.

Zago, A., Connolly, S.A., Spear, P.G., Longnecker, R., 2013. The fusion loops and membrane proximal region of Epstein-Barr virus glycoprotein B (gB) can function in the context of herpes simplex virus 1 gB when substituted individually but not in combination. Virus Res. 171, 227–230.

Zeev-Ben-Mordehai, T., Vasishtan, D., Hernandez Duran, A., Vollmer, B., White, P., Prasad Pandurangan, A., Siebert, C.A., Topf, M., Grunewald, K., 2016. Two distinct trimeric conformations of natively membrane-anchored full-length herpes simplex virus 1 glycoprotein B. Proc. Natl. Acad. Sci. U.S.A. 113, 4176–4181.

Zhang, N., Yan, J., Lu, G., Guo, Z., Fan, Z., Wang, J., Shi, Y., Qi, J., Gao, G.F., 2011. Binding of herpes simplex virus glycoprotein D to nectin-1 exploits host cell adhesion. Nat. Commun. 2, 577.

CHAPTER EIGHT

Entry of betaherpesviruses

Mitsuhiro Nishimura, Yasuko Mori[1]
Division of Clinical Virology, Center for Infectious Diseases, Kobe University Graduate School of Medicine, Kobe, Japan
[1]Corresponding author: e-mail address: ymori@med.kobe-u.ac.jp

Contents

Abstract

In this chapter, we present an overview on betaherpesvirus entry, with a focus on human cytomegalovirus, human herpesvirus 6A and human herpesvirus 6B. Human cytomegalovirus (HCMV) is a complex human pathogen with a genome of 235 kb encoding more than 200 genes. It infects a broad range of cell types by switching

Advances in Virus Research, Volume 104
ISSN 0065-3527
https://doi.org/10.1016/bs.aivir.2019.05.005

its viral ligand on the virion, using the trimer gH/gL/gO for infection of fibroblasts and the pentamer gH/gL/UL128/UL130/UL131 for infection of other cells such as epithelial and endothelial cells, leading to membrane fusion mediated by the fusion protein gB. Adding to this scenario, however, accumulating data reveal the actual complexity in the viral entry process of HCMV with an intricate interplay among viral and host factors. Key novel findings include the identification of entry receptors platelet-derived growth factor-α receptor (PDGFRα) and Netropilin-2 (Nrp2) for trimer and pentamer, respectively, the determination of atomic structures of the fusion protein gB and the pentamer, and the in situ visualization of the state and arrangement of functional glycoproteins on virion. This is covered in the first part of this review. The second part focusses on HHV-6 which is a T lymphotropic virus categorized as two distinct virus species, HHV-6A and HHV-6B based on differences in epidemiological, biological, and immunological aspects, although homology of their entire genome sequences is nearly 90%. HHV-6B is a causative agent of exanthema subitum (ES), but the role of HHV-6A is unknown. HHV-6B reactivation occasionally causes encephalitis in patients with hematopoietic stem cell transplant. The HHV-6 specific envelope glycoprotein complex, gH/gL/gQ1/gQ2 is a viral ligand for the entry receptor. Recently, each virus has been found to recognize a different cellular receptor, CD46 for HHV 6A amd CD134 for HHV 6B. These findings show that distinct receptor recognition differing between both viruses could explain their different pathogenesis.

1. Entry of human cytomegalovirus

1.1 HCMV tropism and entry process

Human cytomegalovirus (HCMV) infects asymptomatically during childhood, although a substantial number of people become infected later in life. This late primary infection of HCMV is of great concern especially to obstetricians, because primary infection with HCMV during early pregnancy is correlated with an increased risk of congenital infections of newborn infants frequently accompanied with birth defects such as neurological abnormality and hearing loss (Davis et al., 2017). Vaccination before pregnancy could prevent congenital HCMV infection, but the complicated nature of the infection, transmission, and pathology of HCMV so far prevented the development of an effective vaccine (Plotkin and Boppana, 2018). Recent advances in our understanding of HCMV infection, especially the entry process, included the identification of HCMV entry receptors and molecular visualization of viral ligands. Although these have been well-documented in several review articles (Nguyen and Kamil, 2018; Sathiyamoorthy et al., 2017; Vanarsdall and Johnson, 2012), an update is timely due to the rapid progress.

HCMV infects a wide variety of human cells such as fibroblasts, epithelial cells, endothelial cells, leukocytes, macrophages, monocytes, and more (Gerna et al., 2004). HCMV possesses multiple viral ligands which serve to recognize host cell receptors. In HCMV, the heterodimer gH/gL which is conserved among all herpesviruses, is extended by being associated with gO (encoded in the UL74 gene) or UL128, UL130 and UL131, forming a trimeric gH/gL/gO or a pentameric gH/gL/UL128/UL130/UL131, respectively. The amount and ratio of trimer and pentamer expression on the virion varies among HCMV strains (Zhang et al., 2018; Zhou et al., 2013, 2015), and is controlled by other viral factors, e.g., HCMV UL148 (suppressing pentamer incorporation) (Li et al., 2015a; Nguyen et al., 2018) and US16 (promoting pentamer incorporation) (Luganini et al., 2017). HCMV passaged in fibroblast culture resulted in a defect of infectivity to endothelial cells (Gerna et al., 2002; Grazia Revello et al., 2001; Kahl et al., 2000; Sinzger et al., 2000), and the HCMV strains lose UL128-UL131 because of the point mutation, frame-shift mutation, or deletion in the genes (Gerna et al., 2005; Hahn et al., 2004; Sinzger et al., 2008), resulting in a lack of pentamer on the virion. Viruses devoid of pentamer, including laboratory strains AD169 and Towne, have a defect in infection of epithelial cells, endothelial cells, monocytes/macrophages, and dendritic cells, while infection of fibroblasts is enhanced (Adler et al., 2006; Gerna et al., 2005; Hahn et al., 2004; Patrone et al., 2005; Ryckman et al., 2008a; Sinzger et al., 2008; Wang et al., 2005a; Wang and Shenk, 2005). Recent data provided a straightforward view about the roles of two viral ligands; the trimer is responsible for fibroblast infection, whereas the pentamer is required for infection of wider range of cells, represented by epithelial and endothelial cells (Ryckman et al., 2008a, b; Vanarsdall et al., 2011) (Fig. 1). However, accessory proteins for forming trimer or pentamer, and it is rationalized from the molecular architecture of these glycoprotein complex. Low resolution structures of HCMV trimer and pentamer were obtained by negative-stain electron microscopy (EM) analysis with single particle reconstitution (Ciferri et al., 2015a, b). Trimer and pentamer have a base density of gH/gL associated with additional densities at the membrane distal side, where gL is located, resulting in longitudinally elongated shapes. The expected densities for the UL128/UL130/UL131 of the pentamer and gO of the trimer apparently occupy a spatially overlapping position, corroborating the mutually exclusive nature of the UL128/UL130/UL131 and the gO in docking on gH/gL. More detailed analyses using mass spectrometry (MS) and site-directed mutagenesis revealed that C144 of gL forms disulfide

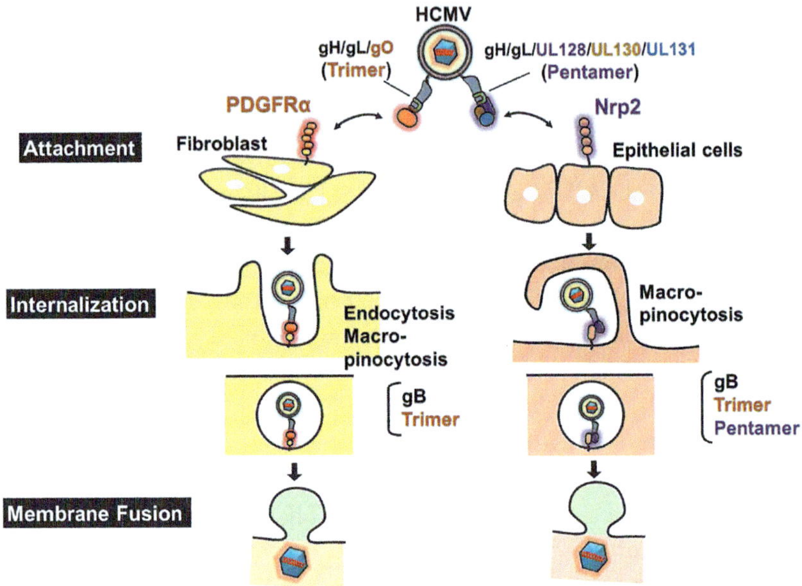

Fig. 1 Distinct processes of HCMV infection to fibroblast and epithelial cells. HCMV entry includes the three steps attachment, internalization, and membrane fusion. In infection of fibroblasts, HCMV uses the trimer gH/gL/gO to interact with PDGFRα, and is internalized by endocytosis. On the other hand, HCMV infects epithelial and endothelial cells via the interaction between Nrp2 and pentameric gH/gL/UL128/UL130/UL131. This interaction triggers micropinocytosis. Membrane fusion occurs in concerted action of gB, trimer and pentamer, as well as host protein CD147. Figures by Vanarsdall et al. (2012) and Vanarsdall and Johnson (2012) were referred to make this illustration.

bonds with C351 of gO in the trimer, and also with C162 of UL128 in the pentamer. Because C144 of gL is conserved in betaherpesviruses, it likely contributes to the formation of tetramers of HHV-6A and HHV-6B as well and may place the gQ1/gQ2 at the distal side (see below). The distal association of gO and UL128/UL130/UL131 near gL is a major different from the gH/gL complex of gammaherpesvirus Epstein-Barr Virus (EBV). Here, within the trimeric gH/gL/gp42 complex gp42 is located at the side position of the longitudinal axis by binding gH (Connolly et al., 2011; Sathiyamoorthy et al., 2014, 2016).

1.2 Crystal structure of the pentamer

The ectodomain structure of HCMV pentamer has been further analyzed by X-ray crystallography at a maximum resolution 3.0 Å (Chandramouli et al., 2017). Similar to previously reported gH structures of other herpesviruses

(Backovic et al., 2010; Chowdary et al., 2010; Matsuura et al., 2010; Xing et al., 2015), the gH within the HCMV pentamer has a tandem domain organization, DI, DII, DIII and DIV from N-to C-terminus, placing DI and DIV at the expected membrane distal and proximal sides, respectively (Fig. 2A and B). The cleft between DI and DII is wide, and gL resides on it. Although the respective domains of HCMV gH are similar to those of other herpesviruses, its overall shape does not match because of differences in the relative arrangement of each domain (Chandramouli et al., 2017; Malito et al., 2018). HCMV gL has a chemokine fold which is located in between the DI and DII of gH, as in other herpesviruses. The crystal structure highlighted the importance of the N-terminal region of gL, which extends from the common chemokine fold and is unique for betaherpesviruses. The N-terminus of HCMV gL forms a helix and resides around the cleft between DI and DII of gH. Two cysteine residues, C47 and C54, on the helix form disulfide bonds with gH C95 and C59, respectively (Fig. 2A). All of these cysteine residues are conserved among betaherpesviruses, and thus the strong covalent connection between gH and gL seems to be a common feature. The following N-terminal part of gL forms a compact fold placed at the distal tip of gH/gL, and serves as a platform for UL128/UL130/UL131. UL130 has two domains: the N-terminal chemokine fold in contact with gL, and the C-terminal part which is combined with the C-terminal part of UL131 forming an α/β core. The N-terminal small fold of UL131 and the α/β core of UL130/UL131 interact with the N-terminal chemokine fold of UL128 which constitutes the distal tip of the pentamer. The C-terminal part of UL128 extends to the platform domain of gL, and cysteine residue C162 forms a disulfide bond with gL C144, consistent with previous observations (Ciferri et al., 2015a, b). The horn-like protrusion containing UL131, UL130, and UL128 represents a right-handed screw toward the tip formed by UL128. The elongated-spindle shape and few contacts among domains separated by large clefts accounts for the intrinsic flexibility of the pentamer observed in the crystallography and EM analyses (Ciferri et al., 2015a, b; Malito et al., 2018). The atomic structure of pentamer also allowed a comprehensive mapping of the antigenic sites of neutralizing antibodies (Ciferri et al., 2015a, b; Ha et al., 2017; Macagno et al., 2010). Some antigenic sites overlap with the binding site for the receptor Neuropilin-2 (Nrp2), explaining the mode of action of the neutralizing antibodies (Martinez-Martin et al., 2018). For the other suggested antigenic sites the mechanism of neutralization is still unclear, and may imply multiple functions of the gH/gL complex.

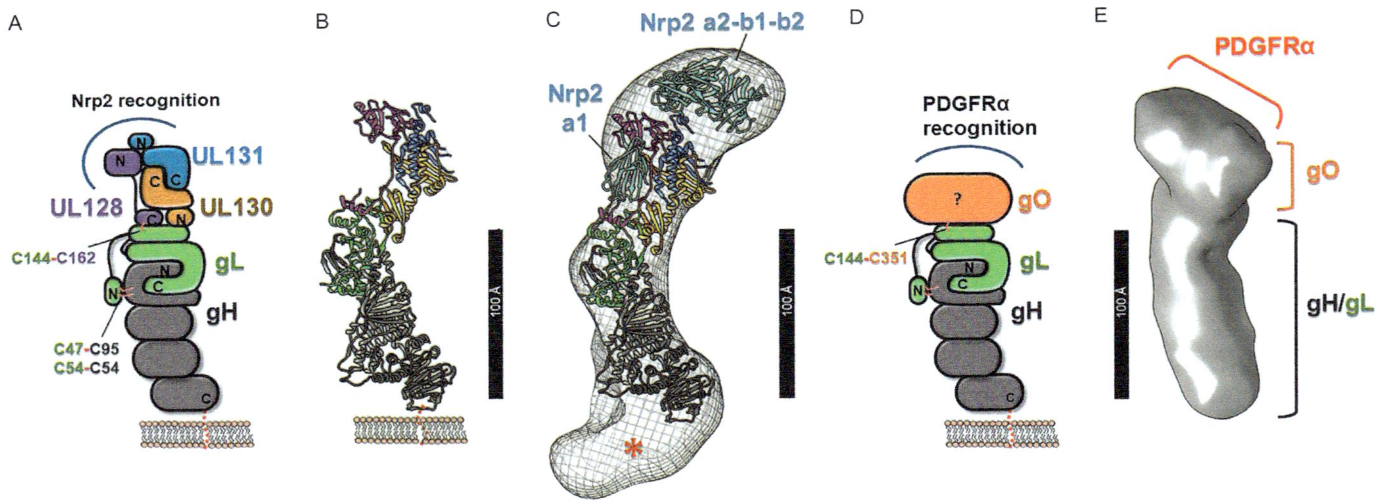

Fig. 2 Structure of pentamer and trimer with their receptors. (A) A schematic illustration and (B) a ribbon model of the ectodomain of HCMV pentamer determined by X-ray crystallography (Chandramouli et al., 2017) (PDB: 5VOB) In (A), the intermolecular disulfide bonds are depicted in red lines. (C) The putative binding model of Nrp2. The density map of the ligand receptor complex (EMDB: 8884) was fitted with pentamer (PDB: 5VOB) and Nrp2 (PDB: 2QQK) structures (Appleton et al., 2007; Chandramouli et al., 2017) by referring to the figure by Martinez-Martin et al. (2018). Red asterisk represents the density of 3G16 Fab. (D) A schematic illustration of the trimer. (E) The density map of the complex between trimer and PDGFRα (EMDB: 3391) (Kabanova et al., 2016). The suggested location of gH/gL, gO, and PDGFRα is indicated. In (B), (C) and (E), the images were produced by UCSF chimera software (Pettersen et al., 2004), and the scale bars stands for 100 Å.

1.3 Entry receptor PDGFRα for HCMV infection of fibroblasts

Although receptors for HCMV gB, such as epidermal growth factor receptor (EGFR) (Wang et al., 2003), integrins (Feire et al., 2004, 2010; Wang et al., 2005b), and platelet-derived growth factor-α receptor (PDGFRα) (Soroceanu et al., 2008) have been identified, the tight correlation between the viral trimer and pentamer, and the cell tropism of HCMV led to the elucidation of specific receptors for these ligands on each cell type. Kabanova et al. re-identified PDGFRα as an entry receptor recognized by the trimer gH/gL/gO (Kabanova et al., 2016). They started from the observation that PDGFRα signaling was triggered only during infection of fibroblasts but not epithelial cells (Kabanova et al., 2016; Wang et al., 2003). By several approaches including genetics and biophysics, they demonstrated that PDGFRα serves as the entry receptor for fibroblast infection by directly interacting with the trimer. The binding affinity between PDGFRα and trimer is quite high. The dissociation constant was evaluated to be 2×10^{-9}M by surface plasmon resonance and biolayer interferometry experiments. The gO component is responsible for the binding to the receptor, and the negative stain EM analysis visualized that a PDGFRα density covers the distal tip of the expected gO density (Fig. 2E) (Kabanova et al., 2016). Consistent with the role of trimer as the tropism determinant, PDGFα is abundantly expressed on fibroblasts, while its expression is hardly detected on epithelial and endothelial cells (Martinez-Martin et al., 2018; Stegmann et al., 2017; Wu et al., 2017).

1.4 Entry receptor Nrp2 for HCMV infection to epithelial cells

Martinez-Martin et al. identified Nrp2 as an entry receptor responsible for the epithelial/endothelial cell infection of HCMV (Martinez-Martin et al., 2018). They utilized a highly sensitive screening method for 1297 single transmembrane receptors of human cells to detect an interaction between Nrp2 and the pentamer, as well as PDGFRα interaction with the trimer. Although they also detected significant interactions between trimer and transforming growth factor beta receptor type 3 (TGFβRIII) or neuroregulin 2 (NRG2), and between pentamer and thrombomodulin (THBD), LILRB3 and FCAR, subsequent in vitro analyses did not support a physiological importance of these interactions. The affinity of Nrp2 to pentamer was evaluated to be 3.3×10^{-7}M, which is lower than that of the PDGFRα-trimer interaction (2×10^{-9}M), but still significant. The binding model of Nrp2 was proposed by a comprehensive approach considering the EM density of the complex, crystal structures of pentamer (Chandramouli et al., 2017) and Nrp2 (Appleton et al., 2007), and analysis of the responsible domain of the Nrp2, crosslinking–MS analysis for the complex, and the binding site of antibodies potent for the binding inhibition (Fig. 2C). The ectodomain of Nrp2 consists of five tandem folds, a1-a2-b1-b2-MAM, with a loose connection between the a1 and the a2-b1-b2 core, and a flexible linker before the MAM domain (Appleton et al., 2007). The a1 domain of Nrp2 occupies a cleft in the pentamer surrounded by gL, UL128, and UL130, and the a2-b1 domain of Nrp2 binds at the distal tip of the pentamer in contact with UL128 and the N-terminal domain of UL131.

1.5 EGFR, integrin, and PDGFRα activation in HCMV infection

In an earlier report, EGFR had been identified as a host receptor recognized by HCMV gB (Wang et al., 2003). Inhibition of EGFR signaling prevented infection of epithelial cells (Kabanova et al., 2016), and decreased infection in monocytes and $CD34^+$ progenitor cells (Chan et al., 2009; Kim et al., 2017; Nogalski et al., 2011), although the contribution of EGFR to HCMV infection was not supported in other studies (Isaacson et al., 2007; Isaacson and Compton, 2009; Vanarsdall et al., 2012). Activation of the integrin Src pathway was suggested to promote HCMV infection to monocytes by regulating paxillin expression (Nogalski et al., 2011). PDGFRα had been suggested as receptor for HCMV gB leading to an activation of phosphoinositide-3-kinase (PI3K) signaling (Soroceanu et al., 2008). However, subsequent experiments did not yield proof of a direct interaction between PDGFRα and gB (Stegmann et al., 2017; Vanarsdall et al., 2012), but gB interaction with PDGFRα was dependent on the trimer (Wu et al., 2017). Varnashdall et al. demonstrated that PDGFRα activation actually causes promotion of HCMV infection but by an abnormal pathway (Vanarsdall et al., 2012). Furthermore, Wu et al. revealed that the interaction between the trimer and PDGFRα does not trigger PDGFRα and the following signaling pathway (Wu et al., 2017, 2018). Taken together, the PDGFRα signaling pathway has a strong influence on HCMV infection, but its contribution in natural infection is still unclear.

1.6 HCMV internalization by micropinocytosis and endocytosis

Following attachment to the target cells via the high affinity ligand-receptor interactions, the HCMV virion is internalized by micropinocytosis or endocytosis. Although pH dependent fusion at the plasma membrane is possible, as reported for HCMV infection of fibroblasts (Compton et al., 1992), endocytosis and micropinocytosis-like engulfment were reported in infection of fibroblasts, epithelial cells and dendritic cells (Haspot et al., 2012; Hetzenecker et al., 2016; Li et al., 2016). Li et al. revealed that the cell surface antigen THY-1 regulates micropinocytosis during HCMV infection (Li et al., 2015b, 2016). Abdellatif et al. visualized a series of HCMV entry processes by scanning transmission electron microscopy tomography with careful specimen preparation (Abdellatif et al., 2018). One interesting observation is the curious morphology of the virion: virions near the target cell membrane showed a protrusion without changing the total volume, named "Pinocchio particles" (Abdellatif et al., 2018), implying a dynamic change in virion structure accompanying virus-host interaction. Consistent with previous reports (Haspot et al., 2012; Hetzenecker et al., 2016), macropinocytosis-like protrusions and ruffles engulfing HCMV virions were observed during infection of fibroblast and endothelial cells, respectively, as well as the resulting HCMV-containing vehicles (Abdellatif et al., 2018).

1.7 Viral ligands and fusion event

For release of the nucleocapsid into the cytosol, gB-mediated membrane fusion must occur. Since virus lacking pentamer can productively infect fibroblasts, trimer alone is competent to trigger gB activation. On the other hand, it has been suggested that pentamer-dependent infection of epithelial cells will be blocked at the fusion step if trimer is not available (Kabanova et al., 2016; Liu et al., 2018; Zhou et al., 2015). Zhou et al. demonstrated that the HCMV Merlin strain, in which pentamer is predominant, was enhanced in its infection to both fibroblasts and epithelial cells by artificial enrichment of trimer, or use of the chemical fusion enhancer polyethylene glycol (PEG) (Zhou et al., 2015), suggesting trimer function in the fusion step and inadequate activation of gB by the pentamer. Several lines of evidence indicated that the requirement of trimer for pentamer-dependent infection is not due to the trimer-PDGFRα interaction (Liu et al., 2018; Stegmann et al., 2017). However, pentamer is also involved in fusion in addition to the trimer, since soluble pentamer impaired HCMV infection to epithelial cells after entering the endosome (Liu et al., 2018). In addition to the described viral ligands and entry receptors, contribution of other host factors has been also reported. HCMV pUL78 is a G protein-coupled receptor homologue expressed on the virion, and specifically required for efficient infection of epithelial and endothelial cells at the internalization (O'Connor and Shenk, 2012). A recently identified entry mediator CD147 influences entry of HCMV into endothelial cells, by promoting fusion in the endosome, while it is not significant in infection of fibroblasts (Vanarsdall et al., 2018). The tetraspanin CD151 is suggested to promote the penetration step during infection of both, fibroblasts and endothelial cells (Hochdorfer et al., 2016). Heparan sulfate glycosaminoglycans (GAG) on target cells was suggested to play a role in HCMV adsorption (Compton et al., 1993). However, a recent report by Liu et al. revealed that heparin sulfate GAG does not contribute to attachment but influences the following internalization and/or fusion steps (Liu et al., 2018).

1.8 Crystal structure of the HCMV gB in the "postfusion form"

The class III fusion protein gB is conserved in all members of the *Herpesviridae* and is essential for herpesviral entry (Connolly et al., 2011; Cooper and Heldwein, 2015; Sathiyamoorthy et al., 2017). Herpesvirus gB forms a homo-trimer like fusion proteins of other viruses (Harrison, 2015). However, gB alone is not fusogenic but requires gH/gL for the regulation which is a unique feature for fusion machineries of herpesviruses. HCMV gB consists of 906 amino acids containing an N-terminal ectodomain, a single pass transmembrane domain, and a cytoplasmic tail domain (endodomain). Not only the ectodomain which is the major part of gB (78–706) but also the transmembrane and the endodomain build a trimeric architecture (Cooper et al., 2018; Rogalin and Heldwein, 2015; Silverman et al., 2012). The detailed structure of the HCMV gB ectodomain has been analyzed by X-ray crystallography (Burke and Heldwein, 2015; Chandramouli et al., 2015). Like the gB structures of alphaherpesvirus HSV-1 (Heldwein et al., 2006) and gammaherpesvirus EBV (Backovic et al., 2009), the isolated gB ectodomain of HCMV also showed a so-called postfusion form, and each monomer of the homo-trimer consists of five domains, DI, DII, DIII, DIV and DV (Fig. 3A–C). DI with the fusion loops is located at a membrane proximal position, and DII forms a knob located on top of DI. DIII includes a long helix which constitutes a helix bundle in the trimer. DIV is located at the distal end and forms a trimeric globular head called "crown." DV is composed of the C-terminal region of the ectodomain running from DIV to the transmembrane domain.

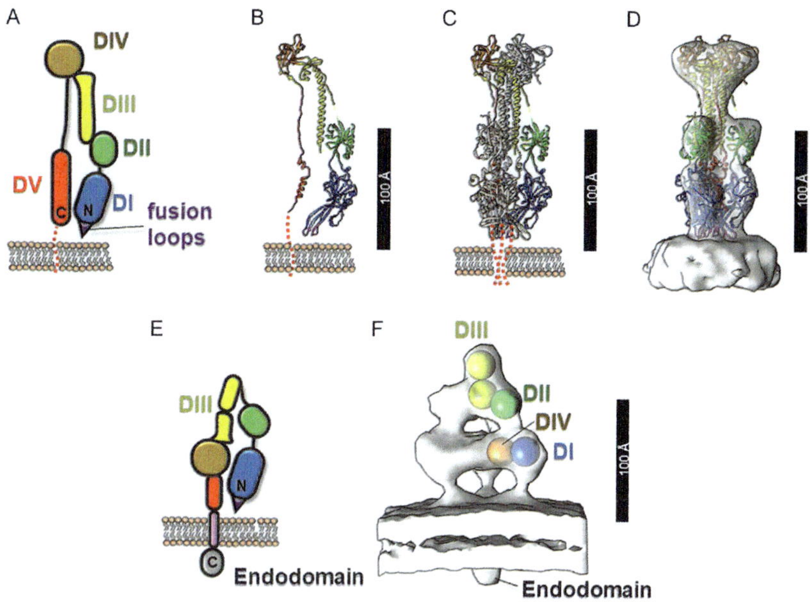

Fig. 3 Structure of HCMV gB. (A) A schematic illustration and (B) a ribbon model of the ectodomain of HCMV gB determined by X-ray crystallography (Burke and Heldwein, 2015; Chandramouli et al., 2015). In (C) the trimeric gB structure assigned as a "postfusion" was depicted. (D) The density map and fitted model of the "Columnar-tree like form" of gB which is well fitted by the crystal structure of the postfusion gB (Burke and Heldwein, 2015; Si et al., 2018). (E) A schematic illustration of prefusion gB and (F) the density map of the "Christmas-tree like form" gB (Si et al., 2018). The suggested location of each domain is indicated as a sphere placed by referring to the original literature (Si et al., 2018). In (A) and (E), the illustrations were prepared by referring to original articles (Burke and Heldwein, 2015; Chandramouli et al., 2015; Si et al., 2018). The images in (B), (C), (E), and (F) were produced by UCSF chimera software (Pettersen et al., 2004) base on the gB crystal structures (PDB: 5CXF) (Burke and Heldwein, 2015) for (B–D), and the published densities of gB postfusion (EMDB: 9328) for (E) and gB prefusion (EMDB: 9329) for (D) and (F).

1.9 In situ visualization of HCMV gB prefusion form

The dynamic feature of HCMV gB was unveiled by advanced cryo-electron tomography (cryoET) with state-of-the-art equipment: energy filter, direct electron detector, and volta phase plate, which improve the signal-to-noise ratio resulting in high contrast without large defocus (Si et al., 2018). They analyzed the HCMV gB structure on purified infection-competent virions in situ. Two forms of gB were observed: a tall Columnar-tree like form, which is similar to the reported crystal structures of gB, (Fig. 3D) and a short

Christmas-tree like form, which has been suggested to represent the prefusion form (Fig. 3F). The abundance of more than 70% of the prefusion gB, makes sense considering the role of fusion proteins which must be fusion competent on infectious virions. These authors carefully validated the methodology, and proposed a model of prefusion gB based on the tomograms, gB crystal structures, molecular dynamics, as well as the expected analogy with other known viral fusion proteins (Fig. 3E and F). In their prefusion gB model, DI faces the fusion loops on the membrane side similar as the "postfusion" form, and the relative arrangement of DII changed only a little. The largest conformational change is imposed on DIII, and authors hypothesized a flip of DIII orientation and a bend in its long helix. As a result, the following DIV occupies a position near the membrane. The arrangement of DI in this model is consistent with previously suggested prefusion models of HSV-1 gB (Fontana et al., 2017; Gallagher et al., 2014), but different from that of the intermediate-like form previously reported for HSV-1 gB, in which DI is located at the top, and the fusion loops are exposed to the solvent side (Zeev-Ben-Mordehai et al., 2015). The in situ tomogram also visualized a density below the lipid bilayer in the prefusion gB, indicating a stable structure of the endodomain (Fig. 3F). Because of the absence of such density in the "postfusion" gB, the stable fold of gB, the endodomain is likely destabilized during or before the conformational change. The ectodomain density is consistent with the proposed existence of a stable fold in the HSV-1 gB endodomain (Rogalin and Heldwein, 2015; Silverman et al., 2012), and the crystal structure of whole HSV-1 gB, which includes not only the ectodomain but also the transmembrane domain and the endodomain stabilized by amphipol to imitate the lipid membrane (Cooper et al., 2018). Although the HSV-1 gB ectodomain has a "postfusion" structure, Cooper et al. suggested the endodomain is likely to take a similar fold also in the prefusion state (Cooper et al., 2018).

1.10 Interaction between gB and gH/gL on the virion

Interaction between gB and gH is of great importance to regulate the fusion activity of gB. Studies on HSV-1 gB revealed a regulatory interaction of gB with the cytoplasmic tail of gH (Rogalin and Heldwein, 2015; Silverman et al., 2012). Vanarsdall et al. found substantial interaction between gH/gL and gB on the HCMV virion (Vanarsdall et al., 2016), indicating the function and importance of the gH/gL dimer without any accessory proteins. Although they did not detected the interaction between gB and the trimer or the pentamer (Vanarsdall et al., 2016), Wu et al. demonstrated

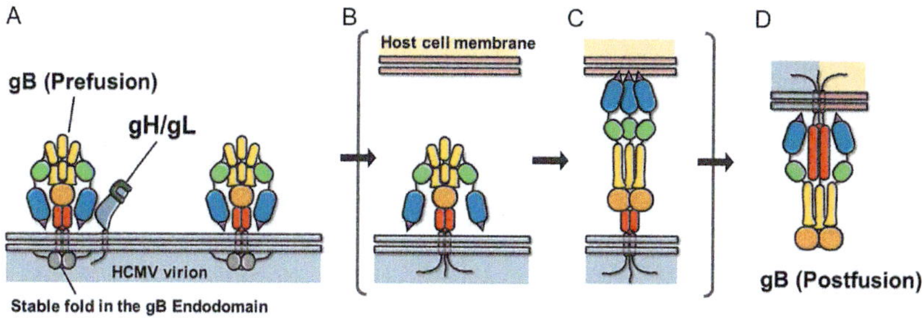

Fig. 4 Proposed conformational change in HCMV gB. A schematic illustration showing the proposed conformational changes of HCMV gB prepared by referring to the article published by Si et al. (2018). In the cryoTM analysis, the gB prefusion state and the complex of gB-gH/gL were observed (A). A density underneath the lipid bilayer represents a stable fold of the gB endodomain. During entry, gB is activated and the fusion loops are exposed (purple triangles) (B) to attack the host cell membrane (C). Drastic conformational changes in the gB to the energetically stable postfusion form drives host-virus membrane fusion (D). Note that the states of (B) and (C) are not observed because of their instability. Several additional states must be involved between (C) and (D), such as a "hemifusion state" (Harrison, 2015; White et al., 2008).

co-precipitation of the trimer, PDGFα, and gB, suggesting that the trimer-PDGFα complex recruits gB (Wu et al., 2017). CryoTM analysis also could visualize the L-shaped gH/gL densities on the virions, and found that 7% of gH/gL is associated with the prefusion gB in a ratio of one gH/gL to one gB trimer, while no such association was found in the "postfusion" form of gB (Si et al., 2018). Their latest diagram regarding the behavior of fusion machineries provided an updated view including the dynamics of gB and its regulation by gH/gL (Fig. 4) (Si et al., 2018). Prefusion gB stably exists on the virion with a stable endodomain; gH/gL directly associates with the prefusion gB and may access the endodomain with the gH cytoplasmic tail which may transmit the receptor binding signal to trigger the conformational change of gB.

2. Entry of human herpesvirus 6A and human herpesvirus 6B

2.1 Characterization of human herpesvirus 6A and human herpesvirus 6B

Human herpesvirus 6 (HHV-6) is an enveloped double stranded DNA virus belonging to the genus *Roseolovirus* within the β-herpesvirus subfamily (Roizmann et al., 1992).

HHV-6 is a T lymphotropic virus which productively replicates in T cells. HHV-6 was initially isolated from the peripheral blood of patients with lymphoproliferative disorders (Salahuddin et al., 1986). Two variants, HHV-6 A and HHV-6B have subsequently been identified, although overall identity of viral genome sequence between HHV-6A and HHV-6B is nearly 90%. (Dominguez et al., 1999; Gompels et al., 1995; Isegawa et al., 1999).

However, these variants were recently categorized as two distinct virus species based on differences in epidemiological, biological, and immunological aspects (Ablashi et al., 2013). Most primary infection and reactivation from latency occur from HHV-6B. Within the first 3 years of life, most children have a primary HHV-6B infection, which becomes latent throughout life. Primary HHV-6B infection causes exanthema subitum (ES) (Yamanishi et al., 1988) which is a childhood febrile illness, sometimes associated with severe encephalopathy. More than 90% of the general population is infected with HHV-6B. HHV-6B reactivation also causes encephalitis in patients with allogeneic hematopoietic stem cell transplant (HSCT). In addition, HHV-6B reactivation is related to drug-induced hypersensitivity syndrome, and might contribute to its severity. However, HHV-6A has not been etiologically related to any specific diseases, although HHV-6A and HHV-7 have been reported to be possibly related to Alzheimer's disease (Haas and Lathe, 2018; Readhead et al., 2018). An association of HHV-6A and HHV-6B with several diseases like chronic fatigue syndrome (Chapenko et al., 2006) and multiple sclerosis (Leibovitch and Jacobson, 2014) has also been suggested. Recently, integration of the HHV-6 genome into host genome has been reported (Arbuckle et al., 2010; Kaufer and Flamand, 2014; Prusty et al., 2013).

2.2 Ligands and receptors of HHV-6A and HHV-6B for entry

HHV-6 infection starts with the attachment of virus particles to target cells. Specific virus-cell interaction begins with the binding of the HHV-6 ligand(s) which is (are) generally expressed on viral envelope, to their cellular receptor(s). Subsequently, the HHV-6 envelope fuses with cellular membrane after endocytosis of the virion (Cirone et al., 1992). Santoro et al. reported that both HHV-6A and HHV-6B use human CD46 as cellular receptor (Santoro et al., 1999), but later reports showed that the HHV-6B strains used in the study (HST strain and the other clinical isolates from infants with ES) could not use CD 46. This discrepancy might be due to the difference of the HHV-6B strains used (Santoro et al. used strains Z29 and PL1 while later group used HST and the other clinical isolates from infants). In 2013, human CD134 was identified cellular receptor of HHV-6B (Tang et al., 2013, 2014; Tang and Mori, 2015).

As for viral ligand, the HHV-6A- or B-specific envelope glycoprotein complex, gH/gL/gQ1/gQ2 (Akkapaiboon et al., 2004; Mori et al., 2003) has been identified as a viral ligand for CD46 (Mori et al., 2004) or CD134 (Tang et al., 2013, 2014; Tang and Mori, 2015).

Beyond viral attachment, HHV-6 also appears to rely on conserved gB and gH/gL for membrane fusion and virus entry (Mori, 2009; Tanaka et al., 2013; Wakata et al., 2018).

2.3 Distinct cellular tropism between HHV-6A and HHV-6B

HHV-6 is T lymphotropic and neurotropic (De Bolle et al., 2005; Lusso et al., 1988). Generally HHV-6 could be propagated in primary activated T cells. However, the usage of T cell lines for HHV-6 virus propagation is different between both viruses. Generally, HSB-2 and JJhan cells are used for HHV-6A, and Molt-3 and MT-4 are for HHV-6B. The different susceptibility for T-cell lines is one of the determinants for their classification into the two species, HHV-6A and HHV-6B (Ablashi et al., 1991), although HHV-6A could infect a broader variety of human cells than HHV-6B.

Both HHV-6A and HHV-6B infect also astrocytes, microglia and neurons (Chan et al., 2001; Cuomo et al., 2001). An association of these viruses with central nervous system diseases has been demonstrated, such as multiple sclerosis (Challoner et al., 1995; Noseworthy et al., 2000) and progressive multifocal leukoencephalopathy (Daibata et al., 2001; Mock et al., 1999). Besides T lymphocytes and nerves system cells, HHV-6 could also infect a broad range of cells, although susceptibility seems to be different between HHV-6A and HHV-6B.

The difference in cell tropism could be explained from the distinct cellular receptor usage of these two viruses. Regarding HHV-6A, its cellular receptor is a ubiquitous molecule, CD46, which is compatible with a broad cell tropism. However, productive HHV-6A infection is limited to a relatively small range of cells, indicating that there should be other cellular factors important for the viral life cycle. Regarding HHV-6B, its entry receptor is CD134 which is mainly expressed in activated T cells (Tang et al., 2013).

2.4 Discovery of HHV-6A unique glycoproteins gQ1 and gQ2 and their complex with gH and gL

Glycoprotein Q (gQ) gene was first described by Pfeiffer et al. in 1993, and is located at the right side of HHV-6A genome. By generating neutralizing antibodies for HHV-6A, they found that the antigen recognized by these neutralizing antibodies was a component of gp82-gp105 (Pfeiffer et al., 1993), and that its mRNA was highly spliced (Pfeiffer et al., 1995). Later, a more abundant mRNA derived from this region in HHV-6A genome were identified by our group, and the gene products were designated as gQ (glycoprotein Q) (Mori et al., 2003) because they were modified by N linked glycans. A form of gQ forms a complex with gH and gL, which interacts with CD46 (Mori et al., 2004). Furthermore, our group identified an another small transcript originating 3′ of this transcript, and a glycoprotein encoded from this small transcript, which was also incorporated into the gH/gL/gQ complex, was designated as gQ2 prompting renaming of the first isolated gQ as gQ1. To form the gH/gL/gQ complex, gQ2 was found to be essential. Therefore the complex was a tetrameric gH/gL/gQ1/gQ2 complex. Subsequently, this tetrameric complex was also identified in HHV-6B (Kawabata et al., 2011).

2.5 Function of gH/gL/gQ1/gQ2 complex in HHV-6A as viral ligands

The most important function of the gH/gL/gQ1/gQ2 complex was eventually defined as the viral ligand for HHV-6A cellular receptor, CD46, in virus entry.

Santoro et al. reported that gH of HHV-6A was the CD46-binding component. They showed that antibody against CD46 immunoprecipitated gH in HHV-6A (GS strain) infected cell lysates, and antibody against CD46 could not immunoprecipitate gH from gH-depleted HHV-6A infected lysates, although it immunoprecipitated gH from gQ1 or gB depleted lysates. CD46 could be detected from immunoprecipitates of HHV-6A infected cells by antibody against gH (Santoro et al., 2003).

Our group showed that not only antibody against gH but also against gQ1 could immunoprecipitate CD46 from HHV-6A (GS strain) infected cell lysates (Mori et al., 2004). Furthermore, monoclonal antibody against gQ1 co-immunoprecipitated CD46 in HHV-6A (GS strain) infected cell lysates, and antibody against CD46 co-immunoprecipitated gH, gL, gQ1 and gQ2 after transient expression of all four genes by transfection (Tang et al., 2011).

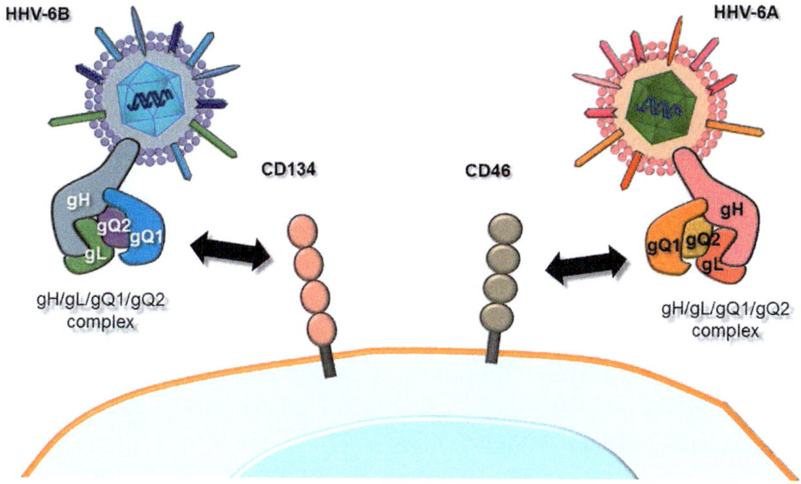

Fig. 5 Viral ligands of HHV-6A and HHV-6B and entry receptors. A schematic illustration of HHV-6A and HHV-6B entry via interaction between gH/gL/gQ1/gQ2 complex and specific entry receptors. The gH/gL/gQ1/gQ2 of HHV-6A recognizes CD46, whereas gH/gL/gQ1/gQ2 of HHV-6B specifically binds to CD134.

Maturation of gQ1, which could be shown by its size shift from 74 KDa (gQ1-74K) to 80 KDa(gQ1-80 K) due to modification with complex N-linked glycans, occurs only when all four components, gH, gL, gQ1 and gQ2 are coexpressed and gQ1-80 K, not gQ1-74 K, was incorporated into gH/gL/gQ1/gQ2 complex (Tang et al., 2011). Tetramer formation was required to bind CD46.

Detailed functional analysis of the HHV-6A gH/gL/gQ1/gQ2 complex showed that all four were needed for complex maturation, transport of the complex from ER to Golgi and its binding to CD46 (Tang et al., 2011). Partial replacement of the complex components with those of HHV-6B demonstrated that HHV-6A gQ1 and gQ2 play the key role for CD46 interaction (Jasirwan et al., 2014). Furthermore, a short amino sequence in HHV-6A gQ1 (494a.a–497a.a) was found to be important for HHV-6A gH/gL/gQ1/gQ2 complex function (Maeki et al., 2013) (Fig. 5).

2.6 Function of CD46 as a receptor

CD46 is a member of the regulators of complement activation (RCA) glycoprotein family (Seya et al., 1999), which are type I transmembrane glycoproteins of 45–67 kDa, expressed on all nucleated cells (Cardone et al., 2011). CD46 consists of four short consensus repeats (SCRs), serine–threonine–proline (STP) rich domain, a small region of unknown significance, a transmembrane domain and a cytoplasmic tail. Most of the extracellular portion is composed of the four SCRs. Four distinct CD46 isoforms generated by alternative RNA splicing are expressed differentially in various cell types. These isomers have the same SCRs but different STP or cytoplasmic domains (Seya et al., 1999). CD46 functions as a co-factor in the factor-I mediated proteolytic cleavage of C3b and C4b and this process protects host from inadvertent lysis by the complement system. Interactions with C3b and C4b involve several regions on domains SCR2, SCR3 and SCR4 (Persson et al., 2010). CD46 is known to have roles not only in innate immunity but also adaptive immunity (Cardone et al., 2011).

CD46 functions as the entry receptor for several viruses (all species of B adenoviruses except 3 and 7 Gaggar et al., 2003, and measles virus Kemper and Atkinson, 2009), and bacteria (Kallstrom et al., 1997, 2001; Rezcallah et al., 2005). As described above, the extracellular domain of CD46 contains four short consensus repeats (SCR) (Seya et al., 1990). Usage of CD46 domains was different among these pathogens. For HHV-6A, SCR2 and SCR3 of CD46 were shown to be required (Greenstone et al., 2002; Mori et al., 2002). In the analysis using CD46 deletions, our group found that SCR4 was also required for HHV-6A infection (Mori et al., 2002). From these findings, the direct binding domains appear to be in SCR2 and SCR3, while SCR4 may be required only for maintaining the conformation of this domain. Because even when SCR4 domain of CD46 was substituted with the corresponding region of DAF, CD46 showed efficient HHV-6 receptor activity, while it was not seen in CD46 lacking SCR4.

For measles virus entry, only SCR1 and SCR2 are required (Greenstone et al., 2002).

Four isoforms of CD46 derive from alternative RNA splicing. Hansen et al. reported that different expression patterns on T cell lines have little effect on the sensitivity of these cells to HHV-6A infection, but only T cell lines with the specific isoforms of CD46 may be susceptible for a particular HHV-6B strain (PL1 strain) (Hansen et al., 2017).

2.7 Discovery of human CD134 as an entry receptor specific for HHV-6B

Although CD46 was first reported as entry receptor for both HHV-6A and HHV-6B, several observations indicated the existence of an alternative receptor for HHV-6B.

HHV-6A could induce cell-cell fusion of CHO (Chinese hamster ovary) cells expressing CD46 even in the absence of viral protein synthesis (fusion from without: FFWO), but this was not observed in HHV-6B (HST strain and other clinical isolates from infants with ES) (Mori et al., 2002; Pedersen et al., 2006). A monoclonal antibody for CD46 could block HHV-6A infection but not HHV-6B infection. In addition, antibody against gH could co-precipitate CD46 from HHV-6A infected cell lysates but not from HHV-6B (Oyaizu et al., 2012).

All these findings led to the identification of a HHV-6B specific receptor, human CD134, and its viral ligand was HHV-6B gH/gL/gQ1/gQ2 complex.

CD134 is a member of the TNF receptor superfamily. It is a type I transmembrane protein, and a co-stimulator for T cell activation (Higgins et al., 1999).

Human CD134 interacted with HHV-6B gH/gL/gQ1/gQ2 complex but not that of HHV-6A. Ectopic expression of CD134 in several HHV-6B non-permissive cell lines rendered these cells susceptible for HHV-6B infection (Tang et al., 2013). Therefore, HHV-6A and -6B recognized distinct receptors which could partially explain their different cell tropism.

Compared with the ubiquitous expression of CD46, CD134 is mainly expressed on activated primary T cells (Weinberg et al., 1996).

Co-expression of only gQ1 and gQ2 was found to be sufficient for CD134 binding of HHV-6B (Tang et al., 2014) while all four components of HHV-6A gH/gL/gQ1/gQ2 complex are required for CD46 binding. These suggests that gQ1/gQ2 complex by itself could bind to CD134, although they are part of the tetrameric complex on the virion. Moreover, an amino acid residue in HHV-6B gQ1 (E127) was identified to be important for CD134 binding by replacement of gQ1 of HHV-6A by HHV-6B (Tang and Mori, 2015).

The ectodomain of CD134 contains four cysteine-rich domains (CRD) and a stalk domain (Marsters et al., 1992) CRD2 is needed for HHV-6B complex binding. Furthermore, two amino acid residues (K79 and W86) within CRD2 were identified to be key for its interaction with the HHV-6B complex, and the positive charge of K79 could associate with the negative charge of E127 of HHV-6B gQ1 and contribute the interaction. Interestingly, mutant CD134 could bind to HHV-6B tetrameric complex when homologous residues in murine CD134 (mCD134) were humanized (mCD134 D75K/Q82W), (Tang et al., 2014; Tang and Mori, 2015).

Discovery of the HHV-6B specific receptor, human CD134 explained several of the differences observed between HHV-6A and HHV-6B infection.

2.8 HHV-6 gH/gL/gO complex which is the other complex with gH/gL

Glycoprotein O (gO) is conserved among betaherpesviruses. gO of HHV-6A and HHV-6B forms a trimeric complex with gH/gL in virus infected cells and is incorporated into the virion. Different forms of HHV-6 gO (120–130KDa gO gO-130 K; 75–80KDa gO gO-80 K) were expressed in infected cells. Only gO-80K was incorporated into HHV-6 virions (Mori et al., 2004). Later study showed that the C-terminal of gO-130K was cleaved during its maturation process, although it is unclear whether the cleavage is important for gO maturation and function (Tang et al., 2015).

Although deletion of HCMV or MCMV gO results in severe growth deficit (Jiang et al., 2008; Scrivano et al., 2010; Wille et al., 2010), gO-deficient HHV-6A showed no obvious growth defect in T cells (Tang et al., 2015), suggesting that HHV-6A gO is non-essential for in vitro virus growth. In case of HHV-6B, it is unknown whether gO is essential for virus infection because so far there is no system for construction of recombinant virus.

The detailed gO function during HHV-6 infection is less clear. It has been already found that HHV-6A- and HHV-6B- gH/gL/gO complexes could not bind CD46 and therefore may associate with an unknown cellular factor and contribute to HHV-6A and HHV-6B infections.

2.9 The roles of gB and gH/gL in the HHV-6 entry process

gB, gH and gL are conserved among herpesviruses and are essential for virus entry (Spear and Longnecker, 2003). Generally, gB and gH are known to be targets for neutralizing antibodies, and also shown to be key for entry of HHV-6A and HHV-6B (Takeda et al., 1996, 1997). gB and gH are required for HHV-6 induced cell-cell fusion (Mori et al., 2002). HHV-6 gB is a ca. 112KDa type I transmembrane protein of 830 amino acids. During its maturation process, gB is cleaved into two subunits of 64 and 58KDa by host furin protease, which are covalently linked by a disulfide bond (Chou and Marousek, 1992; Ellinger et al., 1993). Sequence divergence of HHV-6A and HHV-6B viruses has been reported.

A monoclonal antibody for gB specific for HHV-6A or HHV-6B is used for the differentiation of clinical isolates of HHV-6A and HHV-6B.

The cytoplasmic tail domain (CTD) of gB in other herpesviruses has been demonstrated to be important for regulation of gB function, like enhancing, reducing or abolishing cell fusion (Beitia Ortiz de Zarate et al., 2007; Foster et al., 2001; Garcia et al., 2013). Deletion of the HHV-6A gB CTD abrogated the reconstitution of infectious virus from the mutant virus genome. In addition, HHV-6 gB without CTD diffusely distributed in the cytoplasm while authentic gB accumulated in TGN in transient expression assays. Therefore, HHV-6 gB CTD is thought to be required for intracellular transport of gB (Mahmoud et al., 2016).

2.10 Lipids in both HHV-6 envelope and target cell membrane affect virus entry

Cholesterol both in HHV-6A envelope and target cell is required for HHV-6A entry (Huang et al., 2006; Tang et al., 2008). Moreover, HHV-6A infection induces the re-location of CD46 into the lipid rafts and envelope glycoproteins are associated with their lipid rafts, and these suggest that cellular-membrane lipid rafts are important in viral entry which shows HHV-6A could enter the target cells via the lipid rafts (Tang et al., 2008). Furthermore, it was also demonstrated that HHV-6A envelope contained lipid rafts. Therefore, HHV-6A virion might assemble through lipid rafts (Kawabata et al., 2009). These data demonstrated lipid rafts both in HHV-6A envelope and its target cells are essential role in HHV-6A entry and they may be involved in HHV-6A virus assembly. In the case of HHV-6B, similar events are predicted to occur although there is no report.

2.11 Correlation between HHV-6B reactivation/replication and CD134 expression

As described above, in an immunocompromised host, such as a patient after allogeneic hematopoietic stem cell transplantation (allo-HSCT), HHV-6B can reactivate and sometimes cause encephalitis (Wainwright et al., 2001; Yoshikawa et al., 1991). Recently, a correlation between CD134 upregulation in T cells and HHV-6B replication after allo-HSCT was demonstrated (Nagamata et al., 2018), and the data indicates that CD134 upregulation in T cells is a key player for HHV-6B reactivation/replication after allo-HSCT. The report suggests that latently infected HHV-6B could be reactivated by several conditions, such as immunosuppression, and then reactivated viruses could enter activated-T cells via CD134 and replicate productively. Miyagawa et al. reported that CD134 was upregulated in $CD4^+$ T cells of patients with drug reaction with eosinophilia and systemic symptoms (DRESS)/drug-induced hypersensitivity syndrome (DIHS) (Miyagawa et al., 2016), which has been related to HHV-6 replication (Suzuki et al., 1998). Correlation of HHV-6B reactivation/replication and upregulation of CD134 in T cells has been described in HHV-6B-related diseases. The finding of host receptor(s) of HHV-6 would lead to a step for elucidation of HHV-6B-related pathogenesis.

References

Abdellatif, M.E.A., Sinzger, C., et al., 2018. Investigating HCMV entry into host cells by STEM tomography. J. Struct. Biol. 204 (3), 406–419.
Ablashi, D., Agut, H., et al., 2013. Classification of HHV-6A and HHV-6B as distinct viruses. Arch. Virol. 159 (5), 863–870.
Ablashi, D.V., Balachandran, N., et al., 1991. Genomic polymorphism, growth properties, and immunologic variations in human herpesvirus-6 isolates. Virology 184 (2), 545–552.
Adler, B., Scrivano, L., et al., 2006. Role of human cytomegalovirus UL131A in cell type-specific virus entry and release. J. Gen. Virol. 87 (Pt 9), 2451–2460.
Akkapaiboon, P., Mori, Y., et al., 2004. Intracellular processing of human herpesvirus 6 glycoproteins Q1 and Q2 into tetrameric complexes expressed on the viral envelope. J. Virol. 78 (15), 7969–7983.
Appleton, B.A., Wu, P., et al., 2007. Structural studies of neuropilin/antibody complexes provide insights into semaphorin and VEGF binding. EMBO J. 26 (23), 4902–4912.
Arbuckle, J.H., Medveczky, M.M., et al., 2010. The latent human herpesvirus-6A genome specifically integrates in telomeres of human chromosomes in vivo and in vitro. Proc. Natl. Acad. Sci. U. S. A. 107 (12), 5563–5568.
Backovic, M., DuBois, R.M., et al., 2010. Structure of a core fragment of glycoprotein H from pseudorabies virus in complex with antibody. Proc. Natl. Acad. Sci. U. S. A. 107 (52), 22635–22640.
Backovic, M., Longnecker, R., et al., 2009. Structure of a trimeric variant of the Epstein-Barr virus glycoprotein B. Proc. Natl. Acad. Sci. U. S. A. 106 (8), 2880–2885.
Beitia Ortiz de Zarate, I., Cantero-Aguilar, L., et al., 2007. Contribution of endocytic motifs in the cytoplasmic tail of herpes simplex virus type 1 glycoprotein B to virus replication and cell-cell fusion. J. Virol. 81 (24), 13889–13903.
Burke, H.G., Heldwein, E.E., 2015. Crystal structure of the human cytomegalovirus glycoprotein B. PLoS Pathog. 11 (10), e1005227.
Cardone, J., Le Friec, G., et al., 2011. CD46 in innate and adaptive immunity: an update. Clin. Exp. Immunol. 164 (3), 301–311.
Challoner, P.B., Smith, K.T., et al., 1995. Plaque-associated expression of human herpesvirus 6 in multiple sclerosis. Proc. Natl. Acad. Sci. U. S. A. 92 (16), 7440–7444.
Chan, P.K., Ng, H.K., et al., 2001. Prevalence and distribution of human herpesvirus 6 variants A and B in adult human brain. J. Med. Virol. 64 (1), 42–46.
Chan, G., Nogalski, M.T., et al., 2009. Activation of EGFR on monocytes is required for human cytomegalovirus entry and mediates cellular motility. Proc. Natl. Acad. Sci. U. S. A. 106 (52), 22369–22374.
Chandramouli, S., Ciferri, C., et al., 2015. Structure of HCMV glycoprotein B in the post-fusion conformation bound to a neutralizing human antibody. Nat. Commun. 6, 8176.
Chandramouli, S., Malito, E., et al., 2017. Structural basis for potent antibody-mediated neutralization of human cytomegalovirus. Sci. Immunol. 2 (12). eaan1457.
Chapenko, S., Krumina, A., et al., 2006. Activation of human herpesviruses 6 and 7 in patients with chronic fatigue syndrome. J. Clin. Virol. 37 (Suppl 1), S47–S51.
Chou, S., Marousek, G.I., 1992. Homology of the envelope glycoprotein B of human herpesvirus-6 and cytomegalovirus. Virology 191 (1), 523–528.

Chowdary, T.K., Cairns, T.M., et al., 2010. Crystal structure of the conserved herpesvirus fusion regulator complex gH-gL. Nat. Struct. Mol. Biol. 17 (7), 882–888.

Ciferri, C., Chandramouli, S., et al., 2015a. Structural and biochemical studies of HCMV gH/gL/gO and pentamer reveal mutually exclusive cell entry complexes. Proc. Natl. Acad. Sci. U. S. A. 112 (6), 1767–1772.

Ciferri, C., Chandramouli, S., et al., 2015b. Antigenic characterization of the HCMV gH/gL/gO and pentamer cell entry complexes reveals binding sites for potently neutralizing human antibodies. PLoS Pathog. 11 (10), e1005230.

Cirone, M., Zompetta, C., et al., 1992. Infection by human herpesvirus 6 (HHV-6) of human lymphoid T cells occurs through an endocytic pathway. AIDS Res. Hum. Retroviruses 8 (12), 2031–2037.

Compton, T., Nepomuceno, R.R., et al., 1992. Human cytomegalovirus penetrates host cells by pH-independent fusion at the cell surface. Virology 191 (1), 387–395.

Compton, T., Nowlin, D.M., et al., 1993. Initiation of human cytomegalovirus infection requires initial interaction with cell surface heparan sulfate. Virology 193 (2), 834–841.

Connolly, S.A., Jackson, J.O., et al., 2011. Fusing structure and function: a structural view of the herpesvirus entry machinery. Nat. Rev. Microbiol. 9 (5), 369–381.

Cooper, R.S., Georgieva, E.R., et al., 2018. Structural basis for membrane anchoring and fusion regulation of the herpes simplex virus fusogen gB. Nat. Struct. Mol. Biol. 25 (5), 416–424.

Cooper, R.S., Heldwein, E.E., 2015. Herpesvirus gB: a finely tuned fusion machine. Viruses 7 (12), 6552–6569.

Cuomo, L., Trivedi, P., et al., 2001. Human herpesvirus 6 infection in neoplastic and normal brain tissue. J. Med. Virol. 63 (1), 45–51.

Daibata, M., Hatakeyama, N., et al., 2001. Detection of human herpesvirus 6 and JC virus in progressive multifocal leukoencephalopathy complicating follicular lymphoma. Am. J. Hematol. 67 (3), 200–205.

Davis, N.L., King, C.C., et al., 2017. Cytomegalovirus infection in pregnancy. Birth Defects Res. 109 (5), 336–346.

De Bolle, L., Van Loon, J., et al., 2005. Quantitative analysis of human herpesvirus 6 cell tropism. J. Med. Virol. 75 (1), 76–85.

Dominguez, G., Dambaugh, T.R., et al., 1999. Human herpesvirus 6B genome sequence: coding content and comparison with human herpesvirus 6A. J. Virol. 73 (10), 8040–8052.

Ellinger, K., Neipel, F., et al., 1993. The glycoprotein B homologue of human herpesvirus 6. J. Gen. Virol. 74 (Pt 3), 495–500.

Feire, A.L., Koss, H., et al., 2004. Cellular integrins function as entry receptors for human cytomegalovirus via a highly conserved disintegrin-like domain. Proc. Natl. Acad. Sci. U. S. A. 101 (43), 15470–15475.

Feire, A.L., Roy, R.M., et al., 2010. The glycoprotein B disintegrin-like domain binds beta 1 integrin to mediate cytomegalovirus entry. J. Virol. 84 (19), 10026–10037.

Fontana, J., Atanasiu, D., et al., 2017. The fusion loops of the initial Prefusion conformation of herpes simplex virus 1 fusion protein point toward the membrane. MBio 8 (4).

Foster, T.P., Melancon, J.M., et al., 2001. An alpha-helical domain within the carboxyl terminus of herpes simplex virus type 1 (HSV-1) glycoprotein B (gB) is associated with cell fusion and resistance to heparin inhibition of cell fusion. Virology 287 (1), 18–29.

Gaggar, A., Shayakhmetov, D.M., et al., 2003. CD46 is a cellular receptor for group B adenoviruses. Nat. Med. 9 (11), 1408–1412.

Gallagher, J.R., Atanasiu, D., et al., 2014. Functional fluorescent protein insertions in herpes simplex virus gB report on gB conformation before and after execution of membrane fusion. PLoS Pathog. 10 (9), e1004373.

Garcia, N.J., Chen, J., et al., 2013. Modulation of Epstein-Barr virus glycoprotein B (gB) fusion activity by the gB cytoplasmic tail domain. MBio 4 (1), e00512–e00571.

Gerna, G., Baldanti, F., et al., 2004. Pathogenesis of human cytomegalovirus infection and cellular targets. Hum. Immunol. 65 (5), 381–386.
Gerna, G., Percivalle, E., et al., 2002. The attenuated Towne strain of human cytomegalovirus may revert to both endothelial cell tropism and leuko- (neutrophil- and monocyte-)tropism in vitro. J. Gen. Virol. 83 (Pt 8), 1993–2000.
Gerna, G., Percivalle, E., et al., 2005. Dendritic-cell infection by human cytomegalovirus is restricted to strains carrying functional UL131-128 genes and mediates efficient viral antigen presentation to CD8+ T cells. J. Gen. Virol. 86 (Pt 2), 275–284.
Gompels, U.A., Nicholas, J., et al., 1995. The DNA sequence of human herpesvirus-6: structure, coding content, and genome evolution. Virology 209 (1), 29–51.
Grazia Revello, M., Baldanti, F., et al., 2001. In vitro selection of human cytomegalovirus variants unable to transfer virus and virus products from infected cells to polymorphonuclear leukocytes and to grow in endothelial cells. J. Gen. Virol. 82 (Pt 6), 1429–1438.
Greenstone, H.L., Santoro, F., et al., 2002. Human herpesvirus 6 and measles virus employ distinct CD46 domains for receptor function. J. Biol. Chem. 277 (42), 39112–39118.
Ha, S., Li, F., et al., 2017. Neutralization of diverse human Cytomegalovirus strains conferred by antibodies targeting viral gH/gL/pUL128-131 pentameric complex. J. Virol. 91 (7).
Haas, J.G., Lathe, R., 2018. Microbes and Alzheimer's disease: new findings call for a paradigm change. Trends Neurosci. 41 (9), 570–573.
Hahn, G., Revello, M.G., et al., 2004. Human cytomegalovirus UL131-128 genes are indispensable for virus growth in endothelial cells and virus transfer to leukocytes. J. Virol. 78 (18), 10023–10033.
Hansen, A.S., Bundgaard, B.B., et al., 2017. Divergent tropism of HHV-6AGS and HHV-6BPL1 in T cells expressing different CD46 isoform patterns. Virology 502, 160–170.
Harrison, S.C., 2015. Viral membrane fusion. Virology 479–480, 498–507.
Haspot, F., Lavault, A., et al., 2012. Human cytomegalovirus entry into dendritic cells occurs via a macropinocytosis-like pathway in a pH-independent and cholesterol-dependent manner. PLoS One 7 (4), e34795.
Heldwein, E.E., Lou, H., et al., 2006. Crystal structure of glycoprotein B from herpes simplex virus 1. Science 313 (5784), 217–220.
Hetzenecker, S., Helenius, A., et al., 2016. HCMV induces macropinocytosis for host cell entry in fibroblasts. Traffic 17 (4), 351–368.
Higgins, L.M., McDonald, S.A., et al., 1999. Regulation of T cell activation in vitro and in vivo by targeting the OX40-OX40 ligand interaction: amelioration of ongoing inflammatory bowel disease with an OX40-IgG fusion protein, but not with an OX40 ligand-IgG fusion protein. J. Immunol. 162 (1), 486–493.
Hochdorfer, D., Florin, L., et al., 2016. Tetraspanin CD151 promotes initial events in human Cytomegalovirus infection. J. Virol. 90 (14), 6430–6442.
Huang, H., Li, Y., et al., 2006. Human herpesvirus 6 envelope cholesterol is required for virus entry. J. Gen. Virol. 87 (Pt 2), 277–285.
Isaacson, M.K., Compton, T., 2009. Human cytomegalovirus glycoprotein B is required for virus entry and cell-to-cell spread but not for virion attachment, assembly, or egress. J. Virol. 83 (8), 3891–3903.
Isaacson, M.K., Feire, A.L., et al., 2007. Epidermal growth factor receptor is not required for human cytomegalovirus entry or signaling. J. Virol. 81 (12), 6241–6247.
Isegawa, Y., Mukai, T., et al., 1999. Comparison of the complete DNA sequences of human herpesvirus 6 variants A and B. J. Virol. 73 (10), 8053–8063.
Jasirwan, C., Furusawa, Y., et al., 2014. Human herpesvirus-6A gQ1 and gQ2 are critical for human CD46 usage. Microbiol. Immunol. 58 (1), 22–30.
Jiang, X.J., Adler, B., et al., 2008. UL74 of human cytomegalovirus contributes to virus release by promoting secondary envelopment of virions. J. Virol. 82 (6), 2802–2812.

Kabanova, A., Marcandalli, J., et al., 2016. Platelet-derived growth factor-alpha receptor is the cellular receptor for human cytomegalovirus gHgLgO trimer. Nat. Microbiol. 1 (8) 16082.

Kahl, M., Siegel-Axel, D., et al., 2000. Efficient lytic infection of human arterial endothelial cells by human cytomegalovirus strains. J. Virol. 74 (16), 7628–7635.

Kallstrom, H., Blackmer Gill, D., et al., 2001. Attachment of Neisseria gonorrhoeae to the cellular pilus receptor CD46: identification of domains important for bacterial adherence. Cell. Microbiol. 3 (3), 133–143.

Kallstrom, H., Liszewski, M.K., et al., 1997. Membrane cofactor protein (MCP or CD46) is a cellular pilus receptor for pathogenic Neisseria. Mol. Microbiol. 25 (4), 639–647.

Kaufer, B.B., Flamand, L., 2014. Chromosomally integrated HHV-6: impact on virus, cell and organismal biology. Curr. Opin. Virol. 9, 111–118.

Kawabata, A., Oyaizu, H., et al., 2011. Analysis of a neutralizing antibody for human herpesvirus 6B reveals a role for glycoprotein Q1 in viral entry. J. Virol. 85 (24), 12962–12971.

Kawabata, A., Tang, H., et al., 2009. y Human herpesvirus 6 envelope components enriched in lipid rafts: evidence for virion-associated lipid rafts. Virol. J. 6, 127.

Kemper, C., Atkinson, J.P., 2009. Measles virus and CD46. Curr. Top. Microbiol. Immunol. 329, 31–57.

Kim, J.H., Collins-McMillen, D., et al., 2017. Human cytomegalovirus requires epidermal growth factor receptor signaling to enter and initiate the early steps in the establishment of latency in CD34(+) human progenitor cells. J. Virol. 91 (5), pii: e01206-16.

Leibovitch, E.C., Jacobson, S., 2014. Evidence linking HHV-6 with multiple sclerosis: an update. Curr. Opin. Virol. 9, 127–133.

Li, Q., Fischer, E., et al., 2016. Cell surface THY-1 contributes to human Cytomegalovirus entry via a macropinocytosis-like process. J. Virol. 90 (21), 9766–9781.

Li, G., Nguyen, C.C., et al., 2015a. A viral regulator of glycoprotein complexes contributes to human cytomegalovirus cell tropism. Proc. Natl. Acad. Sci. U. S. A. 112 (14), 4471–4476.

Li, Q., Wilkie, A.R., et al., 2015b. THY-1 cell surface antigen (CD90) has an important role in the initial stage of human Cytomegalovirus infection. PLoS Pathog. 11 (7), e1004999.

Liu, J., Jardetzky, T.S., et al., 2018. The human Cytomegalovirus trimer and pentamer promote sequential steps in entry into epithelial and endothelial cells at cell surfaces and endosomes. J. Virol. 92 (21), pii: e01336-18.

Luganini, A., Cavaletto, N., et al., 2017. Loss of the human Cytomegalovirus US16 protein abrogates virus entry into endothelial and epithelial cells by reducing the virion content of the pentamer. J. Virol. 91 (11), pii: e00205-17.

Lusso, P., Markham, P.D., et al., 1988. In vitro cellular tropism of human B-lymphotropic virus (human herpesvirus-6). J. Exp. Med. 167 (5), 1659–1670.

Macagno, A., Bernasconi, N.L., et al., 2010. Isolation of human monoclonal antibodies that potently neutralize human cytomegalovirus infection by targeting different epitopes on the gH/gL/UL128-131A complex. J. Virol. 84 (2), 1005–1013.

Maeki, T., Hayashi, M., et al., 2013. Identification of the human herpesvirus 6A gQ1 domain essential for its functional conformation. J. Virol. 87 (12), 7054–7063.

Mahmoud, N.F., Jasirwan, C., et al., 2016. Cytoplasmic tail domain of glycoprotein B is essential for HHV-6 infection. Virology 490, 1–5.

Malito, E., Chandramouli, S., et al., 2018. From recognition to execution-the HCMV Pentamer from receptor binding to fusion triggering. Curr. Opin. Virol. 31, 43–51.

Marsters, S.A., Frutkin, A.D., et al., 1992. Identification of cysteine-rich domains of the type 1 tumor necrosis factor receptor involved in ligand binding. J. Biol. Chem. 267 (9), 5747–5750.

Martinez-Martin, N., Marcandalli, J., et al., 2018. An unbiased screen for human Cytomegalovirus identifies Neuropilin-2 as a central viral receptor. Cell 174 (5), 1158–1171, e1119.
Matsuura, H., Kirschner, A.N., et al., 2010. Crystal structure of the Epstein-Barr virus (EBV) glycoprotein H/glycoprotein L (gH/gL) complex. Proc. Natl. Acad. Sci. U. S. A. 107 (52), 22641–22646.
Miyagawa, F., Nakamura, Y., et al., 2016. Preferential expression of CD134, an HHV-6 cellular receptor, on CD4T cells in drug-induced hypersensitivity syndrome (DIHS)/drug reaction with eosinophilia and systemic symptoms (DRESS). J. Dermatol. Sci. 83 (2), 151–154.
Mock, D.J., Powers, J.M., et al., 1999. Association of human herpesvirus 6 with the demyelinative lesions of progressive multifocal leukoencephalopathy. J. Neurovirol. 5 (4), 363–373.
Mori, Y., 2009. Recent topics related to human herpesvirus 6 cell tropism. Cell. Microbiol. 11 (7), 1001–1006.
Mori, Y., Akkapaiboon, P., et al., 2003. The human herpesvirus 6 U100 gene product is the third component of the gH-gL glycoprotein complex on the viral envelope. J. Virol. 77 (4), 2452–2458.
Mori, Y., Akkapaiboon, P., et al., 2004. Discovery of a second form of tripartite complex containing gH-gL of human herpesvirus 6 and observations on CD46. J. Virol. 78 (9), 4609–4616.
Mori, Y., Seya, T., et al., 2002. Human herpesvirus 6 variant A but not variant B induces fusion from without in a variety of human cells through a human herpesvirus 6 entry receptor, CD46. J. Virol. 76 (13), 6750–6761.
Nagamata, S., Nagasaka, M., et al., 2018. Human CD134 (OX40) expressed on T cells plays a key role for human herpesvirus 6B replication after allogeneic hematopoietic stem cell transplantation. J. Clin. Virol. 102, 50–55.
Nguyen, C.C., Kamil, J.P., 2018. Pathogen at the gates: human cytomegalovirus entry and cell tropism. Viruses 10 (12), pii: E704.
Nguyen, C.C., Siddiquey, M.N.A., et al., 2018. Human cytomegalovirus tropism modulator UL148 interacts with SEL1L, a cellular factor that governs endoplasmic reticulum-associated degradation of the viral envelope glycoprotein gO. J. Virol. 92 (18), pii: e00688-18.
Nogalski, M.T., Chan, G., et al., 2011. Human cytomegalovirus-regulated paxillin in monocytes links cellular pathogenic motility to the process of viral entry. J. Virol. 85 (3), 1360–1369.
Noseworthy, J.H., Lucchinetti, C., et al., 2000. Multiple sclerosis. N. Engl. J. Med. 343 (13), 938–952.
O'Connor, C.M., Shenk, T., 2012. Human cytomegalovirus pUL78 G protein-coupled receptor homologue is required for timely cell entry in epithelial cells but not fibroblasts. J. Virol. 86 (21), 11425–11433.
Oyaizu, H., Tang, H., et al., 2012. Complementation of the function of glycoprotein H of human herpesvirus 6 variant a by glycoprotein H of variant B in the virus life cycle. J. Virol. 86 (16), 8492–8498.
Patrone, M., Secchi, M., et al., 2005. Human cytomegalovirus UL130 protein promotes endothelial cell infection through a producer cell modification of the virion. J. Virol. 79 (13), 8361–8373.
Pedersen, S.M., Oster, B., et al., 2006. Induction of cell-cell fusion from without by human herpesvirus 6B. J. Virol. 80 (19), 9916–9920.
Persson, B.D., Schmitz, N.B., et al., 2010. Structure of the extracellular portion of CD46 provides insights into its interactions with complement proteins and pathogens. PLoS Pathog. 6 (9), e1001122.

Pettersen, E.F., Goddard, T.D., et al., 2004. UCSF chimera—a visualization system for exploratory research and analysis. J. Comput. Chem. 25 (13), 1605–1612.
Pfeiffer, B., Berneman, Z.N., et al., 1993. Identification and mapping of the gene encoding the glycoprotein complex gp82-gp105 of human herpesvirus 6 and mapping of the neutralizing epitope recognized by monoclonal antibodies. J. Virol. 67 (8), 4611–4620.
Pfeiffer, B., Thomson, B., et al., 1995. Identification and characterization of a cDNA derived from multiple splicing that encodes envelope glycoprotein gp105 of human herpesvirus 6. J. Virol. 69 (6), 3490–3500.
Plotkin, S.A., Boppana, S.B., 2018. Vaccination against the human cytomegalovirus. Vaccine. Apr 2, pii: S0264-410X(18)30288-3.
Prusty, B.K., Krohne, G., et al., 2013. Reactivation of chromosomally integrated human herpesvirus-6 by telomeric circle formation. PLoS Genet. 9 (12), e1004033.
Readhead, B., Haure-Mirande, J.V., et al., 2018. Multiscale analysis of independent Alzheimer's cohorts finds disruption of molecular, genetic, and clinical networks by human herpesvirus. Neuron 99 (1), 64–82 e67.
Rezcallah, M.S., Hodges, K., et al., 2005. Engagement of CD46 and alpha5beta1 integrin by group A streptococci is required for efficient invasion of epithelial cells. Cell. Microbiol. 7 (5), 645–653.
Rogalin, H.B., Heldwein, E.E., 2015. Interplay between the herpes simplex virus 1 gB cytodomain and the gH cytotail during cell-cell fusion. J. Virol. 89 (24), 12262–12272.
Roizmann, B., Desrosiers, R.C., et al., 1992. The family Herpesviridae: an update. The herpesvirus study Group of the International Committee on taxonomy of viruses. Arch. Virol. 123 (3–4), 425–449.
Ryckman, B.J., Chase, M.C., et al., 2008b. HCMV gH/gL/UL128-131 interferes with virus entry into epithelial cells: evidence for cell type-specific receptors. Proc. Natl. Acad. Sci. U. S. A. 105 (37), 14118–14123.
Ryckman, B.J., Rainish, B.L., et al., 2008a. Characterization of the human cytomegalovirus gH/gL/UL128-131 complex that mediates entry into epithelial and endothelial cells. J. Virol. 82 (1), 60–70.
Salahuddin, S.Z., Ablashi, D.V., et al., 1986. Isolation of a new virus, HBLV, in patients with lymphoproliferative disorders. Science 234 (4776), 596–601.
Santoro, F., Greenstone, H.L., et al., 2003. Interaction of glycoprotein H of human herpesvirus 6 with the cellular receptor CD46. J. Biol. Chem. 278 (28), 25964–25969.
Santoro, F., Kennedy, P.E., et al., 1999. CD46 is a cellular receptor for human herpesvirus 6. Cell 99 (7), 817–827.
Sathiyamoorthy, K., Chen, J., et al., 2017. The COMPLEXity in herpesvirus entry. Curr. Opin. Virol. 24, 97–104.
Sathiyamoorthy, K., Hu, Y.X., et al., 2016. Structural basis for Epstein-Barr virus host cell tropism mediated by gp42 and gHgL entry glycoproteins. Nat. Commun. 7 13557.
Sathiyamoorthy, K., Jiang, J., et al., 2014. Assembly and architecture of the EBV B cell entry triggering complex. PLoS Pathog. 10 (8), e1004309.
Scrivano, L., Esterlechner, J., et al., 2010. The m74 gene product of murine cytomegalovirus (MCMV) is a functional homolog of human CMV gO and determines the entry pathway of MCMV. J. Virol. 84 (9), 4469–4480.
Seya, T., Hara, T., et al., 1990. Complement-mediated tumor cell damage induced by antibodies against membrane cofactor protein (MCP, CD46). J. Exp. Med. 172 (6), 1673–1680.
Seya, T., Hirano, A., et al., 1999. Human membrane cofactor protein (MCP, CD46): multiple isoforms and functions. Int. J. Biochem. Cell Biol. 31 (11), 1255–1260.
Si, Z., Zhang, J., et al., 2018. Different functional states of fusion protein gB revealed on human cytomegalovirus by cryo electron tomography with Volta phase plate. PLoS Pathog. 14 (12) e1007452.

Silverman, J.L., Greene, N.G., et al., 2012. Membrane requirement for folding of the herpes simplex virus 1 gB cytodomain suggests a unique mechanism of fusion regulation. J. Virol. 86 (15), 8171–8184.
Sinzger, C., Hahn, G., et al., 2008. Cloning and sequencing of a highly productive, endotheliotropic virus strain derived from human cytomegalovirus TB40/E. J. Gen. Virol. 89 (Pt 2), 359–368.
Sinzger, C., Kahl, M., et al., 2000. Tropism of human cytomegalovirus for endothelial cells is determined by a post-entry step dependent on efficient translocation to the nucleus. J. Gen. Virol. 81 (Pt 12), 3021–3035.
Soroceanu, L., Akhavan, A., et al., 2008. Platelet-derived growth factor-alpha receptor activation is required for human cytomegalovirus infection. Nature 455 (7211), 391–395.
Spear, P.G., Longnecker, R., 2003. Herpesvirus entry: an update. J. Virol. 77 (19), 10179–10185.
Stegmann, C., Hochdorfer, D., et al., 2017. A derivative of platelet-derived growth factor receptor alpha binds to the trimer of human cytomegalovirus and inhibits entry into fibroblasts and endothelial cells. PLoS Pathog. 13 (4), e1006273.
Suzuki, Y., Inagi, R., et al., 1998. Human herpesvirus 6 infection as a risk factor for the development of severe drug-induced hypersensitivity syndrome. Arch. Dermatol. 134 (9), 1108–1112.
Takeda, K., Haque, M., et al., 1997. Identification of a variant B-specific neutralizing epitope on glycoprotein H of human herpesvirus-6. J. Gen. Virol. 78 (Pt 9), 2171–2178.
Takeda, K., Okuno, T., et al., 1996. Identification of a variant A-specific neutralizing epitope on glycoprotein B (gB) of human herpesvirus-6 (HHV-6). Virology 222 (1), 176–183.
Tanaka, Y., Suenaga, T., et al., 2013. Herpesvirus 6 glycoproteins B (gB), gH, gL, and gQ are necessary and sufficient for cell-to-cell fusion. J. Virol. 87 (19), 10900–10903.
Tang, H., Hayashi, M., et al., 2011. Human herpesvirus 6 glycoprotein complex formation is required for folding and trafficking of the gH/gL/gQ1/gQ2 complex and its cellular receptor binding. J. Virol. 85 (21), 11121–11130.
Tang, H., Kawabata, A., et al., 2008. Human herpesvirus-6 infection induces the reorganization of membrane microdomains in target cells, which are required for virus entry. Virology 378 (2), 265–271.
Tang, H., Mahmoud, N.F., et al., 2015. Maturation of human herpesvirus 6A glycoprotein O requires coexpression of glycoprotein H and glycoprotein L. J. Virol. 89 (9), 5159–5163.
Tang, H., Mori, Y., 2015. Determinants of human CD134 essential for entry of human herpesvirus 6B. J. Virol. 89 (19), 10125–10129.
Tang, H., Serada, S., et al., 2013. CD134 is a cellular receptor specific for human herpesvirus-6B entry. Proc. Natl. Acad. Sci. U. S. A. 110 (22), 9096–9099.
Tang, H., Wang, J., et al., 2014. Detailed study of the interaction between human herpesvirus 6B glycoprotein complex and its cellular receptor, human CD134. J. Virol. 88 (18), 10875–10882.
Vanarsdall, A.L., Chase, M.C., et al., 2011. Human cytomegalovirus glycoprotein gO complexes with gH/gL, promoting interference with viral entry into human fibroblasts but not entry into epithelial cells. J. Virol. 85 (22), 11638–11645.
Vanarsdall, A.L., Howard, P.W., et al., 2016. Human Cytomegalovirus gH/gL forms a stable complex with the fusion protein gB in Virions. PLoS Pathog. 12 (4), e1005564.
Vanarsdall, A.L., Johnson, D.C., 2012. Human cytomegalovirus entry into cells. Curr. Opin. Virol. 2 (1), 37–42.
Vanarsdall, A.L., Pritchard, S.R., et al., 2018. CD147 promotes entry of pentamer-expressing human cytomegalovirus into epithelial and endothelial cells. MBio 9 (3).

Vanarsdall, A.L., Wisner, T.W., et al., 2012. PDGF receptor-alpha does not promote HCMV entry into epithelial and endothelial cells but increased quantities stimulate entry by an abnormal pathway. PLoS Pathog. 8 (9), e1002905.
Wainwright, M.S., Martin, P.L., et al., 2001. Human herpesvirus 6 limbic encephalitis after stem cell transplantation. Ann. Neurol. 50 (5), 612–619.
Wakata, A., Kanemoto, S., et al., 2018. The neutralizing linear epitope of human herpesvirus 6A glycoprotein B does not affect virus infectivity. J. Virol. 92 (5) pii: e02074-17.
Wang, X., Huang, D.Y., et al., 2005b. Integrin alphavbeta3 is a coreceptor for human cytomegalovirus. Nat. Med. 11 (5), 515–521.
Wang, X., Huong, S.M., et al., 2003. Epidermal growth factor receptor is a cellular receptor for human cytomegalovirus. Nature 424 (6947), 456–461.
Wang, D., Shenk, T., 2005. Human cytomegalovirus virion protein complex required for epithelial and endothelial cell tropism. Proc. Natl. Acad. Sci. U. S. A. 102 (50), 18153–18158.
Wang, W., Taylor, S.L., et al., 2005a. Human cytomegalovirus genes in the 15-kilobase region are required for viral replication in implanted human tissues in SCID mice. J. Virol. 79 (4), 2115–2123.
Weinberg, A.D., Bourdette, D.N., et al., 1996. Selective depletion of myelin-reactive T cells with the anti-OX-40 antibody ameliorates autoimmune encephalomyelitis. Nat. Med. 2 (2), 183–189.
White, J.M., Delos, S.E., et al., 2008. Structures and mechanisms of viral membrane fusion proteins: multiple variations on a common theme. Crit. Rev. Biochem. Mol. Biol. 43 (3), 189–219.
Wille, P.T., Knoche, A.J., et al., 2010. A human cytomegalovirus gO-null mutant fails to incorporate gH/gL into the virion envelope and is unable to enter fibroblasts and epithelial and endothelial cells. J. Virol. 84 (5), 2585–2596.
Wu, K., Oberstein, A., et al., 2018. Role of PDGF receptor-alpha during human cytomegalovirus entry into fibroblasts. Proc. Natl. Acad. Sci. U. S. A. 115 (42), E9889–E9898.
Wu, Y., Prager, A., et al., 2017. Human cytomegalovirus glycoprotein complex gH/gL/gO uses PDGFR-alpha as a key for entry. PLoS Pathog. 13 (4), e1006281.
Xing, Y., Oliver, S.L., et al., 2015. A site of varicella-zoster virus vulnerability identified by structural studies of neutralizing antibodies bound to the glycoprotein complex gHgL. Proc. Natl. Acad. Sci. U. S. A. 112 (19), 6056–6061.
Yamanishi, K., Okuno, T., et al., 1988. Identification of human herpesvirus-6 as a causal agent for exanthem subitum. Lancet 1 (8594), 1065–1067.
Yoshikawa, T., Suga, S., et al., 1991. Human herpesvirus-6 infection in bone marrow transplantation. Blood 78 (5), 1381–1384.
Zeev-Ben-Mordehai, T., Weberruss, M., et al., 2015. Crystal structure of the herpesvirus nuclear egress complex provides insights into inner nuclear membrane remodeling. Cell Rep. 13 (12), 2645–2652.
Zhang, L., Zhou, M., et al., 2018. Expression levels of glycoprotein O (gO) vary between strains of human cytomegalovirus, influencing the assembly of gH/gL complexes and virion infectivity. J. Virol. 92 (15).
Zhou, M., Lanchy, J.M., et al., 2015. Human cytomegalovirus gH/gL/gO promotes the fusion step of entry into all cell types, whereas gH/gL/UL128-131 broadens virus tropism through a distinct mechanism. J. Virol. 89 (17), 8999–9009.
Zhou, M., Yu, Q., et al., 2013. Comparative analysis of gO isoforms reveals that strains of human cytomegalovirus differ in the ratio of gH/gL/gO and gH/gL/UL128-131 in the virion envelope. J. Virol. 87 (17), 9680–9690.

Further reading

Heldwein, E.E., 2016. gH/gL supercomplexes at early stages of herpesvirus entry. Curr. Opin. Virol. 18, 1–8.

Ryckman, B.J., Chase, M.C., et al., 2010. Human cytomegalovirus TR strain glycoprotein O acts as a chaperone promoting gH/gL incorporation into virions but is not present in virions. J. Virol. 84 (5), 2597–2609.

CHAPTER NINE

Gammaherpesvirus entry and fusion: A tale how two human pathogenic viruses enter their host cells

Britta S. Möhl[a,*], Jia Chen[b], Richard Longnecker[b]

[a]Institute of Virology, School of Medicine, Technical University of Munich/Helmholtz Zentrum München, Munich, Germany
[b]Department of Microbiology and Immunology, Feinberg School of Medicine, Northwestern University, Chicago, IL, United States
*Corresponding author: e-mail address: britta.moehl@tum.de

Contents

Abstract

The prototypical human γ-herpesviruses Epstein-Barr virus (EBV) and Kaposi Sarcoma-associated herpesvirus (KSHV) are involved in the development of malignancies. Like all herpesviruses, they share the establishment of latency, the typical architecture, and the conserved fusion machinery to initiate infection. The fusion machinery reflects virus-specific adaptations due to the requirements of the respective herpesvirus. For example, EBV evolved a tropism switch involving either the B- or epithelial cell-tropism complexes to activate fusion driven by gB. Most of the EBV entry proteins and their cellular receptors have been crystallized providing molecular details of the initial steps of infection. For KSHV, a variety of entry and binding receptors has also been reported but the mechanism how receptor binding activates gB-driven fusion is not as well understood as that for EBV. However, the downstream signaling pathways that promote the early steps of KSHV entry are well described. This review summarizes the current knowledge of the key players involved in EBV and KSHV entry and the cell-type specific mechanisms that allow infection of a wide variety of cell types.

Advances in Virus Research, Volume 104
ISSN 0065-3527
https://doi.org/10.1016/bs.aivir.2019.05.006

1. Introduction

Members of the *Herpesviridae* are large enveloped double-stranded DNA (dsDNA) viruses with a broad host range including almost all vertebrates and even invertebrates (Fields et al., 2013). Herpesviruses are very successful pathogens that have evolved over the last 60–80 million years together with their hosts due to the establishment of life-long latency (Davison et al., 2002). There are nine different human pathogenic herpesviruses that are grouped as α-, β- and γ-herpesviruses. The characteristic for α- and β-herpesviruses is a lytic infection accompanied by tissue destruction and a variety of diseases such as cold sores, corneal blindness, genital herpes, encephalitis, keratitis, chickenpox, shingles, birth defects through congenital infection, and skin lesions (Fields et al., 2013). However, the prototypical γ-herpesviruses Epstein-Barr virus (EBV) and Kaposi Sarcoma-associated herpesvirus (KSHV) are both associated with quite distinct disease progression, i.e., lymphoproliferative diseases and malignancies.

EBV is the causative agent of infectious mononucleosis in young adults (Longnecker et al., 2013). Infection is associated with epithelial cell malignancies such as gastric carcinoma, nasopharyngeal carcinoma, and B-cell lymphomas such as Burkitt's and Hodgkin's lymphoma. KSHV is involved in endothelial cell malignancies such as Kaposi's sarcoma (KS), primary effusion lymphoma (PEL), and Multicentric Castleman's disease (MCD) which are often associated with suppression of the immune system (Damania and Cesarman, 2013). These EBV- and KSHV-associated malignancies are rare even in immunosuppressed patients compared to the ubiquitous seroprevalence in the human population (Damania and Cesarman, 2013; Longnecker et al., 2013). Nevertheless, the genetic and/or environmental triggers and the interplay with compromised immunity is still poorly understood (Ambinder and Cesarman, 2007). KSHV was first identified in Kaposi sarcoma tumor biopsies (Chang et al., 1994), whereas EBV was first found in Burkitt's lymphoma (Epstein et al., 1964). Both viruses are known to infect divergent tissue types in humans that are the origins for the associated malignancies and the establishment of latency. Infection of new hosts requires reactivation of latent virus that results in secretion of newly produced virions into the saliva and primary infections are commonly asymptomatic. Primary infections are controlled by a strong cellular immune response and infected cells are usually efficiently cleared preventing disease. But to evade the immune system, both KSHV and EBV have evolved numerous strategies to modulate host immune pathways facilitating the establishment and maintenance of latency, a hallmark of herpesviruses (Coscoy, 2007; Thorley-Lawson and Allday, 2008). Currently, only seven human oncogenic viruses are known and among them EBV and KSHV are the only herpesviruses (Schäfer et al., 2015).

Despite having different biological characteristics in human infections, EBV and KSHV have very similar mechanisms of infecting target cells in humans. Contrary to EBV, KSHV initially attaches to heparan sulfate proteoglycans (HSPGs). This relatively non-specific attachment step concentrates the virus onto host cell membranes and favors virus-specific binding to cell type-intrinsic receptors defining the host cell tropism, which then initiate infection. Enveloped viruses such as EBV and KSHV use a fundamental cellular mechanism—membrane fusion—to infect host cells. This process is initiated by receptor binding followed by fusion of the virus envelope with the host cell membrane that requires overcoming the high-energy barrier of fusion of the virion membrane with a host cell membrane. Interestingly, most enveloped viruses require only one protein fulfilling receptor-binding and fusion. In contrast, herpesviruses evolved a fine-tuned and complex entry machinery composed of receptor-binding glycoproteins, the fusion regulatory glycoproteins gH/gL, and the fusogen gB. A specific receptor-binding step enables the interaction of gH/gL with a prefusion form of gB in a metastable conformation on the surface of mature virions. This interaction triggers extensive structural rearrangements within gB resulting in a stable postfusion state that overcomes the high-energy barrier to merge cellular and viral membranes. Full fusion following fusion pore formation ensures the release of the nucleocapsid into the cytoplasm (Connolly et al., 2011; Longnecker et al., 2013). KSHV and EBV have many features in common including the classification as γ-herpesviruses sharing the typical herpesvirus architecture, establishment of latency in lymphocytes and the use of ephrin receptor tyrosine kinase A2 (EphA2) as an entry receptor (Chakraborty et al., 2012a; Chen et al., 2018; Damania and Cesarman, 2013; Fields et al., 2013; Hahn et al., 2012; Longnecker et al., 2013; Zhang et al., 2018). In this chapter, aspects of γ-herpesvirus entry into their host cells will be explored describing EBV and KSHV and their host cell interactions critical for infection and subsequent initiation of the viral life cycle.

2. History of EBV and KSHV

Even though infectious mononucleosis was reported in the late 19th century and described in 1920 (Sprunt and Evans, 1920), it took nearly half a century to identify EBV as the causative agent (Henle et al., 1968). The discovery of EBV began in 1958 with Burkitt's description of a sarcoma affecting the jaws of African children (Burkitt, 1958) and his talk at the Middlesex Hospital on this new sarcoma in 1961. This lecture was attended by Anthony Epstein who established a collaboration with Denis Burkitt. In 1964, Epstein first described EBV using electron microscopy detecting virus particles with the typical morphology of *Herpesviridae* in cultured Burkitt lymphoma cells (Epstein et al., 1964). This allowed the development of an immunofluorescent antibody test, which revealed that patients with Burkitt lymphoma and most Americans had antibodies against EBV (Henle and Henle, 1966). Finally in 1978, strong epidemiological evidence indicated a causal relationship between Epstein-Barr virus and Burkitt's lymphoma (de-The et al., 1978). It was the first time that the causal relationship between a lymphoma and a virus had been described. The lymphoproliferative potential has been supported by the finding that EBV transforms and induces proliferation in human B-lymphocytes in cell culture (Nilsson et al., 1971). Besides B-lymphocytes, epithelial cells are the major site for EBV infection in humans and replication in oropharyngeal epithelial cells was first described in 1984 (Sixbey et al., 1984). Moreover, the lymphoproliferative potential of EBV has been associated with other malignancies such as nasopharyngeal carcinoma in 1970 (zur Hausen et al., 1970), non-Hodgkin lymphoma in patients with acquired immunodeficiency syndrome (AIDS) in 1982 (Ziegler et al., 1982), T-cell lymphoma in 1988 (Jones et al., 1988), and Hodgkin lymphoma in 1989 (Weiss et al., 1989).

In 1872, the Hungarian physician Moritz Kaposi first reported the idiopathic multiple pigmented sarcoma of the skin, which is now called Kaposi sarcoma or KS (Kaposi, 1872). Classic KS is a rare cancer worldwide and described as a slow-growing, chronic tumor confined to the skin mostly in elderly men of the Mediterranean, Eastern Europe, or the Middle East (Chang and Moore, 2014; Robey and Bower, 2015). In the 1950s, a peak of endemic KS, a more fatal form accompanied by lymph node involvement in adults and children, occurred at extraordinary rates in Central and East Africa as one of the most commonly reported cancers in certain cancer registries (Cook-Moẓaffari et al., 1998). In the 1970s, an increasing incidence of

KS was observed among organ transplant recipients involving immunosuppressive therapies and transplantation treatment regiments (Chang and Moore, 2014; Penn, 1979). An unusual cluster of patients with KS and/or pneumocystis pneumonia in New York City and in California heralded the outbreak of the AIDS epidemic in the early 1980s of the last century (Centers For Disease Control, 1981, 1982a). The first report of AIDS described that approximately 37% of the patients had KS (Centers For Disease Control, 1982b), HIV was only identified shortly afterward (Barre-Sinoussi et al., 1983; Gallo et al., 1984). In 1984, herpes-like structures were found in KS lesions using electron microscopy (Walter et al., 1984), but at that time several other viruses were discussed to be causative. It was another decade and more than a century after the first report of KS that Chang and Moore identified that a herpesvirus was present in tumor cells using representational difference analysis using DNA from a KS lesion from an AIDS patient (Chang et al., 1994). More recently, KSHV has been associated with the B-cell lymphoma designated as primary effusion lymphoma, which is often co-infected with EBV and is also observed in AIDS patients (Damania and Cesarman, 2013).

3. Route of infection and host cell tropism

More than 90% of the human population are infected with EBV, thus it is one of the most abundant pathogens among human viruses (Auwaerter, 1999; Balfour et al., 2013). Primary infection with EBV is normally asymptomatic and occurs during childhood via droplet and contact infection with saliva. In young adults or adolescents, 30–60% of primary infected patients can develop infectious mononucleosis that is an acute illness manifested by lymphadenopathy, fever and pharyngitis. In fewer cases (10% of patients), patients may have splenomegaly, palatal petechiae, and hepatomegaly (Cohen, 2000). Primary infection with EBV by saliva contact occurs in the oropharynx, where it enters through the squamous oropharyngeal epithelium that lines the nasopharynx (tonsil). It is still controversial whether EBV directly infects either resting naïve B cells or epithelial cells, which in turn infect naïve B cells. Naïve B cells reside below epithelial cells in Waldeyer's ring that includes adenoids and tonsils. It surrounds the nasopharyngeal region and together with this epithelium is part of the lymphoepithelium. Infected B cells can either produce virus (lytic infection) or express a small subset of viral proteins that facilitate latency. Latently infected B cells become B-cell blasts and mimic a germinal center reaction enabling the virus to persist in the memory B cell compartment (Longnecker et al., 2013).

For spread to naïve individuals, infected memory cells are thought to return to the lymphoepithelium where some of them differentiate into plasma cells and undergo lytic replication (Ambinder and Cesarman, 2007; Laichalk and Thorley-Lawson, 2005; Longnecker et al., 2013). EBV can be amplified in the epithelium for the shedding of virus in saliva or may infect naïve B cells completing the EBV cycle of infection. EBV reactivation is restrained by natural killer (NK) cells and cytotoxic T cells resulting in recognition and destruction of infected cells (Cohen, 2000; Longnecker et al., 2013; Thorley-Lawson and Allday, 2008). This cycle results in a lifelong persistence, which can lead to the development of malignancies originated by B-lymphocytes as in Burkitt and Hodgkin lymphomas or epithelial cells such as nasopharyngeal carcinoma and certain gastric carcinomas (Longnecker et al., 2013). In addition, smooth muscle cells support infection and replication of EBV, which in some cases result in the development of EBV positive smooth muscle tumors, the second most prevalent malignancy of children with AIDS (Dekate and Chetty, 2016; Jenson et al., 1997). EBV can also rarely infect other cell types such as T cells and NK cells in immunosuppressed patients resulting in T/NK-cell lymphoproliferative disorders (Akashi and Mizuno, 2000; Coleman et al., 2015; Isobe et al., 2004; Park and Ko, 2014). Interestingly, the tropism switch, which targets EBV to either epithelial or B cells (Borza and Hutt-Fletcher, 2002) alters its capacity to enter monocytes but blocks monocyte maturation into dendritic cells (Guerreiro-Cacais et al., 2004). EBV infection of monocytes also induces acute maturation into macrophages and infiltration of infected macrophages into cancer tissues associated with EBV are discussed (Shimakage, 2014). The epithelial cell receptor of EBV was identified as EphA2 (Chen et al., 2018; Zhang et al., 2018). EphA2 is known to be involved not only in monocyte differentiation and maturation (Mukai et al., 2017), but also blocks antigen receptor-induced apoptosis by $CD4^+$ T cells (Holen et al., 2008).

Primary infection with KSHV occurs during childhood by salivary transmission (Ambinder and Cesarman, 2007; Henke-Gendo and Schulz, 2004; Pica and Volpi, 2007). In the oropharynx, KSHV infects the tonsillar epithelium that functions as the entrance for the virus to infect underlying B-lymphocytes, which are the site of latency (Chagas et al., 2006; Duus et al., 2004; Pauk et al., 2000; Pica and Volpi, 2007; Vieira et al., 1997; Webster-Cyriaque et al., 2006). Primary infection during childhood is usually asymptomatic but can be associated with febrile illness and maculopapular rash without compromised immunity. But in organ transplant recipients, primary infection can result in fever, lymphoid hyperplasia, splenomegaly and pancytopenia (Ambinder and Cesarman, 2007). In immune-competent individuals, KSHV establishes asymptomatic life-long latency in a subset of B cells, the lambda expressing B cells (CD19+ B cells) that do not circulate in large numbers in peripheral blood (Ambroziak et al., 1995; Hassman et al., 2011; Rappocciolo et al., 2008). The healthy, competent immune system of infected individuals controls lytic reactivation of KSHV and compels latency, thus the immune status is one of the key factors controlling reactivation (Lukac and Yuan, 2007). However, reactivation of KSHV from latency is the critical pathogenic step during progression to KS and the development of other KSHV-associated malignancies such as PEL and MCD. In addition, another key factor in KS progression is the occurrence of co-pathogenic infections that can perturb this regulatory mechanism and thereby influence the pathologic outcome of KSHV infection (Ambinder and Cesarman, 2007; Lukac and Yuan, 2007). Sexual transmission in adults within high-risk groups can result in lymphadenopathy associated with seroconversion (Ambinder and Cesarman, 2007; Henke-Gendo and Schulz, 2004; Lukac and Yuan, 2007; Martin et al., 1998; Pica and Volpi, 2007).

Primary infection with KS in AIDS-KS patients occurs at least 10 years prior to clinically apparent signs of KS lesions (Lukac and Yuan, 2007; Martin et al., 1998). AIDS-KS progression can be followed first by seroconversion and then by appearance of measurable virus DNA in peripheral blood. The viremia of the peripheral blood is accompanied by infection of endothelial cells of the skin resulting in the typical KS lesions. The severity of KS outcome correlates with the viral load in peripheral blood and the risk of developing KS is 10-fold higher in patients with viremia of the peripheral blood mononuclear cells (PBMCs) than in those without this viremia (Lukac and Yuan, 2007). The cell tropism of KSHV is also reflected by the origin of associated malignancies such as KS, which is a slow-growing, highly vascularized tumor of endothelial cell origin influenced by lytic replication, immune modulation and production of multiple cytokines as well as immune cells. KS is differentiated into: (i) the classic rarely aggressive KS in elderly men of Mediterranean and Middle Eastern ancestry, (ii) severe African-endemic KS with significant morbidity and mortality, (iii) iatrogenic KS of pharmacological immunosuppressed patients, (iv) the most aggressive clinical manifestation, endemic AIDS-KS predominantly in Sub-Saharan countries (Thakker and Verma, 2016). In addition, KSHV is associated with two other relatively rare lymphoproliferative disorders including PEL and KSHV-associated MCD. This rare but aggressive immunoblastic or anaplastic large-cell PEL lymphomas are of B cell origin and are characterized by KSHV-infection commonly accompanied by co-infection with EBV. This late manifestation during AIDS progression is described as malignant effusion involving body cavities and occasionally extra-nodal sites (Ambinder and Cesarman, 2007). KSHV is also associated with MCD—an atypical, heterogeneous group of lymphoproliferative disorders that share common histopathological features defined as angiofollicular lymph node hyperplasia. KSHV shows a distinct tropism for both B-cells (CD19+) and T-lymphocytes in KSHV-associated MCD patients (Ambroziak et al., 1995; Hassman et al., 2011; Kikuta et al., 1997; Rappocciolo et al., 2008). Castleman's disease (CD) is classified as unicentric localized or multicentric CD that involves many lymph node groups and spleen (multicentric angiofollicular hyperplasia) and is further subclassified due to presence/absence of KSHV infection (KSHV-associated MCD) (Ambinder and Cesarman, 2007). Based on these findings in human infection, KSHV has a broad host cell tropism in vivo including epithelial cells, endothelial cells, keratinocytes, fibroblasts, B- and T-lymphocytes, monocytes and macrophages as well as dendritic cells (Chandran and Hutt-Fletcher, 2007; Damania and Cesarman, 2013; Kikuta et al., 1997; Pica and Volpi, 2007). Contrary to EBV, KSHV has a very broad tropism regarding susceptible host cell lines. Surprisingly, it can also infect cells across species-barriers such as owl monkey kidney cells, baby hamster kidney fibroblast cells, Chinese hamster ovary cells, and mouse fibroblasts (Chandran and Hutt-Fletcher, 2007; Damania and Cesarman, 2013).

4. The herpesvirus fusion machinery

Herpesviruses require three to six viral envelope proteins forming the fine-tuned complex fusion machinery involving binding to a specific host cell receptor, activation of the fusogen and the virus-driven fusion process. Most enveloped viruses only use one or two proteins driving entry into their host cells. In contrast, the core fusion machinery conserved among *Herpesviridae* consists of gH/gL, which functions as activator, and gB, which is the driver of membrane fusion (Connolly et al., 2011; Heldwein, 2016; Möhl et al., 2016). The heterodimeric complex of gH/gL is the key to the divergent virus-specific tropism among *Herpesviridae* by recruiting additional non-conserved viral proteins engaged in binding cell-specific host receptors (Chandran and Hutt-Fletcher, 2007; Connolly et al., 2011;

Longnecker et al., 2013). For example, the α-herpesviruses herpes simplex virus (HSV) and Pseudorabies virus (PrV) only require gD for determining cell tropism (Connolly et al., 2011). The β-herpesviruses such as human cytomegalovirus (HCMV) utilize gO and UL128/130/131 complexes (Adler, 2015; Revello and Gerna, 2010; Zhou et al., 2015), and human herpesvirus 6 (HHV-6) requires gO and gQ1/Q2 (Jasirwan et al., 2014; Mori, 2009; Tang et al., 2014) that complex with gH/gL. EBV utilizes gp42 that also binds gH/gL to facilitate its cell-specific tropism for B cells, but for epithelial cell entry only gH/gL is required for receptor binding (Borza and Hutt-Fletcher, 2002; Chen et al., 2018; Kirschner et al., 2009; Sathiyamoorthy et al., 2014, 2016; Wang et al., 1998; Zhang et al., 2018). These tropism-determinant complexes form membrane bridging intermediates engaging and activating gB to mediate membrane fusion (Sathiyamoorthy et al., 2017).

The crystal structure of EBV gB revealed similarities to the postfusion form of vesicular stomatitis virus glycoprotein G (VSV G), thus being classified as class III viral fusion protein (Backovic et al., 2009). Crystal structure studies of gB across *Herpesviridae* revealed that the herpesvirus postfusion form exists as spike-like homotrimer formed by five individual domains that adopt an elongated conformation. The EBV gB structure is very similar to the postfusion structures of HSV-1, PrV, and HCMV gB but the domain orientations vary suggesting species-specific functional adaptations (Backovic et al., 2009; Burke and Heldwein, 2015; Chandramouli et al., 2015; Heldwein, 2016; Vallbracht et al., 2017). Besides the presumed role of gH/gL in initiating gB-driven fusion, it is also an important determinant for cell tropism of certain herpesviruses such as EBV, HCMV, and HHV6 (Adler, 2015; Connolly et al., 2011; Möhl et al., 2016; Mori, 2009). The crystal structures of gH/gL across *Herpesviridae* revealed no similarity to known fusogens. However, despite low amino acid conservation, they share a striking structure adopting an elongated rod- or boot-like shape (Backovic et al., 2010; Chowdary et al., 2010; Matsuura et al., 2010; Xing et al., 2015). The overall structure is formed by four structural domains with divergent domain interfaces, whose flexibility is important for gH/gL functioning as a trigger for fusion (Böhm et al., 2015; Chen et al., 2013; Möhl et al., 2015; Wu et al., 2005). Although the crystal structures of KSHV entry proteins have not been solved, the sequence homology between EBV and KSHV glycoproteins, 42% for gB, 26.8% for gH and 27.1% for gL (Gish and States, 1993; Papadopoulos and Agarwala, 2007), allowed homology modeling of KSHV gB, gH and gL. The homology models of KSHV gB, gH and gL (Fig. 1) were generated using the SWISS-MODEL program

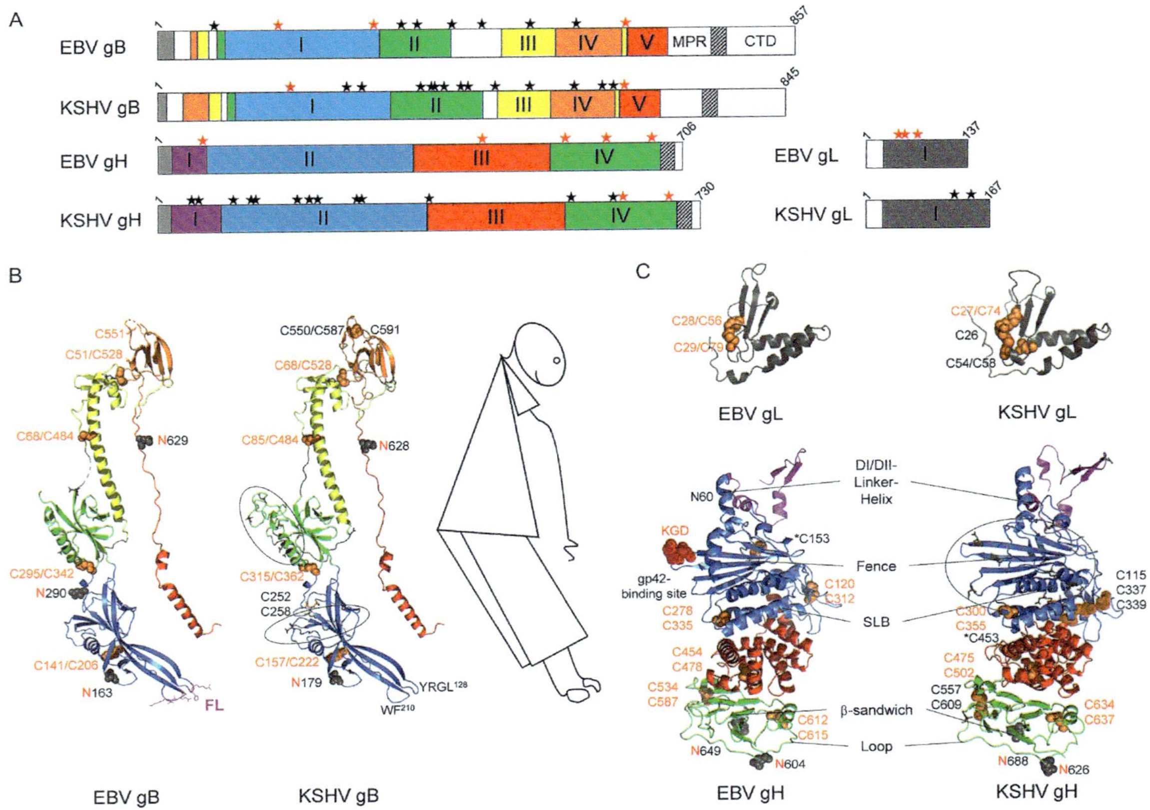

Fig. 1 See legend on opposite page.

(Waterhouse et al., 2018) based on the crystal structures of EBV gB (PDB ID: 3FVC), gH (6C5V) and gL (5T1D) (Backovic et al., 2009; Sathiyamoorthy et al., 2016; Snijder et al., 2018).

The structural sequence alignment of EBV and KSHV gB based on the crystal structure of EBV gB (3FVC) (Backovic et al., 2009) was performed using T-COFFEE Expresso (Notredame et al., 2000). The alignment revealed that the overall domain (D) distribution is similar but that DIV (the "head" of the gB postfusion structure) is enlarged in KSHV (Fig. 1A and B). The hypothetical structure of KSHV gB generated by SWISS-MODEL shows an overall similar elongated rod-like structure as EBV gB, reminiscent of a slouchy and sad stick figure (Fig. 1B). The feet are the fusion loops, the plekstrin-homology modules of DI and DII are the legs. One long α-helix represents the spine that kinks down on the tip giving the impression that the figure is bent over. This impression is supported by the following disordered DIV looking like the slouchy stick figure with head down. The elongated helical DV resembles the loose arms stretching toward its feet. In the homotrimer of gB, this loose extended α-helical arm (DV) reaches into the cavity, formed by the other two monomers together with the α-helical spine of DIII, which wraps around the helices of the other two monomers and stably ties together the homotrimer (Backovic et al., 2009). DIV of EBV gB is more disordered than HSV-1 gB which might result in more flexibility (Backovic et al., 2009). The

Fig. 1 Homology models of KSHV gB, gH and gL. (A) Schematic diagrams of full-length EBV and KSHV gB, gH and gL. The color codes for gB (Connolly et al., 2011) and gH domains (Möhl et al., 2014) are shown as described earlier. Domain I (DI) of gL is colored in gray. N-glycosylation sites are indicated by asterisks (black predicted or red confirmed for EBV). The signal peptide is shown by a gray box, the transmembrane domain by a hatched box, the membrane-proximal region by the label MPR, and the C-terminal domain by CTD. The structure is shown as cartoon for gB (B), gH and gL (C). The color codes correspond to the schema in (A). Disulfide bonds are shown as orange spheres and cysteine's as sticks. The glycan's are indicated as gray spheres (confirmed for EBV) or gray sticks (predicted). Cluster of predicted N-glycan's in KSHV gB and gH are encircled showing a location on one side. (B) Structural view of EBV and KSHV gB with a sketch of the stick figure. The FL of EBV gB are shown in magenta, and the according regions in KSHV gB are labeled by amino acids. (C) Structural view of EBV and KSHV gL and gH. The proposed integrin binding (KGD) motif of EBV gH is indicated as red dots. The homology modeling prediction of KSHV gB and gH/gL were generated using SWISS-MODEL (Waterhouse et al., 2018). The structural views of postfusion EBV gB (PDB ID: 3FVC) (Backovic et al., 2009), gH/gL (3PHF) (Matsuura et al., 2010) and KSHV gB as well as gH and gL were modified using the PyMOL Molecular Graphics System, Version 4.3.0 Schrödinger, LLC.

potentially dynamic structure of DIV could be a common feature of class III viral fusion proteins as suggested by the disordered analogous DIV of baculovirus gp64 (Backovic et al., 2009). DI, with the plekstrin-homology module and the exposed loops (feet) connecting the β-sheets at the extreme bottom of the KSHV gB structure (Fig. 1B), may resemble the fusions loops similar to all gB homologues crystallized to date. A similar plekstrin-homology module is found in DII that is a common feature for a membrane-binding domain but also for ligand-binding epitopes (Backovic et al., 2009). These modules could be involved in gH/gL-binding or interaction with cellular membranes as discussed for HSV-1 gB (Atanasiu et al., 2010a,b; Chandramouli et al., 2015). DII of KSHV gB is not only composed of this module but contains also an exposed helix, and, contrary to EBV gB, carries several predicted N-glycosylation sites (Fig. 1A and B circled). Like HCMV gB, KSHV gB is characterized by several predicted N-glycosylation sites in the interface of DI/DII and in DII around the predominant α-helix structure. HSV-1 gB has only 5 N-glycan's, whereas EBV gB has 10 predicted N-glycosylation sites and 3 (N163, N290, N629) of them showed electron density in the crystal structure (Backovic et al., 2009; Heldwein et al., 2006). Two of the 15 predicted N-glycosylation sites of KSHV gB correspond to N163 and N629 of EBV gB (Fig. 1B). Similar to KSHV gB, HCMV gB carries 18 predicted N-glycosylation sites and 10 could be proven by the crystal structure, whereas the other 6 are located in unstructured regions and 2 outside of the crystallized ectodomain (Chandramouli et al., 2015). As discussed for HCMV gB, the heavy glycosylation around the plekstrin-homology modules might protect these important binding regions from a host antibody response (Chandramouli et al., 2015). It would be interesting to understand why only certain herpesviruses evolved this feature and whether it co-evolved with certain tropism requirements.

The structural sequence alignments of KSHV and EBV gH and gL based on the crystal structures of EBV gH (6C5V) and gL (5T1D) (Chandramouli et al., 2015; Sathiyamoorthy et al., 2016; Snijder et al., 2018) were performed using T-COFFEE Expresso (Notredame et al., 2000). The alignments revealed an overall similar domain distribution (Fig. 1A) but with extended N-terminal domains (DI) for gH and gL that are known to facilitate the intimate complexing between both (Fig. 1C) suggesting a virus-specific adaptation of KSHV. The hypothetical KSHV gH configuration reveals a similar structure of the complex composed of four domains including four conserved disulfide bonds within DII, DIII and DIV and the two predicted, conserved N-glycosylation sites within DIV (Fig. 1C). EBV gL includes two disulfide bonds and three confirmed N-glycans (Matsuura et al., 2010), whereas the hypothetical model of KSHV gL indicates structurally homologous disulfide bonds but distinct N-glycosylation sites (Fig. 1C). KSHV has 15 predicted N-glycosylation sites that differ from EBV gH with its 5 N-glycans. Most of these predicted N-glycans are localized in a region that corresponds to an epitope in EBV gH responsible for gp42-binding (Fig. 1C). Thus, it will be important to understand if this region in KSHV gH also plays a role for binding to another viral entry protein or direct to a specific host cell receptor and if KSHV for this purpose evolved the predicted N-glycans masking the epitope from the host antibody response. In detail, gL and DI of gH are closely linked and connected to the rest of gH by one hinge assembled by the DI/DII-linker helix. DII starts with a striking antiparallel β-sheet structure that shields the rest of gH against DI, the so-called fence. Direct underneath the fence, a bundle of three α-helices resembles a syntaxin-like bundle (SLB) structure. In case of EBV, two disulfide bonds flank this SLB, one is conserved in KSHV gH and a triad of cysteine residues is located around the region corresponding to disulfide bond 1 of EBV gH (Fig. 1C). DIII is α-helical with a stabilizing disulfide bond conserved across gH homologues and followed by DIV with its antiparallel β-sandwich structure and two disulfide bonds that act as hinges connected by a long loop. This loop hides an extended hydrophobic patch that might be exposed during the cascaded fusion process enabling interaction with membranes. The loop and the β-sandwich structure carry the conserved N-glycan's of EBV and KSHV gH (Fig. 1C).

Herpesviruses can engage divergent entry pathways for different target cells (Connolly et al., 2011) despite conserved fusion machinery. An example of this is the observation that EBV infects B-lymphocytes, the site of latency, by endocytosis, but infects epithelial cells by direct fusion with the host cell plasma membrane (Miller and Hutt-Fletcher, 1992; Nemerow and Cooper, 1984). EBV has a distinct tropism for both cell types due to their role in facilitating primary infection, and virions carry two diverse gH/gL complexes, the epithelial cell-tropism complex or the B cell-tropism complex. For epithelial cell entry, only the core fusion machinery of gB and gH/gL and the epithelial cell entry-specific receptor EphA2 are required. Thus, the gH/gL complex is designated as the epithelial cell-tropism complex (Fig. 2A) (Chen et al., 2018; Longnecker et al., 2013; Zhang et al., 2018). For B cell entry, EBV requires an additional glycoprotein (gp)42 that is unique for EBV and other closely related lymphocryptoviruses. EBV gp42 binds to the B cell receptor human leukocyte

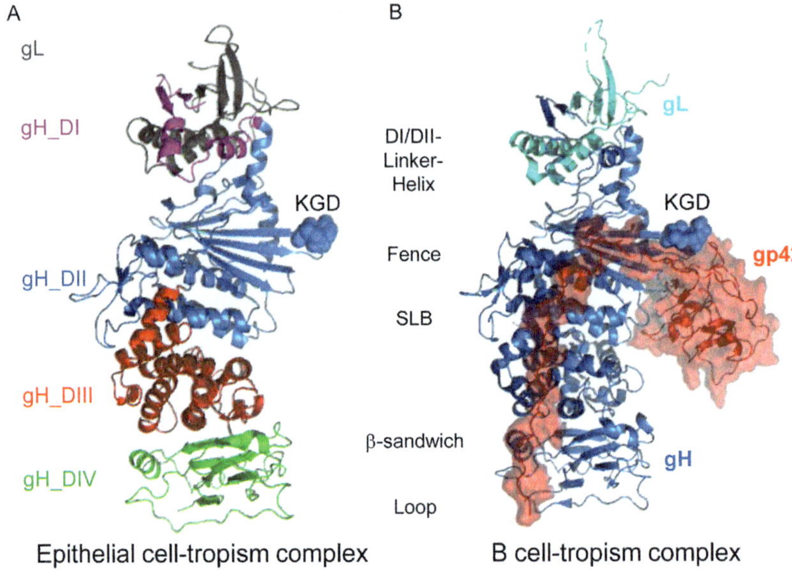

Fig. 2 Tropism switch of EBV. (A) The structural view of the epithelial cell-tropism complex gH/gL is shown as cartoon. The color code is used as described in Fig. 1. (B) The structural view of the B cell-tropism complex gH/gL-gp42 is indicated as cartoon for gH (marine) and gL (cyan), and as cartoon with transparent surface presentation for gp42 (red) with the elongated N-terminus wrapping around on one side of gH. The KGD motif is indicated by marine spheres and labeled. The structural views of gH/gL (3PHF) (Matsuura et al., 2010) and gH/gL-gp42 (5T1D) (Sathiyamoorthy et al., 2016) were modified using PyMOL Molecular Graphics System, Version 4.3.0 Schrödinger, LLC.

antigen (HLA) class II (Longnecker et al., 2013) for B cell entry and the gH/gL-gp42 complex acts as the B cell-tropism entry complex (Fig. 2B). Structural analysis of the gH/gL-gp42 complex revealed that the elongated N-terminus of gp42 wraps around gH involving three of its four external surfaces (Fig. 2B) (Sathiyamoorthy et al., 2016). The receptor binding causes a widening of the hydrophobic pocket within the C-terminus of gp42 and it is proposed that this feature is involved in the cascaded fusion process (Kirschner et al., 2009). The crystal structure of gH/gL-gp42 indicates an intimate interaction of gp42 with gH suggesting that this change is translated to gH/gL activating gB for cell fusion.

5. EBV-driven entry and fusion

B cells are the site of latency for EBV and therefore the major target cells for infection. Infection of B cells is initiated by the interaction of the

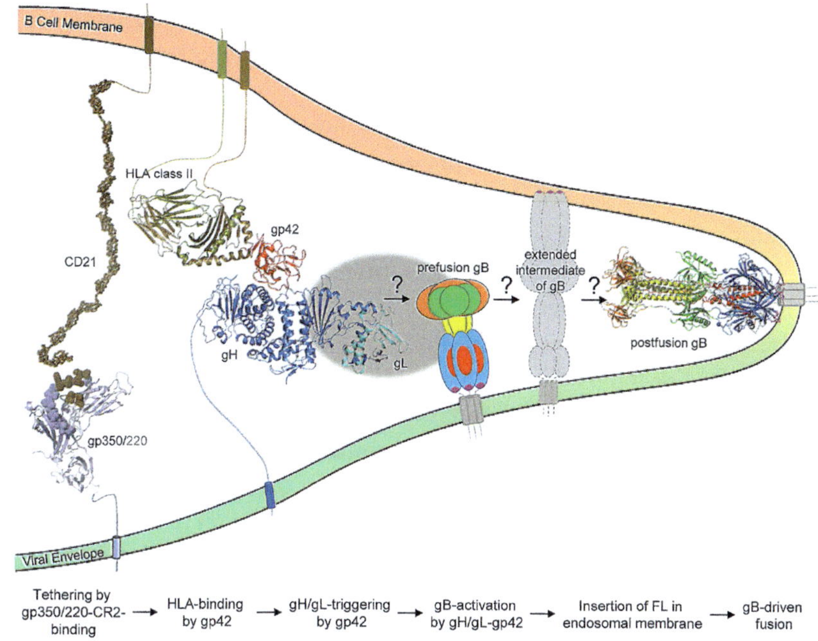

Fig. 3 Endocytosis entry pathway of EBV into B-lymphocytes. First, gp350/220 (only gp350 with the proposed binding region as spheres in corresponding colors is depicted) interacts with CD21 tethering EBV to the B cell membrane. After tethering, gp42 binds to its receptor HLA class II resulting in a widening of the hydrophobic pocket in gp42. This conformational change may activate gp42 by itself or gH/gL enabling the binding to the prefusion form of gB. Extended rearrangements within gB (extended intermediate) enable insertion of the fusion loops (FL) (magenta) into the target cell membrane and refolding of gB forms postfusion gB which mediates merging of the two bilayers. The structural views are shown as cartoons and the color code of the EBV gB domains is according to (Connolly et al., 2011) as well as the color code for the B cell-tropism complex is used as described in Fig. 2. The structural views of CD21 (2GSX) (Gilbert et al., 2006), EBV gp350 (2H6O) (Szakonyi et al., 2006), the complex of gH/gL (3PHF) (Matsuura et al., 2010) and gp42-HLA-DR1 (1KG0) (Mullen et al., 2002; Sathiyamoorthy et al., 2014), as well as postfusion gB (3FVC) (Backovic et al., 2009) were modified using PyMOL Molecular Graphics System, Version 4.3.0 Schrödinger, LLC.

abundant viral attachment protein gp350/220 to complement receptor 2 (CD21 or CR2) or CD35 (Fig. 3) (Ogembo et al., 2013; Tanner et al., 1987). This binding results in capping of CD21 followed by endocytosis of the virion into the host cell (Nemerow and Cooper, 1984; Tanner et al., 1987). This attachment step tethers EBV to its host cell and thereby increases the efficiency of infection, although this step is not essential for B cell entry (Janz et al., 2000). Interestingly, contrary to B-lymphocytes,

B lymphoblastoid cells are infected by direct fusion with the viral envelope (Nemerow and Cooper, 1984). Fusion is mediated by the binding of gp42 in the gH/gL-gp42 complex to the B cell-specific HLA class II receptor (Fig. 3). This binding is accompanied by a widening of the hydrophobic pocket within gp42. This conformational change may expose additional binding sites for engaging other components of the entry machinery or a yet unidentified cellular interaction partner (Kirschner et al., 2009). Structural data indicate that the conformational change within gp42 is involved in triggering the overall fusion process (Kirschner et al., 2009). After receptor binding, the hydrophobic pocket of the C-terminus, including the HLA-binding region, is engaged closer to the complex enabling interaction with the DII/III-interface of gH (Sathiyamoorthy et al., 2016). The stability of the gH/gL-gp42 complex with HLA class II provided the opportunity to discern the overall structure of the B cell-triggering complex required for infection by EBV, a first for herpesviruses (Sathiyamoorthy et al., 2014, 2016). In these electron micrographs, an "open" and "closed" state were observed. The "open" state was characterized by a highly variable orientation of HLA and gH/gL to each other resulting in larger angles, whereas the "closed" state, likely the structural view of the gH/gL-gp42 complex, revealed a more rigid orientation with the C-terminus of gp42 closely engaged to the interface of DII/III of gH (Sathiyamoorthy et al., 2014, 2016). Mutational analysis of the hydrophobic pocket of gp42 indicated that the closed state, which draws the transmembrane domains of gH and HLA closer to each other, is important for activating the cascaded fusion process driven by gB (Sathiyamoorthy et al., 2016). Following HLA-binding and conformational change within gp42, the gH/gL-gp42 complex is then thought to bind to the metastable prefusion gB (Fig. 3). The binding to gB triggers a conformational change of gB from the hypothetical prefusion state to the extended pre-hairpin intermediate enabling the insertion of the fusion loops into the host cell membrane, thereby acting as additional bridge between the two bilayers (Connolly et al., 2011; Heldwein, 2016; Longnecker et al., 2013; Möhl et al., 2016; Sathiyamoorthy et al., 2017). Based on previous studies, a molecular model of herpesvirus gB postulates that the tip of the second fusion loop can further penetrate into the hydrophobic core of the cellular lipid bilayer enabling the side chains forming a rim structure. Thereby, the fusion loops form a stable interaction with the host cell membrane ensuring that the two membranes will fuse (Vallbracht et al., 2017). The collapse and refolding of this extended pre-hairpin intermediate to a trimer of hairpins catalyzes enough free energy to overcome the high kinetic barrier for merging membranes (Connolly et al., 2011; Harrison, 2015; Sathiyamoorthy et al., 2017). The structural analysis of the postfusion form of EBV gB highlighted a long C-terminal arm arranged antiparallel to the coiled-coil core of the trimeric hairpin (Backovic et al., 2009; Connolly and Longnecker, 2012). The formation of this coil-arm complex may facilitate the reorganization of the gB structure into the postfusion form of gB similar to class I fusogens (Connolly and Longnecker, 2012). Thus, the activation of gB triggers the highly orchestrated fusion process of both membranes releasing the nucleocapsid into the cytoplasm of the target cells. B cell-derived virions are gp42-deficient due to g42 being sequestrated by HLA class II during maturation, and thereby acts as tropism switch rendering virions without gp42 capable of infecting epithelial cells. In contrast, virions derived from epithelial cells are enriched in the B cell-tropism complex gH/gL-gp42, which can facilitate entry into B cells (Borza and Hutt-Fletcher, 2002; Longnecker et al., 2013; Wang et al., 1998).

The attachment step tethering EBV to epithelial cells is facilitated by gp350/220 binding to CD21 expressed on tonsillar epithelial cells and BMRF2 interaction with integrins on polarized epithelial cells (Fig. 4) (Jiang et al., 2008; Tugizov et al., 2003; Xiao et al., 2007). The epithelial cell-tropism complex gH/gL binds to the ligand-binding domain of EphA2 similar to KSHV (Chen et al., 2018; Zhang et al., 2018). Interestingly, EBV enters epithelial cells by direct fusion with the plasma membrane, whereas KSHV enters its host cells via an endocytosis pathway involving gH/gL-binding to EphA2, A4 or A5 (Chakraborty et al., 2012a; Chandran, 2010; Chen et al., 2018, 2019; Hahn et al., 2012; Miller and Hutt-Fletcher, 1992; TerBush et al., 2018; Zhang et al., 2018). EBV gH/gL acts as tropism switch by either interacting with gp42 that facilitates B cell receptor binding or directly binding to the epithelial cell receptor EphA2 (Borza and Hutt-Fletcher, 2002; Chen et al., 2018; Matsuura et al., 2010; Möhl et al., 2016; Sathiyamoorthy et al., 2014, 2016; Zhang et al., 2018). Receptor binding by gH/gL enables the transient and dynamic interaction with gB initiating the fusion process (Fig. 4). Similar to the B cell entry-triggering complex, it is likely that the epithelial cell entry-mediating complex requires conformational changes within gH/gL that enable interaction with gB (Longnecker et al., 2013). Compatible with this hypothesis, the domain interfaces of the four domains and their flexibility are required for gH/gL-activated gB-driven fusion process (Böhm et al., 2015; Chen et al., 2013; Möhl et al., 2015; Wu et al., 2005). In addition, the disulfide

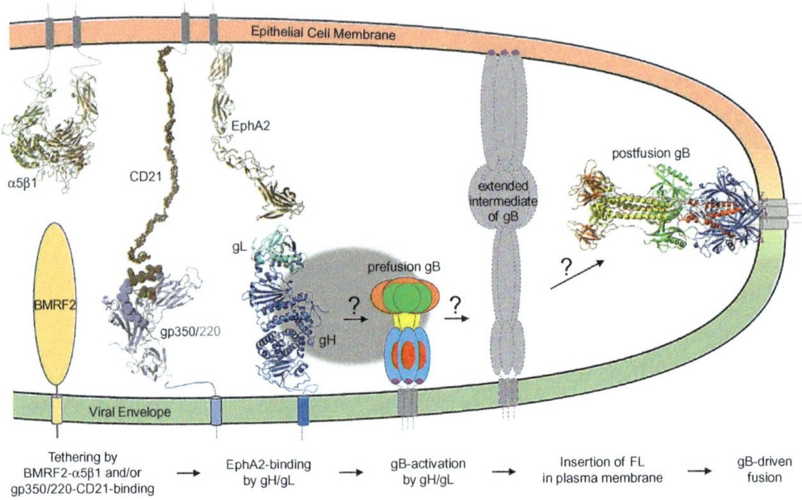

Fig. 4 Entry of EBV via direct fusion into epithelial cells. First, gp350/220 interacts with CD21 and/or BMRF2 with integrin α5β1 tethering EBV to the epithelial cell membrane. After tethering, gH/gL binds to its high-affinity receptor EphA2 enabling the binding to the prefusion form of gB. Extended rearrangements within gB trigger insertion of the FL into the target cell membrane, followed by refolding of gB to postfusion form and thereby mediates merging of the two bilayers. The structural views of integrin α5β1 (3VI4) (Nagae et al., 2012), CD21 (2GSX) (Gilbert et al., 2006), EphA2 (2 × 10) (Seiradake et al., 2010), EBV gp350 (2H6O) (Szakonyi et al., 2006), and gH/gL (3PHF) (Matsuura et al., 2010) as well as postfusion gB (3FVC) (Backovic et al., 2009) were generated using the PyMOL Molecular Graphics System, Version 4.3.0 (Schrödinger, LLC).

bond of DII tightens the syntaxin-like bundles and thereby induces a local rigidity that is required for epithelial cell fusion. In contrast, a larger perturbation of this region next to the DII/III interface that binds to gp42 disturbs the formation of the B cell entry-triggering complex (Möhl et al., 2014; Sathiyamoorthy et al., 2014). Despite the striking structural similarity of the postfusion forms of gB conserved across *Herpesviridae*, the differences in domain orientations indicate virus-specific functional adaptations regarding its receptor-binding partners (Backovic et al., 2009; Burke and Heldwein, 2015; Chandramouli et al., 2015; Connolly et al., 2011; Heldwein et al., 2006; Stampfer and Heldwein, 2012; Vallbracht et al., 2017).

6. KSHV-driven entry and fusion

KSHV predominantly infects endothelial cells that are the origin for development of KS and activated B cells, which are the origin for a rare type of B cell lymphoma or PEL. Besides these two target cell types, KSHV can enter multiple other cell types including PBMCs, fibroblasts, keratinocytes, and epithelial cells (Chandran, 2010; Chandran and Hutt-Fletcher, 2007). The main entry route are diverse pH-dependent endocytosis pathways such as dynamin-dependent, clathrin-mediated endocytosis for fibroblasts and clathrin/caveolin-mediated endocytosis for monocytes (Akula et al., 2003; Chandran, 2010; Damania and Cesarman, 2013; Kerur et al., 2010). In contrast, early stages of KSHV infection targeting endothelial cells involve actin reorganization attended by actin-dependent macropinocytosis and c-Cbl-dependent membrane blebbing (Chandran, 2010; Damania and Cesarman, 2013; Raghu et al., 2009). Activated primary human B cells are infected by endocytosis involving DC-SIGN (dendritic cell-specific intercellular adhesion molecule 3 (ICAM-3)-grabbing non-integrin—CD209) binding (Chandran, 2010; Damania and Cesarman, 2013; Rappocciolo et al., 2008). The broad KSHV cell tropism is supported by the usage of ubiquitous cell surface HSPGs as attachment receptor bound by gH, gB, or the KSHV-specific glycoproteins open reading frame 4 (ORF4) and gpK8.1A (Fig. 5) (Akula et al., 2001a,b;

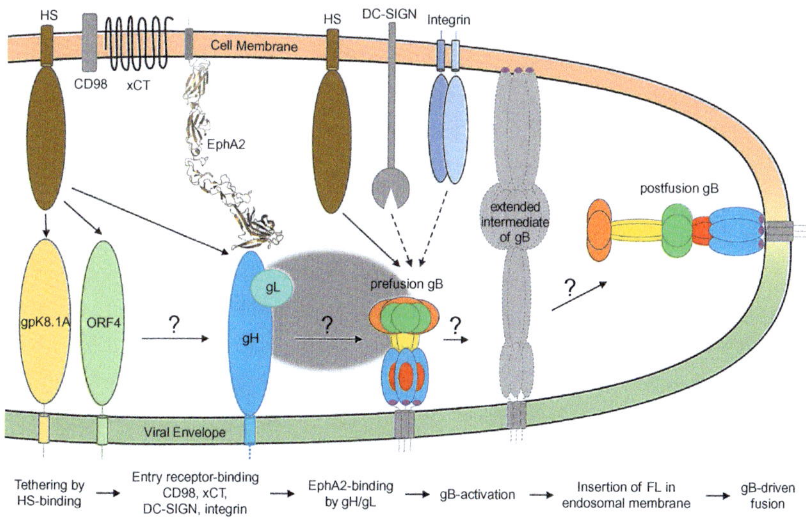

Fig. 5 Endocytosis entry pathway of KSHV into human target cells. First, gH, gB, gpH8.1A and ORF4 interact with HSPG tethering KSHV to human target cells. Tethering may support interaction with following entry receptors CD98, xCT, integrin and DC-SIGN as well as EphA2 (is shown as representative for EphA4, 5 and 7). DC-SIGN and integrin are likely bound by gB, whereas gH/gL binds to its receptor EphA2 that may enable the binding of gH/gL to the prefusion form of gB. Thereby, activated gB triggers merging of the viral membrane with endosome membrane. The structural view of EphA2 (2×10) (Seiradake et al., 2010) were generated using The PyMOL Molecular Graphics System, Version 4.3.0 (Schrödinger, LLC).

Chakraborty et al., 2012b; Hahn et al., 2009; Mark et al., 2006; Wang et al., 2001). Binding to HSPG tethers KSHV to its host cells increasing the efficiency of infection (Chandran, 2010; Damania and Cesarman, 2013). In addition, KSHV uses DC-SIGN as cell-specific receptor on human myeloid dendritic cells, macrophages, and activated B cells, which are the site of latency. A potential interaction partner of DC-SIGN is the highly mannosylated fusogen gB (Fig. 5) (Chandran, 2010; Damania and Cesarman, 2013; Rappocciolo et al., 2008). In addition, other entry receptors have been described including the cysteine transporter xCT and the multifunctional protein CD98 as well as integrins that promote entry into fibroblasts, endothelial and epithelial cells (Fig. 5). Interestingly, KSHV gB is the only gB-homolog with an integrin-binding motif (RGD-sequence). Previous studies demonstrated that several integrins such as αVβ3, αVβ5 and α3β1 are important for infection of human epithelial and endothelial cells as well as esophageal fibroblasts (Akula et al., 2002; Chandran, 2010; Damania and Cesarman, 2013; Wang et al., 2003). More recent studies suggest that KSHV infection of epithelial cells is independent of αν- and β1-family integrin expression (TerBush et al., 2018), but requires HSPGs and either EphA2, EphA4, or EphA5 (Chakraborty et al., 2012a; Chandran, 2010; Chen et al., 2019; Damania and Cesarman, 2013; Hahn et al., 2012; TerBush et al., 2018). Interestingly, EphA7 has recently been found to be a receptor required for KSHV entry into B cells (Großkopf et al., 2019). KSHV attachment to its host cells via HSPG and DC-SIGN is predominately mediated by gB and gpK8.1A (Fig. 5). This binding may promote interaction of viral glycoproteins to the receptors integrin and xCT/CD98 inducing conformational changes within the receptor-binding glycoproteins rendering them accessible to interact with specific entry receptors (Schäfer et al., 2015; Veettil et al., 2014). Interestingly, integrin binding is followed by the interaction of gH/gL with an EphA family member initiating phosphorylation of EphA2 and thereby triggering coordinated integrin-associated downstream signaling that mediates clathrin-mediated endocytosis into fibroblasts (Chakraborty et al., 2012a; Dutta et al., 2013). Activation of cell receptor-associated signaling pathways creates a friendly environment for early events during infection (Kumar et al., 2018). KSHV infection is also known to rely on cell signaling pathways triggered by activation of FAK (focal adhesion kinase), and phosphorylated Src as well as PI3K (phosphoinositide 3-kinase) (Krishnan et al., 2006; Naranatt et al., 2003; Veettil et al., 2006). Moreover, KSHV uses the core fusion machinery of gB and gH/gL as well as KSHV-specific receptor-binding proteins gpK8.1A and ORF4 in divergent settings facilitating entry into a variety of human host cells

(Schäfer et al., 2015; Veettil et al., 2014). KSHV entry into endothelial cells is initiated by tethering KSHV to the target cells by HSPG-binding via gB, gH, gpK8.1A, and ORF4 (Kumar et al., 2018). Tethering may enable the interaction with the specific entry receptors such as integrins, xCT and EphA family members (Fig. 5). Binding of specific entry receptors is thought to induce the activation of a variety of cell signaling pathways involved in actin-rearrangement, acetylation of microtubules, and endocytosis thereby mediating a preferable environment for entry stages (Kumar et al., 2018). For example, KSHV entry into endothelial cells also involves actin-driven formation of blebs facilitating macropinocytosis (Kumar et al., 2018).

The receptor-binding step may activate gH/gL to bind to the metastable prefusion gB (Fig. 5) as indicated for EBV and other herpesviruses such as HSV and PrV. It was previously shown that only gH/gL and gB can mediate cell-cell fusion in a virus-free assay (Pertel, 2002). The conserved structures of the fusogen gB among *Herpesviridae* suggest a fundamental mechanism of gB-driven fusion initiated by the binding of gH/gL to gB. This binding is thought to trigger a conformational change of gB into an extended pre-hairpin intermediate enabling the insertion of the gB fusion loops into the host cell membrane. This bridges both bilayers and subsequent gB refolding brings the membranes together to allow fusion to occur (Connolly et al., 2011; Harrison, 2015; Heldwein, 2016; Longnecker et al., 2013; Sathiyamoorthy et al., 2017).

7. Conclusion

In summary, EBV and KSHV have many features in common including the induction of lymphoproliferation and the expression of conserved key proteins such as gB and gH/gL, which are required for virus-cell fusion (Chakraborty et al., 2012a; Chandran and Hutt-Fletcher, 2007; Chen et al., 2018; Hahn et al., 2012; Zhang et al., 2018). Nevertheless, the molecular details of entry and fusion vary dependent on the diverse cells types that these two herpesviruses infect (Table 1). EBV infection of its major host cells is fine-tuned by two different receptor-binding complexes, which have been characterized by crystal structures of all key players involved in entry except for the prefusion form of gB (Möhl et al., 2016). In contrast, there are no crystal structures for the proteins essential for KSHV entry, but the functions of the KSHV proteins can readily be discerned based on the amino acid sequence homology to EBV. In contrast, the importance of downstream signaling pathways supporting entry and later steps during replication are well

Table 1 Entry proteins and their receptors as well as entry pathways of EBV and KSHV.

	Epstein-Barr virus	**Kaposi Sarcoma-associated herpesvirus**
Entry protein	gB	gB
	gH/gL	gH/gL
	gp42	
	gp350/gp220	
	BMRF2	
		gpK8.1A
		ORF4
Receptor		HSPG (gB, gpK8.1A, gH, ORF4)
	EphA2 (gH/gL)	EphA2, A4, A5 and A7 (gH/gL)
	HLA class II (gp42)	
	CD21/CR2 (gp350/gp220)	
	α5β1 (BMRF1)	Integrins (αVβ3, αVβ5 and α3β1 with proposed gB interaction) DC-SIGN (probably gB) CD98/xCT
Target cell	**Epithelial cells**	Epithelial cells
		Endothelial cells
		Keratinocytes
		Fibroblasts
	Lymphocytes (T, NK and **B cells**)	Lymphocytes (T and **B cells**)
	Monocytes	Monocytes
	Macrophages	Macrophages
		Dendritic cells
	Smooth muscle cells	
Entry pathway	Fusion at the plasma membrane Endocytosis	Endocytosis (clathrin/caveolin- or dynamin-mediated) Macropinocytosis (actin-dependent)
Infection route	Passes through the oropharyngeal epithelium (adenoids and tonsils) to infect underlying naïve B cells	Passes through the tonsillar epithelium to infect underlying lymphocytes sexual transmission in adults of high risk groups

The predominant target cells are shown in bold.

described for KSHV (Chandran, 2010; Kumar et al., 2018). Future research using new techniques such as cryo-electron microscopy will allow a full understanding of the complex entry pathway of these important human pathogens. The IARC-classification of KSHV and EBV as human carcinogens makes it important to understand the molecular details of the diverse entry and fusion pathway adaptions due to cell-type specific requirements to better understand their roles in the development of cancer associated with infections by these two pathogens.

Acknowledgments

The authors thank current and former members of the Longnecker and Jardetzky laboratories for their contributions to the work described. The research was supported by the National Institute of Allergy and Infectious Diseases (grant AI076183 to R.L.) and by the National Cancer Institute (grant CA117794 to R.L.). The DZIF Maternity Leave Stipend (TI 07.005) has been awarded to B.S.M. by the German Center for Infection Research (Deutsches Zentrum für Infektionsforschung—DZIF). J.C. acknowledges funding by the National Institute of Allergy and Infectious Diseases, IRG-15-173-21 American Cancer Society and the Third Coast Center for AIDS Research pilot to J.C. B.S.M. wants to thank especially her little son, who tirelessly tried to contribute to text and figures.

References

Adler, B., 2015. A viral pilot for HCMV navigation? Viruses 7, 3857–3862.

Akashi, K., Mizuno, S., 2000. Epstein-Barr virus-infected natural killer cell leukemia. Leuk. Lymphoma 40, 57–66.

Akula, S.M., Pramod, N.P., Wang, F.Z., Chandran, B., 2001a. Human herpesvirus 8 envelope-associated glycoprotein B interacts with heparan sulfate-like moieties. Virology 284, 235–249.

Akula, S.M., Wang, F.Z., Vieira, J., Chandran, B., 2001b. Human herpesvirus 8 interaction with target cells involves heparan sulfate. Virology 282, 245–255.

Akula, S.M., Pramod, N.P., Wang, F.Z., Chandran, B., 2002. Integrin alpha3beta1 (CD 49c/29) is a cellular receptor for Kaposi's sarcoma-associated herpesvirus (KSHV/HHV-8) entry into the target cells. Cell 108, 407–419.

Akula, S.M., Naranatt, P.P., Walia, N.S., Wang, F.Z., Fegley, B., Chandran, B., 2003. Kaposi's sarcoma-associated herpesvirus (human herpesvirus 8) infection of human fibroblast cells occurs through endocytosis. J. Virol. 77, 7978–7990.

Ambinder, R.F., Cesarman, E., 2007. Clinical and pathological aspects of EBV and KSHV infection. In: Arvin, A., Campadelli-Fiume, G., Mocarski, E., Moore, P.S., Roizman, B., Whitley, R., Yamanishi, K. (Eds.), Human Herpesviruses: Biology, Therapy, and Immunoprophylaxis. Cambridge University Press, Cambridge.

Ambroziak, J.A., Blackbourn, D.J., Herndier, B.G., Glogau, R.G., Gullett, J.H., McDonald, A.R., Lennette, E.T., Levy, J.A., 1995. Herpes-like sequences in HIV-infected and uninfected Kaposi's sarcoma patients. Science 268, 582–583.

Atanasiu, D., Whitbeck, J.C., De Leon, M.P., Lou, H., Hannah, B.P., Cohen, G.H., Eisenberg, R.J., 2010a. Bimolecular complementation defines functional regions of herpes simplex virus gB that are involved with gH/gL as a necessary step leading to cell fusion. J. Virol. 84, 3825–3834.

Atanasiu, D., Saw, W.T., Cohen, G.H., Eisenberg, R.J., 2010b. Cascade of events governing cell-cell fusion induced by herpes simplex virus glycoproteins gD, gH/gL, and gB. J. Virol. 84, 12292–12299.
Auwaerter, P.G., 1999. Infectious mononucleosis in middle age. JAMA 281, 454–459.
Backovic, M., Longnecker, R., Jardetzky, T.S., 2009. Structure of a trimeric variant of the Epstein-Barr virus glycoprotein B. Proc. Natl. Acad. Sci. U. S. A. 106, 2880–2885.
Backovic, M., Dubois, R.M., Cockburn, J.J., Sharff, A.J., Vaney, M.C., Granzow, H., Klupp, B.G., Bricogne, G., Mettenleiter, T.C., Rey, F.A., 2010. Structure of a core fragment of glycoprotein H from pseudorabies virus in complex with antibody. Proc. Natl. Acad. Sci. U. S. A. 107, 22635–22640.
Balfour Jr., H.H., Sifakis, F., Sliman, J.A., Knight, J.A., Schmeling, D.O., Thomas, W., 2013. Age-specific prevalence of Epstein-Barr virus infection among individuals aged 6-19 years in the United States and factors affecting its acquisition. J. Infect. Dis. 208, 1286–1293.
Barre-Sinoussi, F., Chermann, J.C., Rey, F., Nugeyre, M.T., Chamaret, S., Gruest, J., Dauguet, C., Axler-Blin, C., Vezinet-Brun, F., Rouzioux, C., Rozenbaum, W., Montagnier, L., 1983. Isolation of a T-lymphotropic retrovirus from a patient at risk for acquired immune deficiency syndrome (AIDS). Science 220, 868–871.
Böhm, S.W., Eckroth, E., Backovic, M., Klupp, B.G., Rey, F.A., Mettenleiter, T.C., Fuchs, W., 2015. Structure-based functional analyses of domains II and III of pseudorabies virus glycoprotein H. J. Virol. 89, 1364–1376.
Borza, C.M., Hutt-Fletcher, L.M., 2002. Alternate replication in B cells and epithelial cells switches tropism of Epstein-Barr virus. Nat. Med. 8, 594–599.
Burke, H.G., Heldwein, E.E., 2015. Crystal structure of the human Cytomegalovirus glycoprotein B. PLoS Pathog. 11, e1005227.
Burkitt, D., 1958. A sarcoma involving the jaws in African children. Br. J. Surg. 46, 218–223.
Centers For Disease Control, 1981. Kaposi's sarcoma and Pneumocystis pneumonia among homosexual men—New York city and California. MMWR Morb. Mortal. Wkly Rep. 30, 305–308.
Centers For Disease Control, 1982a. A cluster of Kaposi's sarcoma and Pneumocystis carinii pneumonia among homosexual male residents of Los Angeles and Orange counties, California. MMWR Morb Mortal Wkly Rep 31, 305–307.
Centers For Disease Control, 1982b. Update on acquired immune deficiency syndrome (AIDS)—United States. MMWR Morb. Mortal. Wkly Rep. 31, 507–508513–4.
Chagas, C.A., Endo, L.H., Sakano, E., Pinto, G.A., Brousset, P., Vassallo, J., 2006. Detection of herpesvirus type 8 (HHV8) in children's tonsils and adenoids by immunohistochemistry and in situ hybridization. Int. J. Pediatr. Otorhinolaryngol. 70, 65–72.
Chakraborty, S., Veettil, M.V., Bottero, V., Chandran, B., 2012a. Kaposi's sarcoma-associated herpesvirus interacts with EphrinA2 receptor to amplify signaling essential for productive infection. Proc. Natl. Acad. Sci. U. S. A. 109, E1163–E1172.
Chakraborty, S., Veettil, M.V., Chandran, B., 2012b. Kaposi's sarcoma associated herpesvirus entry into target cells. Front. Microbiol. 3, 6.
Chandramouli, S., Ciferri, C., Nikitin, P.A., Calo, S., Gerrein, R., Balabanis, K., Monroe, J., Hebner, C., Lilja, A.E., Settembre, E.C., Carfi, A., 2015. Structure of HCMV glycoprotein B in the postfusion conformation bound to a neutralizing human antibody. Nat. Commun. 6, 8176.
Chandran, B., 2010. Early events in Kaposi's sarcoma-associated herpesvirus infection of target cells. J. Virol. 84, 2188–2199.
Chandran, B., Hutt-Fletcher, L., 2007. Gammaherpesviruses entry and early events during infection. In: Arvin, A., Campadelli-Fiume, G., Mocarski, E., Moore, P.S., Roizman, B., Whitley, R., Yamanishi, K. (Eds.), Human Herpesviruses: Biology, Therapy, and Immunoprophylaxis. Cambridge University Press, Cambridge.

Chang, Y., Moore, P., 2014. Twenty years of KSHV. Viruses 6, 4258–4264.
Chang, Y., Cesarman, E., Pessin, M.S., Lee, F., Culpepper, J., Knowles, D.M., Moore, P.S., 1994. Identification of herpesvirus-like DNA sequences in AIDS-associated Kaposi's sarcoma. Science 266, 1865–1869.
Chen, J., Jardetzky, T.S., Longnecker, R., 2013. The large groove found in the gH/gL structure is an important functional domain for Epstein-Barr virus fusion. J. Virol. 87, 3620–3627.
Chen, J., Sathiyamoorthy, K., Zhang, X., Schaller, S., Perez White, B.E., Jardetzky, T.S., Longnecker, R., 2018. Ephrin receptor A2 is a functional entry receptor for Epstein-Barr virus. Nat. Microbiol. 3, 172–180.
Chen, J., Zhang, X., Schaller, S., Jardetzky, T.S., Longnecker, R., 2019. Ephrin receptor A4 Is a new Kaposi's sarcoma-associated Herpesvirus Virus entry receptor. MBio 10. https://doi.org/10.1128/mBio.02892-18. Epub ahead of print.
Chowdary, T.K., Cairns, T.M., Atanasiu, D., Cohen, G.H., Eisenberg, R.J., Heldwein, E.E., 2010. Crystal structure of the conserved herpesvirus fusion regulator complex gH-gL. Nat. Struct. Mol. Biol. 17, 882–888.
Cohen, J.I., 2000. Epstein-Barr virus infection. N. Engl. J. Med. 343, 481–492.
Coleman, C.B., Wohlford, E.M., Smith, N.A., King, C.A., Ritchie, J.A., Baresel, P.C., Kimura, H., Rochford, R., 2015. Epstein-Barr virus type 2 latently infects T cells, inducing an atypical activation characterized by expression of lymphotactic cytokines. J. Virol. 89, 2301–2312.
Connolly, S.A., Longnecker, R., 2012. Residues within the C-terminal arm of the herpes simplex virus 1 glycoprotein B Ectodomain contribute to its refolding during the fusion step of virus entry. J. Virol. 86, 6386–6393.
Connolly, S.A., Jackson, J.O., Jardetzky, T.S., Longnecker, R., 2011. Fusing structure and function: a structural view of the herpesvirus entry machinery. Nat. Rev. Microbiol. 9, 369–381.
Cook-Mozaffari, P., Newton, R., Beral, V., Burkitt, D.P., 1998. The geographical distribution of Kaposi's sarcoma and of lymphomas in Africa before the AIDS epidemic. Br. J. Cancer 78, 1521–1528.
Coscoy, L., 2007. Immune evasion by Kaposi's sarcoma-associated herpesvirus. Nat. Rev. Immunol. 7, 391–401.
Damania, B., Cesarman, E., 2013. Kaposi's A. Sarcoma-Associated Herpesvirus. Wolters Kluwer Health/Lippincott, Williams & Wilkins, United States, Philadelphiac, 2013.
Davison, A.J., Dargan, D.J., Stow, N.D., 2002. Fundamental and accessory systems in herpesviruses. Antiviral Res. 56, 1–11.
Dekate, J., Chetty, R., 2016. Epstein-Barr virus-associated smooth muscle tumor. Arch. Pathol. Lab. Med. 140, 718–722.
de-The, G., Geser, A., Day, N.E., Tukei, P.M., Williams, E.H., Beri, D.P., Smith, P.G., Dean, A.G., Bronkamm, G.W., Feorino, P., Henle, W., 1978. Epidemiological evidence for causal relationship between Epstein-Barr virus and Burkitt's lymphoma from Ugandan prospective study. Nature 274, 756–761.
Dutta, D., Chakraborty, S., Bandyopadhyay, C., Valiya Veettil, M., Ansari, M.A., Singh, V.V., Chandran, B., 2013. EphrinA2 regulates clathrin mediated KSHV endocytosis in fibroblast cells by coordinating integrin-associated signaling and c-Cbl directed polyubiquitination. PLoS Pathog. 9, e1003510.
Duus, K.M., Lentchitsky, V., Wagenaar, T., Grose, C., Webster-Cyriaque, J., 2004. Wild-type Kaposi's sarcoma-associated herpesvirus isolated from the oropharynx of immune-competent individuals has tropism for cultured oral epithelial cells. J. Virol. 78, 4074–4084.
Epstein, M.A., Achong, B.G., Barr, Y.M., 1964. Virus particles in cultured Lymphoblasts from Burkitt's lymphoma. Lancet 1, 702–703.

Fields, B.N., Knipe, D.M., Howley, P.M., 2013. Fields Virology. Editors-in-Chief, David M. Knipe, Peter M. Howley; associate editors, Jeffrey I. Cohen [et al.]Wolters Kluwer Health/Lippincott, Williams & Wilkins, United States, Philadelphiac2013.
Gallo, R.C., Salahuddin, S.Z., Popovic, M., Shearer, G.M., Kaplan, M., Haynes, B.F., Palker, T.J., Redfield, R., Oleske, J., Safai, B., Et, A., 1984. Frequent detection and isolation of cytopathic retroviruses (HTLV-III) from patients with AIDS and at risk for AIDS. Science 224, 500–503.
Gilbert, H.E., Asokan, R., Holers, V.M., Perkins, S.J., 2006. The 15 SCR flexible extracellular domains of human complement receptor type 2 can mediate multiple ligand and antigen interactions. J. Mol. Biol. 362, 1132–1147.
Gish, W., States, D.J., 1993. Identification of protein coding regions by database similarity search. Nat. Genet. 3, 266–272.
Großkopf, A.K., Schlagowski, S., Hörnich, B.F., Fricke, T., Desrosiers, R.C., Hahn, A.S., 2019. EphA7 functions as receptor on BJAB cells for cell-to-cell transmission of the Kaposi's sarcoma-associated herpesvirus (KSHV) and for cell-free infection by the related rhesus monkey rhadinovirus (RRV). J Virol. https://doi.org/10.1128/JVI.00064-19. Epub ahead of print.
Guerreiro-Cacais, A.O., Li, L., Donati, D., Bejarano, M.T., Morgan, A., Masucci, M.G., Hutt-Fletcher, L., Levitsky, V., 2004. Capacity of Epstein-Barr virus to infect monocytes and inhibit their development into dendritic cells is affected by the cell type supporting virus replication. J. Gen. Virol. 85, 2767–2778.
Hahn, A., Birkmann, A., Wies, E., Dorer, D., Mahr, K., Sturzl, M., Titgemeyer, F., Neipel, F., 2009. Kaposi's sarcoma-associated herpesvirus gH/gL: glycoprotein export and interaction with cellular receptors. J. Virol. 83, 396–407.
Hahn, A.S., Kaufmann, J.K., Wies, E., Naschberger, E., Panteleev-Ivlev, J., Schmidt, K., Holzer, A., Schmidt, M., Chen, J., Konig, S., Ensser, A., Myoung, J., Brockmeyer, N.H., Sturzl, M., Fleckenstein, B., Neipel, F., 2012. The ephrin receptor tyrosine kinase A2 is a cellular receptor for Kaposi's sarcoma-associated herpesvirus. Nat. Med. 18, 961–966.
Harrison, S.C., 2015. Viral membrane fusion. Virology 479–480, 498–507.
Hassman, L.M., Ellison, T.J., Kedes, D.H., 2011. KSHV infects a subset of human tonsillar B cells, driving proliferation and plasmablast differentiation. J. Clin. Invest. 121, 752–768.
Heldwein, E.E., 2016. gH/gL supercomplexes at early stages of herpesvirus entry. Curr. Opin. Virol. 18, 1–8.
Heldwein, E.E., Lou, H., Bender, F.C., Cohen, G.H., Eisenberg, R.J., Harrison, S.C., 2006. Crystal structure of glycoprotein B from herpes simplex virus 1. Science 313, 217–220.
Henke-Gendo, C., Schulz, T.F., 2004. Transmission and disease association of Kaposi's sarcoma-associated herpesvirus: recent developments. Curr. Opin. Infect. Dis. 17, 53–57.
Henle, G., Henle, W., 1966. Immunofluorescence in cells derived from Burkitt's lymphoma. J. Bacteriol. 91, 1248–1256.
Henle, G., Henle, W., Diehl, V., 1968. Relation of Burkitt's tumor-associated herpes-ytpe virus to infectious mononucleosis. Proc. Natl. Acad. Sci. U. S. A. 59, 94–101.
Holen, H.L., Shadidi, M., Narvhus, K., Kjosnes, O., Tierens, A., Aasheim, H.C., 2008. Signaling through ephrin-a ligand leads to activation of Src-family kinases, Akt phosphorylation, and inhibition of antigen receptor-induced apoptosis. J. Leukoc. Biol. 84, 1183–1191.
Isobe, Y., Sugimoto, K., Yang, L., Tamayose, K., Egashira, M., Kaneko, T., Takada, K., Oshimi, K., 2004. Epstein-Barr virus infection of human natural killer cell lines and peripheral blood natural killer cells. Cancer Res. 64, 2167–2174.

Janz, A., Oezel, M., Kurzeder, C., Mautner, J., Pich, D., Kost, M., Hammerschmidt, W., Delecluse, H.J., 2000. Infectious Epstein-Barr virus lacking major glycoprotein BLLF1 (gp350/220) demonstrates the existence of additional viral ligands. J. Virol. 74, 10142–10152.

Jasirwan, C., Furusawa, Y., Tang, H., Maeki, T., Mori, Y., 2014. Human herpesvirus-6A gQ1 and gQ2 are critical for human CD46 usage. Microbiol. Immunol. 58, 22–30.

Jenson, H.B., Leach, C.T., McClain, K.L., Joshi, V.V., Pollock, B.H., Parmley, R.T., Chadwick, E.G., Murphy, S.B., 1997. Benign and malignant smooth muscle tumors containing Epstein-Barr virus in children with AIDS. Leuk. Lymphoma 27, 303–314.

Jiang, R., Gu, X., Nathan, C.O., Hutt-Fletcher, L., 2008. Laser-capture microdissection of oropharyngeal epithelium indicates restriction of Epstein-Barr virus receptor/CD21 mRNA to tonsil epithelial cells. J. Oral Pathol. Med. 37, 626–633.

Jones, J.F., Shurin, S., Abramowsky, C., Tubbs, R.R., Sciotto, C.G., Wahl, R., Sands, J., Gottman, D., Katz, B.Z., Sklar, J., 1988. T-cell lymphomas containing Epstein-Barr viral DNA in patients with chronic Epstein-Barr virus infections. N. Engl. J. Med. 318, 733–741.

Kaposi, M., 1872. Idiopathisches multiples Pigmentsarkom der haut. Arch. Dermatol. Syphilol. 4, 265–273.

Kerur, N., Veettil, M.V., Sharma-Walia, N., Sadagopan, S., Bottero, V., Paul, A.G., Chandran, B., 2010. Characterization of entry and infection of monocytic THP-1 cells by Kaposi's sarcoma associated herpesvirus (KSHV): role of heparan sulfate, DC-SIGN, integrins and signaling. Virology 406, 103–116.

Kikuta, H., Itakura, O., Taneichi, K., Kohno, M., 1997. Tropism of human herpesvirus 8 for peripheral blood lymphocytes in patients with Castleman's disease. Br. J. Haematol. 99, 790–793.

Kirschner, A.N., Sorem, J., Longnecker, R., Jardetzky, T.S., 2009. Structure of Epstein-Barr virus glycoprotein 42 suggests a mechanism for triggering receptor-activated virus entry. Structure 17, 223–233.

Krishnan, H.H., Sharma-Walia, N., Streblow, D.N., Naranatt, P.P., Chandran, B., 2006. Focal adhesion kinase is critical for entry of Kaposi's sarcoma-associated herpesvirus into target cells. J. Virol. 80, 1167–1180.

Kumar, B., Roy, A., Veettil, M.V., Chandran, B., 2018. Insight into the roles of E3 ubiquitin ligase c-Cbl, ESCRT machinery, and host cell signaling in Kaposi's sarcoma-associated herpesvirus entry and trafficking. J. Virol. 92. https://doi.org/10.1128/JVI.01376-17.

Laichalk, L.L., Thorley-Lawson, D.A., 2005. Terminal differentiation into plasma cells initiates the replicative cycle of Epstein-Barr virus in vivo. J. Virol. 79, 1296–1307.

Longnecker, R., Kieff, E., Cohen, J., 2013. Epstein-Barr virus. Wolters Kluwer Health/ Lippincott, Williams & Wilkins, United States, Philadelphiac, 2013.

Lukac, D.M., Yuan, Y., 2007. Reactivation and lytic replication of KSHV. In: Arvin, A., Campadelli-Fiume, G., Mocarski, E., Moore, P.S., Roizman, B., Whitley, R., Yamanishi, K. (Eds.), Human Herpesviruses: Biology, therapy, and immunoprophylaxis. Cambridge University Press, Cambridge.

Mark, L., Lee, W.H., Spiller, O.B., Villoutreix, B.O., Blom, A.M., 2006. The Kaposi's sarcoma-associated herpesvirus complement control protein (KCP) binds to heparin and cell surfaces via positively charged amino acids in CCP1-2. Mol. Immunol. 43, 1665–1675.

Martin, J.N., Ganem, D.E., Osmond, D.H., Page-Shafer, K.A., Macrae, D., Kedes, D.H., 1998. Sexual transmission and the natural history of human herpesvirus 8 infection. N. Engl. J. Med. 338, 948–954.

Matsuura, H., Kirschner, A.N., Longnecker, R., Jardetzky, T.S., 2010. Crystal structure of the Epstein-Barr virus (EBV) glycoprotein H/glycoprotein L (gH/gL) complex. Proc. Natl. Acad. Sci. U. S. A. 107, 22641–22646.

Miller, N., Hutt-Fletcher, L.M., 1992. Epstein-Barr virus enters B cells and epithelial cells by different routes. J. Virol. 66, 3409–3414.
Möhl, B.S., Sathiyamoorthy, K., Jardetzky, T.S., Longnecker, R., 2014. The conserved disulfide bond within domain II (D-II) of Epstein-Barr virus (EBV) gH has divergent roles in membrane fusion with epithelial cells and B cells. J. Virol. 88, 13570–13579.
Möhl, B.S., Schröter, C., Klupp, B.G., Fuchs, W., Mettenleiter, T.C., Jardetzky, T.S., Longnecker, R., 2015. Comparative mutagenesis of pseudorabies virus and Epstein-Barr virus gH identifies a structural determinant within domain III of gH required for surface expression and entry function. J. Virol. 90, 2285–2293.
Möhl, B.S., Chen, J., Sathiyamoorthy, K., Jardetzky, T.S., Longnecker, R., 2016. Structural and mechanistic insights into the tropism of Epstein-Barr virus. Mol. Cells 39, 286–291.
Mori, Y., 2009. Recent topics related to human herpesvirus 6 cell tropism. Cell. Microbiol. 11, 1001–1006.
Mukai, M., Suruga, N., Saeki, N., Ogawa, K., 2017. EphA receptors and ephrin-a ligands are upregulated by monocytic differentiation/maturation and promote cell adhesion and protrusion formation in HL60 monocytes. BMC Cell Biol. 18, 28.
Mullen, M.M., Haan, K.M., Longnecker, R., Jardetzky, T.S., 2002. Structure of the Epstein-Barr virus gp42 protein bound to the MHC class II receptor HLA-DR1. Mol. Cell 9, 375–385.
Nagae, M., Re, S., Mihara, E., Nogi, T., Sugita, Y., Takagi, J., 2012. Crystal structure of alpha5beta1 integrin ectodomain: atomic details of the fibronectin receptor. J. Cell Biol. 197, 131–140.
Naranatt, P.P., Akula, S.M., Zien, C.A., Krishnan, H.H., Chandran, B., 2003. Kaposi's sarcoma-associated herpesvirus induces the phosphatidylinositol 3-kinase-PKC-zeta-MEK-ERK signaling pathway in target cells early during infection: implications for infectivity. J. Virol. 77, 1524–1539.
Nemerow, G.R., Cooper, N.R., 1984. Early events in the infection of human B lymphocytes by Epstein-Barr virus: the internalization process. Virology 132, 186–198.
Nilsson, K., Klein, G., Henle, W., Henle, G., 1971. The establishment of lymphoblastoid lines from adult and fetal human lymphoid tissue and its dependence on EBV. Int. J. Cancer 8, 443–450.
Notredame, C., Higgins, D.G., Heringa, J., 2000. T-Coffee: a novel method for fast and accurate multiple sequence alignment. J. Mol. Biol. 302, 205–217.
Ogembo, J.G., Kannan, L., Ghiran, I., Nicholson-Weller, A., Finberg, R.W., Tsokos, G.C., Fingeroth, J.D., 2013. Human complement receptor type 1/CD35 is an Epstein-Barr virus receptor. Cell Rep. 3, 371–385.
Papadopoulos, J.S., Agarwala, R., 2007. COBALT: constraint-based alignment tool for multiple protein sequences. Bioinformatics 23, 1073–1079.
Park, S., Ko, Y.H., 2014. Epstein-Barr virus-associated T/natural killer-cell lymphoproliferative disorders. J. Dermatol. 41, 29–39.
Pauk, J., Huang, M.L., Brodie, S.J., Wald, A., Koelle, D.M., Schacker, T., Celum, C., Selke, S., Corey, L., 2000. Mucosal shedding of human herpesvirus 8 in men. N. Engl. J. Med. 343, 1369–1377.
Penn, I., 1979. Kaposi's sarcoma in organ transplant recipients: report of 20 cases. Transplantation 27, 8–11.
Pertel, P.E., 2002. Human herpesvirus 8 glycoprotein B (gB), gH, and gL can mediate cell fusion. J. Virol. 76, 4390–4400.
Pica, F., Volpi, A., 2007. Transmission of human herpesvirus 8: an update. Curr. Opin. Infect. Dis. 20, 152–156.

Raghu, H., Sharma-Walia, N., Veettil, M.V., Sadagopan, S., Chandran, B., 2009. Kaposi's sarcoma-associated herpesvirus utilizes an actin polymerization-dependent macropinocytic pathway to enter human dermal microvascular endothelial and human umbilical vein endothelial cells. J. Virol. 83, 4895–4911.

Rappocciolo, G., Hensler, H.R., Jais, M., Reinhart, T.A., Pegu, A., Jenkins, F.J., Rinaldo, C.R., 2008. Human herpesvirus 8 infects and replicates in primary cultures of activated B lymphocytes through DC-SIGN. J. Virol. 82, 4793–4806.

Revello, M.G., Gerna, G., 2010. Human cytomegalovirus tropism for endothelial/epithelial cells: scientific background and clinical implications. Rev. Med. Virol. 20, 136–155.

Robey, R.C., Bower, M., 2015. Facing up to the ongoing challenge of Kaposi's sarcoma. Curr. Opin. Infect. Dis. 28, 31–40.

Sathiyamoorthy, K., Jiang, J., Hu, Y.X., Rowe, C.L., Möhl, B.S., Chen, J., Jiang, W., Mellins, E.D., Longnecker, R., Zhou, Z.H., Jardetzky, T.S., 2014. Assembly and architecture of the EBV B cell entry triggering complex. PLoS Pathog. 10, e1004309.

Sathiyamoorthy, K., Hu, Y.X., Mohl, B.S., Chen, J., Longnecker, R., Jardetzky, T.S., 2016. Structural basis for Epstein-Barr virus host cell tropism mediated by gp42 and gHgL entry glycoproteins. Nat. Commun. 7, 13557.

Sathiyamoorthy, K., Chen, J., Longnecker, R., Jardetzky, T.S., 2017. The COMPLEXity in herpesvirus entry. Curr. Opin. Virol. 24, 97–104.

Schäfer, G., Blumenthal, M.J., Katz, A.A., 2015. Interaction of human tumor viruses with host cell surface receptors and cell entry. Viruses 7, 2592–2617.

Seiradake, E., Harlos, K., Sutton, G., Aricescu, A.R., Jones, E.Y., 2010. An extracellular steric seeding mechanism for Eph-ephrin signaling platform assembly. Nat. Struct. Mol. Biol. 17, 398–402.

Shimakage, M., 2014. Significant role of macrophages in human cancers associated with Epstein-Barr virus (review). Oncol. Rep. 32, 1763–1771.

Sixbey, J.W., Nedrud, J.G., Raab-Traub, N., Hanes, R.A., Pagano, J.S., 1984. Epstein-Barr virus replication in oropharyngeal epithelial cells. N. Engl. J. Med. 310, 1225–1230.

Snijder, J., Ortego, M.S., Weidle, C., Stuart, A.B., Gray, M.D., Mcelrath, M.J., Pancera, M., Veesler, D., McGuire, A.T., 2018. An antibody targeting the fusion machinery neutralizes dual-tropic infection and defines a site of vulnerability on Epstein-Barr virus. Immunity 48799-811.e9.

Sprunt, T., Evans, F., 1920. Mononuclear leukocytosis in reaction to acute infections ("infectious mononucleosis"). Bulletin of the Johns Hopkins Hospital, 31, Johns Hopkins Press, United States, pp. 410–417.

Stampfer, S.D., Heldwein, E.E., 2012. Stuck in the middle: structural insights into the role of the gH/gL heterodimer in Herpesvirus entry. Curr. Opin. Virol. 3, 13–19.

Szakonyi, G., Klein, M.G., Hannan, J.P., Young, K.A., Ma, R.Z., Asokan, R., Holers, V.M., Chen, X.S., 2006. Structure of the Epstein-Barr virus major envelope glycoprotein. Nat. Struct. Mol. Biol. 13, 996–1001.

Tang, H., Wang, J., Mahmoud, N.F., Mori, Y., 2014. Detailed study of the interaction between human herpesvirus 6B glycoprotein complex and its cellular receptor, human CD134. J. Virol. 88, 10875–10882.

Tanner, J., Weis, J., Fearon, D., Whang, Y., Kieff, E., 1987. Epstein-Barr virus gp350/220 binding to the B lymphocyte C3d receptor mediates adsorption, capping, and endocytosis. Cell 50, 203–213.

TerBush, A.A., Hafkamp, F., Lee, H.J., Coscoy, L., 2018. A Kaposi's sarcoma-associated Herpesvirus infection mechanism is independent of integrins alpha3beta1, alphaVbeta3, and alphaVbeta5. J. Virol. 92. https://doi.org/10.1128/JVI.00803-18.

Thakker, S., Verma, S.C., 2016. Co-infections and pathogenesis of Kshv-associated malignancies. Front. Microbiol. 7, 151.

Thorley-Lawson, D.A., Allday, M.J., 2008. The curious case of the tumour virus: 50 years of Burkitt's lymphoma. Nat. Rev. Microbiol. 6, 913–924.

Tugizov, S.M., Berline, J.W., Palefsky, J.M., 2003. Epstein-Barr virus infection of polarized tongue and nasopharyngeal epithelial cells. Nat. Med. 9, 307–314.

Vallbracht, M., Brun, D., Tassinari, M., Vaney, M.C., Pehau-Arnaudet, G., Guardado-Calvo, P., Haouz, A., Klupp, B.G., Mettenleiter, T.C., Rey, F.A., Backovic, M., 2017. Structure-function dissection of the pseudorabies virus glycoprotein B fusion loops. J. Virol. 92. https://doi.org/10.1128/JVI.01203-17.

Veettil, M.V., Sharma-Walia, N., Sadagopan, S., Raghu, H., Sivakumar, R., Naranatt, P.P., Chandran, B., 2006. RhoA-GTPase facilitates entry of Kaposi's sarcoma-associated herpesvirus into adherent target cells in a Src-dependent manner. J. Virol. 80, 11432–11446.

Veettil, M.V., Bandyopadhyay, C., Dutta, D., Chandran, B., 2014. Interaction of KSHV with host cell surface receptors and cell entry. Viruses 6, 4024–4046.

Vieira, J., Huang, M.L., Koelle, D.M., Corey, L., 1997. Transmissible Kaposi's sarcoma-associated herpesvirus (human herpesvirus 8) in saliva of men with a history of Kaposi's sarcoma. J. Virol. 71, 7083–7087.

Walter, P.R., Philippe, E., Nguemby-Mbina, C., Chamlian, A., 1984. Kaposi's sarcoma: presence of herpes-type virus particles in a tumor specimen. Hum. Pathol. 15, 1145–1146.

Wang, X., Kenyon, W.J., Li, Q., Mullberg, J., Hutt-Fletcher, L.M., 1998. Epstein-Barr virus uses different complexes of glycoproteins gH and gL to infect B lymphocytes and epithelial cells. J. Virol. 72, 5552–5558.

Wang, F.Z., Akula, S.M., Pramod, N.P., Zeng, L., Chandran, B., 2001. Human herpesvirus 8 envelope glycoprotein K8.1A interaction with the target cells involves heparan sulfate. J. Virol. 75, 7517–7527.

Wang, F.Z., Akula, S.M., Sharma-Walia, N., Zeng, L., Chandran, B., 2003. Human herpesvirus 8 envelope glycoprotein B mediates cell adhesion via its RGD sequence. J. Virol. 77, 3131–3147.

Waterhouse, A., Bertoni, M., Bienert, S., Studer, G., Tauriello, G., Gumienny, R., Heer, F.T., De Beer, T.A.P., Rempfer, C., Bordoli, L., Lepore, R., Schwede, T., 2018. SWISS-MODEL: homology modelling of protein structures and complexes. Nucleic Acids Res. 46, W296–W303.

Webster-Cyriaque, J., Duus, K., Cooper, C., Duncan, M., 2006. Oral EBV and KSHV infection in HIV. Adv. Dent. Res. 19, 91–95.

Weiss, L.M., Movahed, L.A., Warnke, R.A., Sklar, J., 1989. Detection of Epstein-Barr viral genomes in reed-Sternberg cells of Hodgkin's disease. N. Engl. J. Med. 320, 502–506.

Wu, L., Borza, C.M., Hutt-Fletcher, L.M., 2005. Mutations of Epstein-Barr virus gH that are differentially able to support fusion with B cells or epithelial cells. J. Virol. 79, 10923–10930.

Xiao, J., Palefsky, J.M., Herrera, R., Tugizov, S.M., 2007. Characterization of the Epstein-Barr virus glycoprotein BMRF-2. Virology 359, 382–396.

Xing, Y., Oliver, S.L., Nguyen, T., Ciferri, C., Nandi, A., Hickman, J., Giovani, C., Yang, E., Palladino, G., Grose, C., Uematsu, Y., Lilja, A.E., Arvin, A.M., Carfi, A., 2015. A site of varicella-zoster virus vulnerability identified by structural studies of neutralizing antibodies bound to the glycoprotein complex gHgL. Proc. Natl. Acad. Sci. U. S. A. 112, 6056–6061.

Zhang, H., Li, Y., Wang, H.B., Zhang, A., Chen, M.L., Fang, Z.X., Dong, X.D., Li, S.B., Du, Y., Xiong, D., He, J.Y., Li, M.Z., Liu, Y.M., Zhou, A.J., Zhong, Q., Zeng, Y.X., Kieff, E., Zhang, Z., Gewurz, B.E., Zhao, B., Zeng, M.S., 2018. Author correction: Ephrin receptor A2 is an epithelial cell receptor for Epstein-Barr virus entry. Nat. Microbiol. 3, 1075.

Zhou, M., Lanchy, J.M., Ryckman, B.J., 2015. Human Cytomegalovirus gH/gL/gO promotes the fusion step of entry into all cell types, whereas gH/gL/UL128-131 broadens virus tropism through a distinct mechanism. J. Virol. 89, 8999–9009.

Ziegler, J.L., Drew, W.L., Miner, R.C., Mintz, L., Rosenbaum, E., Gershow, J., Lennette, E.T., Greenspan, J., Shillitoe, E., Beckstead, J., Casavant, C., Yamamoto, K., 1982. Outbreak of Burkitt's-like lymphoma in homosexual men. Lancet 2, 631–633.

zur Hausen, H., Schulte-Holthausen, H., Klein, G., Henle, W., Henle, G., Clifford, P., Santesson, L., 1970. EBV DNA in biopsies of Burkitt tumours and anaplastic carcinomas of the nasopharynx. Nature 228, 1056–1058.